AF389103

Additive Manufacturing Research and Applications

Additive Manufacturing Research and Applications

Editor

Atila Ertas

MDPI • Basel • Beijing • Wuhan • Barcelona • Belgrade • Manchester • Tokyo • Cluj • Tianjin

Editor
Atila Ertas
Texas Tech University
USA

Editorial Office
MDPI
St. Alban-Anlage 66
4052 Basel, Switzerland

This is a reprint of articles from the Special Issue published online in the open access journal *Metals* (ISSN 2075-4701) (available at: https://www.mdpi.com/journal/metals/special_issues/additive_manufacturing_research_applications).

For citation purposes, cite each article independently as indicated on the article page online and as indicated below:

LastName, A.A.; LastName, B.B.; LastName, C.C. Article Title. *Journal Name* **Year**, *Volume Number*, Page Range.

ISBN 978-3-0365-3903-4 (Hbk)
ISBN 978-3-0365-3904-1 (PDF)

© 2022 by the authors. Articles in this book are Open Access and distributed under the Creative Commons Attribution (CC BY) license, which allows users to download, copy and build upon published articles, as long as the author and publisher are properly credited, which ensures maximum dissemination and a wider impact of our publications.
The book as a whole is distributed by MDPI under the terms and conditions of the Creative Commons license CC BY-NC-ND.

Contents

About the Editor

Atila Ertas is the director of the Academy of Transdisciplinary Studies and a professor of Mechanical Engineering at Texas Tech University. Dr. Ertas has many years of experience in teaching transdisciplinary design courses. He is the author or co-author of over 190 technical papers that cover many engineering technical fields. His textbooks include the following: *The Engineering Design Process* (co-author with J. Jones, 1993, 1996, John Wiley & Sons); *Prevention through Design (PtD): Transdisciplinary Process* (2010, ATLAS Publishing); *Engineering Mechanics and Design Applications: Transdisciplinary Engineering Fundamentals* (2011, CRC Press, Taylor & Francis); *Transdisciplinary Engineering Design Process* (2018, John Wiley & Sons); and *Managing System Complexity through Integrated Transdisciplinary Design Tools* (co-author with U. Gulbulak, 2020, ATLAS Publishing). He has edited or co-edited more than 40 proceedings and books. Under his supervision, more than 190 graduate students have received degrees. His professional affiliations are as follows: The American Society of Mechanical Engineers (Fellow); Society for Design and Process Science (Fellow); The Academy of Transdisciplinary Learning & Advanced Studies (honorary member, Fellow); IC2 Institute, The University of Texas at Austin (Fellow, 1996–2019); Luminary Research Institute, Gaya Foundation, Taiwan (Founding Fellow); Center for Transdisciplinary Research (CIRET), France (Honorary Member).

Preface to "Additive Manufacturing Research and Applications"

Additive Manufacturing (AM) has gone through something of a revolution over the last decade, and it has now evolved into a viable industrial manufacturing solution that is capable of creating complex geometries unachievable with traditional manufacturing methods. Thus, AM is becoming more and more capable of redefining the manufacturing landscape. The existing ecosystem for AM has significantly developed in the areas of design digitization, deposition methods, printer capabilities, component geometry re-imagination, and post-processing methods. Big data and machine learning are reaching levels of maturity where they have become capable of assisting in targeted and rapid problem-solving.

However, further research is required to overcome the many challenges additive manufacturing faces today, specifically in the research areas of powder manufacturing technologies in 3D printing, pre-and post-processing technologies and approaches, AM processes and optimizations, inspection processes and quality evaluation, new materials for 3D printing, design and simulation in AM, topology optimization, microstructure design, new materials in 3D printing such as silicone, new AM machine design, and development, etc.

This Special Issue book covers a wide scope in the 3D-printing research field including the following: the use of 3D printing in system design; AM with binding jetting; powder manufacturing technologies in 3D printing; fatigue performance of additively manufactured metals, such as the Ti-6Al-4V alloy; 3D-printing method with metallic powder and a laser-based 3D printer; 3D-printed custom-made implants; laser-directed energy deposition (LDED) process of TiC-TMC coatings; Wire Arc Additive Manufacturing; cranial implant fabrication without supports in electron beam melting (EBM) additive manufacturing; the influence of material properties and characteristics in laser powder bed fusion; Design For Additive Manufacturing (DFAM); porosity evaluation of additively manufactured parts; fabrication of coatings by laser additive manufacturing; laser powder bed fusion additive manufacturing; plasma metal deposition (PMD); as-metal-arc (GMA) additive manufacturing process; spreading process maps for powder-bed additive manufacturing derived from physics model-based machine learning.

Atila Ertas

Editor

Editorial

Additive Manufacturing Research and Applications

Atila Ertas [1,*] and Adam Stroud [2]

[1] Department of Mechanical Engineering, Texas Tech University, Lubbock, TX 79409, USA
[2] Metalworking Company, Ellwood Group, Houston, TX 77060, USA; titaniumwisperer@gmail.com
* Correspondence: Atila.Ertas@ttu.edu; Tel.: +1-806-834-5788

Citation: Ertas, A.; Stroud, A. Additive Manufacturing Research and Applications. *Metals* **2022**, *12*, 634. https://doi.org/10.3390/met12040634

Received: 31 March 2022
Accepted: 1 April 2022
Published: 7 April 2022

Publisher's Note: MDPI stays neutral with regard to jurisdictional claims in published maps and institutional affiliations.

Copyright: © 2022 by the authors. Licensee MDPI, Basel, Switzerland. This article is an open access article distributed under the terms and conditions of the Creative Commons Attribution (CC BY) license (https://creativecommons.org/licenses/by/4.0/).

1. Introduction and Scope

Additive Manufacturing (AM) has undergone somewhat of a revolution over the last decade and it has now evolved into a viable industrial manufacturing solution, able to create complex geometries which are unachievable with traditional manufacturing methods. Thus, AM is becoming more and more capable of redefining the manufacturing landscape. There have been significant advances made in the existing ecosystem for AM in the areas of design digitization, deposition methods, printer capabilities, component geometry re-imagination, and post-processing methods. Big data and machine learning are each reaching a level of maturity whereby they have become capable of assisting in targeted, rapid problem-solving.

However, further research is required to overcome many challenges faced by additive manufacturing today, particularly in the research area of powder manufacturing technologies in 3D printing, pre-and post-processing technologies and approaches, AM processes and optimizations, inspection processes and quality evaluation, as well as new materials for 3D printing, design and simulation in AM (topology optimization, microstructure design), new materials in 3D printing such as silicone, new AM machine design, and development, etc.

2. Contributions

This Special Issue offers a wide scope in the research field around 3D printing, including the following [1–18]: the use of 3D printing in system design, AM with binding jetting, powder manufacturing technologies in 3D printing, fatigue performance of additively manufactured metals such as the Ti-6Al-4V alloy, 3D-printing method with metallic powder and a laser-based 3D printer, 3D-printed custom-made implants, laser-directed energy deposition (LDED) process of TiC-TMC coatings, Wire Arc Additive Manufacturing, cranial implant fabrication without supports in electron beam melting (EBM) additive manufacturing, the influence of material properties and characteristics in laser powder bed fusion, Design For Additive Manufacturing (DFAM), porosity evaluation of additively manufactured parts, fabrication of coatings by laser additive manufacturing, laser powder bed fusion additive manufacturing, plasma metal deposition (PMD), as-metal-arc (GMA) additive manufacturing process, and spreading process maps for powder-bed additive manufacturing derived from physics model-based machine learning.

3. Conclusions and Outlook

This Special Issue of *Metals* was well supported by a diverse range of submissions and consists of a final publication of 18 high-quality peer-reviewed articles. As Guest Editors of this Special Issue, we are very pleased with the final result and hope that the present papers will be useful to researchers and designers working on the area of additive manufacturing. We would like to warmly thank all the authors for their contributions and all reviewers for their efforts to ensure a high-quality publications. At the same time, we would also like to thank the many anonymous reviewers who assisted in the reviewing process. Sincere thanks are also owed to the Editors of *Metals* for their continuous help, and to the *Metals*

Editorial Assistants for their valuable and inexhaustible engagement and support during the preparation of this special issue. In particular, we offer our sincere thanks to Toliver Guo for his help and support.

Conflicts of Interest: The authors declare no conflict of interest.

References

1. Desai, P.; Higgs, C. Spreading Process Maps for Powder-Bed Additive Manufacturing Derived from Physics Model-Based Machine Learning. *Metals* **2019**, *9*, 1176. [CrossRef]
2. Yu, R.; Zhao, Z.; Bai, L.; Han, J. Prediction of Weld Reinforcement Based on Vision Sensing in GMA Additive Manufacturing Process. *Metals* **2020**, *10*, 1041. [CrossRef]
3. Liu, S.; Guo, H. A Review of SLMed Magnesium Alloys: Processing, Properties, Alloying Elements and Postprocessing. *Metals* **2020**, *10*, 1073. [CrossRef]
4. Arévalo, C.; Ariza, E.; Pérez-Soriano, E.; Kitzmantel, M.; Neubauer, E.; Montealegre-Meléndez, I. Effect of Processing Atmosphere and Secondary Operations on the Mechanical Properties of Additive Manufactured AISI 316L Stainless Steel by Plasma Metal Deposition. *Metals* **2020**, *10*, 1125. [CrossRef]
5. Mohd Yusuf, S.; Choo, E.; Gao, N. Comparison between Virgin and Recycled 316L SS and AlSi10Mg Powders Used for Laser Powder Bed Fusion Additive Manufacturing. *Metals* **2020**, *10*, 1625. [CrossRef]
6. Qian, S.; Zhang, Y.; Dai, Y.; Guo, Y. Microstructure and Mechanical Properties of Nickel-Based Coatings Fabricated through Laser Additive Manufacturing. *Metals* **2021**, *11*, 53. [CrossRef]
7. Park, S.; Hong, J.; Ha, T.; Choi, S.; Jhang, K. Deep Learning-Based Ultrasonic Testing to Evaluate the Porosity of Additively Manufactured Parts with Rough Surfaces. *Metals* **2021**, *11*, 290. [CrossRef]
8. Del Prete, A.; Primo, T. Innovative Methodology for the Identification of the Most Suitable Additive Technology Based on Product Characteristics. *Metals* **2021**, *11*, 409. [CrossRef]
9. Haferkamp, L.; Haudenschild, L.; Spierings, A.; Wegener, K.; Riener, K.; Ziegelmeier, S.; Leichtfried, G. The Influence of Particle Shape, Powder Flowability, and Powder Layer Density on Part Density in Laser Powder Bed Fusion. *Metals* **2021**, *11*, 418. [CrossRef]
10. Moiduddin, K.; Mian, S.; Ameen, W.; Alkhalefah, H.; Sayeed, A. Feasibility Study of the Cranial Implant Fabricated without Supports in Electron Beam Melting. *Metals* **2021**, *11*, 496. [CrossRef]
11. Casuso, M.; Veiga, F.; Suárez, A.; Bhujangrao, T.; Aldalur, E.; Artaza, T.; Amondarain, J.; Lamikiz, A. Model for the Prediction of Deformations in the Manufacture of Thin-Walled Parts by Wire Arc Additive Manufacturing Technology. *Metals* **2021**, *11*, 678. [CrossRef]
12. Li, Y.; Zhang, D.; Wang, H.; Cong, W. Fabrication of a TiC-Ti Matrix Composite Coating Using Ultrasonic Vibration-Assisted Laser Directed Energy Deposition: The Effects of Ultrasonic Vibration and TiC Content. *Metals* **2021**, *11*, 693. [CrossRef]
13. Park, J.; Kang, H.; Kim, J.; Kim, H. 3D-Printed Connector for Revision Limb Salvage Surgery in Long Bones Previously Using Customized Implants. *Metals* **2021**, *11*, 707. [CrossRef]
14. Mirea, R.; Biris, I.; Ceatra, L.; Ene, R.; Paraschiv, A.; Cucuruz, A.; Sbarcea, G.; Popescu, E.; Badea, T. In Vitro Physical-Chemical Behaviour Assessment of 3D-Printed CoCrMo Alloy for Orthopaedic Implants. *Metals* **2021**, *11*, 857. [CrossRef]
15. Segurajauregi, U.; Álvarez-Vázquez, A.; Muñiz-Calvente, M.; Urresti, Í.; Naveiras, H. Fatigue Assessment of Selective Laser Melted Ti-6Al-4V: Influence of Speed Manufacturing and Porosity. *Metals* **2021**, *11*, 1022. [CrossRef]
16. Ferreira, A.; Amaral, R.; Romio, P.; Cruz, J.; Reis, A.; Vieira, M. Deposition of Nickel-Based Superalloy Claddings on Low Alloy Structural Steel by Direct Laser Deposition. *Metals* **2021**, *11*, 1326. [CrossRef]
17. Castro-Sastre, M.; García-Cabezón, C.; Fernández-Abia, A.; Martín-Pedrosa, F.; Barreiro, J. Comparative Study on Microstructure and Corrosion Resistance of Al-Si Alloy Cast from Sand Mold and Binder Jetting Mold. *Metals* **2021**, *11*, 1421. [CrossRef]
18. Rauthan, K.; Guzzomi, F.; Vafadar, A.; Hayward, K.; Hurry, A. Experimental Investigation of Pressure Drop Performance of Smooth and Dimpled Single Plate-Fin Heat Exchangers. *Metals* **2021**, *11*, 1757. [CrossRef]

Article

Prediction of Weld Reinforcement Based on Vision Sensing in GMA Additive Manufacturing Process

Rongwei Yu, Zhuang Zhao *, Lianfa Bai and Jing Han *

Jiangsu Key Laboratory of Spectral Imaging and Intelligent Sense, Nanjing University of Science and Technology, Nanjing 210094, China; yrw_njust@163.com (R.Y.); blf@njust.edu.cn (L.B.)
* Correspondence: zhaozhuang3126@gmail.com (Z.Z.); eohj@njust.edu.cn (J.H.); Tel.: +86-151-9599-5329 (Z.Z. & J.H.)

Received: 25 June 2020; Accepted: 29 July 2020; Published: 2 August 2020

Abstract: In the gas-metal-arc (GMA) additive manufacturing process, the shape of the molten pool, the temperature field of the workpiece and the heat dissipation conditions change with the increase of cladding layers, which can affect the dimensional accuracy of the workpiece; hence, it is necessary to monitor the additive manufacturing process online. At present, there is little research about formation-dimension monitoring in the GMA additive manufacturing process; in this paper, weld reinforcement prediction in the GMA additive manufacturing process was conducted, the visual-sensing system for molten pool was established, and a laser locating system was designed to match every frame of the molten pool image with the actual weld location. Extracting the shape and location features of the molten pool as visual features, on the basis of a back-propagation (BP) neural network, we developed the prediction model for weld reinforcement in the GMA additive manufacturing process. Experiment results showed that the model could accurately predict weld reinforcement. By changing the cooling time between adjacent cladding layers, the generalization ability of the prediction model was further verified.

Keywords: GMA additive manufacturing; weld reinforcement; visual features; neural network

---

1. Introduction

Additive manufacturing technology is based on the idea of discreteness and accumulation, using material to stack layer by layer, forming a three-dimensional entity [1]. In recent years, using additive manufacturing technology to obtain a workpiece with high dimensional accuracy and good mechanical properties has become a research hotspot [2]. Gas-metal-arc (GMA) additive manufacturing is the processing technology that uses an electric arc as a heat source to melt metal wire and stacks, forming a metal workpiece [3,4], Yanhu Wang et al. [5,6] pointed out that additive manufacturing based on arc welding has the outstanding advantages of low cost and high efficiency, and can be widely used in many fields. They studied the additive manufacturing of copper-aluminum alloys by adding a small amount of silicon in the cold-metal-transfer (CMT) welding process. In recent years, the visual-sensing method has been widely used in the field of welding manufacturing, Zhuang Zhao et al. [7] proposed an optimal imaging-band selection mechanism for molten pool vision, which has important guiding significance for collecting high-quality molten pool images. Jun Lu et al. [8] obtained the temperature field of the molten pool on the basis of visual imaging in the GMA welding process, and realized the prediction of hump weld bead by monitoring the temperature field distribution of the molten pool.

In the GMA additive manufacturing process, with the increase of cladding layers, heat accumulation of the workpiece is serious, heat dissipation worsens and the shape of the molten pool changes, which ultimately affect the dimensional accuracy of the formed workpiece, so it is very important to monitor the GMA additive manufacturing process online [9]. Ouyang et al. [10,11] produced aluminum-alloy workpieces in a variable-polarity gas-tungsten-arc-welding (GTAW) additive manufacturing process.

They used a charge coupled device (CCD) to monitor the arc length online, pointing out that arc length and preheating temperature have important influence on forming quality. Spencer et al. [12] produced a metal workpiece in a GMA additive manufacturing process, using an infrared thermometer to monitor online the temperature of the welding workpiece. When the temperature of the welding workpiece cooled to a certain value, the next layer could be stacked. This method reduces the time efficiency of the additive manufacturing process, but improves the microstructure and material properties. Kwak et al. [13,14] used two CCD sets to monitor the morphology of the cladding layer in the metal-inert-gas (MIG) welding additive manufacturing process, and used an infrared thermographer to detect the surface-temperature field of the welding workpiece. Jinqian Zhu et al. [15] used an infrared thermographer to detect online the temperature field of the cladding channel in a laser metal-wire additive manufacturing process, realized the prediction of the cladding channel width and the accurate location of welding defects.

The artificial-neural-network model has excellent nonlinear-mapping characteristics, as well as strong learning and induction abilities; the support vector machine (SVM) is a classifier with sparsity and robustness. In recent years, they have been widely used in areas of pattern recognition, and fault diagnosis, prediction, and estimation. Yukang Liu et al. [16] developed a dynamic adaptive neural fuzzy inference system in a GTAW process, and predicted the weld width on the reverse side of the welding seam online by using the visual characteristic parameters of the molten pool. Yanfeng Gao et al. [17,18] developed an SVM to predict the welding-penetration mode through the arc sound signal during the welding process. Jiajia Yang et al. [19] acquired a molten pool image in the near-infrared band in an aluminum-alloy double-wire-welding process, extracted the visual features of molten pool, and established the penetration-state recognition model on the basis of a neural network.

At present, research on GMA additive manufacturing mainly focuses on the feasibility of the forming process, and the mechanical properties of the formed workpiece [20,21]; there is little research on the dimensional accuracy of the formed workpiece [9]. In this paper, we took single-pass multilayer GMA additive manufacturing as the research object, used a color CCD to collect molten pool images and extract the shape and position features of the molten pool in real time; by means of a neural network, established the prediction model for weld reinforcement, and realized the online monitoring of weld reinforcement in the additive manufacturing process, which has important guiding significance for the subsequent control of forming size.

2. Welding Experiment Platform and Approach

2.1. Welding Experiment Platform

Conducting arc additive manufacturing experiment in the CMT welding procedure, which belongs to the GMA welding procedure, the welding experiment platform included a welding power supply (Fronius CMT advanced 4000R), a mobile robot (ABB IRB1400 M2004, ABB, Zurich, Switzerland), a molten pool visual sensor, and a laser locating system, which are shown in Figure 1. The base metal was 304 stainless steel plate, the size was 450 mm × 150 mm × 10 mm, and the welding wire was stainless steel, the diameter of which was 1.2 mm, the chemical composition of welding wire is shown in Table 1, the type of welding torch was Robacta Drive CMT (Pettenbach, Austria). The visual sensor for the molten pool consisted of a color camera (Basler acA640-750uc, Ahrensburg, Germany), computer, and trigger module; the color camera was installed on the welding torch. Furthermore, we developed a laser locating system that consisted of a laser, monochrome camera (Basler ace acA1920-155um, Ahrensburg, Germany), and computer; the laser was installed on the welding torch, and sent out light shining on the upper edge of welding wire. The monochrome camera was fixed on the experiment platform to record the moving trajectory of the laser point. In the GMA additive manufacturing process, the trigger module gave out a fixed frequency signal to control the color and monochrome cameras at the same time; by observing the location of the laser point in every frame of the monochrome image, precisely matching every frame of color molten pool images with the specific weld location.

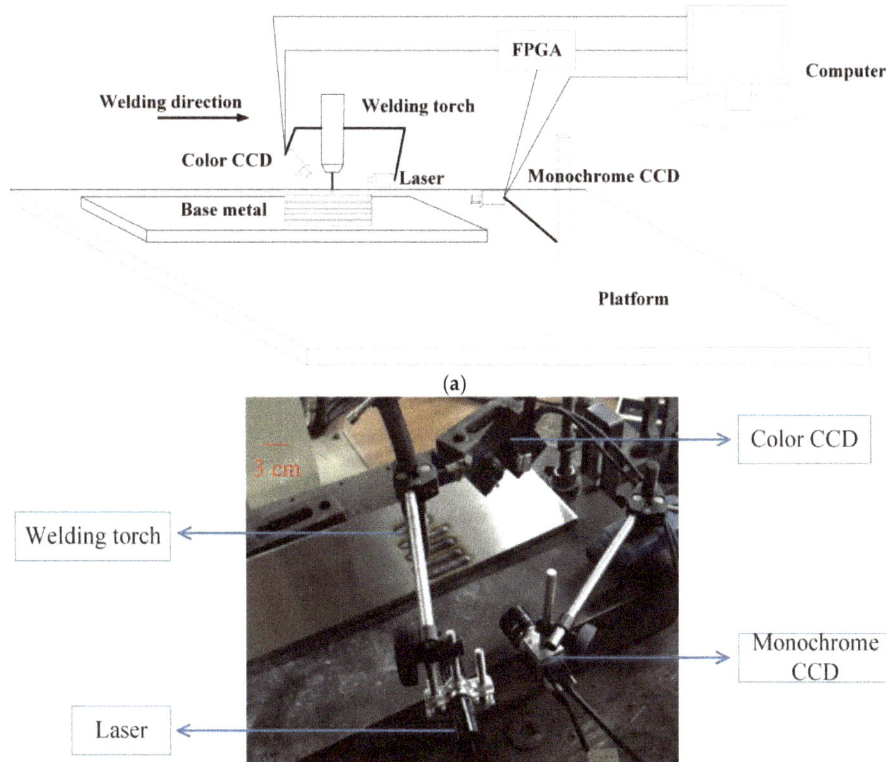

Figure 1. Welding experiment platform. (**a**) Structure diagram; (**b**) Physical diagram.

Table 1. The chemical composition of welding wire.

Elements	C	Mn	Cr	Ni	Mo	Si
Composition	0.012%	2.180%	18.97%	12.70%	2.310%	0.570%

2.2. Experiment Approach

In the GMA additive manufacturing experiment, the welding torch moved horizontally and in the same direction, setting the contact tip to work distance (CTWD) to 1.5 cm, welding current to 130 A, the welding velocity to 5 mm/s, the welding length of each cladding layer to 80 mm, and the number of cladding layers to 10; detailed parameters are shown in Table 2. In the GMA additive manufacturing process, the trigger module gave out fixed frequency signal to control the color and monochrome cameras at the same time, whose frequency was 1000 Hz. We extracted the shape and location features of the molten pool through color molten pool image. By observing the location of the laser point in every frame of the monochrome image, we determined the actual weld location corresponding to each frame collected color molten pool images. After each cladding layer was welded, using three-dimensional scanner to measure the height of the workpiece, we extracted the weld reinforcement. All extracted data constituted the data set of the prediction model.

Table 2. Welding parameters.

Welding Current	Welding Voltage	Welding Speed	Wire Feeding Speed	Shield Gas Flux
129 A	14.4 V	5 mm/s	4.9 m/min	25 L/min

3. Experiment Data Extraction

3.1. Definition and Extraction of Visual Feature Parameters

In the GMA additive manufacturing process, after the color CCD had completed molten pool image acquisition, binary processing was carried out to extract the molten pool outline. Holes were filled up in the contour, and the shape features of the molten pool could be determined. We calculated the molten pool area, length, and width, which were selected as the shape features. Molten pool area was defined as whole pixels within the molten pool outline, molten pool length was defined as the maximal distance of the molten pool contour along the welding direction, and molten pool width was defined as the maximal distance of the molten pool contour perpendicular to the welding direction.

At the arc-striking stage, the cooling rate of the molten pool was relatively fast, the molten pool was difficult to spread out. With the increase of cladding layers, the workpiece tended to be bogged down, which is shown in Figure 2; in addition, the welding procedure is an important factor. The geometric dimensions of the workpiece are shown in Figure 3. The relative position of the molten pool in the image would change; therefore, the location feature of the molten pool is also an important feature. We extracted the wire extension as the location feature of the molten pool. Figure 4 is the schematic diagram of the visual features of the molten pool.

Figure 2. Workpiece formed by additive manufacturing.

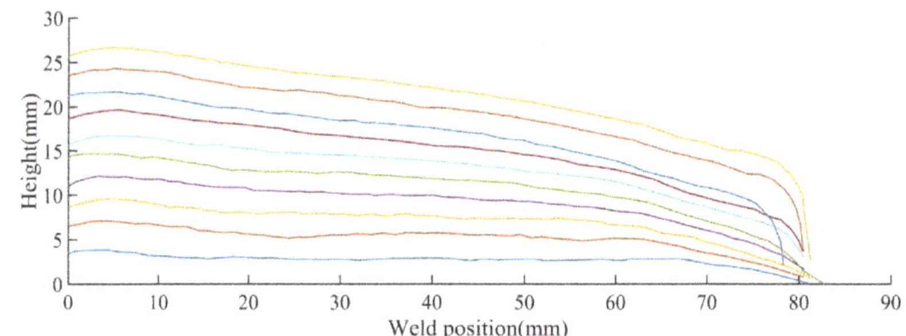

Figure 3. The geometric dimensions of the workpiece.

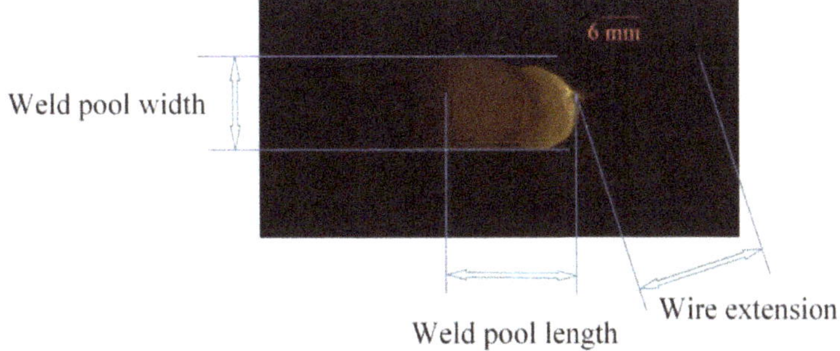

Figure 4. Schematic diagram of visual features of molten pool.

In this paper, the additive manufacturing experiment was conducted in the CMT procedure. The advantages of the CMT welding process are small splash, low heat input, and stable arc. After setting the welding parameters, the current waveform was collected in the actual welding process. Figure 5 [22] shows the welding current waveform of the CMT process under certain welding parameters, while, Figure 6 [22] shows all images collected by color CCD within a CMT cycle.

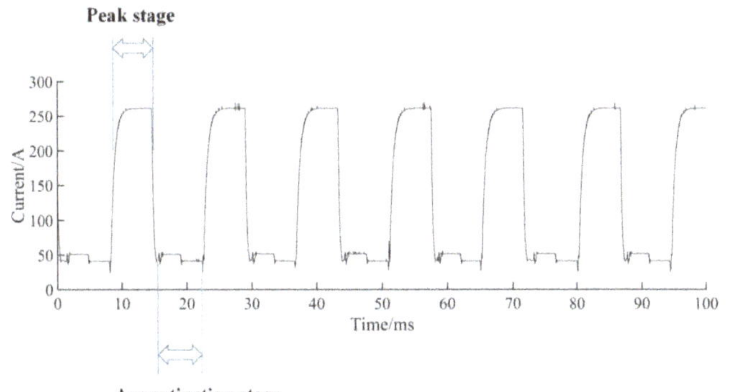

Figure 5. Welding current waveform of cold-metal-transfer (CMT) procedure.

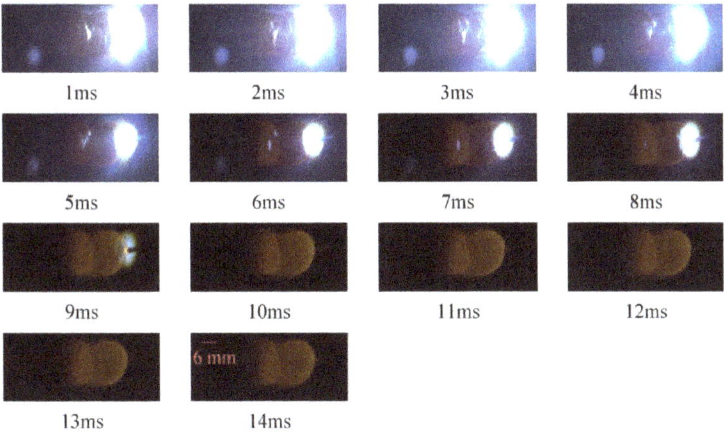

Figure 6. All images collected within CMT cycle.

Figures 5 and 6 show that the CMT cycle was approximately 14 ms, at the peak stage of CMT process. Because of the strong arc light, the collected molten pool image was seriously disturbed, and it was difficult to calculate the visual features of the molten pool. At the base stage of the CMT process, there was almost no arc interference, and the collected molten pool image had high signal-to-noise ratio (SNR), from which the shape and location features of the molten pool could be accurately extracted. In this paper, we chose the first frame image collected at the base stage within every CMT cycle, and extracted the shape and location features of the molten pool.

3.2. Extraction of Weld Reinforcement

In the GMA additive manufacturing process, after each cladding layer was welded, using a 3D scanner (Wiiboox REEYEE 3M) to measure the height of the workpiece (the 3D scanner is shown in Figure 7), getting the relationship between workpiece height and welding seam position, the height difference between adjacent cladding layers was defined as the weld reinforcement of the current cladding layer. Then, by means of the laser locating system, we could calculate the weld reinforcement corresponding to every frame of the molten pool image collected by color CCD.

Figure 7. Three-dimensional scanner.

4. Prediction-Model Establishment for Weld Reinforcement

Artificial neural networks have excellent nonlinear mapping characteristics and strong learning ability. In recent years, they have developed rapidly and are widely used in automatic control, prediction, and other fields. Back-propagation (BP) neural networks are multilayer feedforward neural networks that are trained by error back-propagation algorithms.

Taking the shape and location features of the current frame molten pool as inputs, that is to say, the molten pool area, length, width, and the wire extension as inputs, the weld reinforcement corresponding to the current frame as the output; by means of the BP neural network, the prediction model for weld reinforcement was developed. There were two hidden layers in the neural network. On the basis of the test results, two hidden layers were set with 10 neurons, respectively; the structure of the prediction model is shown in Figure 8.

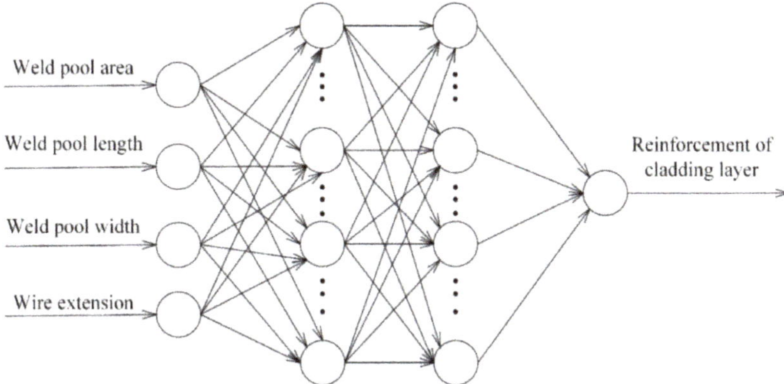

Figure 8. Structure of prediction model.

In the GMA additive manufacturing process, accumulating one layer every 3 min, that is to say, the cooling time between adjacent cladding layers was 3 min, according to the acquired data, we randomly selected 800 groups of data to constitute the training set in each cladding layer, and the other data constituted the test set; detailed data groups are shown in Table 3. Partial data of the shape and location features of the molten pool, and the corresponding weld reinforcement are shown in Table 4.

Table 3. Number of data groups within training and test sets.

Number of Cladding Layers	Training Set	Test Set
3	800	129
4	800	142
5	800	169
6	800	129
7	800	145
8	800	165
9	800	139
10	800	54

Table 4. Partial data of molten pool features and corresponding weld reinforcement.

Molten Pool Area (pixel)	Molten Pool Length (pixel)	Molten Pool Width (pixel)	Wire Extension (pixel)	Weld Reinforcement (mm)
19440	171	146	44	2.1551
25037	212	155	58	2.2781
24453	217	155	58	1.9829
22752	210	140	62	2.0542
21025	198	129	68	1.7656
20227	183	139	76	1.8179
19228	181	131	84	1.6907
16954	152	137	108	1.5000
16036	149	136	131	1.7075

Before training and testing the prediction model, all data needed to be mapped to [0,1], that is normalization, and the detailed calculation formula is:

$$x_{new} = \frac{x_{origin} - x_{min}}{x_{max} - x_{min}} \tag{1}$$

where x_{origin} and x_{new} are the numerical values before and after treatment, respectively; and x_{max} and x_{min} are the maximal and minimal numerical values within the original data set, respectively.

5. Predicted Results and Analysis

By testing the established prediction model, the predicted results of weld reinforcement in the third, fourth, and fifth cladding layers are shown in Figure 9.

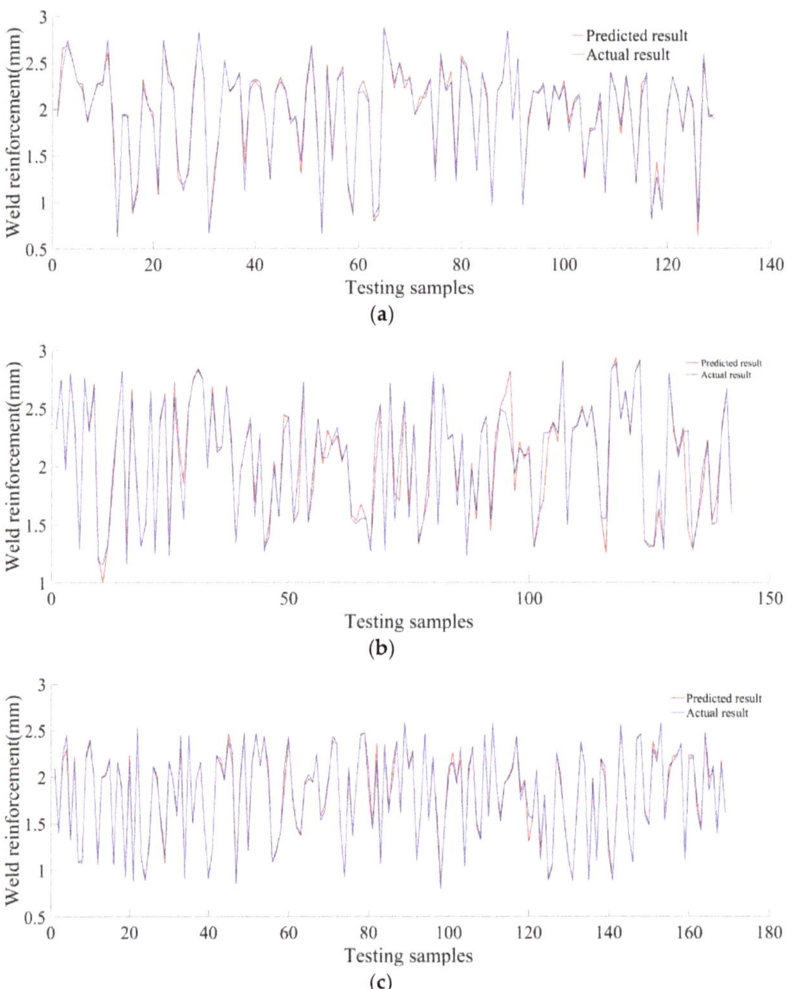

Figure 9. Predicted results of weld reinforcement. (**a**) Third cladding layer; (**b**) Fourth cladding layer; (**c**) Fifth cladding layer.

There are several reasons for the larger prediction error of individual samples, for example, the welding process is a non-linear and multivariable process, and there are some random uncertainties, which leads to the deviation of individual experiment results from the predicted results of the prediction model. In addition, there were some errors in the extraction and calculation of visual features of the molten pool.

The predicted precision of the developed model could be judged through mean absolute error (*MAE*); the calculation equation is as follows:

$$MAE = \frac{1}{m} \cdot \sum_{i=1}^{m} \left| h'(i) - h(i) \right| \tag{2}$$

where h' denotes the predicted weld reinforcement, h denotes the actual weld reinforcement, and m is the whole number of data groups within the test set.

The predicted precision of the established prediction model is shown in Table 5.

Table 5. Predicted precision of established model.

Number of Cladding Layers	MAE (mm)	Number of Cladding Layers	MAE (mm)
3	0.0476	7	0.0412
4	0.0699	8	0.0431
5	0.0388	9	0.0439
6	0.0313	10	0.0353

In the GMA additive manufacturing process, if cooling time between adjacent cladding layers is kept constant, the weld reinforcement of each cladding layer greatly fluctuates; the established prediction model in this paper could accurately predict weld reinforcement. In the GMA additive manufacturing process, the heat-dissipation condition in each cladding layer, and the change rule of the molten pool shape are different; therefore, there is no necessary relationship between MAE and the ordinal number of cladding layers. By reducing the number of data groups in the training set, the test accuracy of the model did not decrease, which indicated that the selected number of data groups in the training set was enough.

When all the data acquired from different cladding layers were combined into one data group, randomly selecting 150 groups of data as the test set, and the other data constituted the training set, the predicted results of weld reinforcement are shown in Figure 10; predicted MAE was 0.1094 mm.

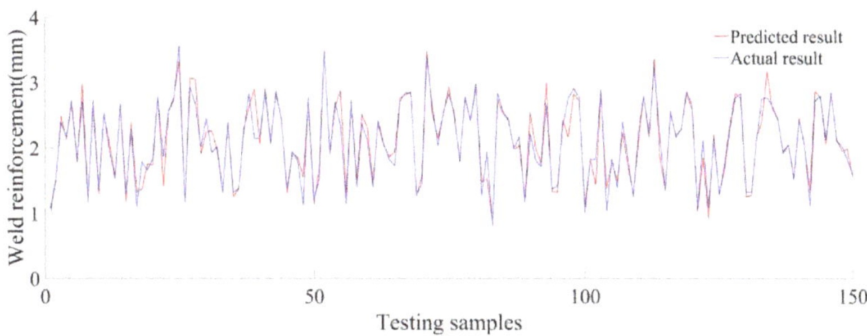

Figure 10. Predicted results of weld reinforcement.

6. Verification of Generalization Ability of Prediction Model

In the GMA additive manufacturing process, cooling time between adjacent cladding layers has great influence on the shape of the molten pool and the precision of the final forming dimension. In order to further verify the generalization ability of the prediction model for weld reinforcement, we changed the cooling time between adjacent cladding layers, with other welding parameters remaining unchanged.

In the case of cooling time between adjacent cladding layers being 4 min, according to the acquired data, we randomly selected 800 groups of data to constitute the training set in each cladding layer, and the other data constituted the test set, as shown in Table 6. The predicted precision of the established prediction model is shown in Table 7.

Table 6. Number of data groups within training and test sets.

Number of Cladding Layers	Training Set	Test Set
3	800	152
4	800	148
5	800	186
6	800	141
7	800	150
8	800	163
9	800	155
10	800	145

Table 7. Predicted precision of the established model.

Number of Cladding Layers	MAE (mm)	Number of Cladding Layers	MAE (mm)
3	0.0312	7	0.0335
4	0.0320	8	0.0304
5	0.0358	9	0.0348
6	0.0228	10	0.0583

When all data acquired from different cladding layers were combined into one data group, randomly selecting 150 groups of data as the test set, and the other data constituted the training set, the predicted results of weld reinforcement are shown in Figure 11; predicted MAE was 0.0777 mm.

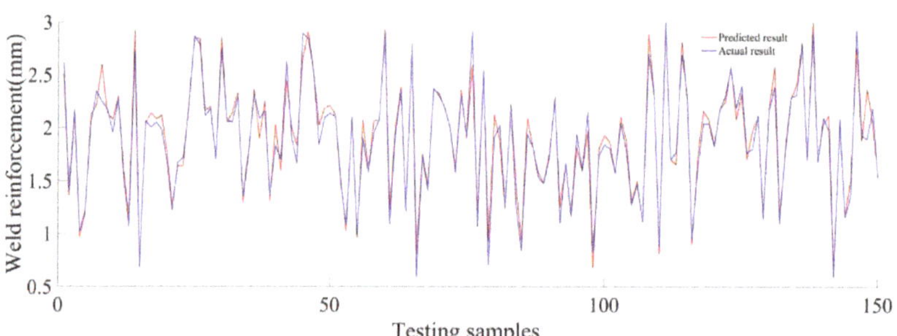

Figure 11. Predicted results of weld reinforcement.

In the case of cooling time between adjacent cladding layers being 2 min, according to the acquired data, we randomly selected 800 groups of data to constitute the training set in each cladding layer, and the other data constituted the test set, as shown in Table 8. The predicted precision of the established prediction model is shown in Table 9.

Table 8. Number of data groups within training and test sets.

Number of Cladding Layers	Training Set	Test Set
3	800	112
4	800	188
5	800	145
6	800	129
7	800	113
8	800	144
9	800	111
10	800	108

Table 9. Predicted precision of established model.

Number of Cladding Layers	MAE (mm)	Number of Cladding Layers	MAE (mm)
3	0.0352	7	0.0549
4	0.0473	8	0.0353
5	0.0365	9	0.0391
6	0.0274	10	0.0563

When all data acquired from different cladding layers were combined into one data group, randomly selecting 150 groups of data as the test set, and the other data constituted the training set, the predicted results of weld reinforcement are shown in Figure 12; predicted MAE was 0.0950 mm.

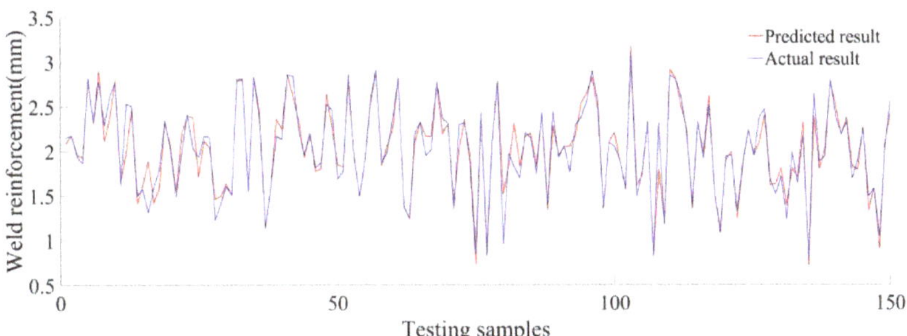

Figure 12. Predicted results of weld reinforcement.

Similarly, due to different heat-dissipation conditions in each cladding layer, the change rule of the molten pool shape was also different; there was no necessary relationship between MAE value and the cooling time of adjacent cladding layers. To sum up, we can conclude that the prediction model for weld reinforcement had high prediction precision and strong generalization ability, which could be applied to the GMA additive manufacturing process.

7. Conclusions

(1) A vision-sensing system for molten pool was established that can extract the shape feature of molten pool in real time. A laser positioning system was also developed to match every frame collected molten pool images with actual weld location.

(2) Taking the shape and location features of the current frame molten pool as input, the weld reinforcement corresponding to current frame as the output, by means of a BP neural network, the prediction model for weld reinforcement was developed; experiment results showed that the predicted MAE of the model was less than 0.11 mm. By changing cooling time between adjacent cladding layers, the generalization ability of prediction model was further verified.

(3) Future work will mainly focus on the study of the online control method for weld reinforcement in the GMA additive manufacturing process.

Author Contributions: Conceptualization, methodology, writing-original draft, R.Y.; investigation, writing-review and editing, Z.Z.; supervision, L.B. and J.H. All authors have read and agreed to the published version of the manuscript.

Funding: This work was supported by the National Natural Science Foundation of China (61727802 and 61901220), the Key Research and Development programs in Jiangsu China (BE2018126), and the Jiangsu postdoctoral research funding program (2019K216).

Conflicts of Interest: The authors declare no conflict of interest.

References

1. Lu, B.-H.; Li, D.-C. Development of the Additive Manufacturing (3D printing) Technology. *Mach. Build. Autom.* **2013**, *42*, 1–4.
2. Xiong, J.; Zhang, G. Adaptive control of deposited height in GMAW-based layer additive manufacturing. *J. Mater. Process. Technol.* **2014**, *214*, 962–968. [CrossRef]
3. Mughal, M.P.; Fawad, H.; Mufti, R.A. Three-Dimensional Finite-Element Modelling of Deformation in Weld-Based Rapid Prototyping. *Proc. Inst. Mech. Eng. Part C J. Mech. Eng. Sci.* **2006**, *220*, 875–885. [CrossRef]
4. Kazanas, P.; Deherkar, P.; De Almeida, P.M.; Lockett, H.; Williams, S. Fabrication of geometrical features using wire and arc additive manufacture. *Proc. Inst. Mech. Eng. Part B J. Eng. Manuf.* **2012**, *226*, 1042–1051. [CrossRef]
5. Wang, Y.; Chen, X.; Konovalov, S.V. Additive Manufacturing Based on Welding Arc: A low-Cost Method. *J. Surf. Investig. X-ray Synchrotron Neutron Tech.* **2017**, *11*, 1317–1328. [CrossRef]
6. Wang, Y.; Chen, X.; Konovalov, S.; Su, C.; Siddiquee, A.N.; Gangil, N. In-situ wire-feed additive manufacturing of Cu-Al alloy by addition of silicon. *Appl. Surf. Sci.* **2019**, *487*, 1366–1375. [CrossRef]
7. Zhao, Z.; Deng, L.; Bai, L.; Zhang, Y.; Han, J. Optimal imaging band selection mechanism of weld pool vision based on spectrum analysis. *Opt. Laser Technol.* **2019**, *110*, 145–151. [CrossRef]
8. Lu, J.; Zhao, Z.; Han, J.; Bai, L.-F. Hump weld bead monitoring based on transient temperature field of molten pool. *Optik* **2020**, *208*, 164031. [CrossRef]
9. Xiong, J.; Zhang, G. Online measurement of bead geometry in GMAW-based additive manufacturing using passive vision. *Meas. Sci. Technol.* **2013**, *24*, 115103. [CrossRef]
10. Wang, H.; Jiang, W.; Ouyang, J.; Kovacevic, R. Rapid prototyping of 4043 Al-alloy parts by VP-GTAW. *J. Mater. Process. Technol.* **2004**, *148*, 93–102. [CrossRef]
11. Ouyang, J.; Wang, H.; Kovacevic, R. Rapid prototyping of 5356-aluminum alloy based on variable polarity gas tungsten arc welding: Process control and microstructure. *Mater. Manuf. Process.* **2002**, *17*, 103–124. [CrossRef]
12. Spencer, J.D.; Dickens, P.M.; Wykes, C.M. Rapid prototyping of metal parts by three-dimensional welding. *Proc. Inst. Mech. Eng. Part B J. Eng. Manuf.* **1998**, *212*, 175–182. [CrossRef]
13. Kwak, Y.-M.; Doumanidis, C. Geometry Regulation of Material Deposition in Near-Net Shape Manufacturing by Thermally Scanned Welding. *J. Manuf. Process.* **2002**, *4*, 28–41. [CrossRef]
14. Doumanidis, C.; Kwak, Y.-M. Geometry Modeling and Control by Infrared and Laser Sensing in Thermal Manufacturing with Material Deposition. *J. Manuf. Sci. Eng.* **2000**, *123*, 45–52. [CrossRef]
15. Zhu, J.; Ling, Z.; Du, F.; Ding, X.; Li, H. Monitoring of laser metal-wire additive manufacturing temperature field using infrared thermography. *Infrared Laser Eng.* **2018**, *47*, 0604002.
16. Liu, Y.K.; Zhang, W.J.; Zhang, Y.M. Estimation of Weld Joint Penetration under Varying GTA Pools. *Weld. J.* **2013**, *92*, 313–321.
17. Gao, Y.; Zhao, J.; Wang, Q.; Xiao, J.; Zhang, H. Weld bead penetration identification based on human-welder subjective assessment on welding arc sound. *Measurement* **2020**, *154*, 107475. [CrossRef]
18. Wang, Q.; Gao, Y.; Huang, L.; Gong, Y.; Xiao, J. Weld bead penetration state recognition in GMAW process based on a central auditory perception model. *Measurement* **2019**, *147*, 106901. [CrossRef]
19. Yang, J.; Wang, K.; Wu, T.; Zhou, X. Welding penetration recognition in aluminum alloy tandem arc welding based on visual characters of weld pool. *Trans. China Weld. Inst.* **2017**, *38*, 49–52.

20. Zhang, Y.M.; Chen, Y.; Li, P.; Male, A.T. Weld deposition-based rapid prototyping: A preliminary study. *J. Mater. Process. Technol.* **2003**, *135*, 347–357. [CrossRef]

21. Xiong, J.; Zhang, G.; Gao, H.; Wu, L. Modeling of bead section profile and overlapping beads with experimental validation for robotic GMAW-based rapid manufacturing. *Robot. Comput. Manuf.* **2013**, *29*, 417–423. [CrossRef]

22. Yu, R.; Han, J.; Zhao, Z.; Bai, L. Real-Time Prediction of Welding Penetration Mode and Depth Based on Visual Characteristics of Weld Pool in GMAW Process. *IEEE Access* **2020**, *8*, 81564–81573. [CrossRef]

© 2020 by the authors. Licensee MDPI, Basel, Switzerland. This article is an open access article distributed under the terms and conditions of the Creative Commons Attribution (CC BY) license (http://creativecommons.org/licenses/by/4.0/).

Review

A Review of SLMed Magnesium Alloys: Processing, Properties, Alloying Elements and Postprocessing

Shuai Liu and Hanjie Guo *

Beijing Key Laboratory of Special Melting and Preparation of High-End Metal Materials, School of Metallurgical and Ecological Engineering, University of Science and Technology Beijing, Beijing 100083, China; lsls1226@163.com
* Correspondence: guohanjie@ustb.edu.cn; Tel.: +86-138-0136-9943

Received: 5 July 2020; Accepted: 31 July 2020; Published: 9 August 2020

Abstract: Selective laser melting (SLM) is an additive manufacturing method with rapid solidification properties, which is conducive to the preparation of alloys with fine microstructures and uniform chemical compositions. Magnesium alloys are lightweight materials that are widely used in the aerospace, biomedical and other fields due to their low density, high specific strength, and good biocompatibility. However, the poor laser formability of magnesium alloy restricts its application. This paper discusses the current research status both related to the theoretical understanding and technology applications. There are problems such as limited processable materials, immature process conditions and metallurgical defects on SLM processing magnesium alloys. Some efforts have been made to solve the above problems, such as adding alloy elements and applying postprocessing. However, the breakthroughs in these two areas are rarely reviewed. Due to the paucity of publications on postprocessing and alloy design of SLMed magnesium alloy powders, we review the current state of research and progress. Moreover, traditional preparation techniques of magnesium alloys are evaluated and related to the SLM process with a view to gaining useful insights, especially with respect to the postprocessing and alloy design of magnesium alloys. The paper also reviews the influence of process parameters on formability, densification and mechanical behavior of magnesium. In addition, the progress of microstructure and metallurgical defects encountered in the SLM processed parts is described. Finally, this article summarizes the research results, and with respect to materials and metallurgy, the new challenges and prospects in the SLM processing of magnesium alloy powders are proposed with respect to alloy design, base material purification, inclusion control and theoretical calculation, and the role of intermetallic compounds.

Keywords: selective laser melting; magnesium alloys; properties

---

1. Introduction

Magnesium is one of the most abundant elements on Earth and represents approximately 2.5% of its composition. Magnesium and its alloys are lightweight metallic structural materials with certain advantages, such as low density, high specific strength and high stiffness [1,2]. Magnesium and its alloys are considered to have great application prospects in the aerospace, transportation, electronics, biomedicine, and energy fields due to their excellent physical and chemical properties, such as low density, good damping performance, biocompatibility, recyclability, large hydrogen storage capacity, and high theoretical specific capacity [3–5]. As the lightest structural material currently available, magnesium alloys have the potential to replace steel and aluminum in many structural applications [6]. Thus, magnesium alloys have already found considerable applications in various fields, including the aerospace, aircraft, automotive, and electronics fields; in particular, magnesium die castings have been widely used in the automotive industry [7]. However, despite the abovementioned advantages, magnesium and its alloys still face many difficulties in large-scale industrial applications. For instance, the poor room temperature plasticity and poor corrosion resistance of these materials still need to be addressed [8,9]. Currently, magnesium alloy development is focused on the production of complex structures with high efficiency and minimal environmental impact; accordingly, many new magnesium alloy preparation technologies have emerged.

Additive manufacturing (AM) is a promising new technology that can dramatically change the way components are manufactured in many different industries and greatly increase manufacturing efficiency. According to different technologies, AM can be divided into electron beam melting (EBM), direct laser forming (DLF) and selective laser melting (SLM). SLM is a technology widely used in the preparation of metal powders. Commonly used metal powders include Fe-based alloys, titanium-based alloys, Al-based alloys, Mg-based alloys, and nickel-based alloys. SLM uses a high-energy laser beam to completely melt metal powder in a protective atmosphere along a defined laser path, and this molten metal rapidly solidifies [10]. By repeating this step and overlapping subsequent layers, a three-dimensional component is eventually formed. SLM provides a means of manufacturing geometrically complex structures, eliminating the need to build molds, which would otherwise require a considerable amount of time and money to manufacture. The cooling rate of the molten pool reaches 10^3–10^8 K/s due to the rapid movement of the laser and the molten metal pool [11]. The characteristics of rapid solidification and layered manufacturing enable the SLM process to produce materials with a more uniform chemical composition, a more concentrated solid solution, a refined microstructure, and better mechanical properties [12,13]. However, due to the inherent heat treatment of the SLM process, wherein each layer is cyclically reheated by the deposition of subsequent layers, the microstructure of the material is also unique [14]. To date, the samples produced by SLM still have some defects, such as pores and cracks, which affect the final use of the produced parts. Therefore, how to control the material density, microstructure and performance by adjusting the SLM process conditions is a research focus.

Due to several characteristics of magnesium alloys, including a low melting point, easy oxidation, and dangerous production, research on the preparation of these materials is still in its infancy worldwide. Research institutions investigating these materials include the Fraunhofer Institute for Laser Technology, Huazhong University of Science and Technology, University of Science and Technology Beijing, Central South University, Université de Technologie de Belfort-Montbéliard, Chongqing University, Hong Kong Polytechnic University, Suzhou University, and Delft University of Technology. Current research is in the small-scale trial production stage in the laboratory, and research and experimental data on the forming characteristics and mechanical properties of selective laser melted (SLMed) magnesium alloys are scarce. On the one hand, owing to the inherent physical properties of magnesium alloys, the research progress of SLMed magnesium alloys is limited. On the other hand, during the production of SLMed magnesium alloys, the generation of defects and the characteristics of microstructures affect the performance of SLMed magnesium alloys. For instance, in the Mg-Al-Zn(AZ) series of magnesium alloys, the presence of the second phase $Mg_{17}Al_{12}$ limits the mechanical properties—especially the elongation—of the alloy, which restricts the further development of SLMed magnesium alloys. At present, many scholars have conducted research and comprehensive reviews on the spheroidization, defects, porosity, and alloying element loss in magnesium alloys and other metals prepared by SLM [15–21]. However, limited processable materials, immature process conditions and metallurgical defects are still problems that magnesium alloys need to face and solve in the SLM process. In the past two years, these three issues of SLMed magnesium alloys have been substantially improved by optimizing process parameters, introducing post treatment and adjusting different alloying elements. However, few review articles have been written in this regard. Therefore, it is necessary to summarize the development of SLMed magnesium alloys from the perspectives of process, element adjustment and post treatment. To improve the machinability of the material, this paper will review the research progress of the addition of alloying elements and the post-treatment to expand the processable magnesium alloy materials. Process conditions and new research progress on relative density, microstructure, mechanical properties and corrosion resistance of SLMed magnesium alloys will be reviewed. In addition, the formation mechanism of metallurgical defects, especially oxidation and cracks, will be discussed and analyzed to provide a reference for the application of SLMed magnesium alloys.

This article also discusses the aspects that need to be addressed in future research on SLMed magnesium alloys, which provides a basis for further research and development of these materials.

2. Formation and Energy Density of SLMed Magnesium Alloys

There are many process parameters that affect the forming of SLMed magnesium alloys. The selection of process parameters directly affects the balling degree, relative density, microstructure, and mechanical properties of magnesium alloys. The selection of the best process parameters is the key to ensuring good sample formation [22–24]. The most commonly studied process parameters are laser power, scanning speed, hatch spacing, and layer thickness. These four process parameters can be expressed by a comprehensive evaluation index: energy density E_v. The definition of E_v is shown in Equation (1):

$$E_v = \frac{P}{vHT}(J/mm^3) \tag{1}$$

where P is the laser power, v is the scanning speed, H is the hatch spacing, and T is the layer thickness. Energy density integrates the key process parameters, which can more intuitively reflect the energy applied by the laser to the powder, to evaluate the heating, melting, and evaporation of the powder. Tables 1 and 2 summarize the research progress of SLMed magnesium alloys in terms of alloy grades, powder characteristics, forming quality, performance, optimal energy density, and research institutions.

As shown in Tables 1 and 2, the research materials involve magnesium and its alloy powders, and different research institutions have studied the mechanical properties and corrosion resistance of SLMed magnesium alloys under best processing parameters. There are differences in the optimal energy input for forming magnesium alloys of the same alloy grade. First, the optimal energy input is related to the properties of the powder. Samples containing different particle sizes and particle size distributions require different amounts of energy to fully melt the powder. Second, there are differences in the research equipment used by different research institutions, which creates differences in process conditions. The introduction of different alloying elements into magnesium powder improves its mechanical properties and corrosion resistance. To further analyze the relationship between forming and magnesium alloy properties in the SLM process, the relationship between the energy density and the magnesium alloy grade in Table 1 is determined, as shown in Figure 1.

We summarized the effects of energy input on the forming, relative density, microhardness, grain size, and mechanical properties of different magnesium alloy series. It can be seen that during the SLM process, the energy input will have certain macroscopic and microscopic effects on the magnesium alloy, including the formation of pores and microstructures. This will be analyzed and discussed in detail in the following chapters.

Table 1. Summary of research progress on SLM magnesium and magnesium alloy powder.

Types	Powder Shape	Partical Size (μm)	Processing	Relative Density (%)	Ultimate Tensile Strength (UTS) (MPa)	Yield Strength (YS) (MPa)	Elongation (EL) (%)	Microhardness (Hv)	Laser Power (W)	Scanning Speed (mm/s)	Energy Density (J·mm^{-3})	Refs.
AZ91	irregular	25–63	SLM	>99	329 ± 6	264 ± 1	3.7 ± 0.8	—	100	800	104	[25]
WE43	irregular	25–63	SLM	>99	307 ± 6	302 ± 3	11.9 ± 1	—	200	700	238	[25]
AZ91D	spherical	59	SLM	99.52	296 ± 2	254 ± 4	1.8 ± 0.2	100	200	333	167	[26]
60	spherical	30	SLM	94.05	-	-	-	78 ± 10	200	300	416	[27]
AZ61	spherical	48	SLM	99.4 ± 0.3	287 ± 3	233 ± 2	3.1 ± 0.1	—	150	400	156	[28]
AZ61	spherical	70	SLM	99 ± 5	-	-	-	93 ± 4	80	3	9609.6	[29]
ZK60	spherical	50	SLM	97.4 ± 2	-	-	-	89.2 ± 5	50	8	600	[30]
ZK30-xAl	spherical	ZK30: 45–74; Al: 5–15	SLM	-	-	-	-	75.7 ± 6	80	3	4004	[31]
Mg-9Al	Mg: irregular Al: spherical	Mg: 42 Al: 17	SLM	82 ± 3	-	-	-	80 ± 7	120	300	93.75	[32]
Mg	spherical	25.8, 43.32	SLM	96.13	-	-	-	52.4	90	10	300	[33]
Mg	irregular	75–150	SLM	-	-	-	-	-			—	[34]
Mg	spherical	5–45	SLM	-	-	-	-	-	20	20	0.99J/mm^2	[34]
Mg	spherical	5–45	SLM	-	-	-	-	0.87 ± 0.13 GPa	13–26	10–200	1.27 × 10^9 J/mm^2 (CW)	[35]
Mg	spherical	5–45	SLM	-	-	-	-	0.95 ± 0.08 GPa	13–26	10–200	1.13 × 10^{12} J/mm^2 (PW)	[35]
Mg	spherical	10–45	SLM	-	-	-	-	0.72 ± 0.07 GPa	18	1	118.2 J/mm^2	[36]
Mg	spherical	1–10	SLM	—	-	-	-	48.3	50	5	100 J/mm^2	[37]
Mg-2Ca	spherical	100–200	LAM	-	111.19	-	-	68	100	10	1200	[38]
Mg-xSn	spherical	Mg: ~10 Sn: ~1	SLM	-	-	-	-	65.7	60	11	107.4 J/mm^2	[39]
Mg-3Zn-xDy	Mg-3Zn spherical Dy: irregular	Mg-3Zn: ~150 Dy: ~130	SLM	-	-	-	-	121.3 ± 3	20	3	360.4 J/mm^2	[40]
AZ61	spherical	48	SLM + HIP	>99	230 ± 2	120 ± 1	-	98.9 ± 5.9	-	-	156	[41]
ZK61-xZn	spherical	50	LAM	69.1 ± 1	-	-	-	106.8 ± 2	90	10	1146	[42]
Mg-Zn	spherical	Mg: ~50 Zn: ~31	SLM	99.35 ± 0.2	148 ± 5	-	11 ± 0.6	50 ± 1	180	700	183.7	[43]
Mg-Gd-Zn-Zr	spherical	44	SLM	99.95 (53.3 J/mm^3)	332 ± 5	325 ± 5	4 ± 0.2	-	80	300	88.9	[44]

Note: CW represents continuous wave, PW represents pulsed wave, and HIP represents hot isostatic pressing.

Table 2. Summary of the biodegradability of SLMed magnesium alloys.

Types	Powder Shape	Particle Size (μm)	Processing	Biodegradation Rate	Refs.
AZ61	spherical	70	SLM	12.26 mg/cm^{-2}	[29]
ZK60	spherical	50	SLM	0.006 mL cm^{-2} h^{-1}	[30]
ZK30-xAl	spherical	ZK30: 45–74; Al: 5–15	SLM	0.17 ± 0.02 mg cm^{-2} day^{-1}	[31]
Mg	irregular	75–150	SLM	Fail	[34]
WE43	spherical	25–60	SLM	0.17 mL/cm^2	[45]
ZK60-Cu	spherical	ZK30: 50; Cu: 80 nm	SLM	Close to 1.01 mm y^{-1}	[46]

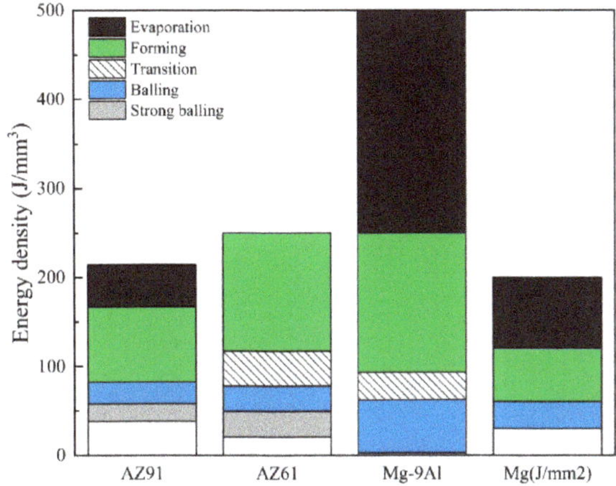

Figure 1. Energy density range of the SLMed magnesium and its alloys [26,28,32,37]. (This picture was drawn by the author; the data in the figure come from citations).

Figure 1 summarizes the energy density required for the formation of several common magnesium alloys during SLM. The forming characteristics of the SLMed magnesium alloy show five feature areas with increasing energy density. These five feature areas can be divided into a strong balling zone, a balling zone, a transition zone, a forming zone and an evaporation zone. Balling refers to some spherical particles existing on the surface of the SLMed sample. There are many mechanisms affecting spheroidization. Severe spheroidization is closely related to the powder characteristics, laser energy density, scanning strategy, wettability between powders, and the Marangoni effect, and there have been many conclusions about spheroidization [16,47–49]. The following summarizes how to control the occurrence of spheroidization from the process point of view and the relationship between the types of spheroidization, pores and other defects and process parameters.

A series of balling particles agglomerate to form a large molten pool of solid–liquid phase or near-liquid phase. The lack of wettability between the molten pool and the previous layer causes a balling effect [19]. The formation of magnesium alloy in the SLM process of each region is summarized hereafter, and the forming characteristics of the SLMed magnesium alloy are shown in Figure 2.

Strong balling zone: SLM forming in the range of high scanning speeds or low laser powers (low energy density), which is characterized by insufficient heating of the magnesium alloy powder, and the temperature of some powders does not reach the melting point temperature and fails to completely melt [26]. Taking AZ61 as an example, when the scanning speed is as high as 1000–1800 mm/s, in fact, the forming is in a solid–liquid state and there is poor bonding between the powder particles, so the sample has a powder accumulation structure without any mechanical strength at this processing parameters; moreover, some samples have a loose metal structure [32]. Balling is a complex metallurgical process that is caused by the instability of the molten pool and the Marangoni effect. The surface characteristics are shown in Table 3 and Figure 2a,e. Liu et al. [28], under an optical microscope, revealed that the pore (size greater than 100 μm) shape of the sample surface in the strong balling area was mostly meniscus-shaped [50]. A similar phenomenon was also found in the preparation of SLMed ZK60 magnesium alloy [27]. At a high speed of 900 mm/s (low energy density), the surface of the magnesium alloy was covered with irregularly shaped pores due to incomplete powder melting, the relative density decreased significantly to 82.25%, and the evaporation was very weak in this region. In addition, too high a scanning speed will cause pores on rough surfaces [30]. A significant heat-affected zone (HAZ) also formed during the orbital melting process. The HAZ was formed by partial melting of particles due to thermal conduction from the center of the molten pool to the adjacent powder. Low energy density also means that the scanning speed was too fast or the laser power was too low. If the scanning speed is too fast, due to the low density and chemical activity of magnesium, more powder is blown upwards and oxidized to form black fog (MgO), contaminating the mold cavity [33]. Balling also occurs at very low scanning speeds because the liquid surface energy and liquid lifetime are reduced in a short length range [51]. Balling occurs due to thermal stress and weak interlayer bonding, resulting in surface deterioration. Therefore, the scanning speed cannot be too high to ensure sufficient energy input.

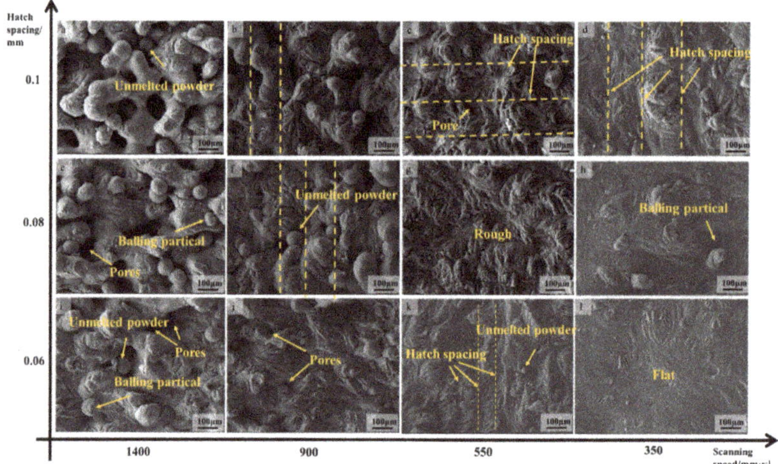

Figure 2. Forming characteristics of SLM magnesium alloy. When the scanning speed is 1400 mm/s, 900 mm/s, 550 mm/s and 250 mm/s respectively, the hatch spacing is 0.06 mm (**a–d**); hatch spacing of 0.08 mm, (**e–h**); and hatch spacing of 0.1 mm, (**i–l**). Reproduction from [28], with permission from Elsevier (Amsterdam, The Netherlands), 2020.

Balling zone: Compared with the strong balling zone, this zone has decreased scanning speeds (800–950 mm/s) or increased laser power (increased energy density to 50–78.13 J/mm^3). Due to the decrease in scanning speed, the interaction time between powder and laser is prolonged, and the temperature of some particles is higher than the melting point, which causes the powder to partially melt. In this case, adjacent particles are sintered together due to the formation of a small amount of liquid [37]; however, the balling zone is still in a liquid–solid state. For SLMed AZ91, the energy density in this range was 66–77 J/mm^3, and the molten pool with a smaller circumference was unstable under relatively low energy input [26]. At the same time, the balling particles tended to grow along the direction of the scanning trajectory, which was attributed to the melting track breaking up into discrete droplets [52]. Therefore, the balling particles parallel to the scanning direction together with the large amount of unmelted powder induced more severe surface deformation (Figure 2f). The number of defects decreased when the scanning speed and hatch spacing decreased (i.e., the energy density E_v increased), because the wettability between the powder and the substrate improved and the balling effect weakened. The decreased scanning speeds (increased energy density) caused the powder to be sufficiently heated and melted, and the adjacent scanning tracks overlapped, thereby reducing the internal pore size. In addition to the reduction in the internal pore size, the surface pore size was also reduced with the decreased scanning speeds [30]. It is note that the size of the pores was also related to the vertical height of the sample, and the size of the pores increased with increasing printing height [18]. The cause of this phenomenon is still related to energy. Based on the characteristics of the layered manufacturing technology, the thin layer produced by the initial printing layer directly contacts the substrate and melts rapidly. As the number of printing layers increases, the heat transferred from the substrate to the powder decreases, and the pore diameter increases. This phenomenon is caused by the energy input being too low, causing the powder to incompletely melt, thereby forming pores. In addition, the adjustment of energy input is achieved by adjusting process parameters. Therefore, the key to controlling the forming quality is to adjust the optimal scanning speed and laser power to provide suitable forming energy density for magnesium alloy.

Transition zone: In this zone, for the AZ61, the scanning speed further decreased (550–750 mm/s). A "hill" formed between the scanning trajectories, which was composed of a series of "melt beads" instead of a completely continuous melting track [34]. Under the effect of decreased scanning speeds, energy input increased, so the instantaneous temperature gradient from the center of the track to the edge increased, the surface tension decreased, the viscosity of the metal droplets decreased, and the surface "melt beads" (ball-shaped particles) were linked together under a smoother flow, which resulted in the formation of scanning trajectories at the "hill" and "valley" positions (Figure 2d,k). Hence, the morphology of the location of the "hill" part was affected by the processing parameters. However, for the Mg-9Al alloy, due to the high scanning speed, even if the laser power is reduced below 30 W (the energy density is small), the forming quality cannot be optimal.

Forming zone: In this region, the powder completely melted into a liquid phase under sufficient and suitable processing parameters (scanning speed of 250–500 mm/s) (Figure 2i). For AZ series magnesium alloys, as the scanning speed decreases, the forming quality becomes better. For Mg-9Al alloy, the laser power is reduced to a lower level (10–20 W), and the best sample can be obtained by adjusting the scanning speed. The temperature of the liquid phase was a certain level above the liquidus temperature. Then, the liquid phase rapidly diffused and solidified, and the melted powders were well combined to form a continuous and smooth trajectory [37]. As the scanning speed decreases or the laser power increases, the interaction time between the powder and the laser increases, and the heat received per unit time increases. Therefore, the energy density in this region increased the temperature of the powder bed and reduced the viscosity of the melt pool, so the molten part can be spread in the layer, thereby promoting more effective densification of the solid powder particles. However, distortion of scanning trajectories caused by flocculated deposition and metal powder evaporation can still be found on some sample surfaces [26].

Evaporation zone: As the laser power continued to increase or the scanning speed continued to decrease, the energy input became excessive and increased the temperature of the molten pool, causing the molten alloy system to reach the temperature of the vapor line, resulting in the evaporation of some alloy components. For Mg-9Al alloy, when the laser power is higher than 60 W, even if the scanning speed increases (~1000 mm/s), the evaporation phenomenon cannot be avoided. Magnesium has a low boiling point (1093 °C) and a high vapor pressure, and the magnesium component was severely evaporated during the forming process when the energy density was too high. This evaporation led to rapid expansion of the system, which caused a recoil pressure on the molten pool [26]. The molten pool was blown away from the surrounding unmelted powder, and the SLMed magnesium alloy failed. In the protective gas chamber, MgO powder and fog were formed. At this time, the Mg:Zn ratio in the sample increased due to evaporation [27]. After each layer of powder was sintered, only an extremely thin metal layer solidified on the substrate, and the shape was irregular [32]. The excessive energy input also reduced the viscosity of the molten pool because of the combined effect of excessive shrinkage of the melting track and high residual stress. The melting track will completely melt or even break, resulting in many visible cracks [24] and deformation.

There is a close relationship between the processing parameters (scanning speed and laser power) and the forming quality of SLMed magnesium alloy samples. The forming characteristics of magnesium alloys vary with respect to different processing parameters. In addition, the optimal energy density for suitable forming of magnesium alloy was 83–250 J/mm^3. From the above summary, it can be seen that the effects of laser power and scanning speed should be considered comprehensively in order to obtain the best quality samples. For magnesium alloys, only changing the laser power or scanning speed may not be able to obtain the best forming quality, and it is necessary to comprehensively adjust these two process parameters. However, research on the influence of laser power on SLMed magnesium alloy forming is still lacking, and needs to be further studied. It can also be seen that there were some differences in the processing parameters of different grades of magnesium alloys during the SLM process. The main consideration was how to adjust the scanning speed and laser power to get the suitable energy density, find the most suitable melting, wetting, spreading, bonding, and solidification conditions, and finally form a magnesium alloy sample with high relative density and good surface quality. These issues are discussed in the following sections.

Table 3. Processing parameters and forming quality of SLMed magnesium alloy [26,28,32].

Surface Feature	Alloy	Scanning Speed (mm/s)	Energy Density (J/mm^3)	Unmelted Powder	Pores	Balling
Strong Balling zone	AZ61	1000–1800	35–45	Lots of	Lots of, ~14%, >100 μm, connected to each other, a network	Spherical: ~100 μm Ellipsoidal: 100–300 μm
	AZ91	1000	38–58	Lots of	Lots of, ~10%	Spherical: ~100 μm Ellipsoidal: ~300 μm
	Mg-9Al	-	-	-	-	-
Balling zone	AZ61	800–950	63–78	Less	Less, ~4%, <50 μm,	Spherical: <50 μm
	AZ91	833	66–77	Lots of	Lots of, ~4%	Spherical: ~100 μm
	Mg-9Al	15–20 W: 160–1000	3–47	-	Poor bond neck	Powder stacking
Transition zone	AZ61	550–750	83–114	None	None	Less/scanning tracks
	AZ91	-	-	-	-	-
	Mg-9Al	10 W: 40 15–30 W: 80	63–94	-	Loose metal structure	-
Forming zone	AZ61	250–500	125–250	None	None	None
	AZ91	333–667	83–167	None	None	None/flocculent depositions, scanning tracks
	Mg-9Al	10 W: 10 15 W: 20–40 20 W: 40	94–250	None	None	-
Evaporation zone	AZ61	-	-	-	-	-
	AZ91	~166	214–429	Fail	Fail	Fail
	Mg-9Al	90–110 W: 0.01–1	~2750	Fail	Fail	Fail

3. Properties of SLMed Magnesium Alloy

3.1. Relative Density of SLMed Magnesium Alloy

Relative density is often used as an indicator of the quality of SLMed parts. The relative density is the ratio of the density obtained by SLM to the theoretical density of the bulk material. The theoretical density of a material can be calculated from the atomic weight and crystal structure of the material. Figure 3 summarizes the trend of the relative density of different series of magnesium alloys as a function of energy density. The relative density ranged from 73 to 99%, and the variation in relative density under different energy densities was large.

Figure 3 shows that the energy density under the SLM process has an important effect on the relative density of the magnesium alloy [26–28,30,32,44]. From the perspective of the AZ61, AZ91, ZK series, Mg-9Al alloys and Mg-Gd-Zn-Zr with increasing energy density, the relative density of the alloy increased.

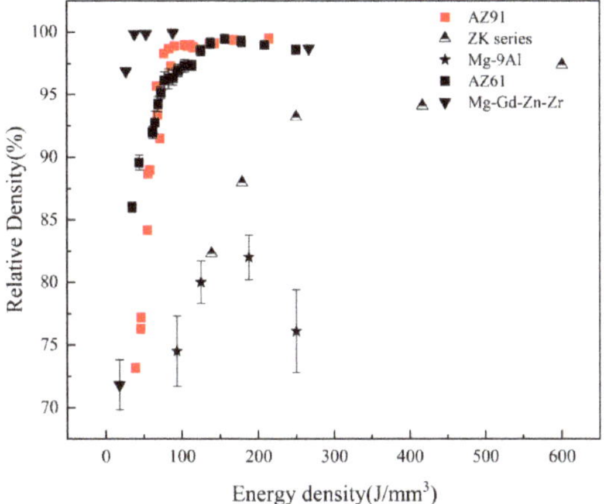

Figure 3. Relationship between energy density and relative density [26–28,30,32,44].

Additionally, Figure 3 shows that in the curve of the relative density of magnesium alloy samples as a function of energy density, the initial slope of the curve was very high. As the energy density increased, it generally increased rapidly at first and then tended to flatten or even slightly decrease. The temperature of the molten pool will change accordingly when the energy density changed [53,54]. The impact of energy density on the relative density of SLMed magnesium alloys can be considered from two aspects: (1) the macroscale effects of the energy density on the melting of the powder; and (2) the microscale effects of the energy density on the solid solution of the elements, which affects the relative density of the SLMed magnesium alloys.

The increase in scanning speed and hatch spacing reduced the peak temperature and temperature gradient of the melt. When the scanning speed and scanning distance were large (lower energy density) [55], heat and mass transfer occurred due to surface tension. Generally, higher surface tension near the edge of the molten pool will pull the molten liquid out of the center of the molten pool, and cracks occur because adjacent melting tracks cannot overlap. Moreover, as the hatch spacing increased, the cracks became wider, which directly affects the final relative density of the sample [56]. The high dynamic viscosity of the laser melt pool was considered to be the main reason for the cracks formed at high scanning speeds (low energy density). The definition of dynamic viscosity is shown in Equation (2) [30]:

$$\mu = \frac{16}{15} \sqrt{\frac{m}{kT}} \lambda \tag{2}$$

where μ is the dynamic viscosity, λ is the surface tension, k is the Boltzmann constant, m is the atomic mass, and T is the temperature of the liquid pool. At low energy density E_v (higher scanning speed), the temperature T in the molten pool was relatively low, and the magnesium alloy molten liquid had a high dynamic viscosity, which was not conducive to the smooth diffusion of the liquid. Additionally, incomplete melting resulted in the formation of large balling particles, and the scanning trajectory was discontinuous. Therefore, it was easy to form pores, and the relative density was low.

As the laser energy density increased, the powder melted more fully, the liquid phase fluidity improved and penetrates into the voids between the particles, so the pores were dispersed and the pore size was reduced, forming a relatively smooth surface, and the density of the magnesium alloy also increased [16].

However, when the hatch spacing was too small, a large amount of molten liquid migrated to the original scanning trajectory at a higher speed, resulting in the accumulation of molten liquid, which affected the density of the magnesium. At high energy densities, the molten pool widened, forming a temperature gradient in the molten pool, exacerbating the change in surface tension, which caused balling and solidification at the edge of the molten pool [57].

The relative density is also related to the solid solution of the elements. A study reported that the relative density increased as the solid solution of Al increased [28]. For the AZ series magnesium alloy, the change in relative density with respect to energy density was no longer obvious after the relative density of the material reached approximately 98%. The relative density of the material decreased when the energy density increased from 100 to 178 J/mm^3. The decrease in relative density was also related to the solid solution of Al. According to the solid solution theory, Al can be dissolved into the α-Mg matrix as a substitutional atom. During rapid solidification, the solid solution of Al will increase in the α-Mg matrix due to the solute retention effect [58,59]. At a suitable energy density, the increase in the melt temperature and the solute retention effect work together to increase the solid solution of Al in the magnesium alloy matrix. However, if the energy density is further increased, the solid solution of Al will decrease due to the weakened solute retention effect. The density of Al is 2.7 g/cm^3, which is higher than the density of pure Mg (1.7 g/cm^3) [60,61], so the solid solution of Al will increase the relative density of the sample.

It can be seen that the energy density has an effect on the relative density of the SLMed magnesium alloy. If the energy density is too low, the powder cannot be fully melted, the system is in a solid–liquid two-phase state, and the surface tension and viscosity of the liquid increase, resulting in the liquid not flowing smoothly and agglomerating into spheres. At the same time, pores are formed, which ultimately results in the sample not being dense. However, the energy density is too high, the powder will evaporate. Moreover, the solute capture effect is weakened, and the relative density of the sample is reduced due to the decrease in solid solution elements. Therefore, the preparation of dense magnesium alloys requires the energy density to be controlled in a suitable range, the solid solution of Al to be enhanced, and the surface tension to be reduced so that the liquid diffuses at a stable rate and fills the pores, thereby eliminating pores and spherical particles between the tracks [57].

According to the current research results, for AZ61 and AZ91, the relative density of the alloy was close to the theoretical density when the energy density was approximately 100–200 J/mm^3. However, the relative density of the ZK series and Mg-9Al alloy was only approximately 75–93% at this energy density. Consequently, the optimal energy density must be determined for the different alloy components in the SLM process.

3.2. Microhardness of SLMed Magnesium Alloy

From the perspective of the SLM preparation process, the energy density also has an important effect on the microhardness of different magnesium alloys, as shown in Figure 4 [26,30,32,33,37,41].

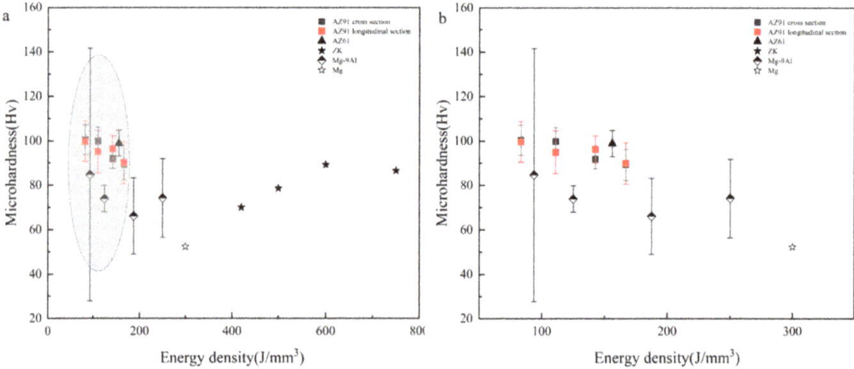

Figure 4. (a) Effect of energy density on microhardness; (b) enlarged image of the gray area in (a) [26,30,32,33,41].

The enlarged view in Figure 4b shows that the microhardness decreases slightly with increasing energy input. This phenomenon was mainly attributed to the growth of grain size and the weakening of the solute capture effect. Figure 5 shows the effect of energy density on grain size. [25,26,28–30,32,37–40,44].

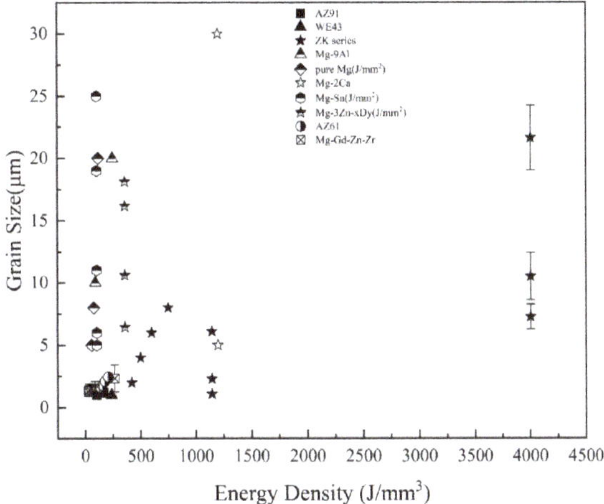

Figure 5. Effect of energy density on grain size [25,26,28–30,32,37–40,44].

First, Figure 5 and Table 4 show that the grain size of magnesium alloys of different alloy series was affected by temperature and increased with increasing energy density. The grain size of SLMed magnesium alloy was mainly 1–15 μm, and the maximum grain size was 30 μm [26]. Compared with as-cast samples, the SLMed samples had significantly smaller grains. According to the Hall–Petch formula, grain refinement can significantly improve microhardness. Conversely, coarsening of the grain size causes a decrease in microhardness. With the introduction of alloying elements, the grain size of the SLMed magnesium alloy changed [31,40]. The grain size of magnesium alloys increased as the Dy content increased, and the microhardness decreased significantly (Figure 6a) [40]. Microhardness is also affected by microstructure. In the same magnesium sample, the microhardness changed with respect to the microstructure at different positions [26,29,62]. This phenomenon is due to different locations in the same sample experiencing different thermal histories. During the SLM process, the microhardness at the center of the molten pool and the microhardness at the edge zone fluctuated, which was caused by the refinement of the microstructure at the center of the molten pool compared with the microstructure of the edge zone [26].

Table 4. Processing parameters and grain size of various SLMed magnesium alloy powders.

Alloy	Energy Density (J/mm^3)	Grain Size (μm)
AZ91 [26]	104–167	1–1.2
WE43 [25]	238	1
ZK60 [30]	420–750	2–8
ZK30-xAl [31]	4004 Al(wt.%): 0–7	21.6 ± 2.6–7.3 ± 1.0
ZK61-xZn [42]	1146 Zn(wt.%): 5–30	1.1–6.1
Mg-9Al [32]	94–250	10–20
Mg [37]	60–120	5–20
Mg-2Ca [38]	1200	5
Mg-xSn [39]	107 Sn(wt.%): 0–7	5–25
Mg-3Zn-xDy [40]	360 Dy(wt.%): 0–5	6.4–18.1
AZ61 [41]	138–208	1.6–2.5
Mg-Gd-Zn-Zr [44]	27–267	1.3 ± 0.4–2.3 ± 1.0

Second, because of the "solute capture" effect caused by rapid solidification, the solid solution of the elements in the matrix was enhanced, so the microhardness increased. Al and Zn, as common alloying elements, were solid-solved into the magnesium alloy matrix. The atomic radii of Al and Zn are 0.1199 and 0.1187 nm, respectively, which are smaller than that of Mg (0.1333 nm) [63]. Therefore, lattice distortion will occur when these elements are dissolved into the Mg matrix. The Al concentration in the SLMed magnesium alloy was much higher than that in the as-cast sample because of the rapid solidification during SLM. According to the solid solution strengthening theory, higher solid solubility results in better mechanical strength [26,64]. With increasing energy density, the solid solubility of the alloying elements increased, and the microhardness of the sample was enhanced by solid solution strengthening. For Mg-Al alloys, the change in microhardness is directly proportional to the Al content in α-Mg [65]. In Mg-Zn-Zr alloys [30], the increase in the solid solubility of Zn in α-Mg is also conducive to increasing hardness. Based on the solid solution strengthening theory, the lattice distortion caused by the solid solution of Zn limits the dislocation slip in the crystal grains, thereby increasing the hardness of the SLMed ZK60 [64]. However, if the energy density is further increased, the temperature of the molten pool will be further increased, the solute trapping effect will be weakened, the solid solution of alloying elements such as Al and Zn will be reduced [28], and the microhardness will be slightly reduced (Figure 4a).

Third, the distribution of the second phase plays a crucial role in the microhardness of SLMed magnesium alloys. Due to the different distribution of intermetallic phases, the microhardness value of magnesium alloys prepared by SLM fluctuated within a certain range, and the amplitude increased as the intermetallic phase volume fraction increased, as shown in Figure 6b. As the Dy content increased, the volume fraction of the Mg-Zn-Dy phase continued to increase, and the dispersed second phase mainly precipitated at the grain boundaries, inhibiting the movement of the grains, which enhanced the microhardness of the alloy [40]. The intermetallic compound in Mg-Al alloys is usually β-Mg$_{17}$Al$_{12}$, and its microhardness is different from that of the α-Mg matrix. It can be seen from the nanoindentation experiments and hardness mapping in Figure 7 that the average hardness of the second phase β-Mg$_{17}$Al$_{12}$ precipitated in the AZ91D magnesium alloy was 174 ± 103 Hv, which was significantly higher than the average hardness of the matrix (123 ± 13 Hv) [62]. The hardness of the β-Mg$_{17}$Al$_{12}$ phase was reduced to 150 ± 60 Hv by applying selective laser surface melting (SLSM) treatment to the magnesium alloy, whereas the hardness of the α-Mg phase was maintained at 126 ± 3 Hv. Although the SLM process reduced the dispersion of the β-Mg$_{17}$Al$_{12}$ phase, it did not change the average hardness of the matrix. This indicates a reduction in element segregation and residual stress. On the other hand, the β-Mg$_{17}$Al$_{12}$ phase was modified by SLSM, which reduced the hardness and dispersion of the β-Mg$_{17}$Al$_{12}$ phase, and the phase distribution of the magnesium alloy was more uniform and continuous [62]. By introducing 0–7 wt.% Al in ZK30 magnesium alloy, the average microhardness gradually increased from 59.7 to 75.7 Hv [31]. An intermetallic hard-brittle β-Mg$_{17}$Al$_{12}$ phase was formed after the addition of Al, which was distributed in the soft α-Mg matrix and acted as a strengthening phase in the alloy [66], resulting in increased strength and microhardness. In the ZK series magnesium alloy, the second phase was Mg$_7$Zn$_3$ [30]. The Mg$_7$Zn$_3$ phase, which has high hardness and a uniform distribution, can strengthen the structure by hindering the dislocation motion between grains.

In addition, the presence of defects also affects microhardness. The porosity generated in the sample at low energy density and the cracks formed due to residual stress in the sample at high energy density will result in reduced hardness [37].

In summary, the microhardness of the magnesium alloy was affected by rapid solidification during the SLM process, which mainly affected the microstructure and the solid solution of the elements. On the one hand, the significant refinement of the microstructure of magnesium alloys was due to the rapid solidification, which made the microhardness of the SLMed magnesium alloy significantly higher than that of the traditional as-cast magnesium alloy. On the other hand, the solute trapping effect in the SLM process resulted in the solid solution of the alloying elements in the matrix, and the different solid solutions of the alloying elements have different strengthening effects. Furthermore, the distribution and amount of the second phase and the presence of defects will also affect the microhardness. The hardness of the second phase was generally higher than that of the matrix, and the microhardness of the magnesium alloy was generally improved by suppressing the dislocation motion between the grains. As the second phase content decreased, the microhardness distribution of the material became more uniform.

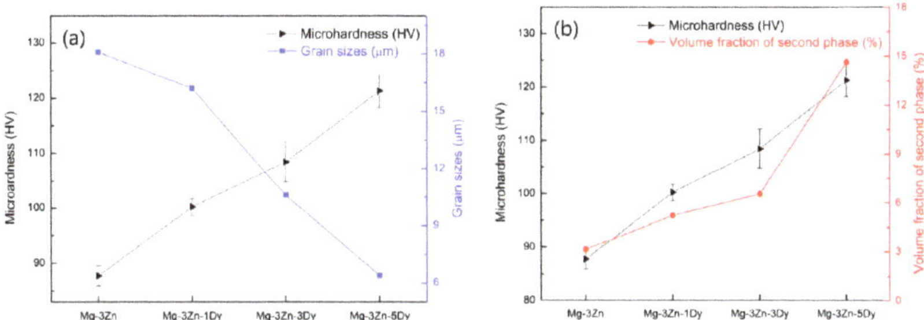

Figure 6. Relationship between grain size, second phase and microhardness of the magnesium alloy: (**a**) grain size and (**b**) second phase volume fraction. Reproduction from [40], with permission from Taylor & Francis (London, UK), 2020.

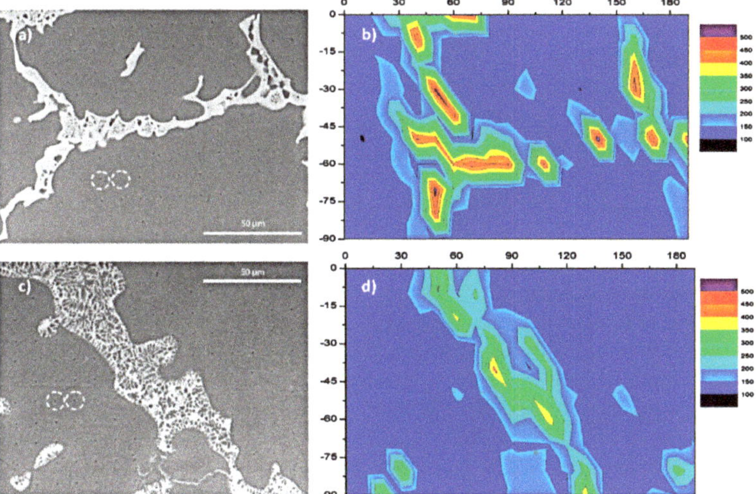

Figure 7. Hardness distribution of the matrix and second phase in AZ91 magnesium alloy. (**a**) Scanning electron microscopy (SEM) image of the as-cast samples and (**b**) corresponding nanoindentation hardness mapping. (**c**) SEM image of the SLSM samples and (**d**) corresponding hardness mapping. Reproduction from [62], with permission from Elsevier, 2020.

3.3. Mechanical Properties of SLMed Magnesium Alloy

Due to their lightweight characteristics, magnesium alloys can reduce fuel costs, and the total life cycle cost of Mg parts is lower than that of parts made of other materials [67]. Therefore, magnesium alloys are widely used in the automotive and aerospace fields. Weight reduction not only saves energy but also reduces greenhouse gas emissions. Reducing the weight of a car by a certain amount will result in a similar improvement in fuel economy [68]. Magnesium alloy parts can usually be used in engines, seat frames, gearboxes, etc. Good strength, ductility and castability are prerequisites for the use of magnesium alloys. However, there are few studies on the mechanical properties of SLMed magnesium alloys, and the amount of experimental data available for reference is limited. According to the existing research, the strength of SLMed magnesium alloys is significantly higher than that of as-cast magnesium [26,28,69] without losing plasticity [25,26,28].

As shown in Figure 8 and Table 5, when the scanning speed was 333–800 mm/s, and energy density was 104–167 J/mm^3, the ultimate tensile strength (UTS) and yield strength (YS) of the SLMed AZ91 magnesium alloy reached 296–330 MPa and 254–264 MPa, respectively. The UTS and YS of SLMed AZ91 were approximately 30% and 50% higher than those of as-cast AZ91 magnesium alloys [25,26], but the elongation of the SLMed samples (1.24–1.83%) was approximately 40% lower than that of as-cast samples (3%), as shown in Figure 9. The AZ61 magnesium alloy prepared by SLM also exhibited the same trend. When the energy density was 156 J/mm^3, the UTS and YS reached 287 MPa and 233 MPa, respectively, which was 93% and 135% higher than those values in the as-cast AZ61 magnesium alloy [69]. However, the elongation was 3.28–2.14%, which was lower than the elongation of the as-cast AZ61 (5.2%). The increase in strength was the result of grain refinement and solid solution strengthening. First, on account of the nature of the SLM process, melting and solidification at a smaller size resulted in fast solidification, which made the microstructure of the entire part more uniform. For the magnesium alloy forming process, segregation of alloying elements occurred on a much smaller scale. The chemical composition of the whole part was more uniform, and the amount of solid solution elements increased, so the strength of the SLMed parts was higher than that of the as-cast parts [70]. Second, Figure 3 shows that the grain size of the SLMed magnesium alloy was significantly refined. The as-cast grain size was usually 200–300 μm [71,72], whereas the grain size after SLM treatment was mostly 1–15 μm, and the average grain size of the AZ series magnesium alloy was approximately 2 μm. According to the Hall–Petch formula, the grains were refined, and the strength of the material was improved. The difference in the strength between different SLMed samples was mainly due to the different SLM process conditions, which led to deviations in density and microstructure, such as defects, structure and grain size [73]. Wei et al. [26] found that at different energy densities, the strength of magnesium alloys increased slightly with increasing energy density. The UTS and YS of the SLMed AZ91 magnesium alloy increased from 274 MPa and 237 MPa (83 J/mm^3) to 296 MPa and 254 MPa (167 J/mm^3), respectively. In fact, compared to the energy density, the laser power and scanning speed (linear energy density) are the main factors that change the characteristics of the melt pool, leading to significant changes in the microstructure and mechanical behavior of the manufactured samples. Therefore, it is necessary to combine these two process parameters for comprehensive analysis. When the scanning speed increases or the laser power decreases (low energy density), the interaction time between the laser and the powder is shortened, the heat accumulation in the molten pool is less and the temperature gradient is smaller, and the crystal grains do not have enough time to grow, which is conducive to grain refinement. As the cooling rate is faster, the higher the cooling rate during solidification, the shorter the time available for grain coarsening. In terms of solid solution elements, the cooling rate of the molten pool is faster at higher scanning speeds, and the solid solution elements in the matrix solidify before they can completely diffuse, resulting in an increase in the solid solution elements in the matrix, achieving the purpose of solid solution strengthening. However, it is still necessary to pay attention to the problem of pores caused by too high scanning speed or too low laser power, which results in the powder being unable to be fully heated and melted. The influence of laser power and scanning speed on microstructure will be discussed in detail in subsequent chapters. An analysis of the fracture behavior of SLMed parts showed that SLMed AZ91 exhibited mixed ductile–brittle fracture. The ultimate compressive strength (UCS) and elastic modulus of SLMed Mg-Ca alloys also increased with increasing laser power (energy density) [38]. This phenomenon was mainly due to the different porosities under different energy inputs. With the increase in laser power, the porosity decreased, resulting in an increase in the UCS, elastic modulus, and plasticity of the porous Mg-Ca alloy.

Table 5. Summary of the processing parameters and properties achieved for various SLMed magnesium alloy powders [26,28,32].

Alloy	Laser Power (W)	Scanning Speed (mm/s)	Energy Density (J/mm^3)	UTS (MPa)	YS (MPa)	EL (%)
AZ91 [26]	200	333–667	83–167	274 ± 7–296 ± 3	227 ± 3–254 ± 4	1.2 ± 0.1–1.8 ± 0.2
AZ91 [25]	100	800	104	329	160	1.8 ± 0.2
AZ61 [28]	150	300–450	138–208	239 ± 3–296 ± 2	217 ± 3–233 ± 2	3.1 ± 0.1
Mg-2Ca [38]	50–100	10	625–1125	5-46 (UCS horizontal) 51-111 (UCS longitudinal)	-	-

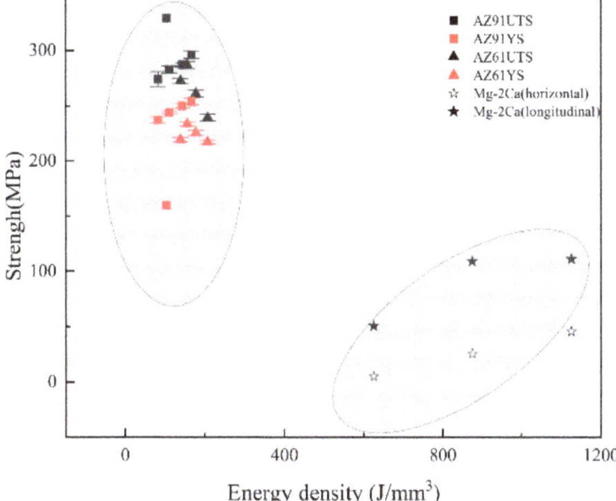

Figure 8. Effect of energy density on strength [26,28,38].

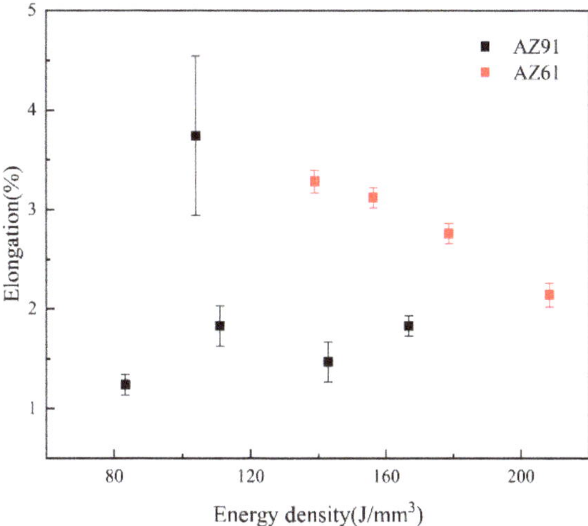

Figure 9. Effect of energy density on elongation [25,26,28].

It was also found that the mechanical properties of magnesium alloys were different in different directions, indicating that the alloy had significant anisotropy [38]. However, the elastic modulus under longitudinal compression and the UCS did not change much when the energy density increased from 875 to 1125 J/mm^3. The longitudinal stress–strain curve at 1125 J/mm^3 had a larger yield platform than the other curves. This variation should mainly be ascribed to the different porosities at different energy inputs and the different distribution modes of the pores under transverse compression and longitudinal compression. The distribution of pores also made the longitudinally compressed skeleton part denser than the laterally compressed skeleton part. Therefore, the longitudinal mechanical properties were better than the transverse mechanical properties. The existence of pores caused stress concentrations, which greatly reduced the mechanical properties of the magnesium alloy; the mechanical properties increased with decreasing porosity. Consequently, the pore size of the porous Mg-Ca alloy can be adjusted by adjusting the laser parameters or energy density, thereby reducing the porosity and ultimately improving the mechanical properties of the porous material.

The fracture morphology of the SLMed magnesium alloy showed that there were obvious cleavage planes around the reserved pores of the porous Mg-Ca alloy, indicating cleavage fracture [38]. River morphology and dimples can be observed on the fracture surface inside the pores of the porous Mg-Ca alloy. The number of longitudinally compressed dimples was significantly greater than the number of transversely compressed dimples. These dimples indicate a certain degree of plastic deformation before fracture, and the plasticity under longitudinal compression was better than that under transverse compression. As the laser power increased, the number and size of dimples increased. This finding indicates that as the energy input increased, the plasticity of the porous Mg-Ca alloy increased. This is because under the increased laser power, the powder receives more sufficient energy per unit time to melt the powder and combine with each other, so the reduction of macroscopic defects leads to an increase in plasticity. In conclusion, the SLMed porous magnesium alloys exhibited a mixed brittle–ductile fracture.

Nevertheless, the elongation of SLMed magnesium alloy is usually low (Figure 9), which may be related to the presence of micropores in the part and the presence and distribution of the second phase. Fractures mostly occur in areas with high porosity. Moreover, for the AZ series magnesium alloy, the research in [28] showed that the existence and distribution of the β-Mg$_{17}$Al$_{12}$ phase had a substantial influence on the plasticity. The α-Mg matrix is relatively soft, and its crystal structure is a close-packed hexagonal structure. The brittle and hard β-Mg$_{17}$Al$_{12}$ phase has a cubic crystal structure. The β-Mg$_{17}$Al$_{12}$ phase in magnesium alloys prepared by SLM usually precipitates along the grain boundaries and is connected to form a network. The crystal structures of the α-Mg and β-Mg$_{17}$Al$_{12}$ phases are incompatible, leading to brittleness of the phase interface (α-Mg/β-Mg$_{17}$Al$_{12}$); moreover, due to the formation of high dislocation density and stress concentrations at the interface of α-Mg and β-Mg$_{17}$Al$_{12}$ [74], cracks nucleated and propagated along the phase interface [75–77]. As shown in Figure 10, the second phase precipitated along the grain boundaries, and the cracks were generated at the interface between the two phases.

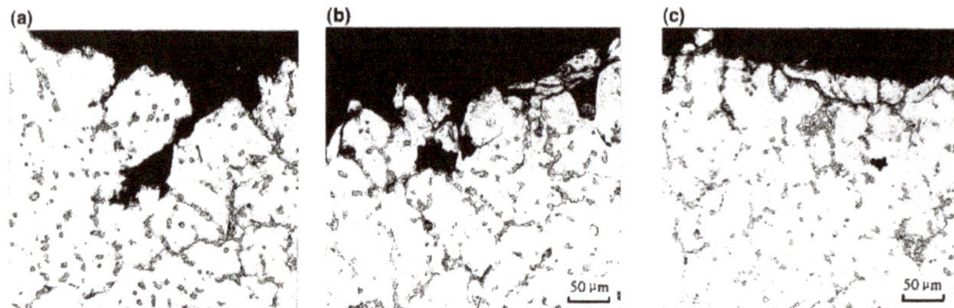

Figure 10. Optical microscopy fractographs showing that the microcracks were prone to initiate near the Mg/Mg$_{17}$Al$_{12}$ interface and the grain boundaries: (**a**) the rear part; (**b**) the middle part; (**c**) the impact part. Reproduction from [75], with permission from Elsevier, 2020.

A side-view of the fracture morphology also confirmed the limitation of the β phase on plasticity. The existence of pores and cracks in the fracture and the second phase at the bottom of the dimple verified the effect of defects and the β phase on plasticity. Furthermore, some fine grains and large grain boundaries formed by small equiaxed grains were observed, for which the fracture occurred along the crystal. This fracture morphology has also been reported in magnesium alloys prepared by other methods [78,79]. After the grains were refined, the stress was dispersed.

During the deformation process, α-Mg grains first undergo plastic deformation, but the deformation is hindered when they encounter the second phase at the grain boundaries, resulting in incomplete deformation and fracture. A similar situation exists in laser-welded Mg alloys, wherein the fracture strain reduced from 13.6 to 3.3% due to the higher number of intermetallic precipitates causing embrittlement of the fusion zone [17]. Therefore, to obtain good plasticity in the Mg-Al alloy, the precipitation of $Mg_{17}Al_{12}$ should be restricted. It is possible to improve the solidification rate by improving the process parameters or by applying post-treatment methods such as solution heat treatment to increase the solid solution of Al, thereby reducing the formation of $Mg_{17}Al_{12}$ at the grain boundaries. Studies have shown that the elongation of Mg alloys is closely related to the content of $Mg_{17}Al_{12}$. When the proportion of $Mg_{17}Al_{12}$ increases to approximately 9%, the ductility decreases [80]. However, under the condition of ensuring the quality of sample forming, higher scanning speed and lower laser power can produce a faster cooling rate, thereby inhibiting the occurrence of phase transition and reducing the precipitation of the second phase along the grain boundary. This part of the content will be in-depth analysis combined with the phase diagram in Section 5.

At present, there is no related research on the effect of heat treatment on the mechanical properties of SLMed magnesium alloys. This is an issue worthy of further exploration in the future. Hot isostatic pressing (HIP) can reduce the number of pores and decrease the anisotropy of the material [81], which has an impact on mechanical properties. In addition, plasticity can be improved by introducing different alloying elements. The literature [82] showed that solute Nd had a significant softening effect on the nucleation critical resolve shear stress (CRSS) of the cone <c + a> slip. When designing a high ductility magnesium alloy, from the perspective of reducing the strength difference between the soft deformation mode and hard deformation mode, the important factors that must be account for include the alloying elements, the solute concentration, and the predeformation strain.

3.4. Corrosion

As a lightweight material, magnesium is widely used for structural materials in various applications, including the automotive and aerospace fields, due to its low density, good heat dissipation, damping and electromagnetic shielding properties. In the biomedical field, osteoconductive implant materials are popular, and the biocompatibility and corrosion resistance of these materials are the research focus in this area. Recently, magnesium-based alloys have attracted interest as possible replacements for some of the metallic alloys in use due to the former being lighter and exhibiting a higher modulus of elasticity [83]. The corrosion behavior of magnesium can be divided into galvanic corrosion, stress corrosion, biocorrosion, etc.

In moist environments, magnesium and its alloys always form a thin surface film. This natural outer layer is not dense because its corresponding Pilling–Bedworth ratio is ~0.81, which indicates that the underlying metal cannot be completely covered. As a result, magnesium and its alloys are highly susceptible to corrosion. Due to the dissolution of magnesium, a loose film consisting mainly of $Mg(OH)_2$ is formed on the surface of magnesium alloy. The poor corrosion resistance of magnesium alloy is mainly caused by the following two reasons. First, the quasipassive hydroxide film formed on the magnesium surface is much less stable than the passive film on the surface of metals such as aluminum and stainless steel. Second, internal galvanic corrosion is caused by intermetallic phases or impurities. Therefore, the improvement of corrosion resistance can also be solved through two aspects. First, the passivation of the surface film is affected by the alloying elements in the magnesium alloy. For example, Hara et al. [84] claimed that adding Al to Mg or increasing the Al content in the Mg matrix can change the composition and structure of the surface film, thereby improving the resistance of the film to local breakdown. Cai et al. [85] asserted that Zn could effectively improve the surface protection film and corrosion resistance of Mg alloys. The effect of alloying elements on corrosion resistance will be discussed in detail in Section 4.

Second, the corrosion resistance can be improved by reducing impurity elements. Elements such as Fe, Ni, and Cu are usually present as intermetallic precipitates, which are harmful because of their low hydrogen overvoltage [86]. The effect of intermetallic compounds on corrosion behavior is complex. The second phase can serve as both a barrier to prevent corrosion of magnesium alloy and as a cathode to accelerate corrosion of magnesium alloy. Hence, the corrosion resistance of Mg alloys is usually related to the type, size and morphology of the intermetallic phase [87,88]. The β phase is expected to be a barrier when there is a small grain size and relatively large β phase fraction, and more importantly, when the β phase is in the form of a continuous network along the Mg grain boundaries. However, microgalvanic corrosion is readily accelerated as the β phase agglomerates and is separately distributed in the Mg matrix with large grain sizes. It is noted that the decreased cathodic and anodic reactions of Mg-Al-Mn-Ca alloy produced by the spinning water atomization process (SWAP) were attributed to the fine dispersion of Al_2Ca particles and the supersaturated Al content in the α-Mg matrix, respectively, resulting in superior corrosion resistance [84]. Therefore, the second phase of the dispersed distribution is beneficial to improve the corrosion resistance of the magnesium alloy.

In terms of biological corrosion, the key point is that an implant must possess the appropriate strength for an adequate period of time to allow healing to take place. During this time, the corrosion rate must be sufficiently slow to not affect the healing process. While it is true that the byproducts of magnesium corrosion are nontoxic, as the metal corrodes, the pH in the localized area increases, and this basic environment may impede healing. Similarly, hydrogen gas evolves during the corrosion process and must be eliminated. When Mg alloy was soaked in an aqueous physiological environment, the following reactions occurred, as shown in Equations (3)–(5):

$$\text{Mg (s)} + 2\text{H}_2\text{O (l)} \rightarrow \text{Mg(OH)}_2 \text{ (s)} + \text{H}_2 \text{ (g)} \tag{3}$$

$$\text{Mg (s)} + 2\text{Cl}^- \text{ (aq)} \rightarrow \text{MgCl}_2 + 2\text{e}^- \tag{4}$$

$$\text{Mg(OH)}_2 \text{ (s)} + 2\text{Cl}^- \text{ (aq)} \rightarrow \text{MgCl}_2 + 2\text{OH}^- \tag{5}$$

It was reported that the anodic dissolution of magnesium in aqueous solutions occurs concurrently with cathodic hydrogen evolution and is accompanied by an anomalous phenomenon called the negative difference effect (NDE), which is characterized by an unexpected increase in the cathodic hydrogen evolution reaction as the anodic overvoltage increases [89].

Rapid solidification has been identified as an effective method for improving the strength and corrosion resistance of magnesium alloys used in structural applications or corrosive environments [90]. The rapid solidification during SLM can refine the microstructure and dissolve the second phase. Moreover, by rapid solidification, the mechanism of corrosion can be changed essentially from pitting corrosion of Mg-Al alloys into overall corrosion [91]. Izumi et al. [92] indicated that the change in cooling rate had a great influence on the corrosion behavior of magnesium alloys. Increasing the cooling rate can delay the occurrence of filamentous corrosion due to grain refinement and the formation of a supersaturated single α-Mg phase solid solution in mg alloys. In addition, laser-induced rapid solidification can also increase the solid solubility of alloying elements, such as manganese and aluminum, and promote the formation of more protective and self-healing thin films, thereby limiting the occurrence of local galvanic corrosion. The corrosion resistance of magnesium alloys prepared by SLM is summarized in Table 2. Shuai et al. [30] studied the corrosion resistance of SLMed ZK60 magnesium alloy at different energy densities. The results indicated that the hydrogen volume evolution rates were in the range of 0.006–0.0019 mL·cm^{-2}·h^{-1} as the laser energy input varied from 420 J/mm^3 to 750 J/mm^3. The optimal energy density for corrosion resistance was 600 J/mm^3. The lowest hydrogen evolution volume rate was 0.006 mL·cm^{-2}·h^{-1}. The refined grains, homogenized microstructure and extended solid solution obtained by rapid solidification contributed to the enhanced corrosion resistance of the SLMed ZK60 alloy. He et al. [29] investigated the corrosion resistance of

SLMed AZ61 magnesium alloy when the laser power was 50–90 W. It was reported that the minimum degradation rate occurred at 80 W, whose values corresponding to 24 and 144 h were approximately 2.4 and 1.2 mm year^{-1}, respectively. This phenomenon occurred due to the elimination of pores and the refinement of the microstructure. When the laser power was 90 W, the microstructure became coarse, and the corrosion resistance decreased.

SLM is a process that can effectively improve the corrosion resistance of magnesium alloys, and the corrosion of magnesium alloys is a complicated process. Therefore, the advantages of the SLM process should be fully used to develop new magnesium alloys or improved alloys to produce finer grains and new phases. Furthermore, research should focus on developing more feasible, reliable, maintainable, and lower-cost protection systems.

3.5. Factors that Affect Performance Defects

3.5.1. Cracking

The existence of cracks in SLMed alloys will affect the subsequent application of the material. Wang et al. [93] researched the reasons for premature material failure. The premature failure schematic diagram is shown in Figure 11. This is due to external defects (pores and unmelted particles) that act as sites for crack initiation. Then, cracks propagated through grain boundaries and/or cellular boundaries that contain continuous brittle second phases. Therefore, cracks are a problem worthy of attention. Mg-Zn binary alloy is a kind of Mg alloy that easily cracks during solidification [94]. This could be explained by the large solidification temperature range, high thermal expansion coefficient, and large solidification shrinkage of these alloys. For the SLM process, if there is sufficient stress at high laser energy density, the melt will crack.

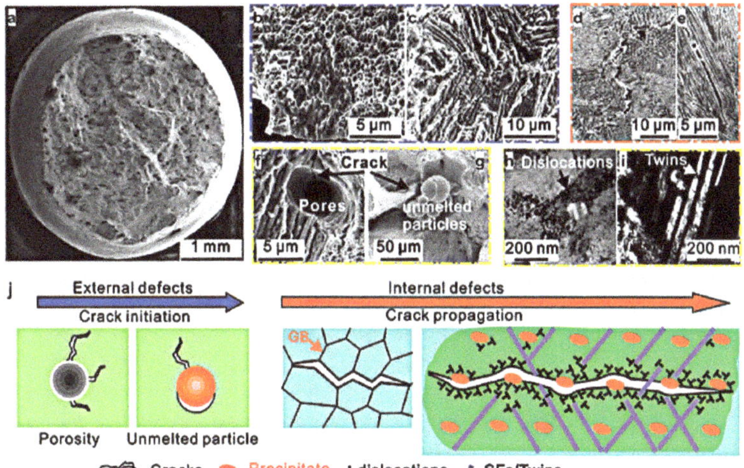

Figure 11. Schematic diagram of SLMed Ag-Cu-Ge alloy premature failure: (**a**) low-magnification SEM image of fracture surface; (**b,c**) dimples and cleavage features; (**d,e**) samples after polishing; (**f,g**) external defects (pores and unmelted particles); (**h,i**) internal defects of TEM micrographs; (**j**) Schematic illustration depicting crack initiation at external defects and crack propagation along internal defects. Reproduction from [93], with permission from Nature Publishing Group (London, UK), 2020.

Shuai et al. [30] found the presence of microcracks in the samples of SLMed ZK60 magnesium alloy. When high *Ev* was used, considerable thermal stress was generated and accumulated in the rapidly solidified SLMed layer. When the thermal stress exceeded a certain threshold, a hot crack formed. Wei et al. [43] observed cracks in Mg-Zn binary alloys. The morphology of the SLMed Mg alloy with different Zn contents is shown in Figure 12.

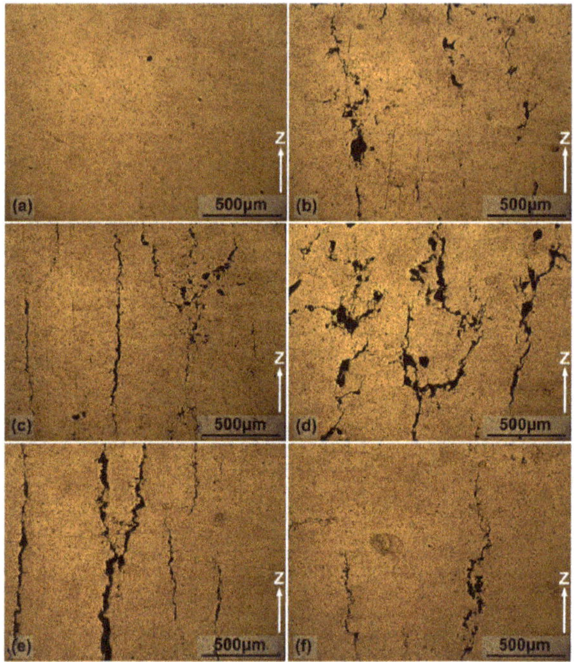

Figure 12. Cracks in SLMed (**a**) Mg-1Zn, (**b**) Mg-2Zn, (**c**) Mg-4Zn, (**d**) Mg-6Zn, (**e**) Mg-8Zn, and (**f**) Mg-10Zn. Reproduction from [43], with permission from Elsevier, 2020.

When the Zn content was different, the cracks in the Mg-Zn binary system were also different. The schematic diagram is shown in Figure 13. At a 1 wt.% Zn content, because the Mg_7Zn_3 eutectic phase was less abundant and mostly granular, the liquid eutectic phase during the final stage of the solidification process was discontinuous and low in content. Therefore, the strength of the mushy zone can resist the solidification shrinkage stress, thereby inhibiting the formation of cracks. As the Zn content increased, the area fraction of the Mg_7Zn_3 eutectic phase increased, which means that in the final stage of SLM pool solidification, the strength of the mushy zone decreased, leading to the formation of solidification cracks. However, at the same time, the fluidity of the residual liquid was significantly enhanced in the final stage of solidification. As a consequence, when the content of Zn was sufficiently high (12 wt.%), the solidification cracks may be backfilled by the residual liquid during the final stage of solidification, thereby alleviating the cracking tendency. The melting point and thermal expansion coefficient of aluminum are similar to those of magnesium. A similar situation usually occurs in SLMed aluminum alloys. The application of 7075 aluminum alloy is limited by cracking [95,96]. Yuki et al. [95] noted that for 7075 alloy, the optimal silicon content was 5%, which is conducive to eliminating cracks and improving mechanical properties. If the Si content was insufficient, the cracks cannot be completely eliminated. In contrast, excessive Si content resulted in brittleness. Microcracks are commonly attributed to solidification cracking and liquation cracking. The cause of solidification cracking is a residual thin film of liquid phase between the primary crystallized grains, and liquation cracking is caused by cyclic heat input, which occurs in multilayer welding and layered fabrication. The reason for the decrease in strength due to the heat input is mainly due to the eutectic phase or partial remelting of the component with a lower melting point. These melting points serve as the starting points for cracking. Generally, the morphology of the top of the sample can be used to distinguish the two kinds of cracks. The surface layer is not affected by thermal conduction from subsequent layers during fabrication; therefore, the solidification structure that is created by single laser irradiation has been preserved. If cracks appeared on the surface, the influence of the cyclic heat input was excluded, indicating that the cracks were solidification cracks. If microcracks appeared in the lower layer, it indicates that the sample were affected by the cyclic heat input, producing fused cracks. Kimura et al. [97] investigated the relationships between the Si content and the mechanical properties of Al-Si binary alloy samples fabricated by SLM. They noted that the additions of silicon enhanced the proof stress and UTS of SLMed samples because the silicon elements enriched the proportion of the secondary phases, providing compositional reinforcement for the aluminum matrix. The Si helped eliminate microcracks in the Al alloys. The addition of Si induced the formation of eutectics with a low melting point and high fluidity, which decreased the crack sensitivity as these eutectics melted and filled the cracks during the final stage of the solidification process [96]. In addition, the selection of process parameters was also very important. As the laser power increased, more low-melting-point eutectic phases melted, which was conducive to the elimination of cracks. In the SLM process, due to the lack of diffusion during nonequilibrium rapid solidification, the actual liquid and solid phase temperatures were further reduced [98]. Therefore, during SLM processing of magnesium alloys, the critical temperature range was larger, and the solidification cracking sensitivity was higher. The addition of alloying elements can narrow the critical solidification range and alter the melt pool composition to avoid cracking [96,98,99]. Therefore, future studies should consider introducing other alloying elements to SLMed Mg-Zn binary alloys to eliminate microcracks. Moreover, it is necessary to study the parameter windows of different magnesium alloys.

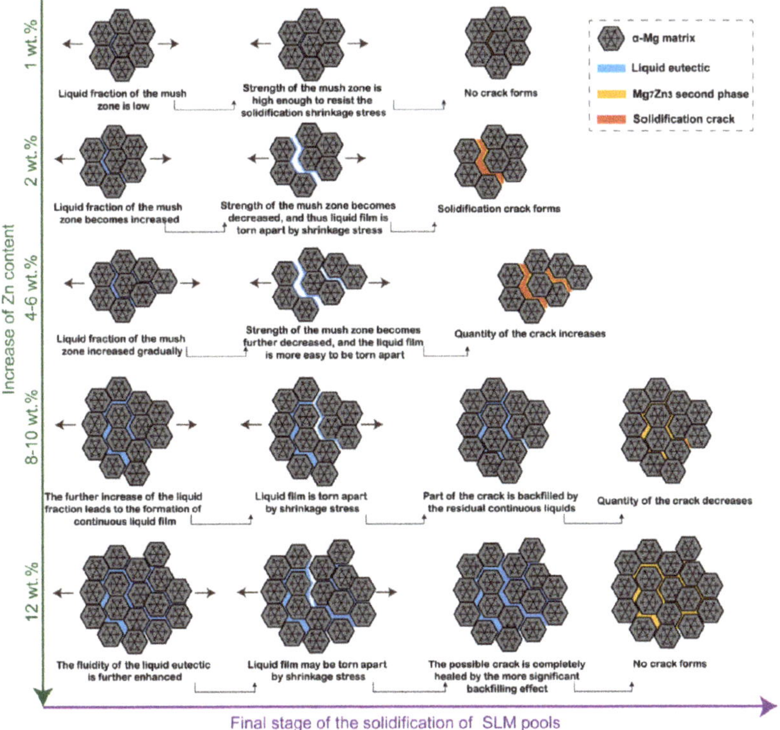

Figure 13. Schematic diagram illustrating the effect of Zn content on the solidification cracking behavior of the SLMed Mg-Zn alloys. Reproduction from [43], with permission from Elsevier, 2020.

3.5.2. Oxide Inclusions

It is well known that magnesium is highly susceptible to oxidation; therefore, elaborate protection from the atmosphere is required [100]. This is achieved using inert gases. Good shielding can avoid burning or porosity. Studies by Ng, CC, and others have shown that there are a large number of oxides in SLMed magnesium, and they tend to form between scan traces rather than between layers. It is also believed that the oxides will form on top of the molten pool, which may slow oxide diffusion, change the wettability and form a porous structure with weak mechanical properties [35]. Liu et al. [28] detected the presence of oxygen in both the Mg matrix and the second phase. The oxygen content in the matrix was 0.58 wt.%, whereas that in the second phase was 0.71 wt.%. Louvis et al. [101] observed that the formation of oxide films on both solid and liquid metal surfaces resulted in oxide films between the laser hatches in every layer of the SLMed Al parts; hence, pores were formed where two oxide films met. For existing oxide films, since the formation of oxide films is completely unavoidable, the SLM process must break these films, which requires high laser power, in order to produce a completely dense part. Therefore, to solve the issues pertaining to the unavoidable oxide films, further research on SLMed Mg should mainly focus on a new method of controlling the oxidation process and destroying oxide films formed within the components.

The oxygen in the SLM process comes from two sources: the oxygen in the molding chamber and the oxygen in the magnesium alloy powder. For the oxygen content in the molding chamber, high-purity protective gas is needed to reduce the presence of oxygen, but it cannot be completely avoided. It is believed that the oxygen content in the powder will lead to pores and reduce the mechanical properties of the sample [102–104]. Dong et al. [102] studied two 12CrNi2 alloy steels prepared via laser melting deposition (LMD) AM: one with high O content (5300 ppm) and the other with lower O content (280 ppm). Studies show that porous alloy steel was additively manufactured using high-O content powder. Using low-O content powder prevents gas pore formation in printed steel. Steel mechanical properties increased when the powder O content decreased [102]. Cao et al. [103] confirmed that γ-Al_2O_3 formed on the surfaces of Al powder particles. Oxidation of an as-gas atomized Al powder prevented the transformation of interparticle boundaries (IPBs) into γ-Al_2O_3 nanoparticle-containing grain boundaries in the bulk Al prepared by spark plasma sintering and led to the formation of continuous γ-Al_2O_3 layers with nanopores at the IPBs, causing a dramatic decrease in the elongation to fracture. Rao et al. [104] asserted that the desired combination of mechanical properties obtained by the use of low oxygen content powder showed its potential for exploring the near net shape advantage of HIP technology for alloy 718 components. In terms of oxygen content control, producing a vacuum in the molding chamber and adopting a powder purification process [12] can be considered. Please note that the presence of oxygen generally leads to the formation of inclusions, which degrade the properties of the material. However, it has also been reported that the introduction of ordered oxygen complexes enhances the plasticity and strength of high-entropy alloys [105]. Therefore, the effect of oxygen on magnesium alloys prepared by SLM and the control of oxygen content in a suitable range still require further research.

3.5.3. Other Factors

There are many other factors affecting the properties of SLMed alloys, such as alloying element loss [26,27], porosity [38,41], and spheroidization [47–49]. These factors, which are discussed in the previous sections, are related to each other. For instance, the consumption of alloying elements will lead to the instability of the scanning track and the high porosity of laser-machined parts [106]. The occurrence of spheroidization is generally accompanied by the appearance of pores [28]. At present, there are no scientific measures to avoid the loss of magnesium alloy during SLM processing. The dominant mechanisms for pore formation during SLM processing of Mg alloys need to be thoroughly investigated. There is no clear relationship between the porosity type and the SLM process parameters. Much work remains to be done in these areas.

4. Effect of Alloying Elements on the Properties of SLMed Magnesium Alloy

Rapid laser melting is an effective way to improve the corrosion resistance of magnesium alloys. The rapid heating and cooling characteristics of laser processing can shorten the processing time to seconds or milliseconds, which is conducive to the refinement of grain size. Grain refinement is the key to improving corrosion resistance [39]. In addition, rapid laser melting can reduce composition segregation, which creates conditions for improving the overall performance of magnesium alloys by adding alloying elements. At present, further adding different alloying elements on the basis of SLM preparation can further optimize the properties of magnesium alloys, thereby obtaining high-performance magnesium alloys. This research direction has been very popular in the past two years. The effects of the addition of alloying elements such as Al, Ca, and Sn on the properties of SLMed magnesium alloys are discussed hereinafter. The number and depth of studies on the effect of adding alloying elements on SLMed magnesium alloys is still limited.

4.1. Regular Elements

4.1.1. Alloying Element: Al

Al is a solid solution strengthening element that will produce precipitation strengthening at low temperatures (<393 K) and has a small impact on corrosion resistance. Al has a large solid solubility in solid Mg, and its ultimate solid solubility is 12.7%. Moreover, with decreasing temperature, the solid solution of Al decreases significantly, and the solid solubility at room temperature is approximately 2.0%. Aluminum can improve the castability and increase the strength of magnesium alloys. The higher the aluminum content, the better the corrosion resistance. However, the stress corrosion sensitivity increases with increasing aluminum content.

The AZ series of Mg alloys is most often used for casted and wrought applications because of its good castability and good mechanical properties [107,108]. The Al concentrations in the AZ series promote the formation of the β-$Mg_{17}Al_{12}$ phase, and the amount of this phase present in the alloys affects the deformation texture. High Al concentrations favor the formation of the $Mg_{17}Al_{12}$ intermediate phase, which weakens the basal texture by inhibiting twin boundary motion and hindering formability. Hence, AZ alloys with high Al concentrations are preferred for engineering applications [109].

Aluminum is often used in different magnesium alloys because it has a significant positive effect on the degradation and mechanical properties of magnesium alloys [110,111]. As an alloying element, aluminum can improve the mechanical properties of magnesium alloys through solid solution strengthening and microstructure refinement [112,113]. In addition, Al is an effective reinforcing element in Mg alloys because it has a significant effect on changing its grain size and intermetallic phase [114]. The effect of Al on the structure and properties of Mg alloy can be attributed to the influence of the Al itself and the intermetallic compound β-$Mg_{17}Al_{12}$ introduced by its addition.

Shuai et al. [31] noted that the grain size and intermetallic phase were the two key factors affecting the biodegradation behavior of magnesium alloys after the introduction of aluminum. Al (0–7 wt.%) was introduced into an SLMed Mg-Zn alloy, and the effect of the Al content on the structure and properties of the SLMed ZK30-xAl magnesium alloy was obtained. The results showed that the grain size and intermetallic phase volume fraction varied with respect to the amount of Al. As the aluminum content increased, the grains were refined, and the intermetallic phase volume fraction increased. When the Al content was less than 3 wt.%, the intermetallic compound β-$Mg_{17}Al_{12}$ precipitated and dispersed, and grain refinement was the main factor affecting the degradation behavior. The finer grains led to an increase in the grain boundary density, the alloy was easily passivated, and the degradation rate was reduced. The Pilling–Bedworth ratio of Mg is less than 1, which means that its oxide layer has high compressive stresses, resulting in the formation of cracks [115]. The grain-refined alloy had a higher grain boundary density, which is beneficial to reduce the compressive stress to compensate for the mismatch between the oxide/base metal. Therefore, the reduction in the mismatch can improve the performance of the oxide film, which offers greater protection from the detrimental action of chloride ions and improves the degradation rate. The grain size was refined as the Al content increased further, and β-$Mg_{17}Al_{12}$ continuously precipitated along the grain boundaries and formed a network when the Al content increased to 7 wt.%. At this time, the intermetallic phase β-$Mg_{17}Al_{12}$ became the main factor affecting the degradation behavior. The larger intermetallic phase volume fraction caused severe galvanic corrosion and accelerated the corrosion process. Therefore, the degradation rate increased when the Al content increased to 5–7 wt.%. ZK30-3Al had the lowest degradation rate (0.17 ± 0.02 mg cm^{-2} day^{-1} in Table 2) and the lowest corrosion current. Therefore, with the introduction of Al, the degree of grain refinement increased and the degradation behavior improved, but the degradation rate gradually deteriorated when the Al content became excessive. In terms of mechanical properties, the microhardness of SLMed magnesium alloy increased with increasing Al content. This phenomenon was mainly caused by grain refinement, solid solution strengthening, and brittle, hard β-$Mg_{17}Al_{12}$ phase strengthening. The reasons for this increase was previously analyzed in detail.

Similarly, we can refer to the influence of alloying elements in the traditional manufacturing process. Homayun et al. [114] studied the effects of different additions of Al on the microstructure, mechanical properties, degradation behavior and biocompatibility of as-cast Mg-4Zn-0.2Ca alloy. Their results showed that to improve the degradation behavior and mechanical properties of the alloy, the Al content should be controlled below 3 wt.%. The addition of Al increased the tensile strength of the alloy from 157 to 198 MPa. The mechanical properties of the magnesium alloys were optimal when the Al content was 3 wt.%. The addition of aluminum had a positive effect on the refinement of the alloy microstructure. However, an excessive amount of aluminum led to a considerable amount of second phase β-$Mg_{17}Al_{12}$ formed at the grain boundaries, which substantially deteriorated the performance of the alloy. When the Al content exceeded 3 wt.%, the intermetallic compounds at the grain boundaries of the alloy formed a network, resulting in deterioration of the mechanical properties and accelerated current corrosion. In fact, although the precipitation of secondary phases in the alloy increased its corrosion potential, the corrosion resistance of the alloy was significantly reduced. Zhao et al. [116] surmised that the intermetallic compound (β-$Mg_{17}Al_{12}$) in AZ91 could be used as a cathode to accelerate the degradation rate or as a barrier to prevent degradation depending on the volume fraction and morphology of the second phase. Lu et al. [117] also found that as the intermetallic volume fraction of ZK30 increased, the degradation resistance decreased. β-$Mg_{17}Al_{12}$ has an important influence on the degradation behavior and corrosion resistance of Mg alloys. For Mg-Al alloys, the β-$Mg_{17}Al_{12}$ phase was discretely distributed along the grain boundaries, which led to severe galvanic corrosion [118].

In addition, the presence of the β-$Mg_{17}Al_{12}$ phase will affect the mechanical properties of Mg. The increase in Al content led to the formation of a considerable amount of β-$Mg_{17}Al_{12}$ phase compounds. The cubic crystal structure of β-$Mg_{17}Al_{12}$ is incompatible with the close-packed hexagonal structure of Mg alloys. The lack of coherence between the crystal structure of the second phase and that of the magnesium alloy resulted in the formation and propagation of cracks under tensile stress. The β-$Mg_{17}Al_{12}$ phase at the grain boundaries interconnected to form a network, which led to a significant reduction in elongation and strain [77]. Liu et al. [28] asserted that the reason for the lower plastic elongation of the SLMed AZ61 magnesium alloy than that of the as-cast sample was due to the precipitation of the second phase along the grain boundaries. As a result, crack sources were generated at the interface of the α-Mg and β-$Mg_{17}Al_{12}$ phases, the plasticity was reduced, and the precipitation of the second phase increased as the Al content increased [31]. The negative effect of β-$Mg_{17}Al_{12}$ on the mechanical properties of alloys has also been reported in other studies [85,112,119]. Therefore, when introducing Al, attention should be paid to the influence of β-$Mg_{17}Al_{12}$ on the properties of SLMed magnesium alloys.

In addition to helping to refine the grains, Al also has a certain effect on the shape of the grains [120]. With the increase in aluminum content in magnesium-based alloys, the grains transformed from columnar to equiaxed morphologies. Increasing the solute content of Al enhanced the component undercooling and promoted heterogeneous nucleation during solidification and grain refinement.

Therefore, the precipitation and distribution of the β-$Mg_{17}Al_{12}$ phase should also be considered when Al is introduced to refine the grains during SLM. The amount of Al added should be controlled within a reasonable range to reduce the precipitation of the second phase β-$Mg_{17}Al_{12}$ to reduce galvanic corrosion and intergranular crack growth.

4.1.2. Alloying Element: Ca

Magnesium alloys have poor room temperature plasticity and poor corrosion resistance, which limits their application and development to a certain extent. Researchers have found that the introduction of Ca into Mg alloys can significantly improve the plasticity, corrosion resistance and overall performance of Mg alloys [121–123]. Ca, an alkaline earth metal, is an element that can notably improve the mechanical properties of Mg-Mn alloys [124]. Ca not only improves the ignition point and high-temperature oxidation resistance of Mg alloys but also plays an important role in the grain refinement of Mg alloys [38,123,125]. Ca has a low solubility in Mg of approximately 0.43 at.% (~0.71 wt.%) at 517 °C [126] (radius ratio is greater than 15%). Ca has a strong affinity for Mg, and Ca can be used as a structural modification element or a dispersed phase to improve the thermal resistance and the yield limit of the alloy [127]. Adding Ca to Mg alloy can form intermetallic compounds such as Mg_2Ca to strengthen Mg-Al and Mg-Zn alloys [128–130]. A previous study [38] investigated the properties of SLMed Mg-Ca porous alloys. The grain size of the porous magnesium–calcium alloys was 5–30 μm, which was significantly lower than the grain size of as-cast magnesium alloys (300–500 μm); in contrast, the grain size of SLMed pure magnesium was 10–20 μm [32]. The microhardness of the SLMed pure magnesium was between 60 Hv and 68 Hv, which was higher than that of as-cast pure magnesium. Grain refinement and solid solution strengthening were the main strengthening mechanisms of Mg-Ca alloys. At an energy density of 1125 J/mm^3, the transverse UCS and elastic modulus reached 45.75 MPa and 0.973 GPa, respectively, and the longitudinal UCS and elastic modulus reached 111.19 MPa and 1.264 GPa, respectively.

Ca is an essential element in the human body that has no adverse biological effects, so it can also be used as an alloying element in orthopedic implant materials. Yang et al. [122] reported that the addition of Ca improved the corrosion behavior of Mg-Al-Mn alloys because the Ca reacted with the original alloy to form a new second phase as a sacrificial anode. Since CaO and Ca act on Mg alloys in a similar manner, CaO introduced in Mg alloys will react with Mg and Al to form a fine $(Mg, Al)_2Ca$ phase [131,132]. However, the potential of the $(Mg, Al)_2Ca$ phase was lower than that of the $Mg_{17}Al_{12}$ phase, which led to reduced potential difference and enhanced corrosion resistance [133].

Shuai et al. [118] introduced 0–12 wt.% CaO into SLMed Mg-Al-Zn (AZ61) alloy. Their results showed that when the content of CaO was less than 6 wt.%, the precipitation of the second phase at the grain boundaries changed from dispersion to a slender branch-like structure. The introduction of CaO changed the discrete $Mg_{17}Al_{12}$ phase into a continuous $(Mg, Al)_2Ca$ phase and then to a coarsened Mg_2Ca phase (9–12 wt.% CaO). The β-$Mg_{17}Al_{12}$ content decreased with the addition of CaO. In the SLM process, CaO reacted with Mg and Al in Mg-Al alloys, as shown in Equations (6) and (7).

$$Mg + 2Al + CaO \rightarrow Al_2Ca + MgO \tag{6}$$

$$3Mg + CaO = Mg_2Ca + MgO \tag{7}$$

The formation of the second phase β-$Mg_{17}Al_{12}$ was suppressed, since reaction (6) consumed a certain amount of Al, and the formation of the $(Mg, Al)_2Ca$ phase was promoted. As shown in reaction (7), the addition of CaO will oxidize Mg, and the resulting Ca will form Mg_2Ca with Mg. Due to the double consumption of Mg, the formation of the $Mg_{17}Al_{12}$ phase was also suppressed to a certain extent.

The corrosion resistance mechanism of Mg alloy after adding different CaO contents is shown in Figure 14. Different phases and microstructures have different corrosion potentials, resulting in different corrosion rates. The introduction of CaO transformed the discrete $Mg_{17}Al_{12}$ phase into a continuous (Mg, Al)$_2$Ca phase and then to the coarsened Mg$_2$Ca phase. The distribution of the second phase in AZ61, AZ61-3CaO and AZ61-6CaO was discontinuous and discrete. At this time, the corrosion behavior was mainly dominated by galvanic corrosion. In the initial stage, a galvanic couple was formed between the second phase (cathode) and the α-Mg matrix (anode) due to a potential difference. α-Mg grains were preferentially dissolved due to their low potential. As the adjacent Mg particles dissolved, most of the second phase disintegrated. The potential of the (Mg, Al)$_2$Ca phase was lower than that of the β-Mg$_{17}$Al$_{12}$ phase, which reduced the potential difference between the α-Mg and the second phase, resulting in weakened galvanic corrosion. Hence, AZ61-3CaO and AZ61-6CaO exhibited better corrosion resistance than AZ61. In AZ61-9CaO Mg alloys, electrochemical corrosion occurs between (Mg, Al)$_2$Ca and α-Mg grains in the initial stage. Once the exposed α-Mg grains were corroded, the (Mg, Al)$_2$Ca phase, which is inert to the corrosion solution and has a continuous network distribution, acted as a barrier layer to prevent further penetration of the corrosion solution. AZ61-9CaO exhibited better corrosion behavior. For AZ61-12CaO, the E$_{corr}$ of Mg$_2$Ca was lower than that of Mg [134]. The Mg$_2$Ca phase replaced the α-Mg phase as the anode and formed a galvanic couple with the (Mg, Al)$_2$Ca phase. Mg$_2$Ca dissolved preferentially, leaving voids along the grain boundaries. When the volume fraction of the Mg$_2$Ca phase was sufficiently high to wrap the α-Mg grains, the enclosed α-Mg grains would fall off once the surrounding Mg$_2$Ca phase was dissolved, resulting in an accelerated corrosion rate. The lowest corrosion rate of 0.031 mg cm^{-2} h^{-1} was observed when the CaO content was 9 wt.% [118].

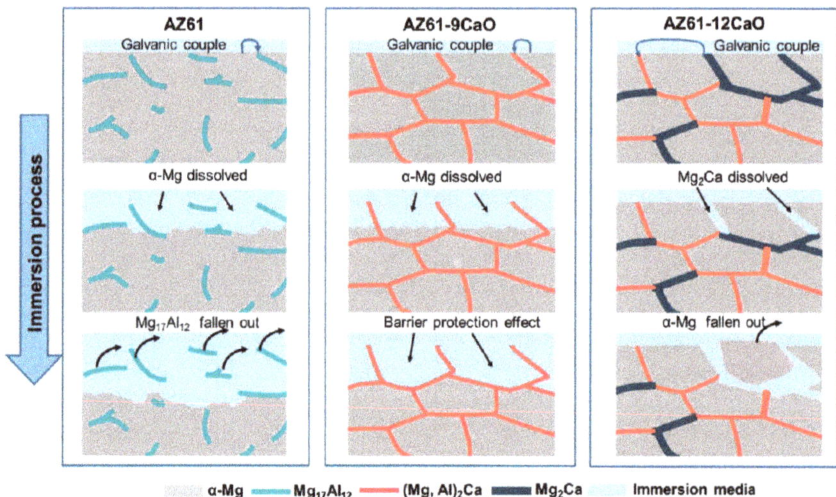

Figure 14. Schematic diagram of the corrosion process of AZ61-CaO. Reproduction from [118], with permission from Taylor & Francis, 2020.

The influence of Ca on the plasticity of SLMed alloys is still lacking. However, this influence can be interpreted from the corresponding mechanism in traditional manufacturing processes. In addition to improving the corrosion resistance of Mg alloys, Ca also has a notable effect on plasticity [135]. The poor formability of magnesium alloys at room temperature limits their wider application. The deformation of magnesium alloys at room temperature is mainly achieved by {0001} < 11$\bar{2}$0> basal slip and {10$\bar{1}$2} <$\bar{1}$011> tension twinning [136,137], whereas the activation of nonbasal slip modes is generally difficult. Studies have shown that the addition of Ca greatly improves the elongation of Mg alloys [138–140]. Pan et al. [138] prepared an extruded Mg-1.0Ca (wt.%) binary alloy with a tensile elongation of 18.0% and a grain size of 2 µm. Zhang et al. [139] found that the extruded Mg-1.0Zn alloy has an elongation of 16.2% and a grain size of 20–50 µm. The grain size of the extruded

Mg-1.0Zn-0.2Ca (wt.%) alloy was refined to 5–20 μm after adding Ca, and the elongation increased to 35.5%. The elongation was 119% higher than that of the extruded Mg-1.0Zn alloy. The high ductility of the Mg-Zn-Ca ternary alloy can be attributed to the texture weakening and grain refinement caused by the segregation of Ca at the grain boundaries [140]. The high ductility of Mg-Al-Ca ternary alloys and Mg-Ca binary alloys indicates that Ca may fundamentally change the deformation behavior of Mg, similar to some rare earth elements. Sandlöbes et al. [141] observed an increase in the activity of the <c + a> slip system in Mg-1Al-0.1Ca. According to molecular dynamics (MD) calculations and density functional theory (DFT) calculations, Ca atoms have higher dislocation binding energy and solid solution strengthening on basal <a> slip than on <c + a> slip, and it is predicted that Ca will promote the basal-to-prismatic cross-slip in Mg [142,143].

Reference [135] studied the mechanism of Ca addition on the tensile strain of Mg alloys and specifically analyzed the reason for the increase in plasticity caused by Ca addition. In a tensile test of extruded Mg-0.47 wt.% Ca alloy, the slip activity in the grains was studied under tensile strains of 1%, 2%, 4%, 8% and 16%. Basal slip was the dominant deformation mechanism in this alloy at all strains, whereas prismatic and pyramidal I <a> slips became important after 2% strain. According to Schmid factor analysis, the CRSSs of prismatic slip and pyramidal <a> slip were approximately twice that of basal slip in this Mg-Ca alloy. The enhancement of nonbasal <a> slip activity improved the ductility of the material. The enhanced nonbasal <a> slip activity can be interpreted as Ca reducing the energy barrier for <a> dislocations to cross-slip to nonbasal planes. According to first-principles calculations, solute Ca atoms will reduce the unstable stacking fault energy of all slip modes. The magnesium alloy prepared by SLM also suffered from low plasticity, and some SLMed magnesium alloys have lower plasticity than as-cast magnesium alloys [26,28]. At present, due to the lack of research on the properties of SLMed Mg alloys containing Ca, the improvement in the plasticity of SLMed magnesium alloy provided by Ca alloying still needs to be further studied because of its complex mechanism.

4.1.3. Alloying Element: Sn

Sn is an essential trace element in the human body. Sn can promote the growth and development of the body, break down hemoglobin, and promote tissue growth and wound healing [144]. The chemical properties of Sn are stable [145]. Several studies have shown that the addition of Sn can improve the properties of Mg alloys and that the resulting alloy exhibits excellent mechanical properties and good corrosion resistance, especially when the Sn content is 1 wt.% [145,146]. Zhou et al. [39] investigated the effect of adding different amounts of Sn on the microstructure and properties of SLMed Mg. The addition of Sn will form the Mg_2Sn phase with Mg. As the Sn content increased, the Mg_2Sn phase hindered the growth of the grains, the grain size decreased from 25 to 5 μm, and the Mg_2Sn phase increased significantly. On the one hand, the reduction of alloying element segregation during the SLM process and grain refinement slowed the degradation rate. On the other hand, the Mg_2Sn phase accelerated the degradation rate due to galvanic corrosion. Hence, reducing the degradation rate of SLMed Mg should balance both the grain size and the volume fraction of the Mg_2Sn phase. The degradation rate of the Mg-Sn alloy was slower than that of Mg when the Sn content was less than 3 wt.%. Grain refinement and segregation reduction were the reasons for this phenomenon. The degradation rate of the alloy increased with increasing Sn content when the Sn content exceeded 3 wt.%. The main reason was that the second phase Mg2Sn caused severe galvanic corrosion and accelerated the degradation of Mg. The corrosion resistance of the Mg alloy was significantly improved when the Sn content was 1 wt.%. In this case, the beneficial effects of grain refinement outweighed the adverse effects of Mg_2Sn on corrosion resistance [39].

In terms of mechanical properties, Mg had the lowest Vickers hardness (37.9 Hv), whereas Mg-7Sn alloy had the highest Vickers hardness (65.7 Hv) [39]. The hardness of the Mg-Sn alloy increased with increasing Sn content, which was caused by grain refinement and second phase strengthening. The rapid solidification characteristics refined the Mg grains. The hardness of Mg_2Sn was greater than that of the Mg phase. Therefore, the strengthening of the second phase was due to the presence of the Mg_2Sn phase. The discrete Mg_2Sn phase precipitated at the grain boundaries, which suppressed the motion of dislocations and strengthened the Mg alloy. The introduction of Sn improved the compressive strength of Mg. With increasing Sn content, the compressive strength of the SLMed Mg-Sn alloy increased to 81 MPa (5 wt.%), and as the Sn content continued to increase, the compressive strength of the alloy decreased. The improvement in compressive strength mainly came from grain refinement and Mg_2Sn phase strengthening. Excessive Sn content led to an increase in the Mg_2Sn phase. The semicontinuous network of Mg_2Sn at the grain boundaries accelerated the initiation and propagation of cracks at the phase interface, leading to a reduction in compressive strength. Therefore, the beneficial and harmful effects should be considered comprehensively when introducing alloying elements

4.1.4. Alloying Element: Zn

Zinc is one of the most abundant nutritionally essential elements in the human body and has basic safety for biomedical applications [147]. Cai et al. [85] showed that Zn could improve the corrosion resistance and mechanical properties of Mg alloys. The improvement in mechanical performance for Mg-Zn alloys with up to 5 wt.% Zn content corresponds to fine grain strengthening, solid solution strengthening, and second phase strengthening. Polarization tests have shown the beneficial effect of Zn on the formation of a protective film on the surface of alloys. However, adding Zn to Mg will make Mg-Zn binary alloys crack easily during solidification [94]. Therefore, the content of Zn should be controlled within an appropriate range. Yang et al. [148] prepared Mg-Zn alloy by powder metallurgy and pointed out that Mg-Zn alloy is suitable for the human body when the Zn content does not exceed 14.5 wt.%. Boehlert [149] studied Mg-Zn alloys containing 0–4.4 wt.% Zn and found that Zn was a potent grain refiner and strengthener for Mg; moreover, they determined that the optimal Zn content was 4 wt.%. Wei et al. [43] discussed the effect of 1–12 wt.% Zn content on the densification, microstructure and mechanical properties of SLMed Mg-Zn binary alloys. Their results showed that the increase in Zn content had a significantly deteriorating effect on the densification response of the SLMed Mg-Zn alloys. At 1 wt.% Zn content, near full-density products could be obtained. As the Zn content (2–10 wt.%) increased, cracks appeared, which severely affected the relative density of the sample. After the Zn content increased to 12 wt.%, the cracks disappeared, but some micropores appeared, which affected the relative density and mechanical properties of the sample. The mechanical properties of the sample were only improved when the Zn content was 1 wt.%, for which the UTS was 148 MPa and the elongation rate was 11%; these values were similar to those of the as-cast Mg. Zhang et al. [42] asserted that with increasing Zn content, the grain size of SLMed ZK61 magnesium alloy was refined, but the surface quality decreased. The precipitated phases experienced successive transitions: $MgZn \rightarrow MgZn + Mg_7Zn_3 \rightarrow Mg_7Zn_3$. With increasing Zn content, the solubility of Zn in the Mg matrix increased. On the basis of the solid solution strengthening theory, high solid solubility will induce excellent properties. In addition, the uniform precipitation of the hard and brittle second phase Mg_7Zn_3 can also promote the mechanical properties. The homogeneously distributed Mg_7Zn_3 phase with higher microhardness could impede the dislocation motion between the grains [150], resulting in increased microhardness. When different manufacturing processes are adopted, the optimal Zn content in the Mg alloy is different. However, it is certain that Zn has a beneficial effect on the properties of SLMed Mg alloys, and the corrosion resistance and mechanical properties of Mg alloys can be improved by grain refinement, solid solution strengthening and precipitation strengthening. When introducing Zn, it is necessary to control the content of Zn and reduce the occurrence of cracking.

4.1.5. Other Regular Elements

Currently, the types of alloying elements introduced in SLMed magnesium alloys are still relatively limited. In traditional manufacturing processes, other alloying elements, such as rare earth elements and Mn, are also introduced to improve the comprehensive performance of as-cast Mg alloys [151,152].

By summarizing the influence of these alloying elements on Mg alloys prepared in the traditional process, more information can be provided for use as a reference for SLMed Mg alloys.

Mg-Mn series alloys are regarded as a new type of Mg alloy that have excellent corrosion resistance, good creep resistance and low cost. The mechanical properties and corrosion resistance of Mg alloys were improved after Mn was introduced [151,153]. Mn addition to Mg-9Al-2Sn alloy led to the formation of $Al_8(Mn, Fe)_5$ phases. The formation of $Al_8(Mn, Fe)_5$ particles in the interior of grains was likely the reason for grain refinement, which was stable in size during the solution treatment. The addition of 0.1 wt.% Mn had an obvious effect on accelerating the aging behavior of the Mg-9Al-2Sn alloy [151,153]. Adjusting the process parameters can improve the performance of the SLMed magnesium alloy to a certain extent. However, adjusting the process parameters had a relatively limited impact on the performance improvement in SLMed magnesium alloys. In addition to controlling the process conditions, alloying elements must be introduced to further improve the performance of SLMed magnesium alloys. Therefore, it is necessary to introduce more alloying elements and expand the SLMed magnesium alloy series. This is also worthy of attention and research in the future.

In summary, the introduction of alloying elements can play a role in refining the grains and improving the corrosion resistance. However, the amount of the new second phase generated after the introduction of different alloying elements will increase with increasing alloying element content and gradually evolve from discrete precipitation to network precipitation along the grain boundaries, which causes crack initiation and propagation at the second phase/matrix interface, limiting the mechanical properties of the SLMed magnesium alloys. A greater number of alloying elements should be introduced in future research, and additional SLMed magnesium alloy systems should be developed. Additionally, more precise studies are needed to determine the optimal number of alloying elements added to SLMed Mg alloys in order to improve corrosion resistance while minimizing the reduction in mechanical properties.

4.2. Rare Earth Elements

The addition of rare earth elements can reduce the grain size and corrosion current of magnesium alloys, which slows hydrogen evolution and improves corrosion resistance [154]. In general, high nucleation rates and low grain growth rates can lead to the formation of fine grains. In addition, the melting point and the diffusion rate of the atoms may also be related to the grain refining effects of the rare earth addition. It has also been reported that the addition of rare earth elements can remarkably improve the mechanical properties of magnesium alloys by solid solution strengthening and precipitation strengthening [155,156]. Therefore, in the preparation of SLMed magnesium alloy, rare earth elements can be added to improve performance.

Studies have shown that the rare earth element Dy has an effect on the microstructure and corrosion resistance of SLMed Mg-Zn alloys [40]. For the alloys containing 1–5 wt.% Dy, the grain size was significantly refined as the Dy content increased. The main reason for the refinement was that the Mg-Zn-Dy phase had a higher melting point, which reduced the temperature difference between the Mg-Zn-Dy phase and the α-Mg phase. In the solidification process, the Mg-Zn-Dy phase was formed by the diffusion of atoms within a lower range, and precipitates first formed in the grain boundaries and then α-Mg solidified. Therefore, the microstructure was effectively refined. Zhang et al. [157] asserted that the addition of rare earth elements reduced the laminar spacing of α-Mg and that rare earth elements acted as grain refiners. In terms of corrosion resistance, the degradation rate significantly decreased when the Dy content was 1 wt.%. The lowest average hydrogen evolution rate was 0.009 mL cm^{-2} h^{-1}. When Dy increased to 5 wt.%, the hydrogen evolution rate increased to 0.068 mL cm^{-2} h^{-1}. When the Dy content was excessive, the abundant heterogeneously distributed Mg-Zn-Dy and $MgZn_2$ phases might serve as a cathode. However, galvanic coupling with the Mg matrix could accelerate the degradation rate of the alloys. Rare earth elements have a double effect on the degradation rate of SLMed magnesium alloy. On the one hand, rare earth elements can refine the grains of magnesium alloys and reduce element segregation, thereby reducing the degradation rate of the alloys; on the other hand, a greater amount of the second phase could cause severe galvanic corrosion, which could accelerate the degradation of the Mg alloys. Therefore, the content of rare earth elements should be controlled within an appropriate range. The influence of the rare earth element content and the determination of the optimum rare earth content still need further study.

Due to the lack of literature on the introduction of rare earth elements into SLMed alloys and their related mechanisms, the research results of the influence of rare earth elements on magnesium alloys produced by the traditional process can be used for reference. In a traditional preparation, Li et al. [158] studied the influence of the rare earth elements Sm and La on the microstructure and mechanical properties of as-cast Mg alloys. The Mg-0.5Zn-0.2Mn-0.2Sm/0.3La-0.4Ca alloy exhibited good ductility with A (elongation) values of 27.5–30.5%, whereas the Mg-0.5Zn-0.2Mn-0.2Sm-0.3La alloy exhibited an excellent balance of strength and ductility with an Rp0.2 (UTS) value of 190 MPa, an Rm (high YS) value of 239 MPa, and an A (elongation) value of 23.9%. After the introduction of Sm and La, the grains were refined by the segregated atoms at the head of the solid–liquid solidification and the aggregated particles near the grain boundaries. Moreover, Sm and La helped weaken the texture, thereby improving the plasticity. The weakened texture with the addition of rare earth elements was due to the dynamic recrystallization (DRX) and the weakened dynamic recrystallized (DRXed) texture. Yamasaki et al. [159] studied the influence of Gd on the mechanical properties of Mg alloys. The Mg-Zn-Gd alloy exhibited high strength (345 MPa) and large elongation (6.9%) due to the refinement of α-Mg grains and the existence of a highly dispersed hard long-period ordered (LPO) structural phase. The LPO phase was also found in the Mg-Zn-Y alloy system [160]. The 18R-type long-period stacking (LPS) phase had a higher hardness than the Mg matrix phase. The most essential characteristic in the microstructures of these Mg-Zn alloys containing rare earth elements (denoted Mg-Zn-RE alloys hereinafter) was a long-period stacking ordered (LPSO) structural phase. The mechanical properties and corrosion resistance of Mg-Zn-RE alloys containing the LPSO phase can be improved by rapid solidification [92]. In the solidification process of Mg alloys, the change in cooling rate had a great influence on the corrosion behavior of Mg-Zn-Y alloys. The rapid solidification technology improved the microstructure and electrochemical uniformity of the Mg-Zn-Y alloys, leading to the passivation of the matrix material. The optimum cooling rate of the Mg-Zn-Y alloys was 3×10^4 K s^{-1}. The cooling rate was slow, and a large number of LPSO phases were formed in the casting process, which deteriorated the corrosion resistance. The increase in cooling rate led to a delay in the occurrence of filamentous corrosion. This was due to the grain refinement in the alloy and the formation of the supersaturated single α-Mg phase solid solution [92].

To date, most reported biomedical magnesium alloys contain rare earth elements. In the application of rare earth elements, attention should be paid to the toxic effects of rare earth elements. Rare earth elements may induce latent toxic and harmful effects on the human body [161,162]. Consequently, alloying elements must be chosen with careful consideration of the possible toxic effects.

Studies on introducing rare earth elements into SLMed alloys are still lacking. Future studies in this area can refer to studies on the addition of rare earth elements in magnesium alloys produced by other forming processes. Combined with the advantages of rapid solidification in the SLM process and the extension of the alloying system used in the SLMed magnesium alloy, it is necessary to explore the influence of the content of rare earth elements on the properties of SLMed magnesium alloys and determine the optimal rare earth element content. However, it is also important to consider the toxicity of rare earth elements in the biological field.

5. Microstructure of SLMed Magnesium Alloy

Because SLM has the characteristics of rapid solidification, the cooling rate can reach 10^4–10^5 K·s^{-1}. The microstructure varied with respect to variations in the cooling rate. The microstructures in the as-cast, sub-rapidly solidified, and rapidly solidified (i.e., SLMed) samples can be compared under different cooling rates.

Figure 15a,b are optical microscopy and SEM images of the as-cast AZ61 alloy [163]. The as-cast alloy exhibited a typical dendritic eutectic network structure. The phase composition in the as-cast AZ61 was composed of an α-Mg matrix and a β-Mg$_{17}$Al$_{12}$ eutectic phase distributed in grain boundaries and grains. The average grain size of the as-cast AZ61 alloy was approximately 320 ± 5 μm. Under normal conditions, the grains were coarse and non-uniform. The microstructure of as-cast AZ91D is shown in Figure 16a,b. As-cast AZ91 also consisted of an α-Mg solid solution and a β-Mg$_{17}$Al$_{12}$ eutectic phase. As shown in the enlarged micrograph of area A (Figure 16b), some of the divorced eutectic β-Mg$_{17}$Al$_{12}$ was surrounded by lamellar eutectic. In the die-cast AZ91D ingot, β precipitates existed in the form of partially divorced eutectic structures. AZ61 and AZ91D are hypoeutectic magnesium aluminum alloys with low zinc content. In hypoeutectic Mg-Al alloys, the morphology of the eutectic phase depends on the cooling rate. A higher cooling rate resulted in a more discrete microstructure [164]. Therefore, the inherently high cooling rate in the SLM process caused a change in β-Mg$_{17}$Al$_{12}$ between the SLMed part and the as-cast part of Mg alloy. For die-cast samples, most of the Al and Zn were concentrated in the β-Mg$_{17}$Al$_{12}$ phase [165]. The reduction in the Al content in the α-Mg solid solution not only reduced the effect of solid solution strengthening but also deteriorated the corrosion behavior.

Sub-rapid solidification is a nonequilibrium solidification process with a solidification rate of 10^3 K/s [69]. The sub-rapidly solidified structure changed considerably from that of the as-cast sample (Figure 15c). The β-Mg$_{17}$Al$_{12}$ phase continuously distributed on the grain boundaries disappeared. The microstructure was obviously refined and was mainly composed of small equiaxed grains. The literature noted that as the thickness of the sheet decreased (i.e., the faster the solidification process), the grain size decreased [72], and a large number of petal-like dendrites appeared in the microstructure. Owing to the high cooling rate, the dendrite structure was very small, so it was difficult to distinguish the dendrite arm spacing in the low-magnification metallographic microstructure. Under sub-rapid solidification, the grain size of the AZ61 magnesium alloy was 13.5 μm, which was much smaller than that of the as-cast AZ61 magnesium alloy.

The solidification rate of SLM was faster than that of sub-rapid solidification. As shown in Figure 15d, the SLMed microstructure was more uniform, the α-Mg grains were refined and equiaxed, and the β-Mg$_{17}$Al$_{12}$ phase precipitated at the grain boundaries. The α-Mg crystal grains and β-Mg$_{17}$Al$_{12}$ in the SLMed microstructure were more refined than those in the sub-rapidly solidified microstructures. The grain size in Figure 15d was 2.46 μm (at an energy density of 208 J/mm^3). Other studies have also found the same microstructural characteristics of SLMed magnesium alloys [26,29,30]. Wei et al. [26] investigated the distribution of Mg and Al in SLMed AZ91D samples. Comparing the SLMed samples with as-cast AZ91D, the distribution of Mg and Al in the SLMed AZ91D was more uniform, and the content of Al in the matrix varied with respect to the energy density. This finding demonstrates that the chemical composition distribution was more uniform under rapid solidification, which is beneficial to reduce the segregation of the components, and the energy input has an effect on the solid solution [166].

The rapid solidification of SLM mainly affects the generation of the second phase in the alloy. Cai et al. [167] compared the microstructure and morphology of the second phase β-Mg$_{17}$Al$_{12}$ in the conventional as-cast part and in the part produced under rapid solidification. Compared with conventional casting, β-Mg$_{17}$Al$_{12}$ in the rapidly solidified AZ91 magnesium alloy had smaller and fewer micropores. The dispersive microporosity will act as crack initiation sites and promote crack propagation under the application of an external force, causing significant deterioration in tensile properties. These micropores were mainly formed by solidification shrinkage and dissolved gas. Liu et al. [28] compared the microstructures under different energy densities under SLM and found that as the energy density increased, the microstructure changed. When the energy density was low, the precipitated second phase was dispersed, and the grain size was small (1.61 μm). As the energy density increased, the second phase gradually precipitated along the grain boundaries, and the grain size slightly increased. Other studies noted that the solid solution was related to the energy input. As the energy input increased, the solid solution of the elements exhibited different monotonic or nonmonotonic changes [28–30]. The solid solution of the elements and the precipitation of the second phase were both related to the energy input and solute capture. Therefore, analysis and discussion need to be combined with the solidification path.

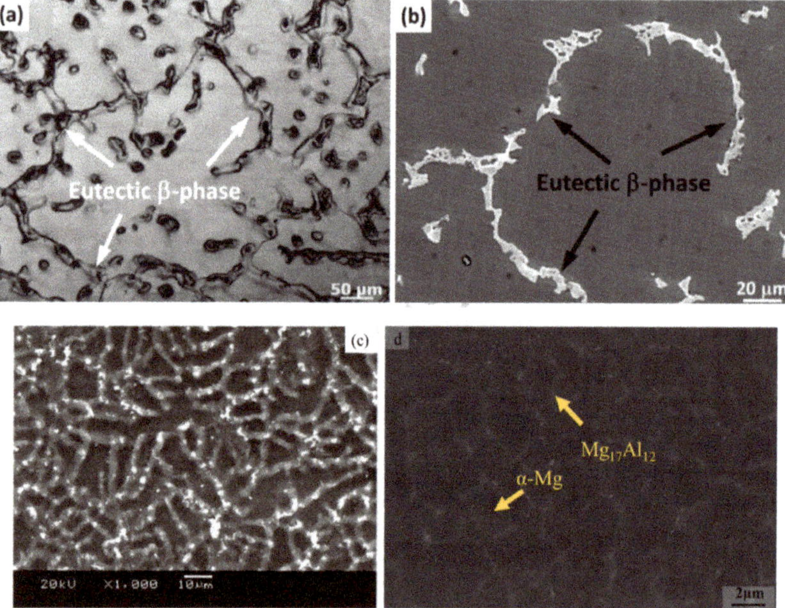

Figure 15. Microstructure of AZ61: (**a**,**b**) as-cast AZ61. Reproduction from [163], with permission fromElsevier, 2020. (**c**) Sub-rapidly solidified AZ61. Reproduction from [72], with permission from Elsevier, 2020. (**d**) SLMed AZ61. Reproduction from [28], with permission from Elsevier, 2020.

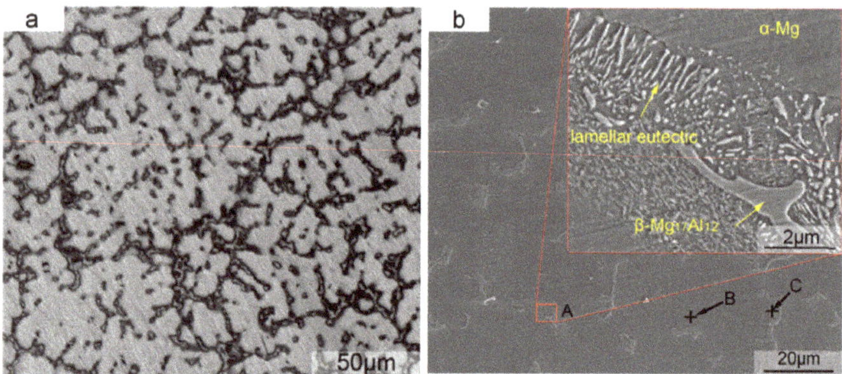

Figure 16. Microstructure of as-cast AZ91. (**a**) microscope picture; (**b**) SEM picture. Reproduction from [26], with permission from Elsevier, 2020.

In Figure 17, the left diagram is the phase diagram of the Mg-Al binary system [168], whereas the right diagram is the phase diagram of the AZ61 Mg alloy [28]. The addition of Zn obviously had a greater impact on the Mg-Al binary system phase diagram. Analysis of the equilibrium cooling process of AZ61 shows that the solidification path of the red line in the figure was L → L + α-Mg → α-Mg → α-Mg + β-Mg$_{17}$Al$_{12}$ → α-Mg + β-Mg$_{17}$Al$_{12}$ + T.

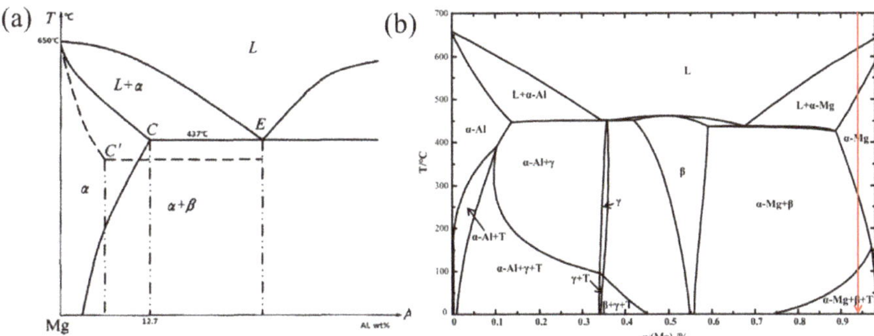

Figure 17. Mg-Al alloy equilibrium phase diagram: (**a**) Mg-Al binary system. Reproduction from [168], with permission from Elsevier, 2020. (**b**) AZ61 (1.27% Zn). Reproduction from [28], with permission from Elsevier, 2020.

Taking the AZ series Mg alloy as an example, the maximum solubility of Al in Mg was 12.7 wt.% at 437 °C during equilibrium solidification, whereas the solubility of Al was only 2 wt.% at room temperature. The reaction L → α-Mg occurred first under slow cooling. The Al atoms diffused sufficiently, and the alloy composition in α-Mg gradually became uniform. As the cooling process progressed, the reaction L → α-Mg ended when the melt temperature dropped to the solid phase line. Subsequently, the second phase β-Mg$_{17}$Al$_{12}$ began to precipitate in the α-Mg solid solution, and the final solidified products in the equilibrium state were α-Mg solid solution and the second phase β-Mg$_{17}$Al$_{12}$, and there was no eutectic phase. However, SLM is a nonequilibrium process under rapid solidification, so the final phase composition of the SLMed part was slightly different from the equilibrium phase diagram. The AZ61 Mg alloy was composed of two phases of α-Mg and β-Mg$_{17}$Al$_{12}$ at room temperature during a relatively slow cooling process. At a faster cooling rate (higher scanning speed and lower laser power), the atoms in the primary α-Mg in the liquid phase did not have time to diffuse sufficiently. From the principle of solute redistribution, it is known that the solid solution of Al was continuously enriched at the solidification front, and the residual liquid phase between the dendrites of the α-Mg reached the eutectic composition in the late solidification period. The two phases grew independently to form a divorced eutectic structure. As a result, the temperature of the molten pool was low, and the cooling rate as fast at low energy input (138.89 J/mm^3), which is equivalent to being extremely cold in the α-Mg region. Due to the lack of sufficient diffusion time, Al dissolved into the matrix. The solid solution of Al in the matrix increased. Hence, the eutectic transformation process L → α-Mg + β-Mg$_{17}$Al$_{12}$ was suppressed. The precipitation of β-Mg$_{17}$Al$_{12}$ was less prominent at low energy input. In contrast, due to the accumulation of heat and the decrease in the solidification speed, the solute capture effect was weakened when the energy input was high (156.25–208.33 J/mm^3). Al diffused more fully, the content of solid solution elements decreased with increasing Ev, the eutectic transformation L → α-Mg + β-Mg$_{17}$Al$_{12}$ occurred, and the crystallinity of β-Mg$_{17}$Al$_{12}$ increased.

The change of microstructure (grain shape, size, second phase distribution) is related to the cooling rate. As expected, various laser parameters resulted in different solidification rates of the molten pool and thermal cycles, leading to variation in microstructures of the melted zone. In the SLM process, the combination of scanning speed and laser power is the key to controlling the cooling rate. In addition, the faster the cooling rate, the finer the microstructure; the lower the cooling rate during solidification, the longer the time available for grain coarsening. Generally speaking, as the scanning speed decreases and the laser power increases (that is, the line energy density increases), the accumulation of heat in the molten pool causes the temperature of the molten pool to rise, the cooling rate slows down, and the grain size gradually becomes coarser [35]. The increase of the laser power or decrease of the laser scanning speed led to the coarsening of the grains in the melted zone because the higher laser power or slower laser scanning speed provided more driving force for grain boundary movement, which then promoted the growth of grains. At lower scanning speeds, the prolonged interaction time between the laser and the powder suppresses the heat dissipation in the molten pool. Therefore, due to the larger heat accumulation, the epitaxial growth kinetic conditions of the grains are enhanced [16]. Similar results were obtained for pulsed laser samples. When the laser power was low or the scanning speed was high, the average gain size became small, since decreasing the laser power in effect acts in the same way as reducing the preheat temperature of the powder when the next pulse hits the base. The relatively high cooling rate at low laser power or high laser scanning speed restrained the growth of α-Mg grain during solidification.

As the line energy density increases (low scanning speed and high laser power), the microstructure follows the evolution of clustered finer dendrites–refined equiaxed grains–coarsened equiaxed grains [26,28]. The grain size of AZ61 magnesium alloy can be refined to 1.61 μm (E_v = 138.89 J/mm^3). Then, the grains changed to an equiaxed shaped (1.79 μm) with an increase in laser energy input to 156 J/mm^3. Further increase in the laser energy input to 178 J/mm^3 and 208 J/mm^3, equiaxed grains of 2.12 μm and 2.46 μm, respectively [28]. This is due to the decrease in cooling rate caused by changes in scanning speed and laser power and accumulation of heat in the molten pool. This same reason can explain the change of the grain size far from the center of the molten pool. The α-Mg grains inside the molten pool show equiaxed crystal morphologies. On the contrary, with the farther away from the molten pool, the α-Mg grains show a transformation from equiaxed crystal to columnar crystal and the grain size increases [38]. The ratio G/R of the temperature gradient (G) and the solidification rate (R: the propulsion rate of the solidification interface in the normal direction) in the crystal direction of the pulsed light spot determines the microstructure morphology after solidification. Inside the molten pool, G is very high, but R tends to 0, so that the value of G/R is high, which causes the solidification structure to be equiaxed crystal. Further away from the molten pool, G decreases but R gradually increases, so that the value of G/R gradually decreases. The influence of scanning speed and laser power on the precipitation of the second phase has been described in detail above.

Consequently, the grain size, the second phase precipitation, and the element solid solution can be achieved by controlling the process parameters, i.e., controlling the scanning speed and laser power (energy density). In terms of grain size, by comparing the microstructure of magnesium alloys at different cooling rates, it can be seen that faster cooling rates are beneficial for refining the microstructure. From the as-cast state to sub-rapid solidification and rapid solidification (SLM), the grain size of the magnesium alloy drops from ~320 μm to ~2 μm, which is a change of two orders of magnitude. Properly increasing the scanning speed and lowering the laser power (i.e., lowering the energy density) is beneficial for the improvement of the cooling rate and the reduction of heat accumulation in the molten pool, thereby refining the microstructure. In terms of second phase precipitation, the increase of the cooling rate is beneficial to suppress the occurrence of the phase transition L $\rightarrow$ α-Mg + β-Mg$_{17}$Al$_{12}$ and the precipitation of the second phase is reduced. In addition, the energy density must be controlled within a suitable range. Too high an energy input will lead to coarse grains, an increase in element segregation and a decrease in the solid solution. However, if the laser power is reduced or the scanning speed is increased blindly (too low an energy density), the powder cannot be fully melted and combined, thus causing serious pores and affecting the sample quality. Therefore, further research must be performed to more accurately control the solidification structure by controlling scanning speed or laser power and the effect of the solidification structure on the defects and mechanical properties of SLMed magnesium alloys.

6. Effect of Heat Treatment on SLMed Magnesium Alloy

6.1. HIP

Some performance breakthroughs have been achieved in SLMed magnesium alloy, but there are still some problems, such as porosity and low elongation. Defects in the sample can be improved by adjusting the process parameters to improve the fusion of the powder. On the other hand, the pores caused by thermal convection in the molten pool are difficult to remove, requiring the introduction of post-treatment processes, such as HIP. The mechanism of HIP is to apply temperature and pressure to the sample at the same time and hold it for a certain time so that the sample is densified during the pressurization process after heating to eliminate the internal pores. There are some studies on collapsing the pores of SLMed parts by HIP [169–171].

According to the literature [172], the strength properties of SLMed samples are considerably inferior to those of the alloy consolidated using HIP technology, suggesting that it is reasonable to expose the complex-shaped SLMed workpieces to additional HIP treatment. Liu et al. [41] believed that the plasticity of the SLMed AZ61 magnesium alloy was greatly improved after 3 h of HIP at 450 °C and 103 MPa. The results showed that the elongation reached 8.2%, the UTS of the material remained unchanged, the elongation increased (160% greater than that of the SLMed part without HIP), and the plasticity was greatly improved. The improvement in plasticity mainly came from the following two parts: the reduction in the original defects (pores) in the SLMed magnesium alloy and the solid solution of the second phase β-$Mg_{17}Al_{12}$. The temperature of the SLMed AZ61 part increased below the melting point during HIP, the magnesium alloy softened and then gradually densified under pressure, eventually eliminating the internal pores, and the density was close to 100%. The elongation was related to the distribution of the second phase. The temperature during HIP was slightly higher than the solidus temperature. As a result, the second phase dissolved into the matrix, which greatly reduced the source of cracks at the interface between the β-$Mg_{17}Al_{12}$ and α-Mg phases. It can be seen that densification and solution treatment are beneficial to improving the plastic elongation of SLMed magnesium alloys.

However, HIP can also adversely affect certain mechanical properties of magnesium alloys. The YS of SLMed magnesium alloys decreased as the grain size increased [41]. Moreover, for SLMed aluminum alloys, HIP post-treatment can have an adverse effect on grain size [173]. HIP postprocessing can lead to grain growth in certain coarser grained areas, probably due to a local imbalance between driving and dragging forces, which corresponds to higher defect density and fewer pinning precipitates. This is a problem that can be solved by appropriately reducing the temperature or shortening the holding time. HIP experiments can verify that the second phase solid solution plays a positive role in improving the plasticity of magnesium alloys. Moreover, the second phase had an impact on the corrosion resistance [174,175]. HIP is a comprehensive heat treatment process that includes closing pores and providing internal heat treatment. Magnesium alloys are affected by the combined effects of temperature, pressure and holding time during HIP, which affect the pores and microstructure. Therefore, different magnesium alloys require different processing parameters, and there is still a lack of experimental data to support the selection of appropriate processing parameters. Hence, more exploration will be needed in order to accurately select the HIP parameters.

6.2. Heat Treatment

Reducing or eliminating the effect of the second phase on the properties of magnesium alloys can also be achieved by heat treatment and other methods. Common magnesium alloy heat treatment methods include solid solution heat treatment (T4), aging heat treatment (T5), and solid solution + aging heat treatment (T6). SLMed magnesium alloy has the problem of poor plasticity, which can be improved by introducing an appropriate heat treatment process.

In AZ series Mg alloys, the presence of the β-phase influences the corrosion behavior to a great extent. The machinability of Mg alloys is also influenced by the amount and distribution of the β-phase. In materials engineering, heat treatment is a promising technique to alter the microstructure to achieve the desired phases and corresponding volume fractions in order to alter the bulk behavior. Chowdary et al. [176] noted that heat treatment of AZ91 Mg alloy greatly affected the amount and distribution of the secondary phase at the grain boundaries and asserted that the machinability of AZ91 Mg alloys can be improved by developing supersaturated grains and reducing the amount of the secondary phase. The corrosion resistance was greatly affected after the heat treatment due to the supersaturated grains, which promoted a higher corrosion rate [176]. The literature [177] showed that the elongation increased from 7.4 to 11.2% in an as-cast magnesium alloy after T4 heat treatment. T4 treatment (solid solution) significantly improved ductility but reduced YS. T6 (solution + aging) increased the UTS but reduces the YS and ductility.

Wang et al. [177] proposed a new heat treatment idea: solution treatment at 413 °C caused the β phase to dissolve into the matrix. In contrast, the design purpose of Tx is to break the β-$Mg_{17}Al_{12}$ phase network, provide better ductility in exchange for a slight decrease in strength, and comprehensively improve strength and elongation. Treatment at a temperature close to the solvus, typically at 365 °C for just 2 h according to the equilibrium phase diagram, caused the β-$Mg_{17}Al_{12}$ phase to dissolve partially, and the β-$Mg_{17}Al_{12}$ network was effectively broken up. This resulted in a composite structure that, with the exception of a few primary α-Mg globules, consisted of fine α-grains and the fine remaining β-$Mg_{17}Al_{12}$-grains located along the grain boundaries, mainly at the triple grain boundary junctions. If the heat treatment temperature was set to Tx, rather than T4, T5 or T6, both strength and elongation can be improved.

In the literature [178], the solution treatment dissolved the β-$Mg_{17}Al_{12}$ phase and coarsened the grains. During the aging process, the β-$Mg_{17}Al_{12}$ phase preferentially and discontinuously precipitated at the grain boundaries, and then a continuous precipitated phase appeared inside the grains. Both the strength and the elongation were improved after solution treatment and aging treatment. Annealing dissolved β-$Mg_{17}Al_{12}$ and increased the concentration of Al solute in the matrix, which effectively reduced the damping capacity. $Mg_{17}Al_{12}$ distributed on the grain boundaries also caused local intergranular cracking.

Jia et al. [179] performed a solution heat treatment on Mg-Zn alloys. The enhancement of UTS was due to the solution strength effects of the alloying elements. In addition, Zn dissolving into the Mg matrix decreased the stacking fault energy of the matrix, which led to a change in the plastic deformation mechanism. Thus, the solution treatment enhanced the UTS and elongation simultaneously. The corrosion resistance of the solution-treated samples was superior to that of the as-cast samples.

It can be seen that heat treatment plays a positive role in improving the strength, plasticity, and corrosion resistance of magnesium alloys. However, there is still a lack of attention in the post-treatment of SLMed magnesium alloys. For the inherent defects in SLMed samples, such as pores, it is necessary to introduce postprocessing steps to eliminate them. Comprehensive improvements in the mechanical properties and corrosion resistance of SLMed magnesium alloys can be considered by adding heat treatment. However, it is necessary to further explore the optimal postprocessing parameters to control the adverse effects of grain size growth on the properties of magnesium alloys to balance the effects of grain size and microstructure.

7. Outlook

The development of SLM of magnesium alloys has been reviewed, and an extensive analysis of the available literature on the preparation of metals via SLM has led to the recognition of the influential process parameters and material properties. At present, the limited processable materials, immature process conditions and metallurgical defects restrict the development and application of SLMed magnesium alloys. Some efforts have been made to solve the above problems, such as adding alloy elements and applying postprocessing. However, the breakthroughs of SLMed magnesium alloys in these two areas have not been reviewed. Therefore, this article gives an overview of these three issues. In this paper, the process parameters, alloying elements, microstructure, properties, and postprocessing steps were systematically summarized. For the purpose of high efficiency, high quality, low cost, and more stable use of SLM to prepare high-performance magnesium alloys, scholars have carried out research on process parameters, alloying elements, and post-treatment steps. Although some progress has been made in terms of the relative density, mechanical properties, and corrosion resistance of magnesium alloys, the literature is still limited to unsystematic performance studies of a few types of magnesium alloys. Hence, current SLMed parts are produced under suboptimal process conditions, wherein it is difficult to control metallurgical defects.

In the future, there are still several issues that need attention in the manufacturing of magnesium alloys via SLM.

First of all, in view of the lack of processable materials, in order to expand more processable SLMed magnesium alloy materials and improve the problem of limited processable materials, we reviewed the progress of SLMed magnesium alloys in the addition of alloying elements, but there are still two problems in this area: the types of added alloying elements are limited, and the research of the addition of alloying elements needs to be in-depth. At present, most studies on SLM fabrication of magnesium alloys focus on the influence of process parameters and simple alloying elements on the microstructure and properties of binary alloys. Aside from research on Mg-Al-Zn alloys, Mg-Zn-Zr alloys, and Mg-Zn alloys, few studies have been reported on other as-cast Mg alloys and wrought Mg alloys, especially the high-strength and highly corrosion-resistant Mg alloys that are in great demand in the aerospace field (e.g., Mg-Mn alloys). We also combined the relevant research on the mature alloy element addition route in the traditional process to provide a reference for the alloy design of SLMed magnesium alloys, such as Mn and Ca, and the addition of these elements in SLMed magnesium alloys has rarely been reported. Therefore, there is a need to broaden the scope of the applicable magnesium alloys for SLM and design new materials whose compositions are suitable for the process characteristics of laser AM. In addition, post-treatment can also be applied to improve the machinability of magnesium alloys in the SLM process. We reviewed the current research progress in the postprocessing part, and the results showed that the post-processing can effectively improve the properties of SLM magnesium alloys, especially to improve the plasticity, but there are few studies at present, which is an area that needs to be further studied in the future. Traditional preparation techniques of magnesium alloys are also evaluated and related to the SLM process with a view to gaining useful insights especially with respect to postprocessing.

Second, in view of the immature process conditions, we comprehensively reviewed the new progress of different magnesium alloy materials in SLM process conditions and forming, and combined process conditions with relative density, grain size, microstructure, mechanical properties, and corrosion resistance. In the review of microstructure, the microstructure under different preparation processes (different cooling rates) was compared longitudinally, and the microstructure under different process conditions in the SLM process was also compared horizontally. In addition, it is different from the previous reviews of SLMed magnesium alloy.

Third, in terms of metallurgical defects, defects such as oxidation and cracks are reviewed. Research on cracks has made progress recently, but there is lack of review of the latest research on cracks. Cracks are a serious defect that limits the application of magnesium alloys and leads to premature failure of materials. Therefore, it is necessary to review the research progress of SLMed magnesium alloy cracks. It is well known that the application of SLMed Al alloy has been limited by cracks, so the paper combines the research progress of SLMed Al alloys on cracks to provide more references for the research of SLMed magnesium alloy cracks. In terms of oxidation, inspired by the purity of molten steel, this review not only looks forward to the development of SLMed magnesium alloys from the perspective of materials and processing, but also considers the future improvement direction of SLMed magnesium alloys from a new perspective—a metallurgical perspective. In addition, this was mentioned in the previous review of SLMed magnesium alloys. In research on steel materials, the harmful effect of nonmetallic inclusions on steel properties has been widely recognized [180,181]. For bearing steel, the deoxidation and purification of liquid steel has always been the focus of attention. Research has found that when the total oxygen content in molten steel is reduced from 0.0026 to 0.001%, the fatigue life of bearing steel is increased by a factor of 10, and when the total oxygen content is reduced from 0.001 to 0.0004%, the fatigue life is increased by an additional factor of 10. Therefore, to improve the performance of steel materials, it is important is to minimize both the dissolved oxygen in liquid steel and the content of oxide inclusions generated during the solidification process [182,183]. There have been many studies on the harmful effects of high oxygen content on material properties in powder metallurgy, such as 718 alloy [104], Al alloy [103], and titanium alloy [184]. Due to the highly oxyphilic nature of magnesium, the oxygen inevitably enters the magnesium alloy base metal and the magnesium alloy powder during the preparation process. However, the amount of oxygen entering the alloy system at various stages, the size and number of oxide inclusions formed in the alloy, and the mechanism of the influence of inclusions on the alloy—especially the influence of inclusions in the performance of magnesium alloys—have rarely been studied. Currently available research is limited to the understanding of the oxide film in the SLM process. It was found that the oxide film is usually located at the top of the molten pool, and the oxide film on the surface will affect the wettability, thereby weakening the interlayer bonding and resulting in the formation of defects. Improving the oxidation also means reducing the defects in the material. It was found that the use of a high-power laser could break the oxide film on the surface, but this increased the burning loss of alloying elements, and balancing the contradiction between these two aspects is a new topic. From the perspective of the preparation process of magnesium alloys, the oxygen that can enter into the alloy mainly comes from the mold cavity and the powder preparation process as well as the smelting of the magnesium base material. Future studies should focus on methods to inhibit the entry of oxygen in these processes and the mechanism of the influence of a small amount of oxygen on SLMed magnesium alloy materials.

In the future, the physical and chemical reactions in the SLMed magnesium alloy process need to be studied theoretically, especially the thermodynamics and kinetics of the formation of intermetallic compounds. For example, the thermodynamic calculation of the enthalpy of fusion of SLMed magnesium alloys with different compositions can theoretically determine the energy required for the magnesium alloy powder of a specific composition, which provides a theoretical basis for setting the laser power for actual experimental production (i.e., this approach can help ensure that the powder is melted under the optimal energy setting). In terms of dynamics research, it is necessary to establish a model of the relationship between the expansion and solidification of magnesium melt to seek the optimal energy input, control the melt temperature, and obtain the relationship between magnesium melt spreading and solidification rate to reduce the occurrence of defects, such as pores and spheroidization.

Furthermore, the influence of intermetallic compounds on SLMed magnesium alloys cannot be ignored. Intermetallic compounds have a dual effect on SLMed magnesium alloys. On the one hand, refining and dispersing intermetallic compounds can play a role in precipitation strengthening; on the other hand, intermetallic compounds can also have an adverse effect on corrosion resistance and mechanical properties. Therefore, controlling the size, shape and content of intermetallic compounds should be studied in further detail in the future. It is necessary to proceed from the theory of metallurgical physical chemistry and seek the conditions and mechanisms for the formation or decomposition of intermetallic compounds from the liquid state to the normal temperature state and finally obtain the best process parameters for controlling the intermetallic compounds in the SLM process, including an appropriate heat treatment method. According to the theoretical basis of postprocessing SLMed magnesium alloys presented above, there is still much research space in this aspect.

8. Conclusions

This review of the published literature on SLMed Mg offers insight into the current knowledge surrounding the influence of processing conditions, alloying elements, and post-treatment on the microstructure, properties, and fracture mechanisms of the produced parts. However, limited processable materials, immature process conditions and metallurgical defects are still problems that magnesium alloys need to face and solve in the SLM process. In the past two years, these three issues of SLMed magnesium alloys have been substantially improved by optimizing process parameters, introducing post treatment and adjusting different alloying elements. Due to the paucity of publications on postprocessing and alloy design of SLMed magnesium alloy powders, we review the current state of research and progress. Moreover, traditional preparation techniques of magnesium alloys are evaluated and related to the SLM process with a view to gaining useful insights especially with respect to postprocessing and alloy design of magnesium alloys. This article also discusses and summarizes the current factors that affect the formability, compactness, and mechanical properties of SLMed magnesium alloys. This article provides a reference for further investigating or controlling the microstructure of SLMed magnesium alloys and improving the densification, mechanical properties, and corrosion resistance of the produced materials. In addition, with respect to materials and metallurgy, new challenges and prospects in the SLM processing of magnesium alloy powders are proposed with respect to alloy design, base material purification, inclusion control and theoretical calculation, and the role of intermetallic compounds.

- With respect to immature process conditions and metallurgical defects, the influence of processing parameters (scanning speed and laser power) on the forming and performance of magnesium alloys cannot be ignored. The relative density of magnesium alloys is closely related to the processing parameters. If the scanning speed is too high or laser power is to low (energy density is too low), the powder cannot be fully melted, and the system is in a state of solid–liquid coexistence. In this state, the surface tension and viscosity of the liquid increase, which inhibits the liquid from flowing smoothly and causing the liquid to agglomerate into spheres and pores, thereby preventing the sample from becoming dense. If the energy density is too high, it will lead to the loss of alloying elements due to the evaporation of elements in the powder. On the other hand, under these conditions, the solute capture effect will be weakened, and the decrease in the solid solution content will cause the relative density of the SLMed sample to decrease. Therefore, the preparation of dense SLMed magnesium alloy requires the energy density to be controlled in a suitable range to enhance the solid solution strengthening and reduce the liquid surface tension, thereby eliminating the pores and spherical particles between the tracks. However, in the SLM process of magnesium alloys, it is difficult to remove the pores by only adjusting the process conditions. Post-treatment methods such as HIP are required to remove these defects. Microcracks are commonly attributed to solidification cracking and liquation cracking. The cause of solidification cracking is a residual thin film of liquid phase between the primary crystallized grains, and liquation cracking is caused by cyclic heat input, which occurs in multilayer welding and layered fabrication.
- The microhardness of the SLMed magnesium alloy is affected by the rapid solidification characteristics in the SLM process, which mainly affect the microstructure and the solid solution of the elements. On the one hand, the rapid solidification characteristics remarkably refine the microstructure of magnesium alloys, and the microhardness of SLMed magnesium alloys is notably higher than that of traditional as-cast magnesium alloys. On the other hand, the solid solution content of the alloying elements is different due to the differences in the solute trapping effects at different energy densities during the SLM process; thus, the strengthening effects are different. Moreover, the distribution and content of the second phase and the presence of defects will also affect the microhardness. Microhardness does not change monotonically with respect to the energy density. It is necessary to comprehensively consider the interaction between the solute capture effect at different molten pool temperatures and the solid solution of the elements under rapid solidification. Hence, while the microstructure is refined, the optimal solid solution is achieved, and the microhardness of the magnesium alloy is improved.
- The grain size of magnesium alloys is increased by increasing the energy input during the SLM process, and the mechanical properties are affected by the grain size and microstructure. In SLMed magnesium alloys, attention should be paid to the dual influence of the second phase on the material properties. The content of the second phase needs to be controlled to balance the strengthening of the second phase and its limitation on plasticity. The mechanical properties and corrosion resistance of SLMed magnesium alloy can be improved by introducing alloying elements.

At present, there are several alloy systems under investigation, and more alloying elements, such as rare earth elements and Mn, need to be introduced to develop different SLMed magnesium alloy systems. The micropores and the second phase generated during the SLM process have certain restrictions on the application of the material. Adjusting the process conditions can reduce the harm of the pores and the second phase to a certain extent, but it cannot be completely avoided. Post-treatments, such as HIP and heat treatment, can be applied to help eliminate inherent porosity and improve the precipitation of the second phase. However, much remains to be done in this area.

- In view of the lack of processable materials, adding alloying elements and post-treatment is an effective way to improve SLMed magnesium alloy. These two methods have played an important role in improving the mechanical properties and corrosion resistance of SLMed magnesium alloy, especially the problem of poor plasticity of SLMed magnesium alloy. In the future, these two aspects need to be further studied to design magnesium alloy materials that are more suitable for SLM process applications.

Author Contributions: Software, S.L.; validation, S.L. and H.G.; formal analysis, S.L.; investigation, S.L.; data Curation, S.L. and H.G.; writing—original draft preparation, S.L.; writing—review & editing, S.L. and H.G.; visualization, S.L.; supervision, H.G. All authors have read and agreed to the published version of the manuscript.

Funding: This research received no external funding.

Acknowledgments: Thanks to the Shanxi United Magnesium Co., Ltd. (Taiyuan City, Shanxi Province, China) for their help.

Conflicts of Interest: The authors declare no conflict of interest.

References

1. Xu, T.; Yang, Y.; Peng, X.; Song, J.; Pan, F. Overview of advancement and development trend on magnesium alloy. *J. Magnes. Alloys* **2019**, *7*, 536–544. [CrossRef]
2. Liu, W.; Zhou, B.; Wu, G.; Zhang, L.; Peng, X.; Cao, L. High temperature mechanical behavior of low-pressure sand-cast Mg–Gd–Y–Zr magnesium alloy. *J. Magnes. Alloys* **2019**, *7*, 597–604. [CrossRef]
3. Ali, Y.; Qiu, D.; Jiang, B.; Pan, F.; Zhang, M.-X. Current research progress in grain refinement of cast magnesium alloys: A review article. *J. Alloys Compd.* **2015**, *619*, 639–651. [CrossRef]
4. Luo, K.; Zhang, L.; Wu, G.; Liu, W.; Ding, W. Effect of Y and Gd content on the microstructure and mechanical properties of Mg–Y–RE alloys. *J. Magnes. Alloys* **2019**, *7*, 345–354. [CrossRef]
5. Yeganeh, M.; Mohammadi, N. Superhydrophobic surface of Mg alloys: A review. *J. Magnes. Alloys* **2018**, *6*, 59–70. [CrossRef]
6. Sanders, P.G.; Keske, J.S.; Leong, K.H.; Kornecki, G. High power Nd:YAG and CO_2 laser welding of magnesium. *J. Laser Appl.* **1999**, *11*, 96–103. [CrossRef]
7. Zhou, Y.; Gui, Q.; Yu, W.; Liao, S.; He, Y.; Tao, X.; Yu, Y.; Wang, Y. Interfacial diffusion printing: An efficient manufacturing technique for artificial tubular grafts. *ACS Biomater. Sci. Eng.* **2019**, *5*, 6311–6318. [CrossRef]
8. Du, J.; Lan, Z.; Zhang, H.; Lü, S.; Liu, H.; Guo, J. Catalytic enhanced hydrogen storage properties of Mg-based alloy by the addition of reduced graphene oxide supported V2O3 nanocomposite. *J. Alloys Compd.* **2019**, *802*, 660–667. [CrossRef]
9. Ning, H.; Zhou, X.; Zhang, Z.; Zhou, W.; Guo, J. Ni catalytic effects for the enhanced hydrogenation properties of Mg17Al12(1 1 0) surface. *Appl. Surf. Sci.* **2019**, *464*, 644–650. [CrossRef]
10. Liu, Y.; Yang, Y.; Mai, S.; Wang, D.; Song, C. Investigation into spatter behavior during selective laser melting of AISI 316L stainless steel powder. *Mater. Des.* **2015**, *87*, 797–806. [CrossRef]
11. Sing, S.L.; An, J.; Yeong, W.Y.; Wiria, F.E. Laser and electron-beam powder-bed additive manufacturing of metallic implants: A review on processes, materials and designs. *J. Orthop. Res.* **2015**, *34*, 369–385. [CrossRef] [PubMed]
12. Zhang, Y.; Zhang, J.; Yan, Q.; Zhang, L.; Wang, M.; Song, B.; Shi, Y. Amorphous alloy strengthened stainless steel manufactured by selective laser melting: Enhanced strength and improved corrosion resistance. *Scr. Mater.* **2018**, *148*, 20–23. [CrossRef]

13. Gokuldoss, P.K.; Eckert, J.; Gokuldoss, P.K. Formation of metastable cellular microstructures in selective laser melted alloys. *J. Alloys Compd.* **2017**, *707*, 27–34.

14. Kürnsteiner, P.; Wilms, M.B.; Weisheit, A.; Barriobero-Vila, P.; Jägle, E.; Raabe, D. Massive nanoprecipitation in an Fe-19Ni- X Al maraging steel triggered by the intrinsic heat treatment during laser metal deposition. *Acta Mater.* **2017**, *129*, 52–60. [CrossRef]

15. Olakanmi, E.O.; Cochrane, R.; Dalgarno, K. A review on selective laser sintering/melting (SLS/SLM) of aluminium alloy powders: Processing, microstructure, and properties. *Prog. Mater. Sci.* **2015**, *74*, 401–477. [CrossRef]

16. Manakari, V.; Parande, G.; Gupta, M. Selective laser melting of magnesium and magnesium alloy powders: A review. *Metals* **2016**, *7*, 2. [CrossRef]

17. Cao, X.; Jahazi, M.; Immarigeon, J.; Wallace, W. A review of laser welding techniques for magnesium alloys. *J. Mater. Process. Technol.* **2006**, *171*, 188–204. [CrossRef]

18. Jahangir, N.; Mamun, M.A.H.; Sealy, M.P. A review of additive manufacturing of magnesium alloys. In Proceedings of the 3rd International Conference on Mechanical Engineering (ICOME 2017), Birmingham, UK, 13–15 October 2017.

19. Makarov, D.; Melzer, M.; Karnaushenko, D.; Schmidt, O.G. Review of selective laser melting: Materials and applications. *Appl. Phys. Rev.* **2016**, *3*, 011101. [CrossRef]

20. Zhang, J.; Song, B.; Wei, Q.; Bourell, D.; Shi, Y. A review of selective laser melting of aluminum alloys: Processing, microstructure, property and developing trends. *J. Mater. Sci. Technol.* **2019**, *35*, 270–284. [CrossRef]

21. Zhang, W.-N.; Wang, L.-Z.; Feng, Z.-X.; Chen, Y.-M. Research progress on selective laser melting (SLM) of magnesium alloys: A review. *Optik* **2020**, *207*, 163842. [CrossRef]

22. Abe, F.; Santos, E.C.; Kitamura, Y.; Osakada, K.; Shiomi, M. Influence of forming conditions on the titanium model in rapid prototyping with the selective laser melting process. *J. Mech. Eng. Sci.* **2003**, *217*, 119–126. [CrossRef]

23. Meier, H.; Haberland, C. Experimental studies on selective laser melting of metallic parts. *Mater. Werkst.* **2008**, *39*, 665–670. [CrossRef]

24. Song, B.; Dong, S.; Zhang, B.; Liao, H.; Coddet, C. Effects of processing parameters on microstructure and mechanical property of selective laser melted Ti6Al4V. *Mater. Des.* **2012**, *35*, 120–125. [CrossRef]

25. Lucas, J.; Meiners, W.; Vervoort, S.; Gayer, C.; Zumdick, N.A.; Zander, D. Selective laser melting of magnesium alloys. *Eur. Cells Mater.* **2015**, *30*, 1.

26. Wei, K.; Gao, M.; Wang, Z.; Zeng, X. Effect of energy input on formability, microstructure and mechanical properties of selective laser melted AZ91D magnesium alloy. *Mater. Sci. Eng. A* **2014**, *611*, 212–222. [CrossRef]

27. Wei, K.; Wang, Z.; Zeng, X. Influence of element vaporization on formability, composition, microstructure, and mechanical performance of the selective laser melted Mg–Zn–Zr components. *Mater. Lett.* **2015**, *156*, 187–190. [CrossRef]

28. Liu, S.; Yang, W.; Shi, X.; Li, B.; Duan, S.; Guo, H.-J.; Guo, J. Influence of laser process parameters on the densification, microstructure, and mechanical properties of a selective Laser melted AZ61 magnesium alloy. *J. Alloys Compd.* **2019**, *808*, 151160. [CrossRef]

29. He, C.; Bin, S.; Wu, P.; Gao, C.; Feng, P.; Shuai, C.; Liu, L.; Zhou, Y.; Zhao, M.; Yang, S.; et al. Microstructure evolution and biodegradation behavior of laser rapid solidified Mg–Al–Zn alloy. *Metals* **2017**, *7*, 105. [CrossRef]

30. Shuai, C.; Shuai, C.; Wu, P.; Lin, X.; Liu, Y.; Zhou, Y.; Feng, P.; Liu, X.; Peng, S. Laser rapid solidification improves corrosion behavior of Mg-Zn-Zr Alloy. *J. Alloys Compd.* **2017**, *691*, 961–969. [CrossRef]

31. Shuai, C.; He, C.; Feng, P.; Guo, W.; Gao, C.; Wu, P.; Yang, Y.; Bin, S. Biodegradation mechanisms of selective laser-melted Mg–xAl–Zn alloy: Grain size and intermetallic phase. *Virtual Phys. Prototyp.* **2017**, *13*, 59–69. [CrossRef]

32. Zhang, B.; Liao, H.; Coddet, C. Effects of processing parameters on properties of selective laser melting Mg–9%Al powder mixture. *Mater. Des.* **2012**, *34*, 753–758. [CrossRef]

33. Hu, N.; Wang, Y.; Zhang, D.; Hao, L.; Jiang, J.; Li, Z.; Chen, Y. Experimental investigation on selective laser melting of bulk net-shape pure magnesium. *Mater. Manuf. Process.* **2015**, *30*, 1298–1304. [CrossRef]

34. Ng, C.C.; Savalani, M.M.; Man, H.; Gibson, I. Layer manufacturing of magnesium and its alloy structures for future applications. *Virtual Phys. Prototyp.* **2010**, *5*, 13–19. [CrossRef]

35. Ng, C.C.; Savalani, M.; Lau, M.; Man, H. Microstructure and mechanical properties of selective laser melted magnesium. *Appl. Surf. Sci.* **2011**, *257*, 7447–7454. [CrossRef]
36. Savalani, M.M.; Pizarro, J.M. Effect of preheat and layer thickness on selective laser melting (SLM) of magnesium. *Rapid Prototyp. J.* **2016**, *22*, 115–122. [CrossRef]
37. Yang, Y.; Wu, P.; Lin, X.; Liu, Y.; Bian, H.; Zhou, Y.; Gao, C.; Shuai, C. System development, formability quality and microstructure evolution of selective laser-melted magnesium. *Virtual Phys. Prototyp.* **2016**, *11*, 173–181. [CrossRef]
38. Liu, C.; Zhang, M.; Chen, C. Effect of laser processing parameters on porosity, microstructure and mechanical properties of porous Mg-Ca alloys produced by laser additive manufacturing. *Mater. Sci. Eng. A* **2017**, *703*, 359–371. [CrossRef]
39. Zhou, Y.; Wu, P.; Shuai, C.; Gao, D.; Feng, P.; Gao, C.; Wu, H.; Liu, Y.; Bian, H.; Shuai, C. The microstructure, mechanical properties and degradation behavior of laser-melted Mg Sn alloys. *J. Alloys Compd.* **2016**, *687*, 109–114. [CrossRef]
40. Long, T.; Zhang, X.; Huang, Q.; Liu, L.; Liu, Y.; Ren, J.; Yin, Y.; Wu, D.; Wu, H. Novel Mg-based alloys by selective laser melting for biomedical applications: Microstructure evolution, microhardness and in vitro degradation behaviour. *Virtual Phys. Prototyp.* **2017**, *13*, 71–81. [CrossRef]
41. Liu, S.; Guo, H.-J. Influence of hot isostatic pressing (HIP) on mechanical properties of magnesium alloy produced by selective laser melting (SLM). *Mater. Lett.* **2020**, *265*, 127463. [CrossRef]
42. Zhang, M.; Chen, C.; Liu, C.; Wang, S. Study on porous Mg-Zn-Zr ZK61 alloys produced by laser additive manufacturing. *Metals* **2018**, *8*, 635. [CrossRef]
43. Wei, K.; Zeng, X.-Y.; Wang, Z.; Deng, J.; Liu, M.; Huang, G.; Yuan, X. Selective laser melting of Mg-Zn binary alloys: Effects of Zn content on densification behavior, microstructure, and mechanical property. *Mater. Sci. Eng. A* **2019**, *756*, 226–236. [CrossRef]
44. Deng, Q.; Wu, Y.; Luo, Y.; Su, N.; Xue, X.; Chang, Z.; Wu, Q.; Xue, Y.; Peng, L. Fabrication of high-strength Mg-Gd-Zn-Zr alloy via selective laser melting. *Mater. Charact.* **2020**, *165*, 110377. [CrossRef]
45. Li, Y.; Zhou, J.; Pavanram, P.; Leeflang, M.; Fockaert, L.; Pouran, B.; Tümer, N.; Schröder, K.-U.; Mol, J.M.C.; Weinans, H.; et al. Additively manufactured biodegradable porous magnesium. *Acta Biomater.* **2018**, *67*, 378–392. [CrossRef]
46. Shuai, C.; Liu, L.; Zhao, M.; Feng, P.; Shuai, C.; Guo, W.; Gao, C.; Yuan, F. Microstructure, biodegradation, antibacterial and mechanical properties of ZK60-Cu alloys prepared by selective Laser melting technique. *J. Mater. Sci. Technol.* **2018**, *34*, 1944–1952. [CrossRef]
47. Li, R.; Liu, J.; Shi, Y.; Wang, L.; Jiang, W. Balling behavior of stainless steel and nickel powder during selective laser Melting process. *Int. J. Adv. Manuf. Technol.* **2011**, *59*, 1025–1035. [CrossRef]
48. Tolochko, N.K.; Mozzharov, S.E.; Yadroitsev, I.A.; Laoui, T.; Froyen, L.; Titov, V.I.; Ignatiev, M.B. Balling processes during selective laser treatment of powders. *Rapid Prototyp. J.* **2004**, *10*, 78–87. [CrossRef]
49. Gu, D.; Shen, Y. Balling phenomena in direct laser sintering of stainless steel powder: Metallurgical mechanisms and control methods. *Mater. Des.* **2009**, *30*, 2903–2910. [CrossRef]
50. Aboulkhair, N.T.; Everitt, N.M.; Ashcroft, I.; Tuck, C.; Tuck, C. Reducing porosity in AlSi10Mg parts processed by selective laser melting. *Addit. Manuf.* **2014**, *1*, 77–86. [CrossRef]
51. Gu, D.; Hagedorn, Y.-C.; Meiners, W.; Meng, G.; Batista, R.J.S.; Wissenbach, K.; Poprawe, R. Densification behavior, microstructure evolution, and wear performance of selective laser melting processed commercially pure titanium. *Acta Mater.* **2012**, *60*, 3849–3860. [CrossRef]
52. Yadroitsev, I.; Gusarov, A.V.; Yadroitsava, I.; Smurov, I.Y. Single track formation in selective laser melting of metal powders. *J. Mater. Process. Technol.* **2010**, *210*, 1624–1631. [CrossRef]
53. Khorasani, A.; Gibson, I.; Awan, U.S.; Ghaderi, A. The effect of SLM process parameters on density, hardness, tensile strength and surface quality of Ti-6Al-4V. *Addit. Manuf.* **2019**, *25*, 176–186. [CrossRef]
54. Xu, W.; Lui, E.W.; Pateras, A.; Qian, M.; Brandt, M. In situ tailoring microstructure in additively manufactured Ti-6Al-4V for superior mechanical performance. *Acta Mater.* **2017**, *125*, 390–400. [CrossRef]
55. Xia, M.; Gu, D.; Yu, G.; Dai, D.; Chen, H.; Shi, Q. Influence of hatch spacing on heat and mass transfer, thermodynamics and laser processability during additive manufacturing of inconel 718 alloy. *Int. J. Mach. Tools Manuf.* **2016**, *109*, 147–157. [CrossRef]
56. Yadroitsev, I.; Bertrand, P.; Smurov, I.; Bertrand, P. Parametric analysis of the selective laser melting process. *Appl. Surf. Sci.* **2007**, *253*, 8064–8069. [CrossRef]

57. Rashid, R.A.R.; Masood, S.; Ruan, D.; Palanisamy, S.; Brandt, M. Effect of scan strategy on density and metallurgical properties of 17-4PH parts printed by selective laser melting (SLM). *J. Mater. Process. Technol.* **2017**, *249*, 502–511. [CrossRef]

58. Danilov, D.; Nestler, B. Phase-field modelling of solute trapping during rapid solidification of a Si–As alloy. *Acta Mater.* **2006**, *54*, 4659–4664. [CrossRef]

59. Aziz, M.J.; Tsao, J.Y.; Thompson, M.; Peercy, P.S.; White, C.W. Solute trapping: Comparison of theory with experiment. *Phys. Rev. Lett.* **1986**, *56*, 2489–2492. [CrossRef]

60. Risha, G.; Son, S.F.; Yetter, R.; Yang, V.; Tappan, B. Combustion of nano-aluminum and liquid water. *Proc. Combust. Inst.* **2007**, *31*, 2029–2036. [CrossRef]

61. Chang, T.-C.; Wang, J.-Y.; Chu, C.-L.; Lee, S. Mechanical properties and microstructures of various Mg–Li alloys. *Mater. Lett.* **2006**, *60*, 3272–3276. [CrossRef]

62. Taltavull, C.; Torres, B.; Lopez, A.J.; Rodrigo, P.; Otero, E.; Rams, J. Selective laser surface melting of a magnesium-aluminium alloy. *Mater. Lett.* **2012**, *85*, 98–101. [CrossRef]

63. Suresh, C.H.; Koga, N. A consistent approach toward atomic radii. *J. Phys. Chem. A* **2001**, *105*, 5940–5944. [CrossRef]

64. Wen, H.; Topping, T.D.; Isheim, D.; Seidman, D.N.; Lavernia, E.J. Strengthening mechanisms in a high-strength bulk nanostructured Cu–Zn–Al alloy processed via cryomilling and spark plasma sintering. *Acta Mater.* **2013**, *61*, 2769–2782. [CrossRef]

65. Caceres, C.; Rovera, D. Solid solution strengthening in concentrated Mg–Al alloys. *J. Light Met.* **2001**, *1*, 151–156. [CrossRef]

66. Yang, Y.; Peng, X.; Wen, H.; Zheng, B.; Zhou, Y.; Xie, W.; Lavernia, E. Influence of extrusion on the microstructure and mechanical behavior of Mg-9Li-3Al-XSr alloys. *Met. Mater. Trans. A* **2012**, *44*, 1101–1113. [CrossRef]

67. Hakamada, M.; Furuta, T.; Chino, Y.; Chen, Y.; Kusuda, H.; Mabuchi, M. Life cycle inventory study on magnesium alloy substitution in vehicles. *Energy* **2007**, *32*, 1352–1360. [CrossRef]

68. Kulekci, M.K. Magnesium and its alloys applications in automotive industry. *Int. J. Adv. Manuf. Technol.* **2007**, *39*, 851–865. [CrossRef]

69. Teng, H.-T.; Li, T.-J.; Zhang, X.-L.; Zhang, Z.-T. Influence of sub-rapid solidification on microstructure and mechanical properties of AZ61A magnesium alloy. *Trans. Nonferrous Met. Soc. China* **2008**, *18*, s86–s90. [CrossRef]

70. Casavola, C.; Campanelli, S.L.; Pappalettere, C. Preliminary investigation on distribution of residual stress generated by the selective Laser melting process. *J. Strain Anal. Eng. Des.* **2008**, *44*, 93–104. [CrossRef]

71. Beausir, B.; Suwas, S.; Tóth, L.; Neale, K.W.; Fundenberger, J.-J. Analysis of texture Evolution in magnesium during equal channel angular extrusion. *Acta Mater.* **2008**, *56*, 200–214. [CrossRef]

72. Teng, H.; Zhang, X.; Zhang, Z.; Li, T.; Cockcroft, S. Research on microstructures of sub-rapidly solidified AZ61 magnesium alloy. *Mater. Charact.* **2009**, *60*, 482–486. [CrossRef]

73. Trevisan, F.; Calignano, F.; Lorusso, M.; Pakkanen, J.; Aversa, A.; Ambrosio, E.P.; Lombardi, M.; Fino, P.; Manfredi, D. On the selective laser melting (SLM) of the AlSi10Mg alloy: Process, microstructure, and mechanical properties. *Materials* **2017**, *10*, 76. [CrossRef]

74. Xu, S.; Matsumoto, N.; Kamado, S.; Honma, T.; Kojima, Y. Effect of Mg17Al12 precipitates on the microstructural changes and mechanical properties of hot compressed AZ91 magnesium alloy. *Mater. Sci. Eng. A* **2009**, *523*, 47–52. [CrossRef]

75. Lü, Y.; Wang, Q.; Ding, W.; Zeng, X.; Zhu, Y. Fracture behavior of AZ91 magnesium alloy. *Mater. Lett.* **2000**, *44*, 265–268. [CrossRef]

76. Lu, Y.; Wang, Q.; Zeng, X.; Ding, W.; Zhu, Y.; Lü, Y.Z.; Wang, Q.D.; Ding, W.J.; Zeng, X.Q.; Zhu, Y.P. Effects of silicon on microstructure, fluidity, mechanical properties, and fracture behaviour of Mg–6Al alloy. *Mater. Sci. Technol.* **2001**, *17*, 207–214.

77. Song, J.-M.; Lui, T.; Chang, H.; Chen, L. The influence of Al content and annealing on vibration fracture properties of wrought Mg–Al–Zn alloys. *Scr. Mater.* **2006**, *54*, 399–404. [CrossRef]

78. Deng, J.; Lin, Y.; Li, S.-S.; Chen, J.; Ding, Y. Hot tensile deformation and fracture behaviors of AZ31 magnesium alloy. *Mater. Des.* **2013**, *49*, 209–219. [CrossRef]

79. Du, X.; Zhang, E. Microstructure and mechanical behaviour of semi-solid die-casting AZ91D magnesium alloy. *Mater. Lett.* **2007**, *61*, 2333–2337. [CrossRef]

80. Dini, H.; Andersson, N.-E.; Jarfors, A.E. Effect of Mg17Al12 fraction on mechanical properties of Mg-9%Al-1%Zn Cast alloy. *Metals* **2016**, *6*, 251. [CrossRef]

81. Shamsaei, N.; Yadollahi, A.; Bian, L.; Thompson, S.M. An overview of direct laser deposition for additive manufacturing; part II: Mechanical behavior, process parameter optimization and control. *Addit. Manuf.* **2015**, *8*, 12–35. [CrossRef]

82. Liu, G.; Zhang, J.; Xi, G.; Zuo, R.; Liu, S. Designing Mg alloys with high ductility: Reducing the strength discrepancies between soft deformation modes and hard deformation modes. *Acta Mater.* **2017**, *141*, 1–9. [CrossRef]

83. Witte, F. The history of biodegradable magnesium implants: A review. *Acta Biomater.* **2010**, *6*, 1680–1692. [CrossRef] [PubMed]

84. Liao, J.; Hotta, M.; Mori, Y. Improved corrosion resistance of a high-strength Mg–Al–Mn–Ca magnesium alloy Made by rapid solidification powder metallurgy. *Mater. Sci. Eng. A* **2012**, *544*, 10–20. [CrossRef]

85. Cai, S.; Lei, T.; Li, N.; Feng, F. Effects of Zn on microstructure, mechanical properties and corrosion behavior of Mg–Zn alloys. *Mater. Sci. Eng. C* **2012**, *32*, 2570–2577. [CrossRef]

86. Nam, N.D.; Kim, W.C.; Kim, J.-G.; Shin, K.; Jung, H. Effect of mischmetal on the corrosion properties of Mg–5Al alloy. *Corros. Sci.* **2009**, *51*, 2942–2949. [CrossRef]

87. Song, G.; Atrens, A.; Dargusch, M. Influence of microstructure on the corrosion of diecast AZ91D. *Corros. Sci.* **1998**, *41*, 249–273. [CrossRef]

88. Ambat, R.; Aung, N.N.; Zhou, W. Evaluation of microstructural effects on corrosion behaviour of AZ91D magnesium alloy. *Corros. Sci.* **2000**, *42*, 1433–1455. [CrossRef]

89. Song, G.L.; Atrens, A. Corrosion mechanisms of magnesium alloys. *Adv. Eng. Mater.* **1999**, *1*, 11–33. [CrossRef]

90. Zhang, H.; Zhang, D.; Ma, C.; Guo, S. Improving mechanical properties and corrosion resistance of Mg-6Zn-Mn magnesium alloy by rapid solidification. *Mater. Lett.* **2013**, *92*, 45–48. [CrossRef]

91. Daloz, D.; Steinmetz, P.; Michot, G. Corrosion behavior of rapidly solidified magnesium-aluminum-zinc alloys. *Corros. Sci.* **1997**, *53*, 944–954. [CrossRef]

92. Izumi, S.; Yamasaki, M.; Kawamura, Y. Relation between corrosion behavior and microstructure of Mg–Zn–Y alloys prepared by rapid solidification at various cooling rates. *Corros. Sci.* **2009**, *51*, 395–402. [CrossRef]

93. Wang, Z.; Xie, M.; Li, Y.; Zhang, W.; Yang, C.; Kollo, L.; Eckert, J.; Prashanth, K.G. Premature failure of an additively manufactured material. *NPG Asia Mater.* **2020**, *12*, 1–10. [CrossRef]

94. Song, J.F.; Pan, F.; Jiang, B.; Atrens, A.; Zhang, M.-X.; Lu, Y. A review on hot tearing of magnesium alloys. *J. Magnes. Alloys* **2016**, *4*, 151–172. [CrossRef]

95. Otani, Y.; Sasaki, S. Effects of the addition of silicon to 7075 aluminum alloy on microstructure, mechanical properties, and selective laser melting processability. *Mater. Sci. Eng. A* **2020**, *777*, 139079. [CrossRef]

96. Li, L.; Li, R.; Yuan, T.; Chen, C.; Zhang, Z.; Li, X. Microstructures and tensile properties of a selective laser melted Al–Zn–Mg–Cu (Al7075) alloy by Si and Zr microalloying. *Mater. Sci. Eng. A* **2020**, *787*, 139492. [CrossRef]

97. Kimura, T.; Nakamoto, T.; Mizuno, M.; Araki, H. Effect of silicon content on densification, mechanical and thermal properties of Al-XSi binary alloys fabricated using selective laser melting. *Mater. Sci. Eng. A* **2017**, *682*, 593–602. [CrossRef]

98. Brandl, E.; Heckenberger, U.; Holzinger, V.; Buchbinder, D. Additive manufactured AlSi10Mg samples using selective laser melting (SLM): Microstructure, high cycle fatigue, and fracture behavior. *Mater. Des.* **2012**, *34*, 159–169. [CrossRef]

99. Kumari, S.S.; Pillai, R.; Rajan, T.; Pai, B. Effects of individual and combined additions of Be, Mn, Ca and Sr on the solidification behaviour, structure and mechanical properties of Al–7Si–0.3Mg–0.8Fe alloy. *Mater. Sci. Eng. A* **2007**, *460*, 561–573. [CrossRef]

100. Abderrazak, K.; Ben Salem, W.; Mhiri, H.; Lepalec, G.; Autric, M. Modelling of CO2 laser welding of magnesium alloys. *Opt. Laser Technol.* **2008**, *40*, 581–588. [CrossRef]

101. Louvis, E.; Fox, P.; Sutcliffe, C.J. Selective laser melting of aluminium components. *J. Mater. Process. Technol.* **2011**, *211*, 275–284. [CrossRef]

102. Dong, Z.; Kang, H.; Xie, Y.; Chi, C.; Peng, X. Effect of powder oxygen content on microstructure and mechanical properties of a laser additively-manufactured 12CrNi2 alloy steel. *Mater. Lett.* **2019**, *236*, 214–217. [CrossRef]

103. Cao, L.; Zeng, W.; Xie, Y.; Liang, J.; Zhang, D. Effect of powder oxidation on interparticle boundaries and mechanical properties of bulk Al prepared by spark plasma sintering of Al powder. *Mater. Sci. Eng. A* **2018**, *742*, 305–308. [CrossRef]

104. Rao, G.A.; Srinivas, M.; Sarma, D. Effect of oxygen content of powder on microstructure and mechanical properties of hot isostatically pressed superalloy inconel 718. *Mater. Sci. Eng. A* **2006**, *435*, 84–99. [CrossRef]

105. Lei, Z.; Liu, X.; Wu, Y.; Wang, H.; Jiang, S.; Wang, S.; Hui, X.; Wu, Y.; Gault, B.; Kontis, P.; et al. Enhanced strength and ductility in a high-entropy alloy via ordered oxygen complexes. *Nature* **2018**, *563*, 546–550. [CrossRef]

106. Das, S. Physical aspects of process control in selective laser Sintering of metals. *Adv. Eng. Mater.* **2003**, *5*, 701–711. [CrossRef]

107. Fintová, S.; Kunz, L. Fatigue properties of magnesium alloy AZ91 processed by severe plastic deformation. *J. Mech. Behav. Biomed. Mater.* **2015**, *42*, 219–228. [CrossRef]

108. Alaneme, K.K.; Okotete, E.A. Enhancing plastic deformability of Mg and its alloys—A review of traditional and nascent developments. *J. Magnes. Alloys* **2017**, *5*, 460–475. [CrossRef]

109. Tahreen, N.; Chen, D.; Nouri, M.; Li, D.Y. Influence of aluminum content on twinning and texture development of Cast Mg–Al–Zn alloy during compression. *J. Alloys Compd.* **2015**, *623*, 15–23. [CrossRef]

110. Zhou, W.R.; Zheng, Y.F.; Leeflang, M.A.; Zhou, J. Mechanical property, biocorrosion and in vitro biocompatibility evaluations of Mg–Li–(Al)–(RE) alloys for future cardiovascular stent application. *Acta Biomater.* **2013**, *9*, 8488–8498. [CrossRef]

111. Kirkland, N.; Lespagnol, J.; Birbilis, N.; Staiger, M. A survey of bio-corrosion rates of magnesium alloys. *Corros. Sci.* **2010**, *52*, 287–291. [CrossRef]

112. Zhang, B.; Hou, Y.; Wang, X.; Wang, Y.; Geng, L. Mechanical properties, degradation performance and cytotoxicity of Mg–Zn–Ca biomedical alloys with different compositions. *Mater. Sci. Eng. C* **2011**, *31*, 1667–1673. [CrossRef]

113. Lai, H.; Li, J.; Li, J.; Zhang, Y.; Xu, Y. Effects of Sr on the microstructure, mechanical properties and corrosion behavior of Mg-2Zn-XSr alloys. *J. Mater. Sci. Mater. Electron.* **2018**, *29*, 87. [CrossRef] [PubMed]

114. Homayun, B.; Afshar, A. Microstructure, mechanical properties, corrosion behavior and cytotoxicity of Mg–Zn–Al–Ca alloys as biodegradable materials. *J. Alloys Compd.* **2014**, *607*, 1–10. [CrossRef]

115. Kirkland, N.; Birbilis, N.; Staiger, M. Assessing the corrosion of biodegradable magnesium implants: A critical review of current methodologies and their limitations. *Acta Biomater.* **2012**, *8*, 925–936. [CrossRef] [PubMed]

116. Zhao, M.-C.; Liu, M.; Song, G.; Atrens, A. Influence of the β-phase morphology on the corrosion of the Mg alloy AZ91. *Corros. Sci.* **2008**, *50*, 1939–1953. [CrossRef]

117. Lu, Y.; Bradshaw, A.; Chiu, Y.; Jones, I. The role of β1′ precipitates in the bio-corrosion performance of Mg–3Zn in simulated body fluid. *J. Alloys Compd.* **2014**, *614*, 345–352. [CrossRef]

118. Shuai, C.; He, C.; Xu, L.; Li, Q.; Chen, T.; Yang, Y.; Peng, S. Wrapping effect of secondary phases on the grains: Increased corrosion resistance of Mg–Al alloys. *Virtual Phys. Prototyp.* **2018**, *13*, 292–300. [CrossRef]

119. Li, Y.; Wen, C.; Mushahary, D.; Sravanthi, R.; Harishankar, N.; Pande, G.; Hodgson, P. Mg–Zr–Sr Alloys as biodegradable implant materials. *Acta Biomater.* **2012**, *8*, 3177–3188. [CrossRef]

120. Niknejad, S.; Liu, L.; Lee, M.-Y.; Esmaeili, S.; Zhou, N.Y. Resistance spot welding of AZ series magnesium alloys: Effects of aluminum content on microstructure and mechanical properties. *Mater. Sci. Eng. A* **2014**, *618*, 323–334. [CrossRef]

121. Zhou, M.; Morisada, Y.; Fujii, H. Effect of Ca addition on the microstructure and the mechanical properties of asymmetric double-sided friction stir welded AZ61 magnesium alloy. *J. Magnes. Alloys* **2020**, *8*, 91–102. [CrossRef]

122. Yang, J.; Peng, J.; Nyberg, E.A.; Pan, F. Effect of Ca addition on the corrosion behavior of Mg–Al–Mn alloy. *Appl. Surf. Sci.* **2016**, *369*, 92–100. [CrossRef]

123. Kondori, B.; Mahmudi, R. Effect of Ca additions on the microstructure, thermal stability and mechanical properties of a cast AM60 magnesium alloy. *Mater. Sci. Eng. A* **2010**, *527*, 2014–2021. [CrossRef]

124. Yuesheng, C.; Zhigang, G.; Kangle, C.; Fang, D. Microstructure and mechanical properties of extruded Mg-Sm-Ca alloys. *Rare Met. Mater. Eng.* **2016**, *45*, 287–291. [CrossRef]

125. Xu, S.; Matsumoto, N.; Yamamoto, K.; Kamado, S.; Honma, T.; Kojima, Y. High temperature tensile properties of As-Cast Mg–Al–Ca alloys. *Mater. Sci. Eng. A* **2009**, *509*, 105–110. [CrossRef]

126. Aljarrah, M.; Medraj, M. Thermodynamic modelling of the Mg–Ca, Mg–Sr, Ca–Sr and Mg–Ca–Sr systems using the modified quasichemical model. *Calphad* **2008**, *32*, 240–251. [CrossRef]

127. Zhang, L.; Deng, K.-K.; Nie, K.; Xu, F.-J.; Su, K.; Liang, W. Microstructures and mechanical properties of Mg–Al–Ca alloys affected by Ca/Al ratio. *Mater. Sci. Eng. A* **2015**, *636*, 279–288. [CrossRef]

128. Oh-Ishi, K.; Watanabe, R.; Mendis, C.L.; Hono, K. Age-hardening response of Mg–0.3at.%Ca alloys with different Zn contents. *Mater. Sci. Eng. A* **2009**, *526*, 177–184. [CrossRef]

129. Xu, S.; Oh-Ishi, K.; Kamado, S.; Uchida, F.; Homma, T.; Hono, K. High-strength extruded Mg–Al–Ca–Mn alloy. *Scr. Mater.* **2011**, *65*, 269–272. [CrossRef]

130. Mendis, C.L.; Oh-Ishi, K.; Ohkubo, T.; Hono, K. Precipitation of prismatic plates in Mg–0.3Ca alloys with In additions. *Scr. Mater.* **2011**, *64*, 137–140. [CrossRef]

131. Kondoh, K.; Fujita, J.; Umeda, J.; Imai, H.; Enami, K.; Ohara, M.; Igarashi, T. Thermo-dynamic analysis on solid-state reduction of CaO particles dispersed in Mg–Al alloy. *Mater. Chem. Phys.* **2011**, *129*, 631–640. [CrossRef]

132. Nam, N.D.; Bian, M.; Forsyth, M.; Seter, M.; Tan, M.Y.; Shin, K. Effect of calcium oxide on the corrosion behaviour of AZ91 magnesium alloy. *Corros. Sci.* **2012**, *64*, 263–271. [CrossRef]

133. Baek, S.-M.; Kang, J.S.; Shin, H.-J.; Yim, C.D.; You, B.S.; Ha, H.-Y.; Park, S.S. Role of alloyed Y in improving the corrosion resistance of extruded Mg–Al–Ca-based alloy. *Corros. Sci.* **2017**, *118*, 227–232. [CrossRef]

134. Zeng, R.; Qi, W.-C.; Cui, H.-Z.; Zhang, F.; Zeng, R.-C.; Han, E.-H. In vitro corrosion of as-extruded Mg–Ca alloys—The influence of Ca concentration. *Corros. Sci.* **2015**, *96*, 23–31. [CrossRef]

135. Zhu, G.; Wang, L.; Zhou, H.; Wang, J.; Shen, Y.; Tu, P.; Zhu, H.; Liu, W.; Jin, P.; Zeng, X. Improving ductility of a Mg alloy via non-basal slip induced by Ca addition. *Int. J. Plast.* **2019**, *120*, 164–179. [CrossRef]

136. Stanford, N.; Barnett, M. Solute strengthening of prismatic slip, basal slip and twinning in Mg and Mg–Zn binary alloys. *Int. J. Plast.* **2013**, *47*, 165–181. [CrossRef]

137. Khosravani, A.; Fullwood, D.T.; Adams, B.L.; Rampton, T.M.; Miles, M.P.; Mishra, R.K. Nucleation and propagation of {101 2} twins in AZ31 magnesium alloy. *Acta Mater.* **2015**, *100*, 202–214. [CrossRef]

138. Pan, H.; Yang, C.; Yang, Y.; Dai, Y.; Zhou, D.; Chai, L.; Huang, Q.; Yang, Q.; Liu, S.; Ren, Y.; et al. Ultra-fine Grain Size and Exceptionally High Strength in Dilute Mg–Ca Alloys Achieved by Conventional One-Step extrusion. *Mater. Lett.* **2019**, *237*, 65–68. [CrossRef]

139. Zhang, B.; Wang, Y.; Geng, L.; Lu, C. Effects of calcium on texture and mechanical properties of hot-extruded Mg–Zn–Ca alloys. *Mater. Sci. Eng. A* **2012**, *539*, 56–60. [CrossRef]

140. Zeng, Z.; Bian, M.; Xu, S.; Davies, C.H.J.; Birbilis, N.; Nie, J. Effects of dilute additions of Zn and Ca on ductility of magnesium alloy sheet. *Mater. Sci. Eng. A* **2016**, *674*, 459–471. [CrossRef]

141. Sandlöbes, S.; Friák, M.; Korte-Kerzel, S.; Pei, Z.; Neugebauer, J.; Raabe, D. A Rare-earth free magnesium alloy with improved intrinsic ductility. *Sci. Rep.* **2017**, *7*, 10458. [CrossRef]

142. Kim, K.-H.; Hwang, J.H.; Jang, H.-S.; Jeon, J.B.; Kim, N.J.; Lee, B.-J. Dislocation binding as an origin for the improvement of room temperature ductility in Mg alloys. *Mater. Sci. Eng. A* **2018**, *715*, 266–275. [CrossRef]

143. Yasi, J.A.; Hector, L.G.; Trinkle, D., Jr. Prediction of thermal cross-slip stress in magnesium alloys from a geometric interaction model. *Acta Mater.* **2012**, *60*, 2350–2358. [CrossRef]

144. Wang, Q.; Li, C.; Zou, Y.; Wang, H.; Yi, T.; Huang, C. A highly selective fluorescence sensor for Tin (Sn4+) and its application in imaging live cells. *Org. Biomol. Chem.* **2012**, *10*, 6740. [CrossRef] [PubMed]

145. Zhao, C.; Pan, F.; Zhao, S.; Pan, H.; Song, K.; Tang, A. Preparation and characterization of as-extruded Mg–Sn alloys for orthopedic applications. *Mater. Des.* **2015**, *70*, 60–67. [CrossRef]

146. Poddar, P.; Kamaraj, A.; Murugesan, A.; Bagui, S.; Sahoo, K.L. Microstructural features of Mg-8%Sn alloy and its correlation with mechanical properties. *J. Magnes. Alloys* **2017**, *5*, 348–354. [CrossRef]

147. Tapiero, H.; Tew, K.D. Trace elements in human physiology and pathology: Zinc and metallothioneins. *Biomed. Pharmacother.* **2003**, *57*, 399–411. [CrossRef]

148. Yan, Y.; Cao, H.; Kang, Y.; Yu, K.; Xiao, T.; Luo, J.; Deng, Y.; Fang, H.; Xiong, H.; Dai, Y. Effects of Zn concentration and heat treatment on the microstructure, mechanical properties and corrosion behavior of As-extruded Mg–Zn alloys produced by powder metallurgy. *J. Alloys Compd.* **2017**, *693*, 1277–1289. [CrossRef]

149. Boehlert, C.J.; Knittel, K. The microstructure, tensile properties, and creep behavior of Mg–Zn alloys containing 0–4.4wt.% Zn. *Mater. Sci. Eng. A* **2006**, *417*, 315–321. [CrossRef]

150. Xie, G.; Ma, Z.; Geng, L. Effect of microstructural evolution on mechanical properties of friction stir welded ZK60 alloy. *Mater. Sci. Eng. A* **2008**, *486*, 49–55. [CrossRef]

151. Zhu, S.; Liu, Z.; Qu, R.; Wang, L.; Li, Q.; Guan, S. Effect of rare earth and Mn elements on the corrosion behavior of extruded AZ61 system in 3.5 Wt% NaCl solution and salt spray test. *J. Magnes. Alloys* **2013**, *1*, 249–255. [CrossRef]

152. Wu, Z.; Ahmad, R.; Yin, B.; Sandlöbes, S.; Curtin, W. Mechanistic origin and prediction of enhanced ductility in magnesium alloys. *Science* **2018**, *359*, 447–452. [CrossRef]

153. Zhu, T.; Fu, P.; Peng, L.; Hu, X.; Zhu, S.; Ding, W. Effects of Mn addition on the microstructure and mechanical properties of cast Mg–9Al–2Sn (wt.%) alloy. *J. Magnes. Alloys* **2014**, *2*, 27–35. [CrossRef]

154. Hort, N.; Huang, Y.; Fechner, D.; Störmer, M.; Blawert, C.; Witte, F.; Vogt, C.; Drücker, H.; Willumeit, R.; Kainer, K.U. Magnesium alloys as implant materials—Principles of property design for Mg–RE alloys. *Acta Biomater.* **2010**, *6*, 1714–1725. [CrossRef]

155. Zhang, J.; Liu, S.; Leng, Z.; Li, M.; Meng, J.; Wu, R. Microstructures and mechanical properties of heat-resistant HPDC Mg–4Al-based alloys containing cheap misch metal. *Mater. Sci. Eng. A* **2011**, *528*, 2670–2677. [CrossRef]

156. Nie, J.; Muddle, B. Characterisation of strengthening precipitate phases in a Mg–Y–Nd alloy. *Acta Mater.* **2000**, *48*, 1691–1703. [CrossRef]

157. Tao, W.; Zhang, M.; Zhongyi, N.; Bin, L. Influence of rare earth elements on microstructure and mechanical properties of Mg-Li alloys. *J. Rare Earths* **2006**, *24*, 797–800. [CrossRef]

158. Fu, L.; Le, Q.; Hu, W.; Zhang, J.; Wang, J. Strengths and ductility enhanced by micro-alloying Sm/La/Ca to Mg–0.5Zn–0.2Mn alloy. *J. Mater. Res. Technol.* **2020**, *9*, 6834–6849. [CrossRef]

159. Yamasaki, M.; Anan, T.; Yoshimoto, S.; Kawamura, Y. Mechanical properties of warm-extruded Mg–Zn–Gd alloy with coherent 14H long periodic stacking ordered structure precipitate. *Scr. Mater.* **2005**, *53*, 799–803. [CrossRef]

160. Itoi, T.; Seimiya, T.; Kawamura, Y.; Hirohashi, M. Long period stacking structures observed in Mg97Zn1Y2 alloy. *Scr. Mater.* **2004**, *51*, 107–111. [CrossRef]

161. Ku, C.-H.; Pioletti, D.P.; Browne, M.; Gregson, P.J. Effect of different Ti–6Al–4V surface treatments on osteoblasts behaviour. *Biomaterials* **2002**, *23*, 1447–1454.

162. El-Rahman, S.S.A. Neuropathology of aluminum toxicity in rats (glutamate and GABA impairment). *Pharmacol. Res.* **2003**, *47*, 189–194. [CrossRef]

163. Jiang, M.; Yan, H.; Chen, R. Microstructure, texture and mechanical properties in an As-Cast AZ61 Mg alloy during multi-directional impact forging and subsequent heat treatment. *Mater. Des.* **2015**, *87*, 891–900. [CrossRef]

164. Dahle, A.K.; Lee, Y.C.; Nave, M.D.; Schaffer, P.L.; StJohn, D. Development of the As-cast microstructure in magnesium–aluminium alloys. *J. Light Met.* **2001**, *1*, 61–72. [CrossRef]

165. Wen, Z.; Wu, C.; Dai, C.; Yang, F. Corrosion behaviors of Mg and its alloys with different Al contents in a modified simulated body fluid. *J. Alloys Compd.* **2009**, *488*, 392–399. [CrossRef]

166. Chen, J.; Wei, J.; Yan, H.; Su, B.; Pan, X. Effects of cooling rate and pressure on microstructure and mechanical properties of sub-rapidly solidified Mg–Zn–Sn–Al–Ca alloy. *Mater. Des.* **2013**, *45*, 300–307. [CrossRef]

167. Cai, J.; Ma, G.; Liu, Z.; Zhang, H.; Xu, M. Influence of rapid solidification on the microstructure of AZ91HP alloy. *J. Alloys Compd.* **2006**, *422*, 92–96. [CrossRef]

168. Zhao, Y.-C.; Zhao, M.-C.; Xu, R.; Liu, L.; Tao, J.-X.; Gao, C.; Shuai, C.; Atrens, A. Formation and characteristic corrosion behavior of alternately lamellar arranged α and β in As-cast AZ91 Mg alloy. *J. Alloys Compd.* **2019**, *770*, 549–558. [CrossRef]

169. Tradowsky, U.; White, J.; Ward, R.; Read, N.; Reimers, W.; Attallah, M.M. Selective laser melting of AlSi10Mg: Influence of post-processing on the microstructural and tensile properties development. *Mater. Des.* **2016**, *105*, 212–222. [CrossRef]

170. Zhao, X.; Li, S.; Zhang, M.; Liu, Y.; Sercombe, T.B.; Wang, S.; Hao, Y.; Yang, R.; Murr, L.E. Comparison of the microstructures and mechanical properties of Ti–6Al–4V fabricated by selective laser melting and electron beam melting. *Mater. Des.* **2016**, *95*, 21–31. [CrossRef]

171. Yan, X.; Lupoi, R.; Wu, H.; Ma, W.; Liu, M.; O'Donnell, G.; Yin, S. Effect of hot isostatic pressing (HIP) treatment on the compressive properties of Ti6Al4V lattice structure fabricated by selective laser melting. *Mater. Lett.* **2019**, *255*, 126537. [CrossRef]

172. Kaplanskii, Y.; Sentyurina, Z.A.; Loginov, P.; Levashov, E.; Korotitskiy, A.; Travyanov, A.Y.; Petrovskii, P. Microstructure and mechanical properties of the (Fe,Ni)Al-based alloy produced by SLM and HIP of spherical composite powder. *Mater. Sci. Eng. A* **2019**, *743*, 567–580. [CrossRef]

173. Spierings, A.; Dawson, K.; Dumitraschkewitz, P.; Pogatscher, S.; Wegener, K. Microstructure characterization of SLM-processed Al-Mg-Sc-Zr alloy in the heat treated and HIPed condition. *Addit. Manuf.* **2018**, *20*, 173–181. [CrossRef]

174. Feng, H.; Liu, S.; Du, Y.; Lei, T.; Zeng, R.; Yuan, T. Effect of the second phases on corrosion behavior of the Mg-Al-Zn alloys. *J. Alloys Compd.* **2017**, *695*, 2330–2338. [CrossRef]

175. Song, G.-L.; Bowles, A.L.; StJohn, D.H. Corrosion resistance of aged die cast magnesium alloy AZ91D. *Mater. Sci. Eng. A* **2004**, *366*, 74–86. [CrossRef]

176. Dumpala, S.C. Influence of heat treatment on the machinability and corrosion behavior of AZ91 Mg alloy. *J. Magnes. Alloys* **2018**, *6*, 52–58.

177. Wang, Y.; Liu, G.; Fan, Z. Microstructural Evolution of Rheo-Diecast AZ91D Magnesium Alloy During Heat Treatment. *Acta Mater.* **2006**, *54*, 689–699. [CrossRef]

178. Zhao, D.; Wang, Z.; Zuo, M.; Geng, H. Effects of heat treatment on microstructure and mechanical properties of extruded AZ80 magnesium alloy. *Mater. Des.* **2014**, *56*, 589–593. [CrossRef]

179. Jia, H.; Feng, X.; Yang, Y. Influence of solution treatment on microstructure, mechanical and corrosion properties of Mg-4Zn alloy. *J. Magnes. Alloys* **2015**, *3*, 247–252. [CrossRef]

180. Zhang, L. State of the art in the control of inclusions in tire cord steels—A review. *Steel Res. Int.* **2006**, *77*, 158–169. [CrossRef]

181. Park, J.H.; Kang, Y. Inclusions in stainless steels—A review. *Steel Res. Int.* **2017**, *88*, 1700130. [CrossRef]

182. Shi, C.; Chen, X.-C.; Guo, H.-J.; Zhu, Z.-J.; Sun, X.-L. Control of MgO·Al2O3 spinel inclusions during protective gas electroslag remelting of die steel. *Met. Mater. Trans. A* **2012**, *44*, 378–389. [CrossRef]

183. Shi, C.; Zheng, D.; Guo, B.; Li, J.; Jiang, F. Evolution of oxide–sulfide complex inclusions and its correlation with steel cleanliness during electroslag rapid remelting (ESRR) of tool steel. *Met. Mater. Trans. A* **2018**, *49*, 3390–3402. [CrossRef]

184. Yan, M.; Xu, W.; Dargusch, M.; Tang, H.P.; Brandt, M.; Qian, M. Review of effect of oxygen on room temperature ductility of titanium and titanium alloys. *Powder Met.* **2014**, *57*, 251–257. [CrossRef]

 © 2020 by the authors. Licensee MDPI, Basel, Switzerland. This article is an open access article distributed under the terms and conditions of the Creative Commons Attribution (CC BY) license (http://creativecommons.org/licenses/by/4.0/).

Article

Effect of Processing Atmosphere and Secondary Operations on the Mechanical Properties of Additive Manufactured AISI 316L Stainless Steel by Plasma Metal Deposition

Cristina Arévalo [1], Enrique Ariza [2], Eva Maria Pérez-Soriano [1,*], Michael Kitzmantel [2], Erich Neubauer [2] and Isabel Montealegre-Meléndez [1]

[1] Escuela Politécnica Superior, Universidad de Sevilla, 41011 Sevilla, Spain; carevalo@us.es (C.A.); imontealegre@us.es (I.M.-M.)
[2] RHP-Technology GmbH, 2444 Seibersdorf, Austria; e.ar@rhp.at (E.A.); m.ki@rhp.at (M.K.); e.ne@rhp.at (E.N.)
* Correspondence: evamps@us.es; Tel.: +34-954482278

Received: 31 May 2020; Accepted: 17 August 2020; Published: 21 August 2020

Abstract: Plasma metal deposition (PMD) is an interesting additive technique whereby diverse materials can be employed to produce end parts with complex geometries. This study investigates not only the effects of the manufacturing conditions on the final properties of 316L stainless steel specimens by PMD, but it also affords an opportunity to study how secondary treatments could modify these properties. The tested processing condition was the atmosphere, either air or argon, with the other parameters having previously been optimized. Furthermore, two standard thermal treatments were conducted with the intention of broadening knowledge regarding how these secondary operations could cause changes in the microstructure and properties of 316L parts. To better appreciate and understand the variation of conditions affecting the behavior properties, a thorough characterization of the specimens was carried out. The results indicate that the presence of vermicular ferrite (δ) varied slightly as a consequence of the processing conditions, since it was less prone to appear in specimens manufactured in argon than in air. In this respect, their mechanical properties suffered variations; the higher the ferrite (δ) content, the higher the mechanical properties measured. The degree of influence of the thermal treatment was similar regardless of the processing conditions, which affected the properties based on the heating temperature.

Keywords: plasma metal deposition; additive manufacturing; 316L; processing conditions; mechanical properties; microstructure

1. Introduction

Nowadays, several additive manufacturing (AM) processes are employed in diverse industrial sectors as recognized techniques to produce specimens with determined and specific geometries [1,2]. These well-known techniques involve layer-by-layer shaping and building components [3] in opposition to the conventional manufacturing techniques, in which the parts are built in bulk [4]. In the framework of the AM processes in the field of metals, their general classification could be carried out while considering the two different feedstock systems: powder–bed and injection deposition techniques [5]. The technology of the powder–bed processes enables parts to be built with complex geometric shapes; however, the building volumes may be limited to small components [6]. Components of a larger volume can be produced by injection deposition techniques, whereby shapes and geometries often demand a lower degree of complexity [7]. The technologies of injection deposition can be broadly classified into powder and wire systems, in which the starting materials could be fed or blown [1]. The technology of a wire-fed system is faster than the other techniques, thereby allowing a high production rate [8]. However, this process is limited by the availability of raw materials. Laser and plasma metal deposition (LMD & PMD) are included in the category of blown powder techniques.

The PMD manufacturing process is a rapidly rising AM technology thanks to its advantages over laser-based AM processes. In this technique, a plasma is used as the energy source to melt the starting materials [9,10]. The plasma spot, larger than the laser spot, presents a high feeding rate (1–10 kg/h) and, as a consequence, a high production rate, which is interesting for today's industrial sector [11,12]. Moreover, it is not limited by the feedstock material weight or component size as in powder–bed systems since the plasma torch is attached to a Gantry system and the deposition is performed on a workbench of the desired size. Currently, PMD is used in the prototype scale [13], as well as for structural components [14]. Due to the high energy delivered to the feedstock deposited on a substrate, a melt pool is generated; simultaneously, an inert gas is blown. Such an inert gas aids in the injection of the powder feed and also protects the formed components from oxidation. Thus, the specimen is produced layer-by-layer. Finally, the three-dimensional component is built. Concerning the deposition process and the feedstock melting, the consolidated material on the substrate may be involved in the reheating cycles, which could promote variations on the microstructure and changes in the final component behavior. Several authors describe the importance of a thorough understanding of the relevant parameters, not only regarding the starting powder (flow rate, size, shape, composition, etc.), but also the processing conditions (transferred arc, speed, atmospheric conditions, etc.) [10,15]. Therefore, the influence of these significant parameters should be considered on the final properties of the components.

In AM technology, 316L stainless steel powder is commonly employed to develop complex geometries since this alloy presents attractive properties for many industrial sectors, such as highly resistant corrosion behavior and toughness [16]. Previous authors indicate that the microstructure of this austenitic stainless steel could suffer variations as a consequence of the manufacturing process. Retained δ-ferrite phase with various morphologies and intermetallic σ triggered by the replacement of δ, could be present in the microstructure in addition to the expected γ-austenite phase. This may be related to the processing parameters and ambient conditions in which the parts are built; in this regard, the final mechanical and tribological properties of the components could be affected [17–20].

The peculiarities of this innovative PMD process, such as its fast heating/cooling rates, could result in remarkably different microstructures compared with traditional fabrication routes, such as casting. In other additive techniques, such as selective laser melting (SLM), many authors state the benefits of AM and how the optimization of fabrication parameters presents a positive effect on the final properties of the specimens [21–24]. Moreover, these investigations reveal that there would be a direct relationship between the final microstructure and the properties with the number of layers; it is suggested that the microstructure could be influenced by the accumulated heat which would lead to the non-equilibrium microstructure. Secondary thermal treatments are carried out in order to improve the final properties [25], and their effect on the microstructure is studied.

In this work, the importance of the ambient condition during the PMD process is studied through the manufacture of two walls made from powders of 316L stainless steel, built in air and argon, respectively. Furthermore, two standard thermal treatments are performed in selected specimens extracted from each wall, to evaluate possible variations in their final properties. In this context, this research is intended to deepen and broaden scientific knowledge on how ambient conditions and the secondary treatments affect the properties and the microstructure of the PMD specimens.

2. Materials and Methods

2.1. Materials

The starting powder studied in this research was AISI 316L stainless steel, produced by the water-atomization method, leading to a particle size range of 75–150 µm. This powder was supplied by the Hoeganaes Corporation (Radevorwald, Germany).

The feedstock was chemically analyzed to confirm the composition and compare it to the standard composition. For this purpose, a scanning electron microscope equipped with an energy dispersion X-ray spectroscopy detector (SEM/EDS Hitachi TM1000, Tokyo, Japan) was used; the results are listed in Table 1. Additionally, a particle morphology analysis was carried out via optical microscopy (Figure 1a, OM, Zeiss ICM 405, ZEISS Invers LM, Oberkochen, Germany) and SEM (Figure 1b,c). The morphology appeared to be irregular with rounded edges. In Figure 1a, some internal porosity can be observed.

Table 1. Chemical composition of the standard 316L and the starting powder employed as feedstock.

	Composition in Mass%								
SOURCE INFO	**Fe**	**Cr**	**Ni**	**Mo**	**Mn**	**Si**	**P**	**S**	**C**
Hoeganaes Corporation	Bal.	16.37	12.78	2.22	0.07	0.84	0.019	0.002	0.015
EDS Analysis	Bal.	23.5	6.8	0.6	–	0.6	–	–	–
Standard [26]	Bal.	18–16	14–10	3–2	max. 2	max. 0.75	max. 0.045	max. 0.03	max. 0.03

Figure 1. Microscopy analysis of the feedstock. (**a**) optical microscopy (OM) image of the cross-section of the starting particles of 316L; (**b,c**) SEM images of the powder particles.

X-ray diffraction (XRD) analysis was conducted by Bruker D8 Advance A25 equipment (Bruker, Billerica, MA, USA) with Cu-K$_\alpha$ radiation. In Figure 2, the X-ray pattern reveals the significant peaks of the γ phase of the 316L stainless steel.

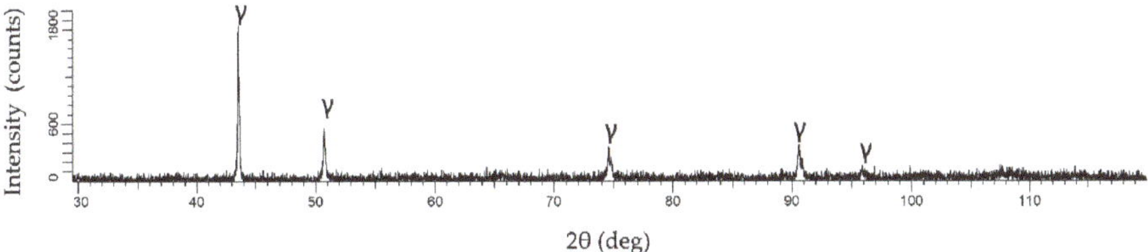

Figure 2. XRD analysis of the starting powder.

2.2. Plasma Metal Deposition Process

Samples were manufactured via a PMD technique which employed an argon plasma as an energy source for melting the starting material in powder, injected directly in the plasma torch. The PMD device was mounted on a 2.5-axis CNC machine; both types of equipment are the property of RHP-Technology GmbH.

A commercial AISI 1015 steel plate with dimensions of 200 mm × 200 mm × 10 mm was selected as substrate. As on previous authors' work [9,15], the feasibility of manufacturing the raw material powders by PMD processing was tested by optimizing the following welding parameters: main arc current (A), pilot current (A), welding speed (mm/min), material feeding rate (g/min), pilot gas flow (L/min) and shielding gas flow (L/min). The parameter optimization was carried out in order to avoid cracks, delamination between layers due to a lack of input energy and porosity as a result of gas trapped in the matrix when high energy is used.

Two walls of 150 mm × 40 mm × 15 mm in size were produced by an oscillation movement: one wall was manufactured in open atmosphere condition (see Figure 3a), using a local shielding gas only as a protective agent against oxidation; while the second wall was fabricated inside a hermetic box filled with argon as the inert atmosphere (see Figure 3b), which protected the entire wall from oxidation during the whole process. In this regard, the influence of the ambient condition was evaluated on the final behavior of the specimens.

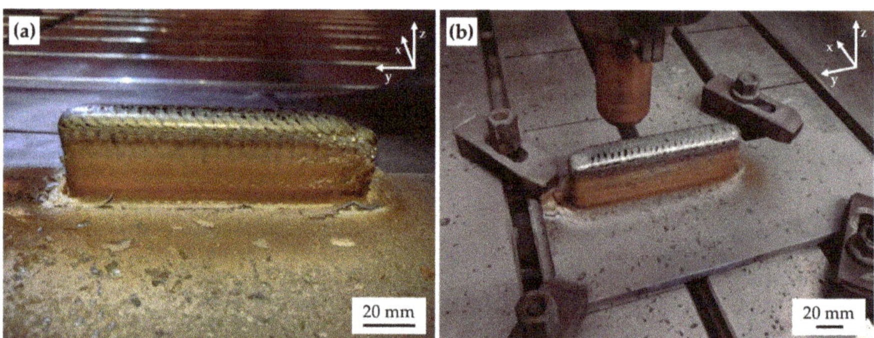

Figure 3. Images of the fabricated walls. (**a**) W1, produced in air atmosphere; (**b**) W2, produced in argon protective atmosphere.

The optimized manufacturing parameters selected to fabricate both walls are shown in Table 2.

Table 2. Processing parameters of plasma metal deposition (PMD) 316L.

Main Current (A)	Pilot Current (A)	Oscillation Welding Speed (mm/min)	Material Feeding Rate (g/min)	Pilot Gas Flow (L/min)
130	35	600	16.2	1.5

In the context of the manufacturing conditions, for the open atmosphere wall (W1), an argon shielding gas flow of 15 L/min was used; on the other hand, for the inert atmosphere test wall (W2), no shielding gas was employed since the hermetic box was filled with argon to control the oxygen below 50 ppm.

2.3. Sample Extraction and Thermal Treatments

Three different sets of tensile test samples were extracted by an electro discharge machining (EDM) Mitsubishi DWC 110 (Mitsubishi, Tokyo, Japan). The dimensions of the extracted specimens were in accordance with the standard [27]. Their geometry and their location in the walls are shown in Figure 4.

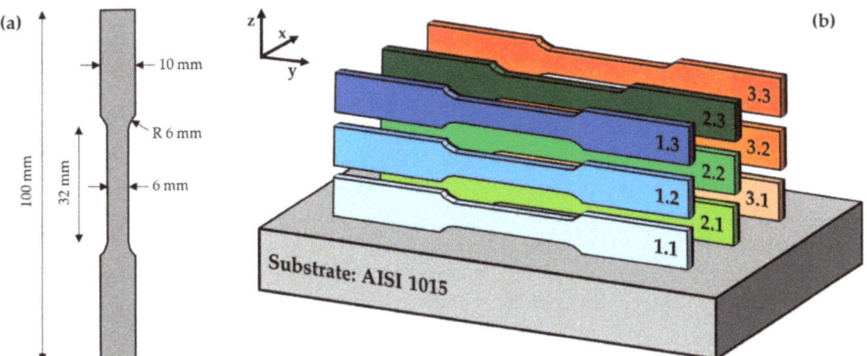

Figure 4. Schematic images of (**a**) a tensile test sample with its dimensions and (**b**) the numbering of the samples for the tensile tests and their location in the walls: (**b.1**) 1.1, 1.2, 1.3 as-built specimens; (**b.2**) 2.1, 2.2, 2.3 specimens treated by TT1; and (**b.3**) 3.1, 3.2, 3.3 specimens treated by TT2.

Following the investigation of the as-built specimens, the thermal treatments were performed using a high vacuum vertical tube furnace model APF-0716-MM (Thermal Technology GmbH, Bayreuth, Germany) under argon atmosphere. The thermal treatment parameters were selected based on the existing literature [26,28]; these are listed in Table 3. The treatment named TT1 involved an enhancement of the temperature only up to 600 °C (stress relief); in the treatment named TT2, in contrast, the heating was carried out at 950 °C (austenitization). The postprocessing parameters of 10 °C/min as the heating rate, 2 h for holding time and 10 °C/min as the cooling rate were the same for both processes; the heating temperature was the crucial factor to induce possible changes in the treated specimens.

Table 3. Thermal treatments performed on the two sets of 316L tensile test specimens, for both cases of air and argon atmosphere manufacturing, along with the sample identification.

Thermal Treatment	Thermal Treatment Cycle	Samples Location in Wall (Figure 4)
As built	–	1.1, 1.2, 1.3
TT1	10 °C/min, 600 °C, 2 h, 10 °C/min, RT	2.1, 2.2, 2.3
TT2	10 °C/min, 950 °C, 2 h, 10 °C/min, RT	3.1, 3.2, 3.3

2.4. Microstructural Characterization and Mechanical and Tribological Behavior

A detailed characterization was conducted in this research. The microstructural study was performed, after metallographic preparation, with a Nikon Epiphot optical microscope (Nikon, Tokyo, Japan) and a scanning electron microscope (FEI Teneo, Hillsboro, OR, USA) equipped with an EDS system to carry out compositional analysis and elemental mapping.

The Archimedes' method [29] enabled the density to be determined for the different sets of specimens. Image analysis (IA) was performed with a Nikon Epiphot optical microscope (Nikon, Tokyo, Japan) coupled with a Jenoptik Progres C3 camera (Jenoptik, Jena, Germany) and then processed using Image-Pro Plus 6.2 software (Media Cybernetics, Bethesda, MD, USA). Struers-Duramin A300 (Ballerup, Germany) was employed to reach the Vickers hardness (HV2). The hardness measurements took place on the polished cross-section of both walls.

Macro mechanical characterization was carried out through tensile tests by employing an Instron 5505-load unit (Norwood, MA, USA) using a strain rate of 0.5 mm·min^{-1} at room temperature. The mechanical properties under study were ultimate tensile strength (UTS), yield strength (YS, σ_y), Young's modulus (E) and elongation (ε).

An instrumented microindentation device, MTR3/50-50/NI (Microtest, Madrid, Spain), was employed on samples to evaluate their micromechanical behavior. The Vickers pyramid diamond tip was set in the simple loading/unloading mode. The resulting load–penetration (P–h) curves were measured using the Oliver and Pharr analysis [30], and the elastic relaxation values were calculated. Six indentations were made for each sample: three for a maximum load of 1 N and three for 5 N (load rate of 1 N/min and 5 N/min, respectively), with a holding time of 40 s.

After macro- and micromechanical characterization, the tribological behavior of the samples was ascertained by a ball-on-disc tribometer (Microtest MT/30/NI, Madrid, Spain). Ceramic alumina balls of 6 mm in diameter were employed. Moreover, for running the tests, the specimens were grounded, polished, cleaned with acetone in an ultrasonic bath, and, as a last step, dried. The tests were carried out under the following parameters: room temperature, a normal load of 5 N on the ball and a 200 rpm sliding speed for 15 min on a circular path of 2-mm radius. The worn surface morphology was studied by optical microscopy (OM) with a Leica Zeiss DMV6 (Leica Microsystems, Heerbrugg, Switzerland).

3. Results

The results were studied and compared in accordance with the manufacturing atmosphere condition, air and argon, in addition to the as-built specimens or the thermal treatments, TT1 and TT2, performed on specific specimens, as described above.

3.1. Microstructure

The microstructural characterization confirmed the existence of an austenite (γ) matrix in all the specimens, regardless of the processing conditions and the thermal treatment, as expected [25]. The retained ferrite (δ) was also present and exhibited a fine vermicular morphology. There were slight differences in how this phase appeared depending on the atmosphere conditions and the postprocessing treatment carried out (Figure 5).

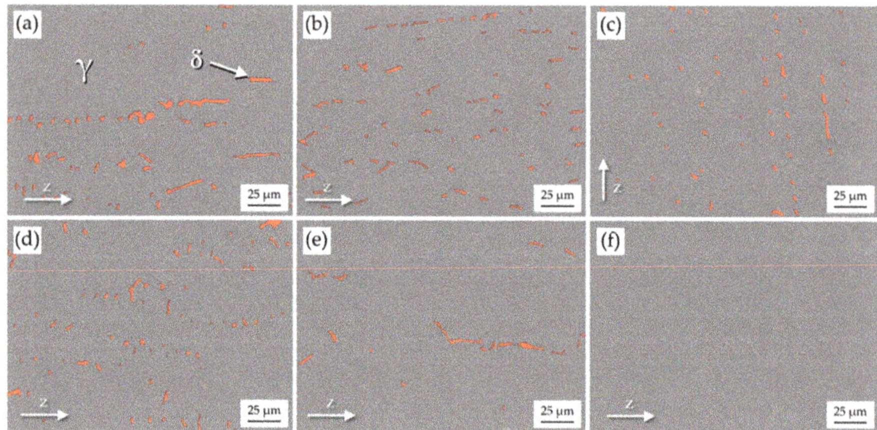

Figure 5. Processed OM images from the following specimens. (**a**) air-1.1; (**b**) air-2.2; (**c**) air-3.3; (**d**) argon-1.1; (**e**) argon-2.2; and (**f**) argon-3.3.

Observing the microstructure of the processed specimens, δ grew as a dendritic morphology [18] along with the vertical direction of the wall (z). The austenite phase was formed around the boundaries of the ferrite; therefore, γ seems to be oriented within the dendrites. This orientation (z) was more noticeable in air (see Figure 5a) than in argon (see Figure 5d). In the case of the specimens manufactured in argon, there was a soft phase orientation. Furthermore, the content of this ferrite was slightly reduced when the specimens were processed in argon, regardless of the thermal treatment, as can be qualitatively appreciated in Figure 5.

From the point of view of the postprocessing, both thermal treatments resulted in decreasing the vermicular phase while maintaining a similar tendency in its occurrence on the basis of the aforementioned air and argon conditions. TT2 caused a greater decrement of the ferrite than did TT1. This trend was more significant in specimens processed in argon than in specimens fabricated in the air condition, as can clearly be observed when the microstructures shown in Figure 5c,f are compared.

In a detailed evaluation of the microstructure of specimens located in determinate positions, several differences could be appreciated. Regarding the specimens that were extracted from the top and the base of the walls, argon-1.3 and argon-1.1, respectively, the content of the vermicular phase was similar; however, the specimens cut from the center of the walls, argon-1.2, presented a slightly lower content of ferrite (δ). This suggests that the heat dissipation on the center was less pronounced than in the areas close to the substrate and the top. In this regard, a low rate of heat dissipation during the manufacturing process contributed towards the disaggregation and transformation of the retained vermicular phase [18,31,32]. This phenomenon is shown in Figure 6.

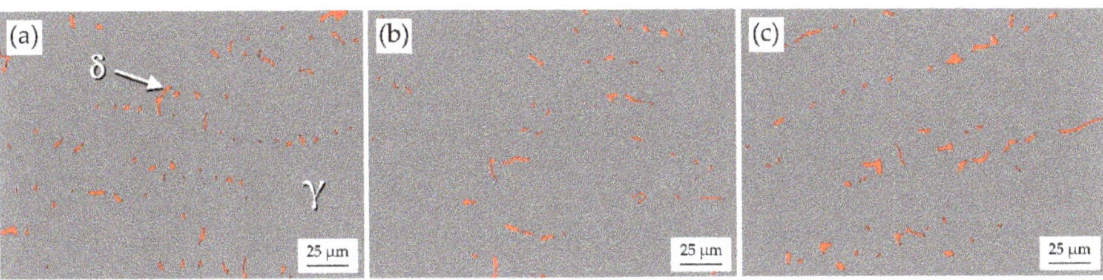

Figure 6. Processed OM images from the following specimens. (**a**) argon-1.1; (**b**) argon-1.2; and (**c**) argon-1.3.

In order to confirm the elemental content of the various phases, three specific points were analyzed using a circular backscatter (CBS-) SEM with EDS for sample air-1.1 (Figure 7). Spot 1, marked in Figure 7, shows the lowest content of chromium, with this point set in the austenite matrix. For Spot 3, the concentration of chromium lightly increased; and the highest content of chromium was measured at Spot 2; in both these latter spots, the vermicular morphology could be identified. These EDS analyses suggest that the vermicular phase is rich in chromium and also in molybdenum, while the matrix (Spot 1) presents more nickel content than in the vermicular phase.

Furthermore, a compositional mapping was carried out in sample argon-2.2 in order to verify the elemental distribution. Figure 8 confirms the presence of chromium and molybdenum in the vermicular areas; both elements appeared in the δ-ferrite phase in a higher content than in the austenite matrix.

It should be borne in mind that the vermicular δ phase may embed the intermetallic σ phase that commonly possesses the highest content of chromium and molybdenum [25]. In accordance with previous authors, this σ phase arises from the decomposition of the δ phase. Therefore, the identification of this intermetallic may relate to the presence of the vermicular areas. Figure 9a shows a higher-magnification SEM image that enabled the detection of the intermetallic σ. There were small amounts of this phase embedded in the vermicular ferrite areas and preferably placed in the gamma-delta ferrite interphase. More intermetallic sigma phase can be found in as-built samples, in the central position of the walls [18]. These regions can also be observed in OM (Figure 9b).

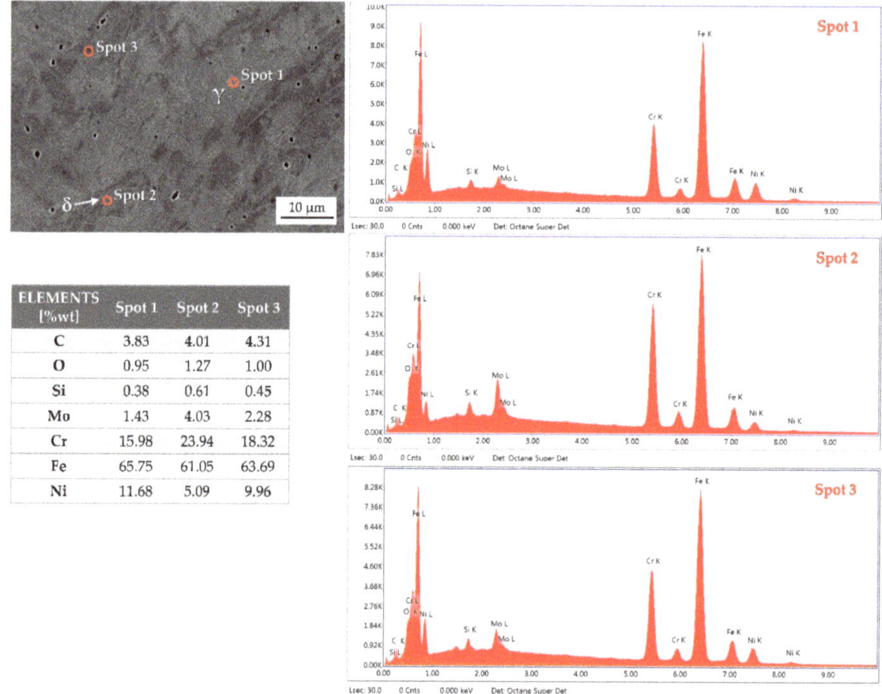

ELEMENTS [%wt]	Spot 1	Spot 2	Spot 3
C	3.83	4.01	4.31
O	0.95	1.27	1.00
Si	0.38	0.61	0.45
Mo	1.43	4.03	2.28
Cr	15.98	23.94	18.32
Fe	65.75	61.05	63.69
Ni	11.68	5.09	9.96

Figure 7. EDS–SEM image and the analysis of three spots in sample air-1.1.

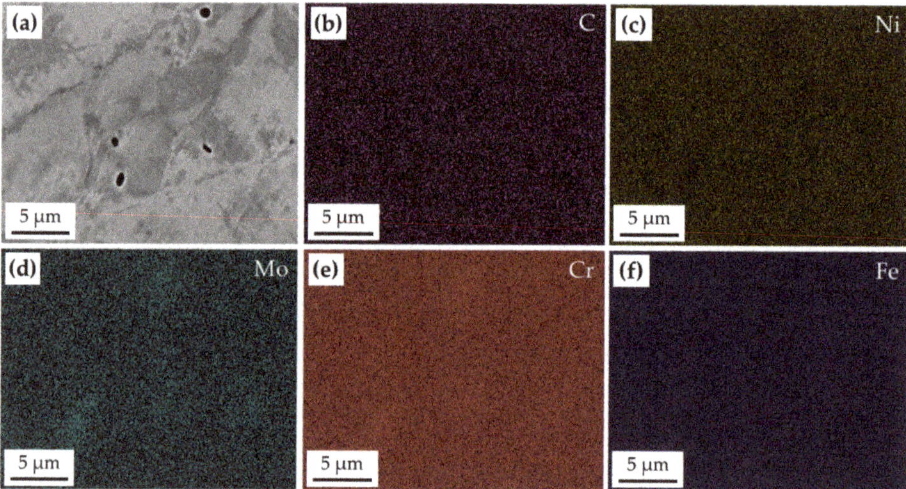

Figure 8. Compositional mapping of sample argon-2.2. (**a**) SEM image; (**b**) carbon; (**c**) nickel; (**d**) molybdenum; (**e**) chromium; and (**f**) iron.

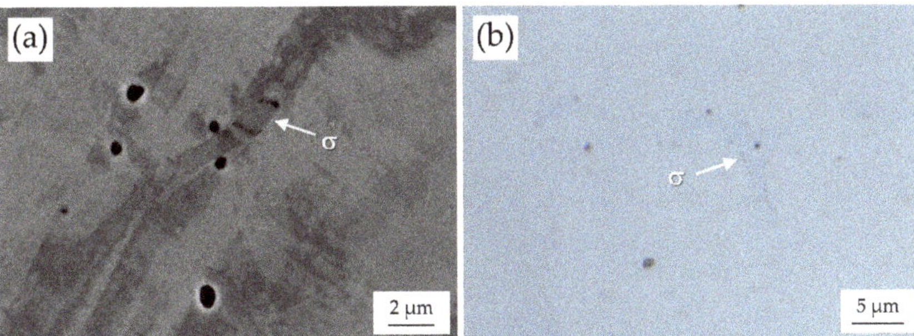

Figure 9. Identification of intermetallic σ phase. (**a**) CBS–SEM image from sample air-1.2; and (**b**) OM image from sample argon-1.2.

3.2. Mechanical Properties

In general terms, AISI 316L stainless steel specimens processed by PMD were dense with porosity values lower than 1%, obtained by IA and densification by the Archimedes' method.

Concerning hardness, a whole wall section was measured along with the height. The obtained results were similar within each wall. The average hardness values were 150 HV in air and 140 HV in argon. Therefore, in both cases, the achieved hardness was slightly reduced from the nominal values, but similar to the typical results from metal deposition techniques (140–175 HV) [18,28,33,34].

In the context of the mechanical properties measurements, the tensile test results were studied and compared on the basis of: (i) ultimate tensile strength (UTS), (ii) elongation (ε), (iii) yield strength (YS) and (iv) Young's modulus (E). A comparative study was performed, not only to achieve the effect of the manufacturing conditions in the final behavior of the specimens (air and argon), but also to comprehend the influence of the thermal treatment applied. Figures 10–12 provide data on the average calculated values after the testing of three specimens located in the three different positions shown in Figure 4.

A graphic depiction of the average properties of the as-built specimens studied can be observed in Figure 10. The goodness of selected manufacturing parameters was verified since the UTS, ε, and YS achieved values close to the standard values (450 MPa, 40% and 170 MPa, respectively) [35], regardless of the air or argon conditions. The influence of the ambient conditions during the PMD process involved differences on the mechanical properties between specimens processed in air and in argon, especially in the UTS and YS. It was observed how the average UTS value increased by 9% when specimens were manufactured in air; the YS also showed increments of 13%. Young's modulus and elongation showed similar values, with 161 GPa and 44%, respectively, for specimens processed in air (see Figure 10).

Regarding the effect of the thermal treatment, Figures 11 and 12, the specimens processed in air suffered mild decrements of their UTS, of approximately 7% after TT1 and 10% after TT2. When the specimens were made in argon conditions, their UTS after TT1 and TT2 remained closer, at 446 MPa and 453 MPa, respectively, which means that there was a negligible variation of UTS behavior after the postprocessing treatments when the specimens were performed in argon.

However, the YS values diminished after both treatments. Its tendency remained similar to the as-built specimens and resulted in higher values in air than in argon. Comparing the effect of the heating temperature of the postprocessing treatment: the higher the temperature, the lower the YS measured.

Young's modulus values also presented attenuation with the thermal treatments. These measured lower values after TT2 than after TT1. In this respect, the influence of the two treatments was similar, independently of whether the specimens were fabricated in air or in argon.

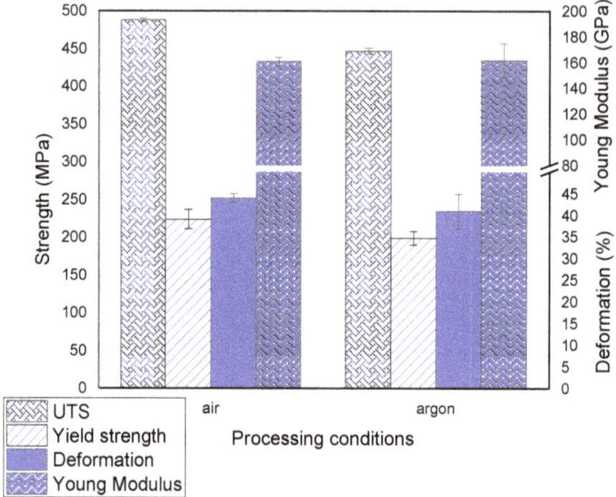

Figure 10. Mechanical properties of PMD as-built specimens vs. the processing conditions of air and argon.

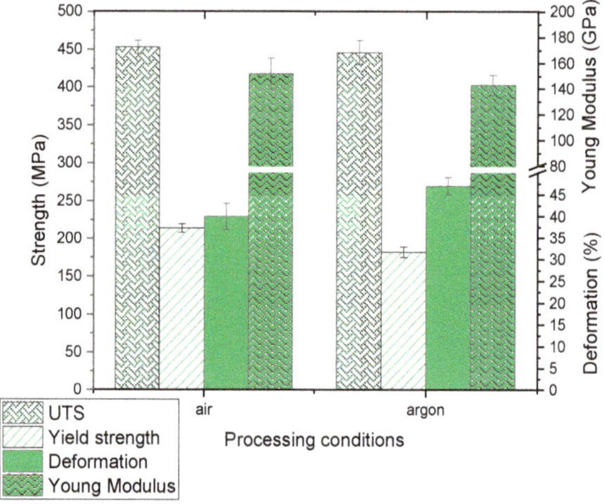

Figure 11. Mechanical properties of PMD specimens after TT1 vs. the processing conditions of air and argon.

On studying the results of the YS behavior, the application of the TT1 led to a decrease of 5% and 9% in specimens manufactured in air and in argon, respectively. Similarly, the TT2 contributed towards reducing the YS by 15% and 18% regarding the as-built specimens. It seems that both treatments more strongly affected the specimens fabricated in argon; in general, TT2 caused a greater decrement in the YS. The Young's modulus values were affected in the same way by each of the thermal treatments.

The deformation of the specimens after each of the thermal treatments varied slightly. In specimens fabricated in argon, the enhancement of the elongation reached from 41% of the as-built specimens up to approximately 47%, after TT1 and TT2. In the case of the specimens manufactured in air, the deformation remained closer to 44% independent of the treatment carried out. Therefore, the effect of the postprocessing was more significant in specimens fabricated in argon.

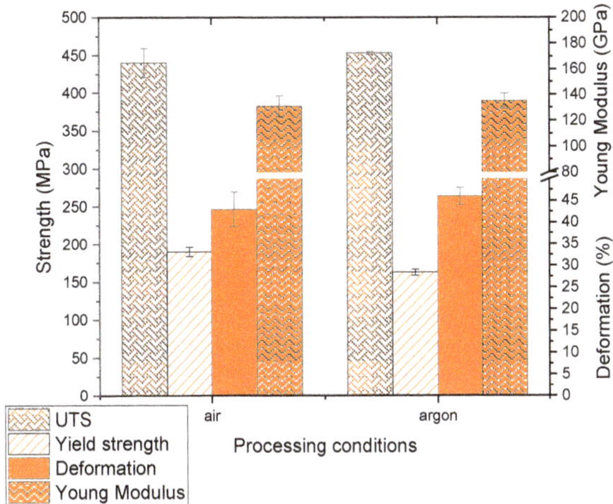

Figure 12. Mechanical properties of PMD specimens after TT2 vs. the processing conditions of air and argon.

In general, it becomes apparent that the YS and Young's modulus present lower values as the heat treatment temperature increases, and therefore these samples offer a lower resistance to permanent plastic deformation.

Tables 4 and 5 show the values of the mechanical properties of each of the measured specimens. In the context of the sample locations in the manufactured wall, there was only a negligible trend among the specimens, with the exception of Young's modulus. Even after the thermal treatments, the variation of these properties considering the sample position remained unclear.

Table 4. Ultimate tensile strength and elongation results for the specimens studied.

	W1 (Built in Air)			W2 (Built in Argon)	
Sample	**UTS (MPa)**	**Elongation (%)**		**UTS (MPa)**	**Elongation (%)**
1.1 as built	491	44	1.1 as built	450	42
1.2 as built	483	45	1.2 as built	442	37
1.3 as built	489	43	1.3 as built	449	44
As built Average	488 ± 3 (0.7%)	44 ± 1 (1.9%)		447 ± 4 (0.8%)	41 ± 4 (8.6%)
ASTM A 240	485	40		485	40
[12]	450	40		450	40
2.1 TT1	465	44	2.1 TT1	465	47
2.2 TT1	451	36	2.2 TT1	448	44
2.3 TT1	444	40	2.3 TT1	426	49
TT1 Average	453 ± 9 (1.9%)	40 ± 3 (8.2%)		446 ± 16 (3.6%)	47 ± 2 (4.4%)
3.1 TT2	416	49	3.1 TT2	461	49
3.2 TT2	440	42	3.2 TT2	442	47
3.3 TT2	463	39	3.3 TT2	455	43
TT2 Average	440 ± 19 (4.4%)	43 ± 4 (9.7%)		453 ± 8 (1.8%)	46 ± 2 (5.4%)

Table 5. Young's modulus and yield strength results for the specimens studied.

	W1 (Built in Air)			W2 (Built in Argon)	
Sample	Young's Modulus (GPa)	Yield Strength (MPa)		Young's Modulus (GPa)	Yield Strength (MPa)
1.1 as built	160	238	1.1 as built	161	205
1.2 as built	159	205	1.2 as built	149	205
1.3 as built	165	230	1.3 as built	176	188
As built Average	161 ± 3 (1.9%)	224 ± 13 (5.8%)		162 ± 13 (8.3%)	199 ± 9 (4.3%)
ASTM A 240	200	170		200	170
[12]		170			170
2.1 TT1	144	218	2.1 TT1	139	190
2.2 TT1	143	205	2.2 TT1	136	183
2.3 TT1	168	218	2.3 TT1	154	173
TT1 Average	152 ± 12 (7.6%)	214 ± 6 (2.9%)		143 ± 8 (5.5%)	182 ± 7 (3.8%)
3.1 TT2	127	196	3.1 TT2	137	168
3.2 TT2	124	193	3.3 TT2	127	162
3.3 TT2	143	182	3.6 TT2	141	159
TT2 Average	131 ± 8 (6.4%)	190 ± 6 (3.2%)		135 ± 6 (4.4%)	163 ± 4 (2.3%)

On observing the Young's modulus results shown in Table 5, the specimens extracted from the central areas of both walls showed lower values than the samples located on the top and the bottom of the walls. This phenomenon was also detected after each of the thermal treatments.

3.3. Microindentation and Wear Behavior

The microindentation tests were conducted in each of the specimens and employed two different maximum loads of 1 N and 5 N; three runs were performed for each load. Figure 13 depicts the P–h curves, loading and unloading, in which the load vs. indentation depth is represented using average values after the tests. Concerning the applied thermal treatments, there were no significant variations in the penetration depth curves. Regardless of the thermal treatment, the penetration was higher in samples processed in argon (Figure 13c,d) than in specimens manufactured in air (Figure 13a,b). This trend is clearly visible under the high applied load.

In Table 6 (1 N) and Table 7 (5 N), values of the elastic recovery, absolute and relative, are listed. In general, the relative elastic recovery was greater in specimens fabricated in argon than in air. It should be borne in mind how as-built specimens and specimens after TT1 fabricated in argon exhibited similar values. However, after TT2 there was an enhancement of the elastic recovery; it can be inferred that the 316L stainless steel became softer than for as-built specimens. Results were more noticeable at five newtons maximum load than at one newton.

Table 6. Elastic recovery values employing a load of 1 N.

	W1 (Built in Air)			W2 (Built in Argon)	
Sample	Absolute Elastic Recovery (μm)	Relative Elastic Recovery (%)		Absolute Elastic Recovery (μm)	Relative Elastic Recovery (%)
1.1 as built	0.85	22.2	1.1 as built	1.00	26.0
2.2 TT1	1.09	26.7	2.2 TT1	1.08	28.4
3.3 TT2	0.97	23.8	3.3 TT2	1.23	34.3

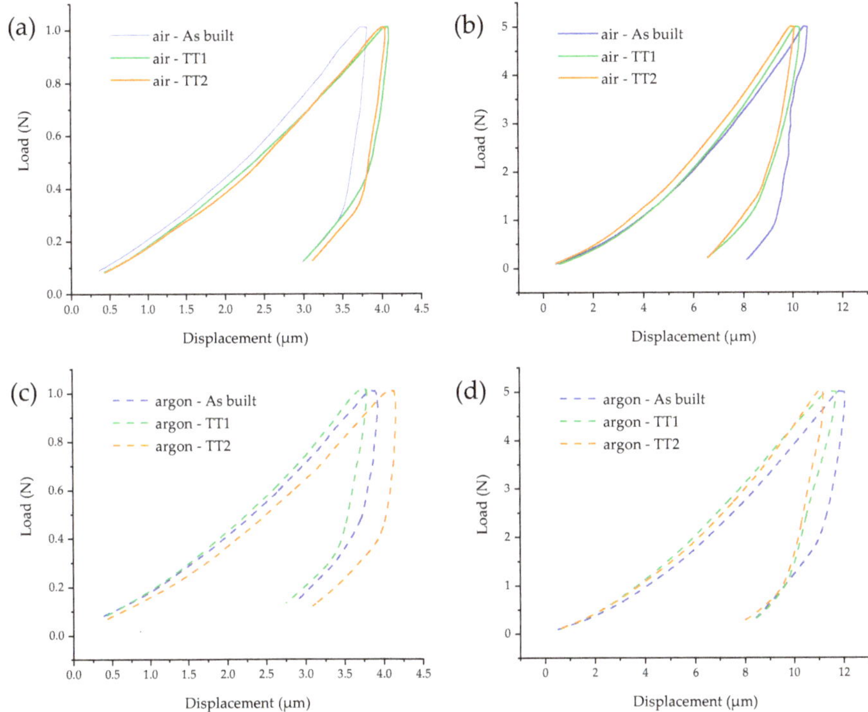

Figure 13. P–h curves for micromechanical behavior of samples. (**a**) Air atmosphere, 1 N; (**b**) air atmosphere, 5 N; (**c**) argon atmosphere, 1 N; and (**d**) argon atmosphere, 5 N.

Table 7. Elastic recovery values employing a load of 5 N.

	W1 (Built in Air)			W2 (Built in Argon)	
Sample	Absolute Elastic Recovery (µm)	Relative Elastic Recovery (%)		Absolute Elastic Recovery (µm)	Relative Elastic Recovery (%)
1.1 as built	2.37	22.3	1.1 as built	3.04	30.1
2.2 TT1	3.73	36.2	2.2 TT1	3.52	30.0
3.3 TT2	3.55	35.3	3.3 TT2	3.82	36.5

The tribological characterization revealed slight variations on the wear behavior depending on the processing conditions. On observing the resulting coefficient of friction (COF), there was a small increase in specimens processed in argon, as can be appreciated in Figure 14.

Regarding the effect of the thermal treatment on the wear behavior, there was no distinguishable trend. There was an initial stage named accommodation period, in which the COF enhancement sharply took place. Subsequently, a second tranche showed a slight increase, and, finally, the stabilization occurred, which is nearly constant in its variation vs. time [36]. Moreover, as can be observed in Figure 14, the COF suffered a visible rise after 200 s; this enhancement was greater in specimens with a low COF than in specimens with high values thereof. It can be presupposed that such an increase may be related to a possible adhesion mechanism that corresponds to the beginning of the third stage of the curves.

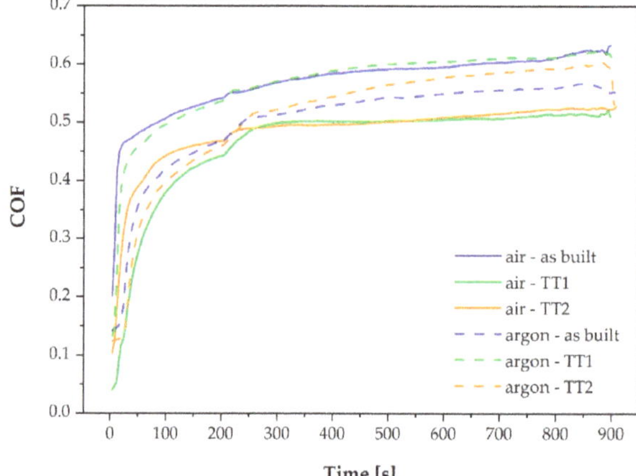

Figure 14. Representation of the coefficient of friction vs. analysis time.

Figure 15 reveals OM images of the worn tracks. Abrasion mechanisms can be appreciated; parallel marks in the same sliding direction can easily be identified. The material removal can clearly be observed as debris in these worn tracks.

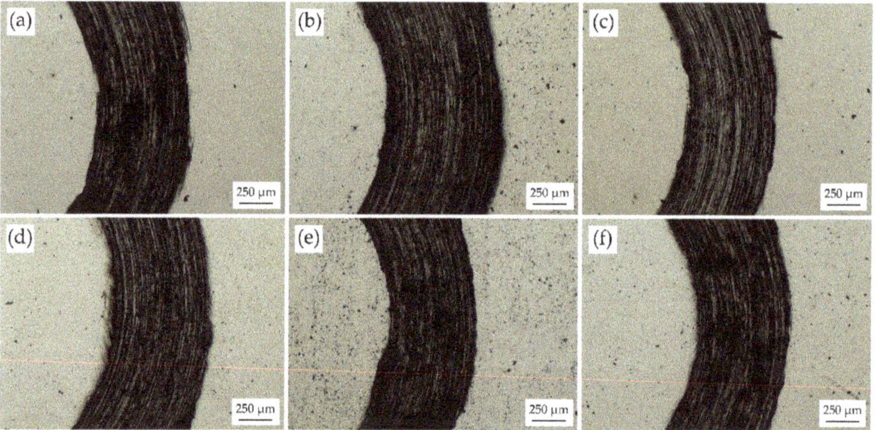

Figure 15. OM images from the worn tracks of the following specimens. (**a**) air-1.1; (**b**) air-2.2; (**c**) air-3.3; (**d**) argon-1.1; (**e**) argon-2.2; and (**f**) argon-3.3.

The worn track areas of the specimens are depicted in Figure 16; these values were calculated by means of the difference between the external and inner circular areas. The tendency was inverse regarding the COF. On comparing the effect of the processing conditions, the values obtained for worn tracks were slightly larger in specimens processed in air than those fabricated in argon. As an additional observation, the influence of thermal treatment was of no significance on the worn tracks, which made difficult to determine a clear trend.

Considering the short period of testing time and distance, in addition to the wear mechanisms that resulted in debris, no significant findings were found regarding the weight loss and the wear rate.

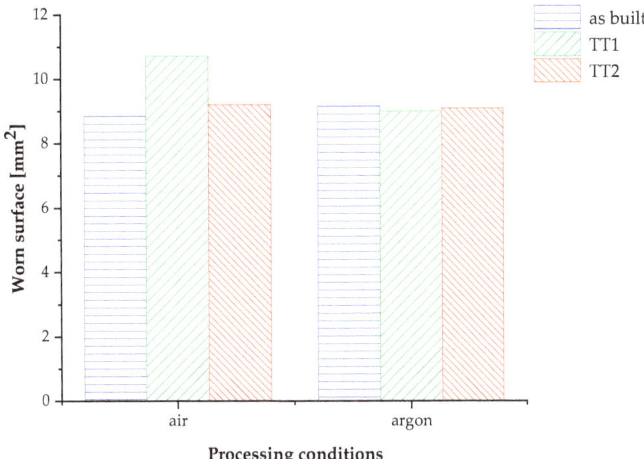

Figure 16. Worn track area values obtained after the tribological tests.

4. Discussion

An austenite (γ) matrix with a retained vermicular ferrite phase (δ) constituted the microstructure of the specimens, regardless of the processing conditions and the thermal treatment.

As previously described, the δ-phase grew as a dendritic morphology. Subsequently, around the boundaries of the ferrite, the austenite phase emerged due to a peritectic transformation during the cooling process with γ oriented within the dendrites.

It was confirmed that the retained δ-phase was rich in chromium and molybdenum. One possible explanation for this composition is related to the inability of these larger and heavier elements to complete their diffusion in rapid solidification since they may have insufficient time to dissolve.

In previous studies regarding the development of 316L stainless steel by AM, in which the built part presented a great height, the analysis of the heat-affected zones showed specimens with significant variations in the ferrite phase morphology depending on the analyzed area [18,25]. In these studies the microstructural evolution started from the top with reticular morphology of the δ phase; it then followed on to a vermicular structure in the center of the wall; and, finally, in the studied microstructure drawn of the base of the built part, this δ phase manifested a fine vermicular structure. Regarding the specimens extracted from the manufactured walls in this research, the observed microstructures revealed only a vermicular morphology and a slight refining trend, particularly in the center of the built part, as can be observed in Figure 6.

Relating the observed microstructure and the study of the macro- and micromechanical and tribological results, the content of the retained ferrite was linked to the properties measured. There were slight variations in the ferrite occurrence in terms of the processing atmosphere. Thus, in specimens manufactured in air, more ferrite could be appreciated and the values of UTS and YS were higher than those manufactured in argon. This retained ferrite phase, located mainly at the grain boundaries [18,32], was responsible for the enhancement of these mechanical properties, regardless of the thermal treatment, making samples stiffer. Other researchers have established that in welded and cast steel, δ phase acts as a strengthener [25]. These differences were less significant when the specimens were postprocessed through thermal treatment.

When specimens were produced in air with a posterior thermal treatment, this led to a decrement of the UTS, YS and Young's modulus. It seems that the effect of the thermal treatment depended on the property evaluated, whereby the changes in UTS more noticeable in air than in argon and the YS, Young's modulus and elongation more affecting in argon than in air. Considering the UTS, the possible appearance of inner porosity, as well as additional microstructural defects, could diminish the UTS of the specimens manufactured in air. It should be borne in mind that the specimens processed in argon showed results that were more homogenous than specimens fabricated in air. An increase in elongation produced strain-hardening as observed in argon-built specimens [25].

This evolution was also related to the retained ferrite and the austenite phases; when the percentage of austenite phase increased, the deformation of the 316L stainless steel reached higher values. By evaluating how the thermal treatment affected the behavior of the specimens, the main answer was found in the transformation of the ferrite into austenite. TT2 involved heating at 950 °C for two hours, causing not only a possible enlargement of the grain size, but also a diminution of the quantity of ferrite phase. Thus, the TT2 promoted the softening of the materials, which can be inferred from the results of the tensile tests that show lower YS, more elongation and more elastic recovery measured after TT2 than after TT1. Moreover, observing the microstructure, the presence of austenite phase was more notorious in the specimens heated at the highest temperature.

Related to the micromechanical findings, the resulting trend remained unclear. The relative elastic recovery, as well as penetration depth, appeared to be higher in specimens processed in argon than in air. The thermal treatment would enhance this elastic behavior. The presence of the austenite, in addition to less content of the ferrite located at grain boundaries, promoted the softening of the specimen.

At five newtons of maximum load, the results were more evident than at one newton. When loading at five newtons, the indentation seemed to encompass a more comprehensive microstructure response. In contrast, loading at one newtons made the trend unclear. This could be related to the formation of oxide layers during the manufacturing of the specimens. Furthermore, it could be due to the microplasticity phenomenon in local areas that is probably caused by the low load.

According to the determined COF, there was a singular relation between specimens with high elongation and high COF. This fact could be linked to the mechanisms that took place during the tribological test. Therefore, the more deformable the material is, the more shear force and surface roughness appear, as well as the adherence of the loose particles on the worn surface [37]. Furthermore, the greater the deformability of the specimens, the more debris appears which in turn generates a higher COF. A smaller groove width is generated in these specimens due to the adhesion of the debris to the ball, thereby generating instability in the formation of the groove, which is appreciable in the exterior of the worn tracks.

5. Conclusions

The key conclusions are presented in terms of the two main factors of influence studied herein: processing conditions and thermal treatment.

(i)   Processing conditions:

- The influence of the inert atmosphere generated by argon gas promoted slight effects on the microstructure and final behavior of the specimens;
- Relating to the retained ferrite (δ), specimens processed in argon showed a lower content in this phase than specimens manufactured in air. This phenomenon was reflected in the mechanical properties measured in as-built specimens processed in argon;
- The elastic recovery was lower in specimens manufactured in air than in argon;
- The COF presented higher values when the specimens were processed in argon than in air and the results suggested that there were abrasive mechanisms with more debris.

(ii)   Thermal treatments:

- These affected the specimens to a lower degree than expected; the specimens treated at a higher temperature (TT2) suffered a higher variation of their mechanical properties than specimens treated at TT1. Therefore, the higher the thermal treatment temperature, the lower the YS and Young's modulus measured;

- Both thermal treatments resulted in a decrease the vermicular phase. TT2 caused greater transformation of the ferrite (δ) than TT1;
- The relation of the elastic recovery with the thermal treatment was unclear, although TT2 appeared to be more influential in the enhancement of the elastic recovery.

Author Contributions: All the authors have cooperated to obtain high-quality research. Conceptualization, C.A. and E.N.; methodology, I.M.-M., E.M.P.-S. and C.A.; formal analysis, I.M.-M.; investigation, C.A., I.M.-M., E.A. and E.M.P.-S.; resources, M.K.; data curation, E.A. and E.M.P.-S.; writing—original draft preparation, I.M.-M., E.M.P.-S., C.A. and E.A.; writing—review and editing, E.A. and I.M.-M.; visualization, E.M.P.-S. and C.A.; and funding acquisition, E.N., E.A., M.K., I.M.-M. and C.A. All authors have read and agreed to the published version of the manuscript.

Funding: This work has received funding from the European Union Horizon 2020 Program (H2020) under grant agreement no 768,612.

Acknowledgments: The authors want to thank the Universidad de Sevilla for the use of experimental facilitates at CITIUS, Microscopy and X-ray Laboratory Services (VI PPIT-2019-I.5 Cristina Arévalo Mora and VI PPIT-2019-I.5 Isabel Montealegre Meléndez). The authors also wish to thank the technicians Jesús Pinto, Mercedes Sánchez and Miguel Madrid for their experimental assistance.

Conflicts of Interest: The authors declare no conflict of interest.

References

1. Gibson, I.; Rosen, D.W.; Stucker, B. Post-processing. In *Additive Manufacturing Technologies*, 2nd ed.; Springer: New York, NY, USA, 2015; pp. 346–348.
2. Guo, N.; Leu, M.C. Additive manufacturing: Technology, applications and research needs. *Front. Mech. Eng.* **2013**, *8*, 215–243. [CrossRef]
3. Gideon, N.L. The role rand future of the Laser Technology in the Additive Manufacturing environment. *Phys. Procedia* **2010**, *5*, 65–80.
4. Bartolomeu, F.; Buciumeanu, M.; Pinto, E.; Alves, N.; Carvalho, O.; Silva, F.S.; Miranda, G. 316L stainless steel mechanical and tribological behavior—A comparison between selective laser melting, hot pressing and conventional casting. *Addit. Manuf.* **2017**, *16*, 81–89. [CrossRef]
5. Wong, K.V.; Hernandez, A. A Review of Additive Manufacturing. *ISRN Mech. Eng.* **2012**, *12*, 1–10. [CrossRef]
6. Gu, D.D.; Meiners, W.; Wissenbach, K.; Poprawe, R. Laser additive manufacturing of metallic components: Materials, processes and mechanisms. *Int. Mater. Rev.* **2012**, *57*, 133–164. [CrossRef]
7. Mukherjee, M. Effect of build geometry and orientation on microstructure and properties of additively manufactured 316L stainless steel by laser metal deposition. *Materialia* **2019**, *7*, 100359. [CrossRef]
8. Lervåg, M.; Sørensen, C.; Robertstad, A.; Brønstad, B.M.; Nyhus, B.; Eriksson, M.; Aune, R.; Ren, X.; Akselsen, O.M.; Bunaziv, I. Additive Manufacturing with Superduplex Stainless Steel Wire by CMT Process. *Metals* **2020**, *10*, 272. [CrossRef]
9. Pérez-Soriano, E.M.; Ariza, E.; Arévalo, C.; Montealegre-Meléndez, I.; Kitzmantel, M.; Neubauer, E. Processing by Additive Manufacturing Based on Plasma Transferred Arc of Hastelloy in Air and Argon Atmosphere. *Metals* **2020**, *10*, 200. [CrossRef]
10. Mercado Rojas, J.G.; Wolfe, T.; Fleck, B.A.; Qureshi, A.J. Plasma transferred arc additive manufacturing of Nickel metal matrix composites. *Manuf. Lett.* **2018**, *18*, 31–34. [CrossRef]
11. Savolainen, J.; Collan, M. How Additive Manufacturing Technology Changes Business Models?—Review of Literature. *Addit. Manuf.* **2020**, *32*, 101070. [CrossRef]
12. Najmon, J.C.; Raeisi, S.; Tovar, A. Review of additive manufacturing technologies and applications in the aerospace industry. In *Additive Manufacturing for the Aerospace Industry*, 1st ed.; Elsevier: Amsterdam, The Netherlands, 2019; pp. 7–31.
13. Neubauer, E.; Ariza Galvan, E.; Meuthen, J.; Bielik, M.; Kitzmantel, M.; Baca, L.; Stelzer, N. *Analysis of the Anisotropy of Properties in Titanium Alloys made by Plasma Metal Deposition, Proceedings of the Euro PM2019, Maastricht, The Netherlands, 13–16 October 2019*; European Powder Metallurgy Association (EPMA): Shrewsbury, UK, 2019.
14. Norsk Titanium delivers FAA approved AM part to Boeing. *Met. Powder Rep.* **2017**, *72*, 279. [CrossRef]

15. Ariza-Galván, E.; Montealegre-Meléndez, I.; Pérez-Soriano, E.M.; Arévalo Mora, C.; Meuthen, J.; Neubauer, E.; Kitzmantel, M. *Processing of 17-4PH by Additive Manufacturing Using a Plasma Metal Deposition (PMD) Technique, Proceedings of the Euro PM2019, Maastricht, The Netherlands, 13–16 October 2019*; European Powder Metallurgy Association (EPMA): Shrewsbury, UK, 2019; p. 4348497.

16. Song, R.; Xiang, J.; Hou, D. Characteristics of mechanical properties and microstructure for 316L austenitic stainless steel. *J. Iron Steel Res.* **2011**, *18*, 53–59. [CrossRef]

17. Saboori, A.; Piscopo, G.; Lai, M.; Salmi, A.; Biamino, S. An investigation on the effect of deposition pattern on the microstructure, mechanical properties and residual stress of 316L produced by Directed Energy Deposition. *Mater. Sci. Eng. A* **2020**, *780*, 139179. [CrossRef]

18. Chen, X.; Li, J.; Cheng, X.; He, B.; Wang, H.; Huang, Z. Microstructure and mechanical properties of the austenitic stainless steel 316L fabricated by gas metal arc additive manufacturing. *Mater. Sci. Eng. A* **2017**, *703*, 567–577. [CrossRef]

19. Cherry, J.A.; Davies, H.M.; Mehmood, S.; Lavery, N.P.; Brown, S.G.R.; Sienz, J. Investigation into the effect of process parameters on microstructural and physical properties of 316L stainless steel parts by selective laser melting. *Int. J. Adv. Manuf. Technol.* **2014**, *76*, 869–879. [CrossRef]

20. Fayazfar, H.; Salarin, M.; Rogalsky, A.; Sarker, D.; Russo, P.; Paserin, V.; Toyserkani, E. A critical review of powder-based additive manufacturing of ferrous alloys: Process parameters, microstructure and mechanical properties. *Mater. Des.* **2018**, *144*, 98–128. [CrossRef]

21. Debroy, T.; Wei, H.L.; Zuback, J.S.; Mukherjee, T.; Elmer, J.W.; Milewski, J.O.; Beese, A.M.; Wilson-Heid, A.; De, A.; Zhang, W. Additive manufacturing of metallic components-Process, structure and properties. *Prog. Mater. Sci.* **2018**, *92*, 112–224. [CrossRef]

22. Suryawanshi, J.; Prashanth, K.G.; Ramamurty, U. Mechanical behaviour of selective laser melted 316L stainless steel. *Mater. Sci. Eng. A* **2017**, *696*, 113–121. [CrossRef]

23. Prashanth, K.G.; Scudino, S.; Chatterjee, R.P.; Salman, O.O.; Eckert, J. Additive Manufacturing: Reproducibility of Metallic Parts. *Technologies* **2017**, *5*, 8. [CrossRef]

24. Salman, O.O.; Gammer, C.; Chaubey, A.K.; Eckert, J.; Scudino, S. Effect of heat treatment on microstructure and mechanical properties of 316L steel synthesized by selective laser melting. *Mater. Sci. Eng. A* **2019**, *748*, 205–212. [CrossRef]

25. Chen, X.; Li, J.; Cheng, X.; Wang, H.; Huang, Z. Effect of heat treatment on microstructure, mechanical and corrosion properties of austenitic stainless steel 316L using arc additive manufacturing. *Mater. Sci. Eng. A* **2018**, *715*, 307–314. [CrossRef]

26. ASTM. *A240: Standard Specification for Chromium and Chromium-Nickel Stainless Steel Plate, Sheet, and Strip for Pressure Vessels and for General Applications*; ASTM International: West Conshohocken, PA, USA, 2019.

27. ASTM. *E-8 Standard Test Methods for Tension Testing of Metallic Materials*; ASTM International: West Conshohocken, PA, USA, 2016.

28. Montero Sistiaga, M.L.; Nardone, S.; Hautfenne, C.; Van Humbeeck, J. Effect of heat treatment of 316L stainless steel produced by selective laser melting (SLM). In Proceedings of the 27th Annual International Solid Freeform Fabrication Symposium, Austin, TX, USA, 8–10 August 2016.

29. ASTM. *C373-88: Standard Test Method for Water Absorption, Bulk Density, Apparent Porosity, and Apparent Specific Gravity of Fired Whiteware Products*; ASTM International: West Conshohocken, PA, USA, 2006.

30. Oliver, W.C.; Pharr, G.M. An improved technique for determining hardness and elastic modulus using load and displacement sensing indentation experiments. *Mater. Res.* **1992**, *7*, 1564–1583. [CrossRef]

31. Pasang, T.; Kirchner, A.; Jehring, U.; Aziziderouei, M.; Tao, Y.; Jiang, C.P.; Wang, J.C.; Aisyah, I.S. Microstructure and Mechanical Properties of Welded Additively Manufactured Stainless Steels SS316L. *Met. Mater. Int.* **2019**, *25*, 1278–1286. [CrossRef]

32. Wang, C.; Tan, X.; Liu, E.; Tor, S.B. Process parameter optimization and mechanical properties for additively manufactured stainless steel 316L parts by selective electron beam melting. *Mater. Des.* **2018**, *147*, 157–166. [CrossRef]

33. Lanzutti, A.; Marin, E.; Tamura, K.; Morita, T.; Magnan, M.; Vaglio, E.; Andreatta, F.; Sortino, M.; Totis, G.; Fedrizzi, L. High temperatura study of the evolution of the tribolayer in additively manufactured AISI 316L steel. *Addit. Manuf.* **2020**, *34*, 101258.

34. Wang, L.; Xue, J.; Wang, Q. Correlation between arc mode, microstructure, and mechanical properties during wire arc additive manufacturing of 316L stainless steel. *Mater. Sci. Eng. A* **2019**, *751*, 183–190. [CrossRef]

35. ASTM. *A473-15: Standard Specification for Stainless Steel Forgings*; ASTM International: West Conshohocken, PA, USA, 2015; pp. 1–5.
36. Fellah, M.; Labaïz, M.; Assala, O.; Dekhil, L.; Zerniz, N.; Iost, A. Tribological behavior of biomaterial for total hip prosthesis. *Mater. Tech.* **2014**, *102*, 601. [CrossRef]
37. Rominiyi, A.L.; Shongwe, M.B.; Ogunmuyiwa, E.N.; Babalola, B.J.; Lepele, P.F.; Olubambi, P.A. Effect of nickel addition on densification, microstructure and wear behaviour of spark plasma sintered CP-titanium. *Mater. Chem. Phys.* **2020**, *240*, 122130. [CrossRef]

© 2020 by the authors. Licensee MDPI, Basel, Switzerland. This article is an open access article distributed under the terms and conditions of the Creative Commons Attribution (CC BY) license (http://creativecommons.org/licenses/by/4.0/).

Article

Spreading Process Maps for Powder-Bed Additive Manufacturing Derived from Physics Model-Based Machine Learning

Prathamesh S. Desai and C. Fred Higgs III *

Mechanical Engineering Department, Rice University, 6100 Main St, Houston, TX 77005, USA; pratnsai@gmail.com
* Correspondence: higgs@rice.edu

Received: 26 September 2019; Accepted: 18 October 2019; Published: 31 October 2019

Abstract: The powder bed additive manufacturing (AM) process is comprised of two repetitive steps—spreading of powder and selective fusing or binding the spread layer. The spreading step consists of a rolling and sliding spreader which imposes a shear flow and normal stress on an AM powder between itself and an additively manufactured substrate. Improper spreading can result in parts with a rough exterior and porous interior. Thus it is necessary to develop predictive capabilities for this spreading step. A rheometry-calibrated model based on the polydispersed discrete element method (DEM) and validated for single layer spreading was applied to study the relationship between spreader speeds and spread layer properties of an industrial grade Ti-6Al-4V powder. The spread layer properties used to quantify spreadability of the AM powder, i.e., the ease with which an AM powder spreads under a set of load conditions, include mass of powder retained in the sampling region after spreading, spread throughput, roughness of the spread layer and porosity of the spread layer. Since the physics-based DEM simulations are computationally expensive, physics model-based machine learning, in the form of a feed forward, back propagation neural network, was employed to interpolate between the highly nonlinear results obtained by running modest numbers of DEM simulations. The minimum accuracy of the trained neural network was 96%. A spreading process map was generated to concisely present the relationship between spreader speeds and spreadability parameters.

Keywords: powder-bed additive manufacturing (AM); powder spreading; spreading process map; discrete element method (DEM); machine learning

---

1. Introduction

Powder-bed additive manufacturing has been rapidly evolving over the last decade, due in part to the design freedom it offers [1]. It is defined as the process of joining materials to make objects from 3D model data, usually layer upon layer [2]. The process consists of layer by layer spreading of a powdered material and selective fusion or binding of the spread layer to produce 3D objects. State of the art 3D printed parts have microscopic defects, at layer-level, like porosity, roughness and over or under fused/bound particles and macroscopic, at part level, defects like poor surface finish, voids, dimensional inaccuracy and shear-induced deformation. Few of these defects are directly related to the spreading step in powder-bed additive manufacturing (AM) [3–14].

1.1. Powder Spreading Studies

Authors of this study have previously defined powder spreadability as the ease with which a powder will spread under a set of load conditions [15] and have introduced four spread layer properties which can be used to quantify spreadability—mass of powder retained in the sampling region after spreading, spread throughput, roughness and porosity of the spread layer [4]. Spread throughput refers to the volume of powder spread per unit width of spreader per unit time and is indicative of speed of printing with respect to spreading [4]. Roughness of the spread layer is based on the RMS value of the topography of the spread layer [4]. Porosity of the spread layer represents the ratio of volume of pores to the volume of the spread layer cuboid or alternatively one minus the ratio of volume of powder particles to the volume of the spread layer cuboid [4]. Snow et al. [3] have developed a similar spreadability metric comprising of the percent build plate coverage, the powder deposition rate and the rate of change of the avalanche angle. There is a consensus in the literature on the applicability of the Discrete Element Method (DEM) to study the problem on AM powder spreading [5–9,11–14]. DEM was used by Haeri in 2017 [10] to optimize the shape of blade-like spreaders to minimize spread layer defects. Few studies [16–18] have tried to use high energy X-rays to visualize the powder spreading in situ, however these studies had to make use of a thin slice of powder in direction perpendicular to the X-rays due to the low penetration depth of view offered by X-rays. Thereby these studies were unable to study the full 3D nature of the powder spreading problem.

1.2. AM Machine Learning Studies

Recent advances in the field of data science and machine learning (ML) have resulted in works combining ML and AM. Everton et al. [19] review in situ analysis for AM processes and highlight key technologies for process control in metal AM. The big data from sensors is mentioned to be the bottleneck for developing effective closed-loop process control. Bumann et al. [20] provide a review of state-of-the-art AM ML studies. ML studies are classified into studies used to optimize AM process parameters, to improve object properties, to enable process monitoring and closed-loop process control and to improve digital security. An overview of AM informatics—science of managing AM data across its lifecycle is provided in Reference [21]. A review of descriptive, predictive and prescriptive ML studies in continuum materials mechanics—process parameters, microstructure, mechanical properties and performance is provided in Reference [22].

Kamath in 2016 [23] introdced an iterative synergistic approach involving different levels of complexity in experiments and simulations to optimize the process parameters of a selective laser melting machine to print dense 316L SS parts (>99% dense). Data mining and statistical inference were used in this study.

A data mining approach in real-time measurement for polymer additive manufacturing was studied in Reference [24]. Brian et al. [25] developed an autonomous objective system devoid of subjective human judgement, to implicitly characterize powder micrographs as a distribution of local image features. The system was capable of classifying powders having different distributions of particle size, shape and surface texture with an accuracy of over 95%. The ML system was also useful in quantitatively identifying representative and atypical powder images. A hybrid ML algorithm comprised of clustering (unsupervised learning) and SVM (supervised learning) was developed to manage design knowledge and help inexperienced designers to explore AM-enabled design freedoms in Reference [26]. Uhlmann et al. [27] provided data driven predictions for condition monitoring and clustering of data from an SLM machine.

Gu et al. [28] proposed design of hierarchical materials using machine learning and finite element analysis. Additionally, this work showed that machine learning can be used as an alternative method of coarse-graining, with the ability to analyze and design materials without the use of full microstructural data. A random forests-based predictive modeling approach to predict surface roughness of parts produced using FDM (fused deposition modeling) was proposed in Reference [29]. Gorbert et al. [30] developed in situ defect detection strategy for PBF AM using in situ sensor data from a DSLR camera and ex situ ground truth defect data from CT images of the 3D printed part. The lighting condition dependent strategy was shown to detect defects of incomplete fusion, cracks and porosity with 80% accuracy. Bessa and Pellegrino [31] developed a data-driven framework comprising of Design of Experiments (DoE), computational analyses, machine learning and multi-objective optimization to develop design charts and find optimal design of Triangular Rollable And Collapsible (TRAC) boom. This type of structure consists of two thin cylindrical shells of uniform thickness and has an approximately triangular cross-section, consisting of single-thickness transversely curved flanges and a double-thickness flat web. A multi-scale convolutional neural network was used to detect anomalies in laser powder bed fusion systems in Reference [32].

Machine learning techniques were investigated for the continuous printing speed modeling, selection and optimization for Continuous Liquid Interface Production (CLIP) a vat photopolymerization based additive manufacturing in Reference [33]. Siamese Network was shown to be more effective at capturing features for identifying the optimum manufacturing speed. In Reference [34], a data-driven and ensemble learning-based predictive model was developed to predict surface roughness of parts printed using fused filament fabrication (FFF). The condition monitoring data, which served as input to the predictive model, was obtained using multiple sensors. Zhang et al. [35] used deep learning in the form of LSTM NN (long short-term memory neural network) and layer-wise relevance propagation to predict tensile strength of dog-bone specimens printed using fused deposition modeling (FDM). The tensile strength of the final specimen was predicted as a function of the contribution from each of the printing layers, measured using in-process sensing of temperature and vibration data, fused together with process parameters and material property [35].

However, none of these studies provide intelligent closed-loop control of 3D printers with regards to the spreading step. The present computational study proposes to solve this by using ML to develop physics model-based surrogate model for a high-fidelity, computationally expensive physics-based model of the spreading process. To the best of our knowledge, no study exists that tries to elucidate the spreading process to obtain process maps by following the methodology described in Section 2. Predictions from the machine learning model are discussed in Section 3, followed by conclusions in Section 4.

2. Methodology

It is challenging to study the spreading step of powder-bed AM, not only experimentally but also computationally. It is difficult to study this experimentally as the in situ measurements of layer defects are challenging. Physics-based computational modeling of the spreading step is highly expensive due to the need to simulate hundreds of thousands of powder particles. Therefore, a synergistic, three-step approach as shown in Figure 1 was employed to study the powder spreading step in powder-bed AM. This approach is the same as the one proposed by authors in an earlier study [36] but improves the fidelity of the physics-based model by accounting for realistic particle size distribution (PSD) of an industry-grade Ti-6Al-4V and roughness of the spreading substrate. The first step involves the characterization of the AM powder using a powder rheometer developed by Freeman Technology, U.K. and using the data for calibrating a 'virtual powder' which behaves similar to the real AM powder as discussed in the previous work of authors [15]. The second step involves the spreading simulation study of this rheometry-validated virtual powder and comparison to the real spreading, as discussed in [4]. Finally, referring to Figure 1, the simulation data is used to develop a physics model-based surrogate model of the physics-based model. This model is then used to generate, what the authors refer to as, a spreading process map which maps the spreader speeds to the spread layer defects and spread throughput. The present study is directed towards elaborating parts of the second and third step of this methodology and discussing the results obtained using these steps.

2.1. Physics-Based Polydispersed DEM Model

The authors have developed an in-house multiphysics computational modeling platform known as the Particle-Surface Tribology Analysis Code (P-STAC), which has a DEM module as its particle dynamics simulation tool. The DEM model is based on the one developed by Cundall and Strack [37] and has been validated in an earlier work by the authors [15]. It was shown by the authors of this study that it is important to account for a realistic particle size distribution while performing virtual powder spreading [4]. This validated polydispersed model will be used in the present study.

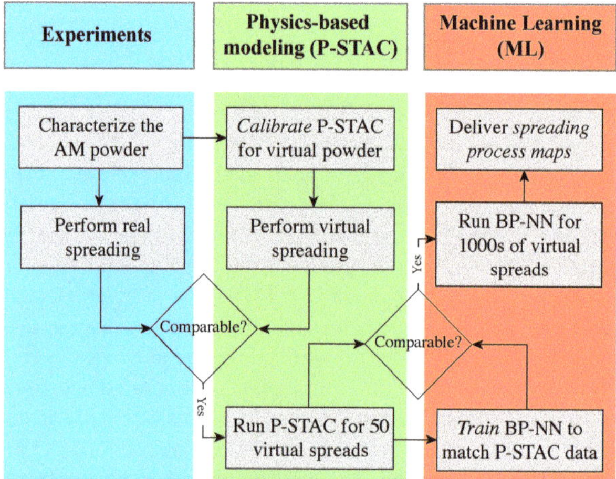

Figure 1. Synergy between experiments, physics-based modeling and physics model-based machine learning enabled surrogate modeling. Reproduced from [36], with permission from the Solid Freeform Fabrication Symposium—An Additive Manufacturing Conference, 2017.

P-STAC is used to simulate single layer spreading as seen in a retrofit to a commercial powder-bed 3D printer, developed by ExOne, USA, refer Figure 2a. The spreading substrate as seen in the inset in Figure 2a was additively manufactured using a powder-bed fusion printer and the spreading of a single layer of powder was carried out over it. Sample simulation output is shown in Figure 2b. Here the front half of the particles has been color-coded by particle size and the back half has been color-coded by particle speed. The DEM simulations had particles of Ti-6Al-4V with an assumed normal of Gaussian particle size distribution (PSD) in the range of 100–250 μm. The real AM powder also had particles in the range of 100–250 μm. The topography of the spreading substrate is also shown. The roughness of the spreading substrate, S_q is 79 microns. The spreader is able to spread a dense layer of powder everywhere except where the spreading substrate has steep protrusions. These defects of partial coverage result in the porosity of the spread layer and can eventually result in voids in the final 3D-printed part. These voids will act as stress concentration points and thereby a source of part failure [38]. The polydispersed DEM simulations can be used to correct such defects by altering the spreader speeds.

Spread Layer Characterization

The polydispersed DEM model allows one to predict spread layer properties which otherwise might be difficult to measure experimentally. These properties include mass of the powder in the sampling region M_s, spread throughput Q_s, roughness of the spread layer R_q and porosity of the spread layer ϕ. The sampling region is a square area of 25 mm by 25 mm located in the central region of the spread layer. The mathematical description of these properties, as reproduced from Reference [4] is as follows:

Figure 3 shows a generalized schematic of the side view of a spread layer i. The heights of the previously 3D-printed (i.e., spread and fused) layer $(i - 1)$ are denoted by b's while the heights of the post-spread surface (consisting of fresh powder particles) are denoted by h's. In other words, b's are the heights measured by a virtual profilometer before the fresh powder was spread and h's are the heights measured by a virtual profilometer after the powder was spread. Notice the point of layer $(i - 1)$ denoted by b_3 where no powder was spread due to a steep peak. This point will be also registered as an h point (i.e., h_3) for layer i while calculating its roughness. In the present study, points at the intersection of randomly distributed 1000×1000 lines over the surface area of the sampling region were used for the measurements of $h's$ and $b's$.

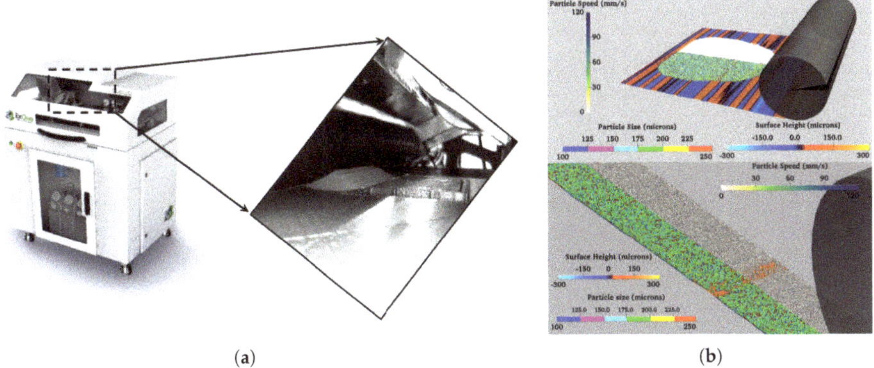

(a)   (b)

Figure 2. Synergy between experiments and physics-based modeling (**a**) A powder-bed 3D printer with inset showing single layer spreading experiment on an additively manufactured spreading substrate, refer [4] for details (**b**) Particle-Surface Tribology Analysis Code (P-STAC) simulation images—full view (**top**), zoomed-in view showing only a part of the spread layer with a visible spread layer defect of partial coverage (**bottom**).

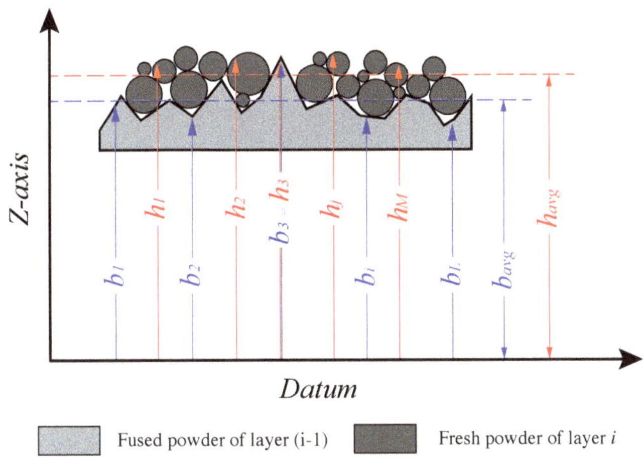

Figure 3. Schematic to understand computations of spread layer properties.

The average height of layer $(i-1)$, denoted by b_{avg} and that of layer i denoted by h_{avg}, can be given by:

$$b_{avg} = \frac{\sum_{i=1}^{L} b_i}{L} \tag{1}$$

$$h_{avg} = \frac{\sum_{j=1}^{M} h_j}{M} \tag{2}$$

The average spread layer thickness can then be given by:

$$thk_{avg} = h_{avg} - b_{avg} \tag{3}$$

The quantities which answer the above four questions can be mathematically obtained as follows. The mass of powder particles in sampling region, M_s can be expressed as

$$M_s = \frac{\pi \rho}{6} \sum_{k=1}^{N} \phi_k^3 \tag{4}$$

Here N is the number of powder particles in the sampling region of L_x by L_y. While ϕ_k is the diameter of the kth particle and ρ is the material density of powder particles.

Volume of powder spread per unit time per unit width of spreader, Q_s can be expressed as

$$Q_s = U \cdot thk_{avg} \tag{5}$$

Where, thk_{avg} is given by (3) and U is the translation speed of the spreader.

Roughness of the spread layer, R_q can be expressed as

$$R_q = \left[\frac{\sum_{j=1}^{M} (h_j - h_{avg})^2}{M} \right]^{\frac{1}{2}} \tag{6}$$

And finally, porosity of the spread layer, Φ can be expressed as

$$\Phi = \left[1 - \frac{M_s}{\rho L_x L_y (4R_q)} \right] \cdot 100\% \tag{7}$$

Spread layer sampling regions for a constant spreader translation speed and varying rotational speed are shown in Figure 4. The variation in spread layer properties is not easily seen however the spread layer properties do exhibit quite a lot of variation for these three spreader speeds. Refer to Section 3 for details.

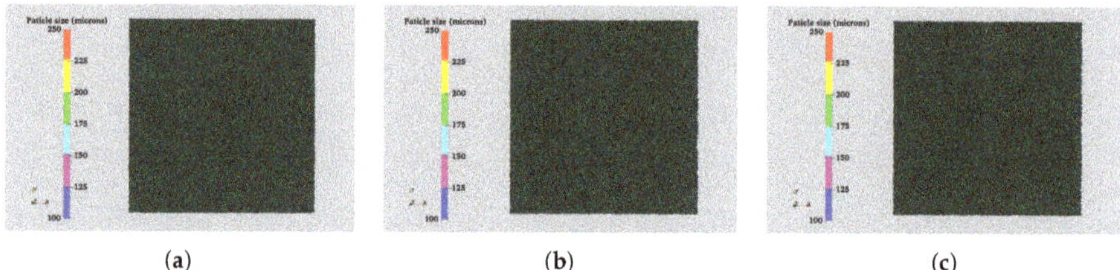

(a)   (b)   (c)

Figure 4. Sampling regions for the characterization of the spread layer for spreader speeds (**a**) U = 100 mm/s and ω = −5 rad/s (**b**) U = 100 mm/s and ω = 0 (**c**) U = 100 mm/s and ω = 5 rad/s.

2.2. Design of Simulations for Virtual Spreading

Physics-based simulations with the polydispersed DEM module of P-STAC can be used to provide predictions about the spread layer characteristics by varying the translation and rotational speeds of the spreader. However these simulations are computationally expensive. It can take on an average 2 h to run a single layer spreading physics-based simulation with 435,000 DEM particles having sizes in the range of 100–250 μm. Thus it is not feasible to cover the entire process parameter space at refined intervals using the polydispersed DEM model. However, a Design of Simulations (DoS), akin to Design of Experiments (DoE), can be carried out by sampling the process parameter coarsely and then training a machine learning model to regress between the results obtained by the physics-based model. The different parameters for such a DoS of spreading simulations are summarized in Table 1. Spreading cases with no roller rotation and roller rotating in clockwise direction (negative values), resulted in deposition of 2–3 powder layers over a single pass of spreader as opposed to a single layer of particles with single pass of anticlockwise rotating roller. The physics-based simulation results, as seen in Figures 6–9 (blue and black points), are highly non-linear. To gain a better understanding of the effect of these speeds on the spread layer properties M_s, Q_s, R_q and Φ, machine Learning (ML) enabled surrogate modeling can be employed to regress between these highly non-linear data points as described in the next sub-section.

Table 1. Design of Simulations (DoS) for virtual spreading of polydispersed particulate media.

Parameter	Range of Values
Spreader diameter (mm)	30
Spreader length (mm)	70
Spreader translation speed, U (mm/s)	40, 55, 70, 85, 100
Spreader rotation speed, ω (rad/s)	0, 5, 10, 15, 20, −5, −10, −15, −20
Substrate roughness S_q (µm)	79

2.3. Physics Model-Based Machine Learning Enabled Surrogate Model

A schematic of a Back Propagation Neural Network (BP-NN) with 2 inputs and 4 outputs, is shown in Figure 5. A single hidden layer was used in this study. The topology of BP-NN used in the present study is shown in Table 2. This topology is based on the neural network used in an earlier study by the authors [36]. This neural network creates a map between the spreader speeds as the inputs and spread layer properties as the outputs. Sigmoid function is used as the activation function for the hidden layer and L-2 regularization is used to avoid overfitting. The data for training and testing the neural network is obtained by running the polydispersed DEM model by following the DoS described in Section 2.2. The data from physics-based simulations is devoid of noise which can occur in data obtained using physical sensors and thus the topology of the neural network is much simpler and shallower than what might be required for a neural network trained on the data from physical experiments. The results from the ML surrogate model are discussed in the next section.

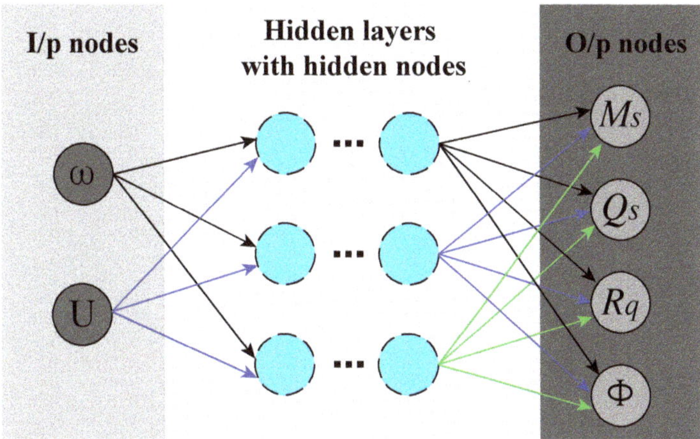

Figure 5. Schematic of a general neural network (NN) having 2 input nodes and 4 output nodes, with multiple hidden layers with each hidden layer having multiple hidden nodes. NN used in this study was comprised of a single hidden layer.

Table 2. Back Propagation Neural Network (BP-NN) parameters.

Parameter	Value	Parameter	Value
Number of training samples	35	Activation function for hidden layer	Sigmoid
Number of test samples	10	Activation function for output layer	Linear
Number of hidden layers	1	Learning rate α	0.0001
Number of hidden nodes	200	L2-regularization parameter λ	0.1

3. Results and Discussion

Regression results from the final BP-NN with parameters listed in Table 2 are shown in Figures 6–9 for spreading simulations over a rough substrate (S_q = 79 μm). Throughout, blue dots represent training data and black dots represent test data. The numerical values of the spread layer properties for the 3 sampling regions shown in Figure 4, can be seen in Figures 6–9 (points on back-most edge of the surface corresponding to U = 100 mm/s). These properties show a non-linear dependence on the speeds of the spreader. Such shearing flow of powders with polydispersity results in percolation-type segregation [39,40].

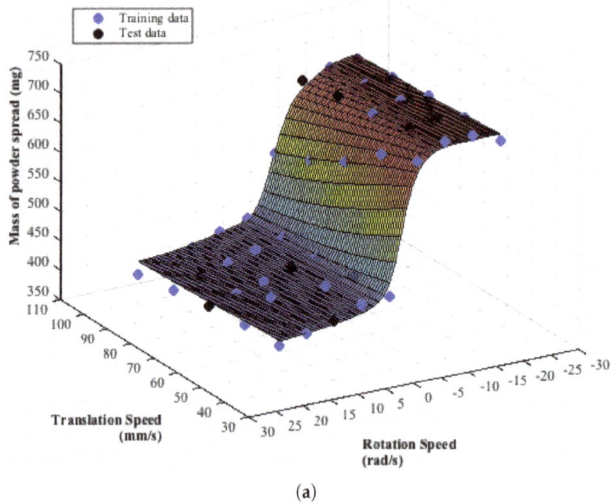

(**a**)

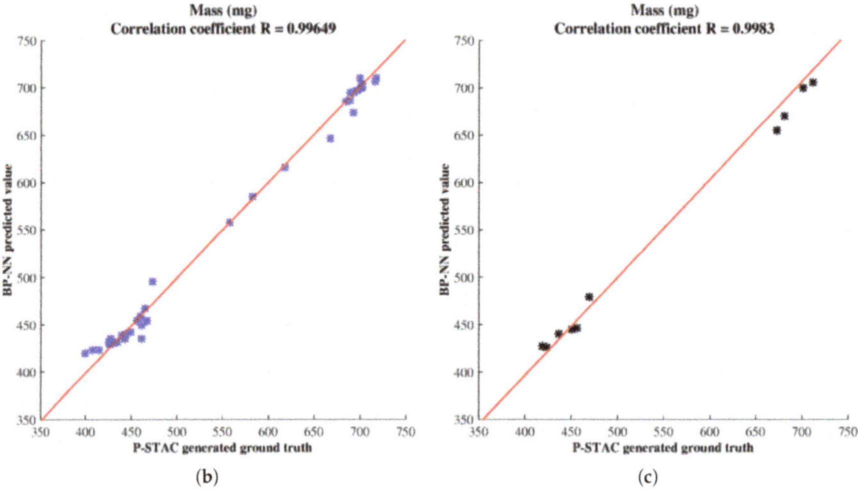

(**b**)   (**c**)

Figure 6. BP-NN predictions for the spreadability metric of mass of powder in the sampling region, M_s for the case of spreading Ti-6Al-4V powder on an additively manufactured spreading substrate with S_q = 79 μm (**a**) Regressed surface (**b**) Training data (**c**) Test data.

The surfaces predicted by this BP-NN, refer sub-figure (*a*) of Figures 6–9, nicely blanket the simulation data points, both training and test data points, generated via DoS. In these surface plots, dark blue corresponds to the lowest values and dark red corresponds to the highest values. Also shown is the correlation coefficient R between P-STAC simulation results and BP-NN predicted results (*b*) and (*c*) sub-figures of Figures 6–9. The near unity value of R for both training and test data points for each of the layer properties M_s, Q_s, R_q and Φ, suggests a near perfect regression. BP-NN trained for spreading of Ti-6Al-4V powder on 79 µm substrate is able to predict results with at least 96% accuracy.

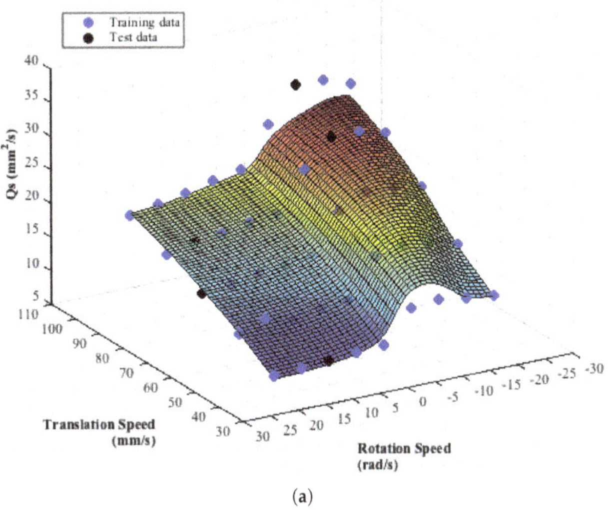

(a)

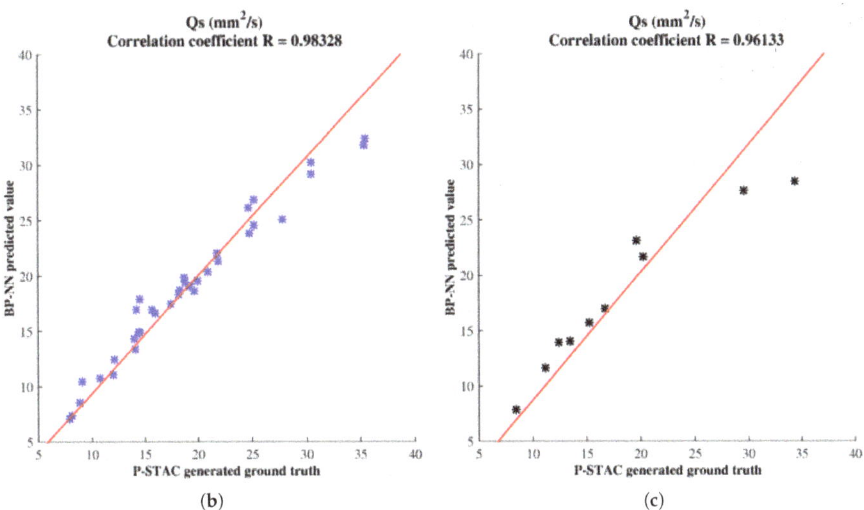

(b)   (c)

Figure 7. BP-NN predictions for the spreadability metric of spread throughput, Q_s for the case of spreading Ti-6Al-4V powder on an additively manufactured spreading substrate with S_q = 79 µm (**a**) Regressed surface (**b**) Training data (**c**) Test data.

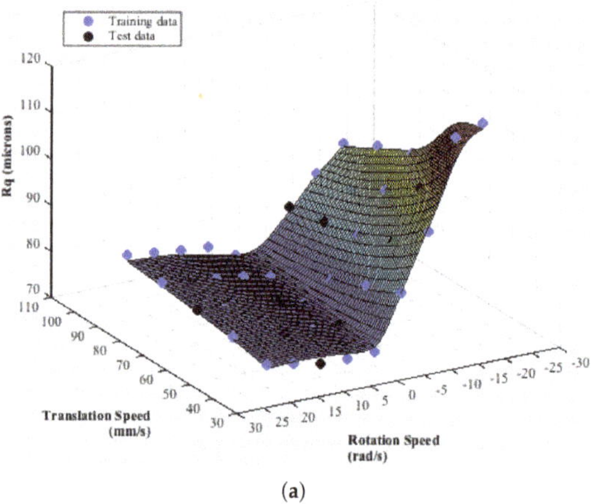

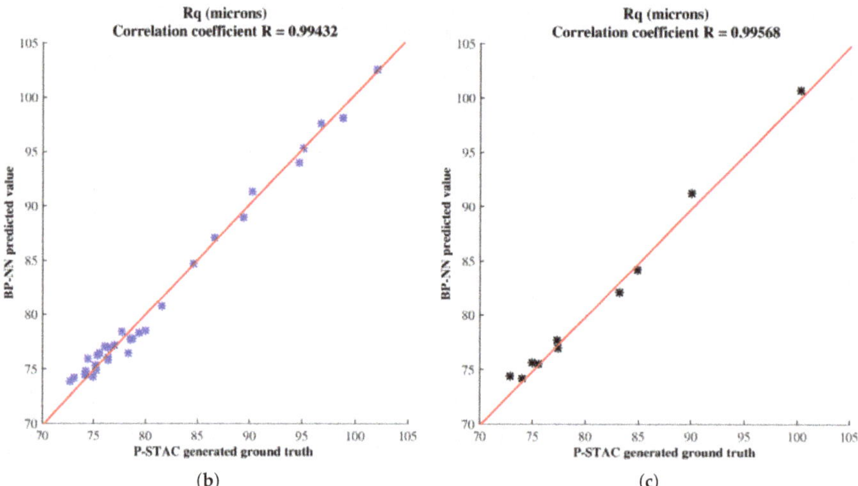

(**b**)　　　　　　　　　　(**c**)

Figure 8. BP-NN predictions for the spreadability metric of spread layer roughness, R_q for the case of spreading Ti-6Al-4V powder on an additively manufactured spreading substrate with S_q = 79 µm (**a**) Regressed surface (**b**) Training data (**c**) Test data.

Figure 6 shows the variation of mass of powder, M_s in the sampling region with spreader speeds. As the roller rotational speed decreases from a positive value to zero, M_s increases. This is due to the decrease of energy being transferred from the roller to the powder. Then as the rotational speed changes to negative values, there is a sharp increase in M_s. This is due to the deposition of more than one layer of powder. As the negative rotation speed increases in magnitude, M_s first increases then decreases. The flip occurs when the translation speed matches the rotational speed at the bottom-most point of roller (i.e., when the effective speed at the bottom-most point is zero). The dependence of M_s on translation speed of roller is maximum only for the rotational speed of -5 rad/s because for this rotational speed, the velocity of the bottom-most point of the roller varies from a negative value (-35 mm/s for translation speed of 40 mm/s) to a positive value (25 mm/s for translation speed of 100 mm/s). For the remaining rotational speeds, the variation of M_s with translation speed is minimum as the values of velocity of the bottom-most point do not vary much; they stay either all positive or all negative for a constant rotational speed.

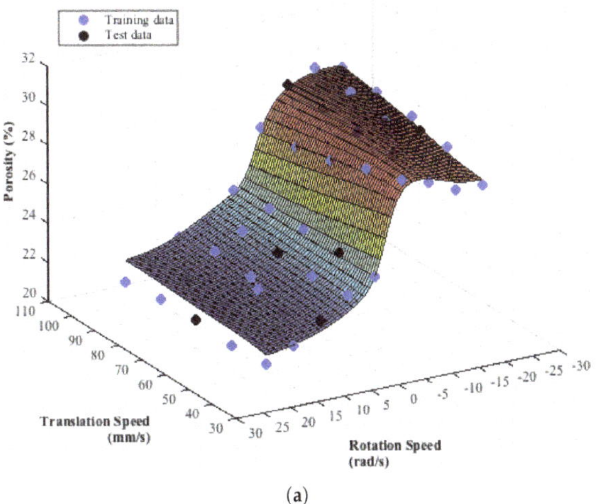

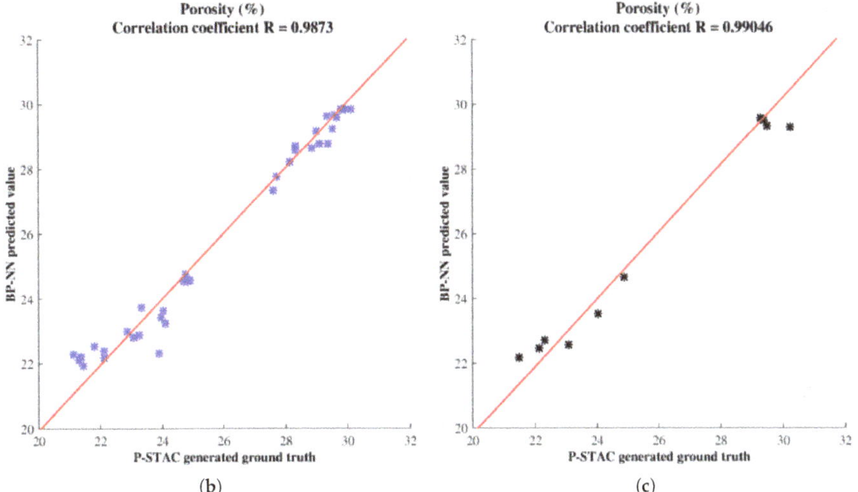

Figure 9. BP-NN predictions for the spreadability metric of spread layer porosity, Φ for the case of spreading Ti-6Al-4V powder on an additively manufactured spreading substrate with S_q = 79 μm (**a**) Regressed surface (**b**) Training data (**c**) Test data.

Figure 7 shows the variation of spread throughput, Q_s with spreader speeds. Similar to M_s, there is an increase in Q_s as the rotational speed of roller decrease from a positive value to zero. There is a sudden jump in the value of Q_s as the direction of rotation of roller changes from positive (i.e., anti-clockwise) to negative (i.e., clockwise). This is again, as was seen in the case of M_s, due to the fact that spreading with negative roller rotational speeds results in the deposition of multiple powder layers while spreading with positive roller rotational speed results in the deposition of a single layer of powder. Once again, as in M_s, as the negative rotation speed increases in magnitude, Q_s first increases then decreases. A monotonic increase in spread throughput can be seen with increasing roller translation speed. This is expected as spread throughput is directly proportional to the translation speed of roller (see Equation (5)).

Figure 8 shows the variation of spread layer roughness, R_q with spreader speeds. The spread layer roughness, for a constant translation speed, is minimum for the case where roller has zero rotation speed. The spread layer roughness increases with increase in the magnitude of the rotational speed. The increase in R_q is much more drastic for negative rotational speeds than for positive rotational speeds.

Figure 9 shows the variation of spread layer porosity, *Phi* with spreader speeds. Φ increases monotonically with increasing translation speed, at a constant rotation speed of roller. Similar to the previous spread layer properties, there is a sudden increase in layer porosity as the rotation of roller changes from positive to negative. The maximum spread layer porosity occurs when the net speed at the bottom-most point of the roller is zero.

Depending on the application of the part being 3D printed, it might be desirable to have least porosity or least roughness but always maximum spread throughput. Once a 3D printer operator knows the desired minimum porosity (or roughness) he/she can use these blanket plots to determine the U-ω pair which has the maximum throughput.

It is difficult to present the information shown in the blankets plots of Figures 6–9 on a single process map due to the higher dimensionality of this data. However, it is possible to present the information from the four layer properties on a single plot by creating hybrid properties like ((R_q normalized by the maximum (R_q)) * ((Φ normalized by maximum (Φ)) and ((Q_s normalized by the maximum (Q_s)) * ((M_s normalized by maximum (M_s)). The first hybrid property is indicative of layer defects while the second is indicative of speed of printing. Most of the applications will need minimum of the first hybrid property and maximum of the second hybrid property. A process map with these hybrid properties is shown in Figure 10. For a given R_q, Φ pair, the rightmost point on the process map corresponds to the maximum throughput or print speed. A 3D printer operator can perform a spread at a U-ω pair and examine the roughness (R_q) and the porosity (Φ) of the layer and also find the corresponding U-ω point on the spreading process map. Now, if a lower porosity and roughness is desired then he/she can slide down along the blue contour corresponding to the U value. But by doing so, the throughput decreases. The operator can sustain the spread throughput by following a vertical line through the U-ω pair and moving downwards. Thus he/she can determine the next U-ω pair at which he/she can conduct the spreading which will result in a smoother and denser layer while sustaining the throughput. Alternatively, such a process map can be installed on a 3D printer thereby providing a layer-wise control on defects as a function of spreader speeds.

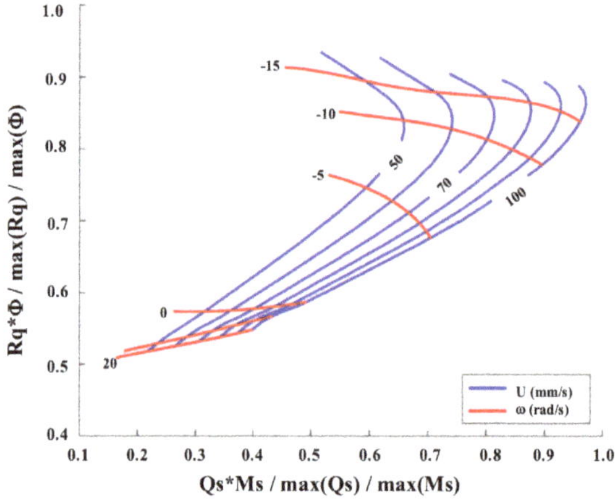

Figure 10. Spreading process map relating hybrid spread layer properties or spreadability metrics to the translation speed U and rotational speed ω of the spreader for polydispersed Ti-6Al-4V powder.

4. Conclusions

A modeling framework, comprised of physics-based modeling and physics model-based ML modeling, has been presented in this study to provide spreading predictions in powder-bed AM. The physics-based polydispersed DEM model is successfully applied to simulate the spreading of virtual Ti-6Al-4V powder, which is modeled as a collection of 235,000 smooth, spherical and cohesionless particles. This spreading model is further used to extract spread layer properties such as the mass of the powder in the sampling region, spreading throughput, layer roughness and layer porosity. A back propagation neural network is used to regress between the highly non-linear spread layer properties, with a minimum accuracy of 96%, to provide a spreading process map which can be installed on a 3D printer to offer layerwise control over spreading defects.

Author Contributions: Conceptualization, P.S.D. and C.F.H.III; methodology, P.S.D.; formal analysis, P.S.D.; investigation, P.S.D. and C.F.H.III; data curation, P.S.D.; writing—original draft preparation, P.S.D.; writing—review and editing, P.S.D. and C.F.H.III; visualization, P.S.D.; supervision, C.F.H.III; project administration, C.F.H.III; funding acquisition, C.F.H.III

Funding: This research was funded by Carnegie Mellon University and Rice University.

Acknowledgments: Authors of this work would like to acknowledge Akash Mehta for his help on P-STAC coding. We also thank Wentai Zhang for his inputs on Back Propagation Neural Network.

Conflicts of Interest: The authors declare no conflict of interest. The funders had no role in the design of the study; in the collection, analyses, or interpretation of data; in the writing of the manuscript, or in the decision to publish the results.

References

1. Frazier, W.E. Metal additive manufacturing: A review. *J. Mater. Eng. Perform.* **2014**, *23*, 1917–1928. [CrossRef]
2. ASTM International. *ASTM Committee F42 on Additive Manufacturing Technologies*; Subcommittee F42. 91 on Terminology. Standard Terminology for Additive Manufacturing Technologies; ASTM International: West Conshohocken, PA, USA, 2012.
3. Snow, Z.; Martukanitz, R.; Joshi, S. On The Development of Powder Spreadability Metrics and Feedstock Requirements for Powder Bed Fusion Additive Manufacturing. *Addit. Manuf.* **2019**, *28*, 78–86. [CrossRef]
4. Desai, P.S. Tribosurface Interactions involving Particulate Media with DEM-calibrated Properties: Experiments and Modeling. Ph.D. Thesis, Carnegie Mellon University, Pittsburgh, PA, USA, 2017.
5. Herbold, E.; Walton, O.; Homel, M. *Simulation of Powder Layer Deposition in Additive Manufacturing Processes Using the Discrete Element Method*; Technical report; Lawrence Livermore National Lab (LLNL): Livermore, CA, USA, 2015.
6. Parteli, E.J.; Pöschel, T. Particle-based simulation of powder application in additive manufacturing. *Powder Technol.* **2016**, *288*, 96–102. [CrossRef]
7. Mindt, H.; Megahed, M.; Lavery, N.; Holmes, M.; Brown, S. Powder bed layer characteristics: The overseen first-order process input. *Metall. Mater. Trans. A* **2016**, *47*, 3811–3822. [CrossRef]
8. Steuben, J.C.; Iliopoulos, A.P.; Michopoulos, J.G. Discrete element modeling of particle-based additive manufacturing processes. *Comput. Methods Appl. Mech. Eng.* **2016**, *305*, 537–561. [CrossRef]
9. Haeri, S.; Wang, Y.; Ghita, O.; Sun, J. Discrete element simulation and experimental study of powder spreading process in additive manufacturing. *Powder Technol.* **2017**, *306*, 45–54. [CrossRef]
10. Haeri, S. Optimisation of blade type spreaders for powder bed preparation in Additive Manufacturing using DEM simulations. *Powder Technol.* **2017**, *321*, 94–104. [CrossRef]
11. Mindt, H.W.; Desmaison, O.; Megahed, M.; Peralta, A.; Neumann, J. Modeling of powder bed manufacturing defects. *J. Mater. Eng. Perform.* **2018**, *27*, 32–43. [CrossRef]
12. Markl, M.; Körner, C. Powder layer deposition algorithm for additive manufacturing simulations. *Powder Technol.* **2018**, *330*, 125–136. [CrossRef]
13. Nan, W.; Pasha, M.; Bonakdar, T.; Lopez, A.; Zafar, U.; Nadimi, S.; Ghadiri, M. Jamming during particle spreading in additive manufacturing. *Powder Technol.* **2018**, *338*, 253–262. [CrossRef]
14. Nan, W.; Ghadiri, M. Numerical simulation of powder flow during spreading in additive manufacturing. *Powder Technol.* **2019**, *342*, 801–807. [CrossRef]

15. Desai, P.S.; Mehta, A.; Dougherty, P.S.; Higgs, F.C., III. A rheometry based calibration of a first-order DEM model to generate virtual avatars of metal Additive Manufacturing (AM) powders. *Powder Technol.* **2019**, *342*, 441–456. [CrossRef]

16. Escano, L.I.; Parab, N.D.; Xiong, L.; Guo, Q.; Zhao, C.; Fezzaa, K.; Everhart, W.; Sun, T.; Chen, L. Revealing particle-scale powder spreading dynamics in powder-bed-based additive manufacturing process by high-speed X-ray imaging. *Sci. Rep.* **2018**, *8*, 15079. [CrossRef] [PubMed]

17. Escano, L.I.; Parab, N.D.; Xiong, L.; Guo, Q.; Zhao, C.; Sun, T.; Chen, L. Investigating Powder Spreading Dynamics in Additive Manufacturing Processes by In-situ High-speed X-ray Imaging. *Synchrotron Radiat. News* **2019**, *32*, 9–13. [CrossRef]

18. Guo, Q.; Zhao, C.; Escano, L.I.; Young, Z.; Xiong, L.; Fezzaa, K.; Everhart, W.; Brown, B.; Sun, T.; Chen, L. Transient dynamics of powder spattering in laser powder bed fusion additive manufacturing process revealed by in-situ high-speed high-energy X-ray imaging. *Acta Mater.* **2018**, *151*, 169–180. [CrossRef]

19. Everton, S.K.; Hirsch, M.; Stravroulakis, P.; Leach, R.K.; Clare, A.T. Review of in-situ process monitoring and in-situ metrology for metal additive manufacturing. *Mater. Des.* **2016**, *95*, 431–445. [CrossRef]

20. Baumann, F.W.; Sekulla, A.; Hassler, M.; Himpel, B.; Pfeil, M. Trends of machine learning in additive manufacturing. *Int. J. Rapid Manuf.* **2018**, *7*, 310–336. [CrossRef]

21. Mies, D.; Marsden, W.; Warde, S. Overview of additive manufacturing informatics: "A digital thread". *Integr. Mater. Manuf. Innov.* **2016**, *5*, 6. [CrossRef]

22. Bock, F.E.; Aydin, R.C.; Cyron, C.J.; Huber, N.; Kalidindi, S.R.; Klusemann, B. A Review of the Application of Machine Learning and Data Mining Approaches in Continuum Materials Mechanics. *Front. Mater.* **2019**, *6*, 110. [CrossRef]

23. Kamath, C. Data mining and statistical inference in selective laser melting. *Int. J. Adv. Manuf. Technol.* **2016**, *86*, 1659–1677. [CrossRef]

24. Zhao, X.; Rosen, D.W. A data mining approach in real-time measurement for polymer additive manufacturing process with exposure controlled projection lithography. *J. Manuf. Syst.* **2017**, *43*, 271–286. [CrossRef]

25. DeCost, B.L.; Jain, H.; Rollett, A.D.; Holm, E.A. Computer vision and machine learning for autonomous characterization of am powder feedstocks. *Jom* **2017**, *69*, 456–465. [CrossRef]

26. Yao, X.; Moon, S.K.; Bi, G. A hybrid machine learning approach for additive manufacturing design feature recommendation. *Rapid Prototyp. J.* **2017**, *23*, 983–997. [CrossRef]

27. Uhlmann, E.; Pontes, R.P.; Laghmouchi, A.; Bergmann, A. Intelligent pattern recognition of a SLM machine process and sensor data. *Procedia Cirp* **2017**, *62*, 464–469. [CrossRef]

28. Gu, G.X.; Chen, C.T.; Richmond, D.J.; Buehler, M.J. Bioinspired hierarchical composite design using machine learning: Simulation, additive manufacturing, and experiment. *Mater. Horiz.* **2018**, *5*, 939–945. [CrossRef]

29. Wu, D.; Wei, Y.; Terpenny, J. Surface Roughness Prediction in Additive Manufacturing Using Machine Learning. In Proceedings of the ASME 2018 13th International Manufacturing Science and Engineering Conference, June 18–22, College Station, TX, USA, 2018.

30. Gobert, C.; Reutzel, E.W.; Petrich, J.; Nassar, A.R.; Phoha, S. Application of supervised machine learning for defect detection during metallic powder bed fusion additive manufacturing using high resolution imaging. *Addit. Manuf.* **2018**, *21*, 517–528. [CrossRef]

31. Bessa, M.; Pellegrino, S. Design of ultra-thin shell structures in the stochastic post-buckling range using Bayesian machine learning and optimization. *Int. J. Solids Struct.* **2018**, *139*, 174–188. [CrossRef]

32. Scime, L.; Beuth, J. A multi-scale convolutional neural network for autonomous anomaly detection and classification in a laser powder bed fusion additive manufacturing process. *Addit. Manuf.* **2018**, *24*, 273–286. [CrossRef]

33. He, H.; Yang, Y.; Pan, Y. Machine learning for continuous liquid interface production: Printing speed modelling. *J. Manuf. Syst.* **2019**, *50*, 236–246. [CrossRef]

34. Li, Z.; Zhang, Z.; Shi, J.; Wu, D. Prediction of surface roughness in extrusion-based additive manufacturing with machine learning. *Robot. Comput.-Integr. Manuf.* **2019**, *57*, 488–495. [CrossRef]

35. Zhang, J.; Wang, P.; Gao, R.X. Deep learning-based tensile strength prediction in fused deposition modeling. *Comput. Ind.* **2019**, *107*, 11–21. [CrossRef]

36. Zhang, W.; Mehta, A.; Desai, P.S.; Higgs, C.F., III. Machine Learning enabled Powder Spreading Process Map for Metal Additive Manufacturing (AM). In Proceedings of the 28th Annual International Solid Freeform Fabrication Symposium, Austin, TX, USA, 7–9 August 2017; pp. 1235–1249.

37. Cundall, P.A.; Strack, O.D. A discrete numerical model for granular assemblies. *Geotechnique* **1979**, *29*, 47–65. [CrossRef]
38. Zeltmann, S.E.; Gupta, N.; Tsoutsos, N.G.; Maniatakos, M.; Rajendran, J.; Karri, R. Manufacturing and security challenges in 3D printing. *Jom* **2016**, *68*, 1872–1881. [CrossRef]
39. Tang, P.; Puri, V. Methods for minimizing segregation: A review. *Part. Sci. Technol.* **2004**, *22*, 321–337. [CrossRef]
40. Dougherty, P.S.; Marinack, M.C., Jr.; Sunday, C.M.; Higgs, C.F., III. Shear-induced particle size segregation in composite powder transfer films. *Powder Technol.* **2014**, *264*, 133–139. [CrossRef]

© 2019 by the authors. Licensee MDPI, Basel, Switzerland. This article is an open access article distributed under the terms and conditions of the Creative Commons Attribution (CC BY) license (http://creativecommons.org/licenses/by/4.0/).

Article

Comparison between Virgin and Recycled 316L SS and AlSi10Mg Powders Used for Laser Powder Bed Fusion Additive Manufacturing

Shahir Mohd Yusuf, Edmund Choo and Nong Gao *

Materials Research Group, Faculty of Engineering and Physical Sciences, University of Southampton, Southampton SO17 1BJ, UK; symy1g12@soton.ac.uk (S.M.Y.); eztc1a15@soton.ac.uk (E.C.)
* Correspondence: N.Gao@soton.ac.uk; Tel.: +44-023-8059-3396

Received: 18 November 2020; Accepted: 1 December 2020; Published: 3 December 2020

Abstract: In this study, the comparison of properties between fresh (virgin) and used (recycled) 316L stainless steel (316L SS) and AlSi10Mg powders for the laser powder bed fusion additive manufacturing (L-PBF AM) process has been investigated in detail. Scanning electron microscopy (SEM), electron-dispersive X-ray spectroscopy (EDX), and X-ray diffraction (XRD) techniques are used to determine and evaluate the evolution of morphology, particle size distribution (PSD), circularity, chemical composition, and phase (crystal structure) in the virgin and recycled powders of both materials. The results indicate that both recycled powders increase the average particle sizes and shift the PSD to higher values, compared with their virgin powders. The recycled 316L SS powder particles largely retain their spherical and near-spherical morphologies, whereas more irregularly shaped morphologies are observed for the recycled AlSi10Mg counterpart. The average circularity of recycled 316L SS powder only reduces by ~2%, but decreases ~17% for the recycled AlSi10Mg powder. EDX analysis confirms that both recycled powders retain their alloy-specific chemical compositions, but with increased oxygen content. XRD spectra peak analysis suggests that there are no phase change and no presence of any undesired precipitates in both recycled powders. Based on qualitative comparative analysis between the current results and from various available literature, the reuse of both recycled powders is acceptable up to 30 times, but re-evaluation through physical and chemical characterizations of the powders is advised, if they are to be subjected for further reuse.

Keywords: virgin; recycled; metal powders; laser powder bed fusion; additive manufacturing

---

1. Introduction

Laser powder bed fusion (L-PBF) is one of the main categories of metal additive manufacturing (AM) technology, in which selective laser melting (SLM) is a widely used technique under this category. In the SLM process, a focused laser beam is used to selectively melt multiple layers of powder bed in succession and produce complex, functional metallic components according to the initial computer-aided-design (CAD) file [1]. To date, a wide range of metals and alloys have been successfully fabricated by SLM, including 316L SS, 304L SS, 15-5 PH SS, AlSi10Mg, Scalmalloy (AlMgSc), CoCr, IN 718, Ti6Al4V, and CuSn10 for various functional use, especially in the aerospace, biomedical, and automotive industries.

To date, a significant amount of research on metal AM has focused on the process–microstructure –property relationship. In particular, emphasis is given on the influence of various processing parameters such as laser power, scan speed, scan line (hatch) spacing, and layer thickness on the microstructures and the resulting properties such as fracture, yield, and tensile strengths, fatigue life, as well as corrosion and tribological properties [2–8]. Studies have shown that AM-fabricated metallic components can possess similar or even enhanced properties, compared to conventionally manufactured (CM) parts [9–12]. Such improvements in these properties, e.g., strength, have been generally attributed to the unique multi-scale microstructures, much finer grain sizes, and the presence of ultrafine/nano-sized precipitates [13–20]. Furthermore, efforts for the standardization and qualification of metal AM parts have been actively carried out by organizations such as the American Society for Testing and Materials (ASTM) and the National Institute of Standards and Technology (NIST) by publishing various standards pertaining to additively manufactured metallic components [21–27].

While research into the process–microstructure–property relationship of AM metallic parts is certainly important, another equally essential aspect that has been given less attention is the properties of the metal powder used for AM processing, i.e., the feedstock materials. So far, available studies in the literature have revealed that the quality of these feedstock materials play a major role in determining the quality of the as-built AM parts, particularly densification levels and porosity contents [28–30]. In fact, the quality of metal powders used for AM is often assessed through their physical properties such as particle size, particle size distribution (PSD), morphology, packing density and flowability, as well as their chemical composition. In turn, the quality of metal powders is significantly determined by the powder manufacturing method, in which water atomization (WA), gas atomization (GA), and the plasma rotating electrode process (PREP) are prevalent techniques [28,31–35]. Nevertheless, for L-PBF AM, high packing density and high flowability are desired to ensure optimal powder melting and homogeneous deposition of the powder bed layers [36]. Thus, in terms of morphology, smooth, spherical particles with as little satellites/defects as possible are desirable [29,37–39]. In addition, Hajnys et al. [37], Liu et al. [40], and Benson et al. [41] explained that a wide PSD with a mix of large and fine particle sizes can improve the packing density of the powder bed through the accumulation of the fine powders within the interstitial spaces that exist around the larger powder particles. However, some studies have also shown slightly reduced flowability when a wide PSD is used for AM fabrication due to increased inter-particle friction, thereby suggesting that some compromise is needed when selecting the particle sizes to attain the highest possible packing density while maximizing flowability [42–44]. On the other hand, several researchers recommend different nominal powder particle sizes to be used as the starting feedstock for L-PBF AM processes, e.g., 10–60 [28], 10–45 [29], and 5–80 μm [45]. Nonetheless, regardless of the particle size range chosen for AM processing, the powder particles need to be spherical and free from defects as mentioned previously.

On the other hand, back in 2016, Honeywell projected that the metal AM industry will experience an immense growth amounting to $3.1 billion over the next 5–10 years [46]. However, the relatively high cost of virgin powders for L-PBF AM (~$100–$200/kg) and high usage volume per build (~40–50 kg is needed to fill the powder dispenser of the L-PBF machine) could be a major obstacle that could increase the overall production cost along the supply chain [47–52]. Thus, it is often economically viable to reuse/recycle the powders for further build cycles. However, the recycling of used powders may induce alterations in their physical properties and chemistry, which could affect the eventual mechanical properties of the as-built parts [27]. In fact, studies have shown that the recycled powders will almost always have larger particle sizes, rougher surface areas, reduced particle circularity, increased contamination, and a higher uptake of oxygen, nitrogen, or other inert gases regardless of the number of reuse times [53–61].

For example, Zhang et al. [47] attributed the rougher texture with the increased number of satellites in recycled 15-5 PH SS powder used for L-PBF to the repeated heating/cooling cycles during processing and increased inter-particle friction during powder bed recoating, as compared to its virgin counterpart. Galicki et al. [56] and Heiden et al. [57] both reported increased uptakes of oxygen in recycled 316L SS powder due to the oxidation that occurs during the manufacturing and the presence of residual oxygen in the build chamber. Similarly, Gorji et al. [38] detected an increase in the metal oxide content in recycled 316L SS powder, while the initially spherical particles evolved into a more irregularly shaped morphology, together with the presence of agglomerates and spatter. However, Mellin et al. [62] and Lutter-Gunther et al. [63] discovered an insignificant change in oxidation level in recycled 316L SS powders, and hypothesized that the elements that have high affinity to oxygen such as

Cr, Mn, and Fe diffuse out of the powder surface during the AM fabrication process. Nevertheless, any increased oxygen absorption and oxide formation need to be taken into consideration before reusing the recycled powders as they may adversely affect the performance of the eventual as-built parts during service [28,29]. On the other hand, Hajnys et al. [37] observed an increase in the PSD of recycled 316L SS powder but considered further reuse to be acceptable. Similarly, although Nezhadfar et al. [64] observed more of satellites and porosity on the surface of recycled 316L SS powder after 12 times of reuse, they reported an inconclusive trend in the ductility and mechanical strength in the solidified parts. Interestingly, some studies have also shown that a phase change could occur in recycled 316L SS and other SS alloys after >10 times of reuse without any considerable changes in the physical and chemical properties, as well as apparent density and hardness, implying that the recycled powders can still be reused for further build cycles [57,65].

Moreover, other direct investigations on the packing density and flowability of virgin and recycled powders for L-PBF based on established ASTM standards have shown some interesting results. For example, Seyda et al. [66] observed an increased flowability of recycled Ti6Al4V powder after 12 reuse times, which was attributed to the reduced number of satellites and fine particles despite the coarsening of the powder particles, which become increasingly irregularly shaped with increasing PSD values [66]. Quintana et al. [55] reported similar results in recycled Ti6Al4V powder with increasing content of oxygen from 0.09 wt.% to 0.13 wt.% after 31 reuse times. They even reported comparatively higher tensile strength in the parts built using the recycled powder due to the oxide strengthening effect. However, Tang et al. [67] revealed that while the Ti6Al4V powder particles became rougher and less spherical with fewer satellites after 21 build cycles, the PSD actually became narrower but still exhibited improved flowability without any compromise on the tensile properties in the as-built parts. In addition, Cordova et al. [36] studied the physical and chemical properties, as well as packing density and flowability, of virgin and recycled Ti6Al4V, AlSi10Mg, Scalmalloy, and IN 718 for various reuse times. In general, the recycled powders displayed an increased flowability and maintained a high packing density due to the reduced number of satellites and fine particles. They concluded that these would enable optimal powder melting and subsequently homogeneous deposition of the powders on the build platform.

Hence, these studies have shown the importance of evaluating and understanding the changes that occur in powders after undergoing numerous build cycles in terms of quality control and consideration for further reuse of the recycled powders for AM fabrication. However, comprehensive studies are still limited in terms of the correlation of the changes in the physical and chemical properties of virgin and recycled metal powders for AM processing, and their feasibility for further reuse [24,36,55]. Therefore, in this study, the physical and chemical properties of virgin and recycled 316L SS and AlSi10Mg powders used particularly for SLM processing were characterized and compared via the following approaches: (i) Microstructural observations through scanning electron microscopy (SEM), (ii) chemical composition analysis through electron-dispersive X-ray spectroscopy (EDX), and (iii) phase analysis through X-ray diffraction (XRD). The results from this study were then compared with those in the available literature to qualitatively assess the suitability of these recycled powders for further reuse.

2. Materials and Methods

Four batches of gas-atomized 316L SS and AlSi10Mg powders (2 fresh sets (virgin), and 2 used sets (recycled)) were used in this study. The fresh powders were obtained in their as-received state from the supplier, Concept Laser, while the recycled powders were taken from the powder dispenser container in the Laser Cusing M2 SLM machine. The following are the chemical compositions of the virgin powders as listed by Concept Laser (in wt.%): (i) 316L SS; Cr: 16.5–18.5; Ni: 10.0–13.0; Mo: 2.0–2.5; Mn: 0–2.0; Si: 0–1.0; P: 0–0.045; C: 0–0.03; S: 0–0.03; Fe: *Bal.* and (ii) AlSi10Mg; Si: 10.5–13.5; Mg: 0–0.05; Fe: 0–0.55; Mn: 0–0.35; Ti: 0–0.15; Cu: 0–0.05; Zn: 0–0.10; C: 0–0.05; Ni: 0–0.05; Pb: 0–0.05; Sn: 0–0.05, Al: *Bal.*

The recycled powders have been used about 20 times prior to collection. It should be noted that during the SLM process, the build chamber was purged with nitrogen to limit contamination and oxidation as much as possible. In addition, the recycled powders were also sieved for particles larger than 80 µm for 316L SS, and 50 µm for AlSi10Mg prior to characterization.

The size, morphology, circularity, and distribution of the virgin and recycled powder particles were evaluated via JSM-JEOL 9500 scanning electron microscopy (SEM, JEOL Tokyo, Japan) observations combined with analysis of the SEM images using ImageJ software (V 1.52, NIH and LOCI, Wisconsin, WI, USA). The changes in chemical composition between the virgin and recycled powders were assessed using EDX analysis, while phase evolution was determined through X-ray diffraction (XRD) analysis using a Bruker D2 Phaser powder diffraction machine (θ/2θ configuration, range: 40–100°, Bruker, Billerica, MI, USA).

3. Results and Discussion

3.1. Microstructural Analysis

3.1.1. Morphology and Distribution

The morphology and cumulative particle size distribution (PSD) of the virgin and recycled 316L SS and AlSi10Mg powders are shown in Figure 1. It is clear that the powder particles for both virgin powders are largely spherical or near-spherical (Figure 1a,d), but some irregularly shaped particles, such as cylindrical and tear-drop morphologies, can be observed for the virgin AlSi10Mg powders (Figure 1d). The surfaces of the recycled powders also appear rougher compared to the initially smooth virgin powders. In addition, fine particles called satellites (<5 µm) are attached to the surface of the virgin 316L SS and AlSi10Mg powders (e.g., dashed arrows of the zoomed areas in Figure 1a,d), which is common in virgin metallic powders produced by the gas atomization process [37,42,68]. However, fewer satellites are observed in both recycled powders as shown in Figure 1b,d, often attributed to their detachment from the surface of unmelted powder particles when they are being blown off from the as-built AM parts [27,55,67].

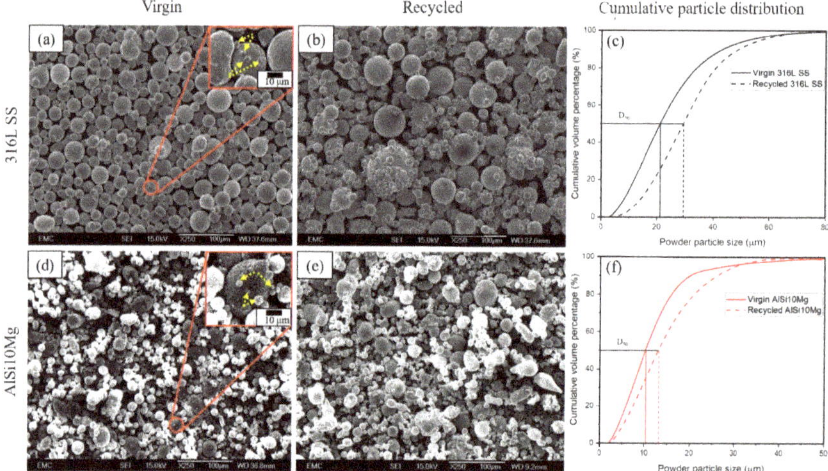

Figure 1. (**a**) Virgin, (**b**) recycled, and (**c**) cumulative particle size distribution of 316L SS powders; (**d**) virgin, (**e**) recycled, and (**f**) cumulative particle distribution of AlSi10Mg powders. Dashed arrows of the zoomed areas inset of (**a,d**) show satellites on the surface of the virgin powders.

In addition, the agglomeration of small powder particles on the larger ones could be observed for the recycled 316L SS, as shown in Figure 1b. Such agglomerates are typically caused by the spattering of unmelted powders (powder spatter) from the powder bed upon contact with the laser heat source during SLM fabrication, which are then attached to other unconsolidated or semi-consolidated powder particles [36–38,47]. Similarly, other studies have shown that such agglomerates could also arise from melted powder particles that are ejected from the powder bed (droplet spatter), which eventually solidify during the flight and then either adhere to the as-fabricated part or impinge on other powder particles [69–72].

Although the recycled 316L SS powders largely retain their spherical/near-spherical shape (Figure 1b), the recycled AlSi10Mg powders generally deform into more irregular shapes (Figure 1e). Interestingly, the powder particles are apparently larger for both sets of recycled powders (Figure 1b,e) compared to their virgin counterparts (Figure 1a,d). This is also quantified by the cumulative distribution function showing the average particle size (D_{50}) of the virgin and recycled powders (Figure 1c,f). The values of D_{50} for virgin and recycled 316L SS powders are ~22 μm and ~29 μm, respectively, which are higher than those of the virgin and recycled AlSi10Mg powders: ~11 μm and ~13 μm, respectively. These could be ascribed to the melting of smaller powder particles during the previous SLM processing [47]. The higher D_{50} values for the recycled powders in this study are similar to those obtained for various recycled AM metal powders observed in other studies, which often show a wider distribution and variation in particle sizes [36,58,59]. Nevertheless, Figure 2 indicates that the 316L SS powder consists of a wide range of particle sizes with broad size distribution (0–80 μm) when compared with the AlSi10Mg powder that comprises relatively finer particle sizes with narrower size distribution (0–50 μm).

On the other hand, several noteworthy observations can be made from the morphologies of the recycled 316L SS and AlSi10Mg powders shown by the high-magnification SEM images in Figure 2. For example, Figure 2a,d show the deformation of powders into more irregular shapes and visibly rougher surface texture compared to the initially smooth virgin powder particles. A number of factors have been related to these occurrences, including: (i) Continuous melting and cooling cycles of AM processing, (ii) remelting of powders upon exposure to the laser heat source, (iii) friction among powder particles during the spreading of successive powder bed layers, and long exposure times to high temperatures during the fabrication process [29,36,67,73].

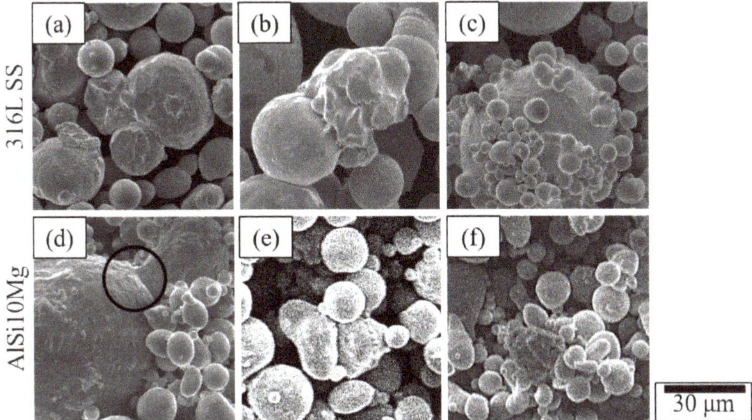

Figure 2. Examples of powder morphologies for recycled 316L SS (**a–c**) and AlSi10Mg (**d–f**): (**a,d**) Deformed powders, rough surface, and sintering necks (circled area in Figure 2d); (**b,e**) incomplete powder melting and fusion; and (**c,f**) powder agglomeration due to spatter.

In addition, a sintering neck can be observed to form between two adjacent powder particles of the recycled AlSi10Mg powders, as shown by the circled area in Figure 2d. Similarly, Figure 2b exhibits the fusion of a partially melted powder particle with an unmelted one for the recycled 316L SS, while an example of the incomplete fusion of differently sized powders in recycled AlSi10Mg is shown in Figure 2e. Altogether, the occurrence of these phenomena can be attributed to the insufficient heat input and considerable temperature difference between loose powders and powders that are already fused [74–76]. They have been identified as undesirable defects that contribute not only to the surface roughness of as-built AM parts, but also, more importantly, to the porosity that could compromise their strength and service properties [6,67,68,77].

Furthermore, the agglomeration of powder particles due to droplet and powder spatter, previously observed in Figure 1b,e, are shown in more detail in Figure 2c,f for recycled 316L SS and AlSi10Mg powders, respectively. In both cases, the size of the agglomerates is measured as >45 μm. For 316L SS, much smaller powder particles ranging from 5 to 15 μm are impinged on larger powder particles of >45 μm. As for AlSi10Mg, particles with sizes of 5–20 μm are observed to be clustered together, whereas larger particles remain either free from such agglomerates, or only have fewer satellites attached onto them. Interestingly, the clumped particles are of different morphologies that include spherical, near-spherical, and irregularly shaped ones. This is in contrast with the agglomerates in recycled 316L SS, in which much smaller spherical or near-spherical powder particles are attached to the larger powder particles with similar morphologies. These can be ascribed to the higher tendency of fine particle sizes with narrow size distribution to agglomerate, compared to those possessing comparatively larger particle sizes and wider size distribution, as is the case for the AlSi10Mg and 316L SS powders in this study, respectively [78,79].

Moreover, the histograms in Figure 3 reveal the PSD of the virgin and recycled 316L SS and AlSi10Mg powders in more detail. Both virgin and recycled 316L SS powders (Figure 3a,b) exhibit a comparatively wider size distribution than those of AlSi10Mg powders (Figure 3c,d), ranging from 0 to 80 μm and 0 to 50 μm, respectively. A clear shift toward higher PSD values, i.e., increasing percentages of larger particle sizes, can be observed: >25 μm for 316L SS (Figure 3b) and >10 μm for AlSi10Mg (Figure 3d). At the same time, fewer fine particles ranging from 0 to 5 μm can be observed in the recycled powders (Figure 3b,d). This confirms the higher D_{50} values observed for the recycled powders shown by the cumulative PSD graphs Figure 1c,f, which can be attributed to the presence of agglomerates and the removal of satellites during AM fabrication [36,67].

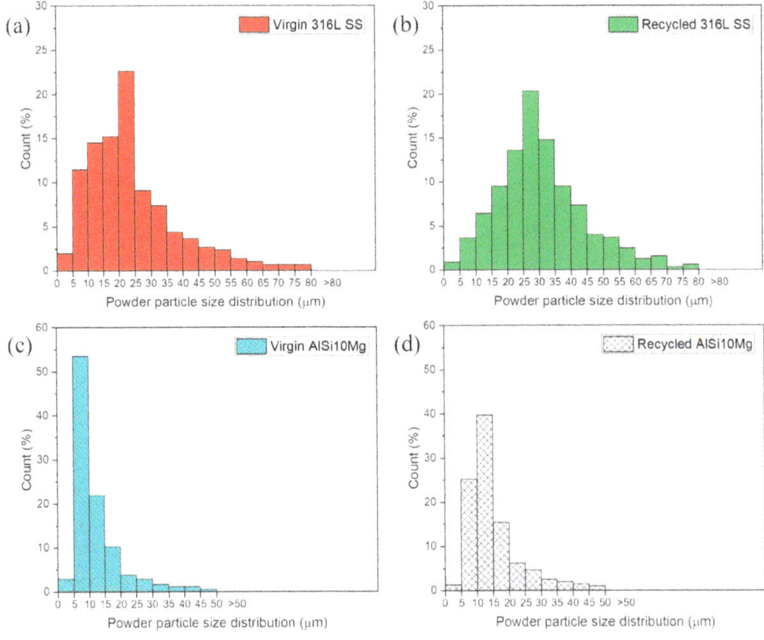

Figure 3. Powder particle size distribution of virgin and recycled 316L SS (**a,b** respectively) and AlSi10Mg (**c,d** respectively) powders.

3.1.2. Circularity

In addition to SEM observations that provide visual and qualitative assessments of the powder particle shapes, the circularity factor, *C*, is introduced to evaluate the morphology of the virgin and recycled 316L SS and AlSi10Mg powders, given by the formula [80]:

$$C = 4\pi A/P^2 \tag{1}$$

where *P* is the perimeter (μm) and *A* is the pore area (μm^2), both determined from the ImageJ software. *C* values close to 1 indicate near-perfect circular particles, while *C* values close to 0 demonstrate increasingly irregularly shaped particles. The results of the calculations are shown in Table 1.

Table 1. Average circularity of virgin and recycled 316L SS and AlSi10Mg powders.

Powder	Circularity		Percentage Reduction (%)
	Virgin	Recycled	
316L SS	0.89 ± 0.11	0.87 ± 0.13	2.3
AlSi10Mg	0.83 ± 0.17	0.69 ± 0.31	16.9

It is clear that the virgin powders possess higher average circularity values than the recycled powders, with a slightly higher value in 316L SS (0.89 ± 0.11) than AlSi10Mg (0.83 ± 0.17). This suggests rounder and more regular shapes in the virgin 316L SS compared to the virgin AlSi10Mg, confirming the SEM observations in Figure 1a,d. The average circularity of recycled 316L SS powder remains virtually unchanged at 0.87 ± 0.13, while that of recycled AlSi10Mg powder decreases to 0.69 ± 0.31. This correlates well with the observations in Figure 1b,d, in which the 316L SS powder particles largely retain their spherical or near-spherical shapes, whereas the AlSi10Mg powder particles tend to deform into more irregular shapes upon reuse for SLM fabrication. On the other hand, representative visual examples of powder particle circularity are shown in Figure 4. A range of morphologies can be observed with the corresponding circularity, with the highest circularity, C = 1, obtained from virgin 316L SS powder (Figure 4a), and the lowest circularity was evaluated as C = 0.38 from recycled AlSi10Mg powder (Figure 4e).

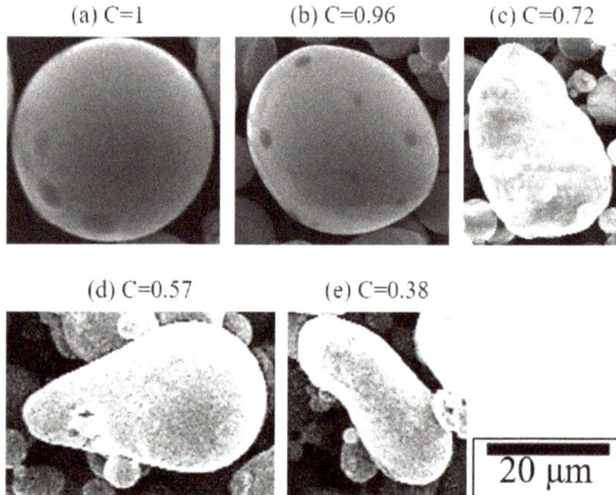

Figure 4. Examples of particle circularity for the metal powders used in this study. Representative images of (**a**,**b**) are taken from 316L SS, while those of (**c–e**) are taken from AlSi10Mg.

3.2. Compositional Analysis

Semi-quantitative EDX analysis was conducted on the surface of the virgin and recycled 316L SS and AlSi10Mg powders to determine their chemical compositions, and the results are presented in Table 2.

Table 2. Chemical composition of virgin and recycled powders via electron-dispersive X-ray spectroscopy (EDX) analysis (wt.%).

Element	316L SS		AlSi10Mg	
	Virgin	**Recycled**	**Virgin**	**Recycled**
Fe	61.71 ± 0.08	60.38 ± 0.05	-	-
Ni	12.92 ± 0.04	11.89 ± 0.05	-	-
Cr	18.43 ± 0.06	17.55 ± 0.03	-	-
Al	-	-	86.97 ± 0.07	86.16 ± 0.05
Si	0.94 ± 0.01	1.11 ± 0.02	11.09 ± 0.04	10.43 ± 0.03
Mg	-	-	0.45 ± 0.03	0.40 ± 0.04
Mn	1.96 ± 0.03	2.18 ± 0.01	-	-
Mo	2.46 ± 0.02	2.21 ± 0.03	-	-
O	1.42 ± 0.07	4.68 ± 0.09	1.49 ± 0.08	3.01 ± 0.08

It can be observed that the composition of the main elements in both virgin powders determined via EDX analysis (Table 2) corresponds well and is within the range of the supplier-listed compositions detailed in Section 2. In addition, the composition of most elements for the virgin and recycled powders only fluctuates slightly and remains stable, except for oxygen. For 316L SS, the oxygen content in the recycled powder increases by ~3 times compared to the virgin powder, whereas it doubles that of the virgin powder for the recycled AlSi10Mg. The increase in oxygen content in the recycled powders is expected and has been observed in various metal powders that have undergone AM processing in other studies [36,47,55,67]. The higher percentage of oxygen in recycled AM powders can be attributed to a number of factors, including: (i) Oxygen pick-up due to contaminants, moisture, and residual oxygen that might still be present within the build chamber despite being purged with protective inert gases such as argon and nitrogen (this study); (ii) vaporization of oxygen (a trace element) from the powder surface that might already exist during powder production upon contact with the laser heat source; and (iii) the exposure of powder to ambient air during their removal from the as-received part, during the sieving process, as well as during powder handling and storage [36,38,67].

In particular, several studies in the field of powder metallurgy have also shown the strong affinity of major alloying elements such as Mn and Si to oxygen and the selective oxidization on the powder surface [81–83], which might also be the case for the 316L SS and AlSi10Mg powders subjected to AM fabrication in this study, respectively. In fact, Simonelli et al. [69] found that the high affinity of Mn and Si to oxygen was the reason that both elements were selectively oxidized on the surface of molten material spatter-ejected from the powder bed, which then solidify during flight in the build chamber. Nevertheless, further study could include EDX map analysis for visual confirmation of the presence of these oxides on the surface of these powders. In addition, an important thing to note is that EDX is commonly used as a semi-quantitative tool to determine the chemical composition of a material, but the exact values detected might not be accurate, particularly for trace elements such as oxygen [42,47,61]. Thus, more detailed surface chemical analysis is necessary, such as by using X-ray photoemission spectroscopy (XPS) and inductively coupled plasma mass spectrometry (ICP-MS), to obtain a more accurate reading of the chemical compositions, as well as to detect thin surface oxide layers that may be present on the surface of the powders [47,70]. Nonetheless, the EDX analysis in this study adequately captures the trend of increasing oxygen content in the recycled powders, and suggests that the deviations in chemical compositions are within the acceptable range for both materials [84,85].

3.3. Phase Analysis

The phase identification of the virgin and recycled 316L SS and AlSi10Mg powders is shown by the XRD spectra in Figure 5.

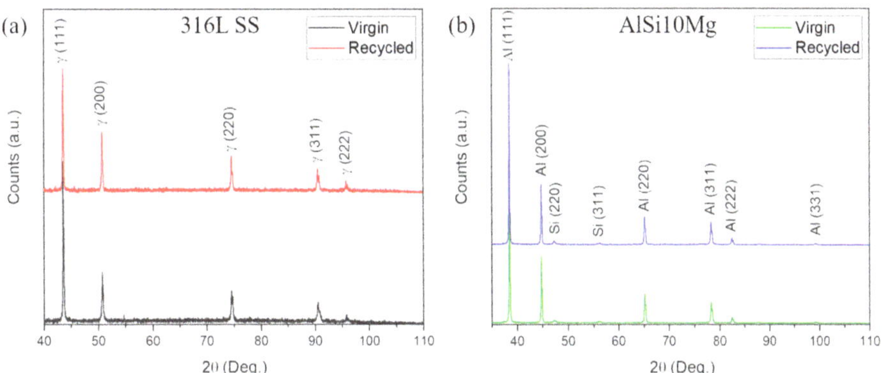

Figure 5. XRD spectra of virgin and recycled 316L SS (**a**) and AlSi10Mg (**b**) powders.

Figure 5a shows no phase change for both virgin and recycled 316L SS powder, such that they remain a single γ-austenite FCC structure. Similarly, the recycled AlSi10Mg powder exhibits identical peaks of Al and Si as the virgin counterpart (Figure 5b). No additional peaks can be observed, suggesting that any change in phase (if any) is below the XRD detection limit and represents very little contamination within the build chamber, and therefore, it is negligible [38,70]. However, the hypothesis of insignificant contamination during the AM build cycle might seem paradoxical, seeing that the EDX results show increasing oxygen content in the recycled powder that is often associated with the presence of contaminants inside the build chamber. Thus, the more reasonable explanation is that even though no phase change can be observed in the XRD spectra, it could still occur but below the limit of detection of the XRD equipment used in this study. Furthermore, nitrogen is known as an austenite stabilizer and, especially for 316L SS, the presence of nitrogen in the build chamber during AM processing might even suppress the formation of other phases despite the three-fold increase in oxygen content detected by EDX analysis in the recycled powder.

In fact, Heiden et al. [57] conducted electron-backscattered diffraction (EBSD) and XRD analyses on virgin and recycled 316L SS and observed the presence of the δ-ferrite BCC phase in the recycled powder (reused for 30 times). Jacob et al. [65] observed a progressive increase in ferritic-martensitic BCC phase in 17-4 PH stainless-steel powder via XRD analysis after a minimum use of four times. In addition, Simonelli et al. [69] detected an enrichment of thin Mn and Mg oxide films on the surface of recycled 316L SS and AlSi10Mg powders through EDX maps analysis, respectively. They attributed the growth of these oxide films to the volatile nature of Mn and Mg and their high affinity toward oxygen [5,60,86]. Therefore, future investigations could include carrying out EDX mapping, electron-backscattered diffraction (EBSD), and transmission electron microscopy (TEM) analyses to determine sub-micron-scale phase and microstructures, which can definitively confirm whether other phases can be observed in the recycled powders.

4. Discussion

It is known that flowability and packing density are two of the most important qualities of metal powders that are to be used for PBF AM processing as they have considerable influence on the quality of as-built parts, together with the processing parameters applied [42,87,88]. Altogether, these two properties are determined by the particle size, PSD, morphology, and chemical composition of the powders [61,66,70]. However, with increasing number of uses, the virgin powders with initially favorable properties will undergo physical and chemical changes that require further evaluation before a decision can be made about whether they can be reused. Based on the guidelines proposed by Cordova et al. [36], it is recommended to consider the following five key aspects in assessing the suitability of reusing the powders, including: (i) Particle size distribution (PSD), (ii) powder morphology, (iii) chemical composition, (iv) flowability, and (v) packing density.

First, the nominal powder PSD typically used for SLM is in the range of 10–45 μm, which is much finer than those used for other PBF processes, such as electron beam melting (EBM) (45–106 μm) and directed energy deposition (DED) (20–200 μm) [28,29,89]. This is because of the architecture of SLM machines that only allows for fine powder distributions to reduce surface roughness by utilizing lower layer thicknesses, as compared with those of EBM and DED counterparts that enable the use of higher layer thicknesses and larger particle sizes [90]. Based on the results in this study, the PSD for both recycled powders fall within this range, up to 80 μm for 316L SS and up to 50 μm for AlSi10Mg. In particular, the recycled 316L SS powder can be further sieved to remove the agglomerates and unmelted particle sizes of >45 μm.

Secondly, powder morphology has been identified as another important aspect that needs to be examined as it significantly influences flowability and packing density. Smooth, spherical, and dry particles possess the best flowability characteristics due to the low inter-particle friction that enable the deposition of densely solidified layers and produces homogeneous microstructures [42,91]. On the other hand, powder particles with rough surfaces and more irregular shapes reveal larger specific surfaces that increase surface friction between them, typically reducing flowability and leaving voids and gaps that lead to porosity in the as-built parts [47,68]. Therefore, to be considered for further reuse, the guidelines provided by Cordova et al. [36] recommended that the circularity factor of the recycled powders needs to be as close to 1 as possible, or at least ≥7.0. In this study, there are no issues with the recycled 316L SS powders as the average circularity is ~0.87, which does not differ much from its virgin counterpart (~0.89). However, the average circularity of recycled AlSi10Mg is ~0.69, which is on the borderline value suggested. Thus, the recycled AlSi10Mg might still be considered for a few more uses, but only for noncritical components outside of high-risk industries, e.g., in the aerospace sector and some biomedical applications [38]. Furthermore, reuse after >30 build cycles is generally not recommended, due to severe degradation that deteriorates the quality of the metal powders, as has often been observed in powders of various metallic materials, including considerable agglomerates and satellites, and noticeable deformation and distortion into highly irregular shapes [55,57,68,92].

Thirdly, the chemical composition of the recycled powders must be measured to ensure that they remain within alloy-specific compositions, and to address the issues of losses due to evaporation such as spatter, contamination within the build chamber and during powder handling and recovery, as well any pick-up of oxygen or other trace elements such as nitrogen or argon during the AM build process [29,36]. In addition, no (significant) phase change should occur in the recycled powder, as any changes in the crystal structure, e.g., through additional peaks in the XRD spectra, or any precipitates and oxide layers detected from TEM or XPS, would result in considerable contamination that could negatively impact the structural and mechanical properties of the fabricated parts [38,70]. In the present investigation, EDX analysis shows a 2–3 times increase in oxygen content for both recycled powders, which could be caused by residual oxygen present in the chamber, or evaporation of entrapped oxygen within the powders (from the gas atomization powder production process) due to powder and droplet spatter [1,93]. Nevertheless, the XRD spectra analysis does not show any unexpected peaks in the recycled powders, and their chemical compositions remain within the acceptable range of both 316L SS and AlSi10Mg alloys. Therefore, based on the EDX and XRD results, at least, the further reuse of both recycled powders is considered acceptable. However, if more detailed analyses such as through XPS, EBSD, EDX mapping, and TEM reveal undesired precipitates, oxides, or phases in the recycled powders, then the recycled powders should not be used for any structural or other critical end-use applications.

The last two guidelines from Cordova et al. [36] recommend the flowability and packing density of the recycled powders to be directly measured through analytical tools following the various ASTM or other international standards. High flowability and high packing density are desired to ensure homogeneous deposition of powder bed layers and optimal powder melting, which would lead to high densification and low porosity levels in the as-built parts. In particular, the packing density of the powder bed has been found to be more important than processing parameters such as laser power (P), scan speed (S), hatch spacing (d), and layer thickness (h) in determining part density and porosity [36,42]. Smooth and spherical particles with minimal defects are known to improve packing density, while irregularly shaped particles with numerous defects such as agglomerates and satellites reduce the packing density of the powder bed, similar to their influences on powder flowability [29,39,94]. In addition, the inclusion of fine powder particles (fines) also contributes to high packing density due to the fine particles filling the interstitial spaces that exist among the larger particles [40,44,90]. However, these fine particles tend to result in wider PSD and have been found to reduce powder flowability [30,36,55,67]. Therefore, as a compromise, it is suggested that smooth, spherical particles with as minimum defects as possible with narrower PSD should be used for AM fabrication to attain optimized flowability with relatively high packing density [29,37,42].

In this study, direct measurements of flowability and packing density were not carried out. However, comparative analysis between the current results and discussions with the first three recommendations from Cordova et al. [36], as well as with established literature, can still be made to provide qualitative assessments for the viability of reusing the recycled powders in this study (after 20 times of use). The properties of the recycled powders analyzed in this study, powder characteristics that are suitable for reuse as suggested by available literature, and recommendations from the authors regarding the possible further reuse of the recycled powders are listed in Table 3. Based on the comparative analysis, the authors suggest that the reuse of the recycled 316L SS and AlSi10Mg powders is acceptable up until 30 build cycles overall, but further reuse beyond that should require re-evaluation through further detailed physical and chemical characterizations.

Table 3. Qualitative analysis on the recommendation of reuse of the recycled powders in the present study based on results established in available literature.

Powder Properties	Recommendation from Literature	Properties of Recycled Powders in This Study		Overall Recommendation from Current Authors
		316L SS	AlSi10Mg	
PSD	10–45 µm; Narrow PSD; reduced fine particles [29,36,37,42]	0–80 µm (Wide PSD); reduced fine particles of sizes < 5 µm	0–50 µm (Narrower PSD); reduced fine particles sizes <5 µm	PSD for both powders are acceptable. Particles >45 µm can be further sieved, as they are mostly agglomerates anyway.
Morphology and circularity	Smooth, spherical, with minimum defects (agglomerates and satellites); circularity as close to 1 as possible or at least ≥0.7 [36,37,42,92]	Mostly spherical and near-spherical; agglomerates present; fewer satellites: circularity: ~0.87 (virgin: ~0.89)	Mix of spherical and increasingly irregular-shaped particles; agglomerates present; fewer satellites; circularity: ~0.69 (virgin: 0.83)	No issues for 316L SS powder. Exercise caution when reusing AlSi10Mg powder, probably not more than 30 times overall, and not for critical components.
Chemical composition and phase (crystal structure)	Maintain within alloy-specific compositions; minimum pick-up of oxygen or other trace elements; No change from initial powders and no undesired precipitates should form [29,36,38,70,91]	Within alloy-specific composition; oxygen content triples from that of initial powders; No phase change from initial powders can be detected through XRD	Within alloy-specific composition; oxygen content doubles from that of initial powders; No phase change from initial powders can be detected through XRD	Reuse of both recycled powders is acceptable based on the EDX and XRD results, but there are concerns on the oxygen pick-up. Detailed analysis through XPS, EDX mapping, EBSD, and TEM could be carried out in the future to definitively confirm the presence of any undesired phases/precipitates.

5. Conclusions

In this study, the properties of 316L SS and AlSi10Mg powders in the virgin condition and recycled state after 20 times of reuse are analyzed and compared by using SEM, EDX, and XRD techniques. Qualitative assessments on the viability of using the recycled powders for further build cycles are made based on the results in this study and those recommended in the available literature. The following conclusions can be made:

(1) Compared with virgin powders, a shift to higher particle size distribution (PSD) values are observed, resulting in higher average particle sizes for both 316L SS and AlSi10Mg recycled powders.

(2) SEM observations indicate the presence of agglomerates, but fewer satellites are detected on the surface of the recycled 316L SS and AlSi10Mg powders.

(3) The recycled 316L SS powder particles retain their spherical and near-spherical morphologies, whereas that of the AlSi10Mg counterpart deforms into more irregularly shaped morphologies.

(4) The average circularity of recycled 316L SS powder only reduces slightly by ~2%, but a higher reduction in circularity is attained for the recycled AlSi10Mg powder (~17%).

(5) EDX analysis suggests the chemical composition of both recycled powders remains within alloy-specific values, but an increase in oxygen pick-up is detected in both cases.

(6) XRD spectra peak analyses indicate no change in crystal structures or presence of any undesired precipitates in both recycled powders.

(7) Based on the present results and qualitative comparative analysis with available literature, the reuse of both recycled powder for further build cycles is acceptable up to 30 overall build cycles. However, these are preliminary assessments, and extensive re-evaluation through physical and chemical characterizations is necessary if they are to be further reused.

Author Contributions: S.M.Y. contributes primarily to the writing and compilation of manuscript and data analysis; E.C. contributes primarily to carrying out the main experiments and data collection; N.G. is heavily involved in the project conceptualization, administration, and planning, as well as the review and editing of the manuscript. All authors have read and agreed to the published version of the manuscript.

Funding: This research received no external funding.

Conflicts of Interest: The authors declare no conflict of interest.

References

1. Yusuf, S.M.; Gao, N. Influence of energy density on metallurgy and properties in metal additive manufacturing. *Mater. Sci. Technol.* **2017**, *33*, 1269–1289. [CrossRef]
2. Frazier, W.E. Metal additive manufacturing: A review. *J. Mater. Eng. Perform.* **2014**, *23*, 1917–1928. [CrossRef]
3. Herzog, D.; Seyda, V.; Wycisk, E.; Emmelmann, C. Additive manufacturing of metals. *Acta Mater.* **2016**, *117*, 371–392. [CrossRef]
4. Manakari, V.; Parande, G.; Gupta, M. Selective laser melting of magnesium and magnesium alloy powders: A review. *Metals* **2016**, *7*, 2. [CrossRef]
5. Olakanmi, E.O. Selective laser sintering/melting (SLS/SLM) of pure Al, Al–Mg, and Al–Si powders: Effect of processing conditions and powder properties. *J. Mater. Process. Technol.* **2013**, *213*, 1387–1405. [CrossRef]
6. Olakanmi, E.O.; Cochrane, R.F.; Dalgarno, K.W. A review on selective laser sintering/melting (SLS/SLM) of aluminium alloy powders: Processing, microstructure, and properties. *Prog. Mater. Sci.* **2015**, *74*, 401–477. [CrossRef]
7. Sun, Y.; Moroz, A.; Alrbaey, K. Sliding wear characteristics and corrosion behaviour of selective laser melted 316L stainless steel. *J. Mater. Eng. Perform.* **2013**, *23*, 518–526. [CrossRef]
8. Bartolomeu, F.; Buciumeanu, M.; Pinto, E.; Alves, N.; Carvalho, O.; Silva, F.; Miranda, G. 316L stainless steel mechanical and tribological behavior—A comparison between selective laser melting, hot pressing and conventional casting. *Addit. Manuf.* **2017**, *16*, 81–89. [CrossRef]
9. Lewandowski, J.J.; Seifi, M. Metal additive manufacturing: A review of mechanical properties. *Annu. Rev. Mater. Res.* **2016**, *46*, 151–186. [CrossRef]

10. Gorsse, S.; Hutchinson, C.; Gouné, M.; Banerjee, R. Additive manufacturing of metals: A brief review of the characteristic microstructures and properties of steels, Ti-6Al-4V and high-entropy alloys. *Sci. Technol. Adv. Mater.* **2017**, *18*, 584–610. [CrossRef] [PubMed]

11. Boisselier, D.; Sankaré, S. Influence of powder characteristics in laser direct metal deposition of SS316L for metallic parts manufacturing. *Phys. Procedia* **2012**, *39*, 455–463. [CrossRef]

12. Song, B.; Zhao, X.; Li, S.; Han, C.; Wei, Q.; Wen, S.; Liu, J.; Shi, Y. Differences in microstructure and properties between selective laser melting and traditional manufacturing for fabrication of metal parts: A review. *Front. Mech. Eng.* **2015**, *10*, 111–125. [CrossRef]

13. Saeidi, K.; Gao, X.; Zhong, Y.; Shen, Z. Hardened austenite steel with columnar sub-grain structure formed by laser melting. *Mater. Sci. Eng. A* **2015**, *625*, 221–229. [CrossRef]

14. Chen, N.; Ma, G.; Zhu, W.; Godfrey, A.; Shen, Z.; Wu, G.; Huang, X. Enhancement of an additive-manufactured austenitic stainless steel by post-manufacture heat-treatment. *Mater. Sci. Eng. A* **2019**, *759*, 65–69. [CrossRef]

15. Tucho, W.M.; Cuvillier, P.; Sjolyst-Kverneland, A.; Hansen, V. Microstructure and hardness studies of Inconel 718 manufactured by selective laser melting before and after solution heat treatment. *Mater. Sci. Eng. A* **2017**, *689*, 220–232. [CrossRef]

16. Tucho, W.M.; Lysne, V.H.; Austbø, H.; Sjolyst-Kverneland, A.; Hansen, V. Investigation of effects of process parameters on microstructure and hardness of SLM manufactured SS316L. *J. Alloys Compd.* **2018**, *740*, 910–925. [CrossRef]

17. Olakanmi, E.O.; Cochrane, R.F.; Dalgarno, K.W. Densification mechanism and microstructural evolution in selective laser sintering of Al–12Si powders. *J. Mater. Process. Technol.* **2011**, *211*, 113–121. [CrossRef]

18. Hadadzadeh, A.; Baxter, C.; Amirkhiz, B.S.; Mohammadi, M. Strengthening mechanisms in direct metal laser sintered AlSi10Mg: Comparison between virgin and recycled powders. *Addit. Manuf.* **2018**, *23*, 108–120. [CrossRef]

19. Hadadzadeh, A.; Amirkhiz, B.S.; Mohammadi, M. Contribution of Mg2Si precipitates to the strength of direct metal laser sintered AlSi10Mg. *Mater. Sci. Eng. A* **2019**, *739*, 295–300. [CrossRef]

20. Chen, B.; Moon, S.; Yao, X.; Bi, G.; Shen, J.; Umeda, J.; Kondoh, K. Strength and strain hardening of a selective laser melted AlSi10Mg alloy. *Scr. Mater.* **2017**, *141*, 45–49. [CrossRef]

21. Seifi, M.; Christiansen, D.; Beuth, J.; Harrysson, O.; Lewandowski, J.J. Process mapping, fracture and fatigue behaviour of Ti-6Al-4V produced by EBM additive manufacturing. In *Proceedings of the 13th World Conference on Titanium*; Venkatesh, V., Pilchak, A.L., Allison, J.E., Ankem, S., Boyer, R.R., Christodoulou, J., Fraser, H.L., Imam, M.A., Kosaka, Y., Rack, H.J., Eds.; The Minerals, Metals & Materials Society: Pittsburgh, PA, USA, 2016; pp. 1373–1378.

22. Seifi, M.; Salem, A.; Beuth, J.; Harrysson, O.; Lewandowski, J.J. Overview of materials qualification needs for metal additive manufacturing. *JOM J. Manag.* **2016**, *68*, 747–764. [CrossRef]

23. Seifi, M.; Gorelik, M.; Waller, J.; Hrabe, N.; Shamsaei, N.; Daniewicz, S.; Lewandowski, J.J. Progress towards metal additive manufacturing standardization to support qualification and certification. *JOM* **2017**, *69*, 439–455. [CrossRef]

24. Slotwinski, J.A.; Garboczi, E.; Hebenstreit, K.M. Porosity measurements and analysis for metal additive manufacturing process control. *J. Res. Natl. Inst. Stand. Technol.* **2014**, *119*, 494–528. [CrossRef] [PubMed]

25. Slotwinski, J.; Moylan, S. *Applicability of Existing Materials Testing Standards for Additive Manufacturing Materials*; NIST National Institute for Standards and Technology: Gaithersburg, MD, USA, 2014.

26. Hrabe, N.; Barbosa, N.; Daniewicz, S.; Shamsaei, N. *Findings from the NIST/ASTM Workshop on Mechanical Behavior of Additive Manufacturing Components*; NIST/ASTM: Gaithersburg, MD, USA, 2016. [CrossRef]

27. Brien, M.O. Existing standards as the framework to qualify additive manufacturing of metals. In Proceedings of the IEEE Aerospace Conference, Big Sky, MT, USA, 3–10 March 2018; pp. 1–10.

28. DebRoy, T.; Wei, H.L.; Zuback, J.S.; Mukherjee, T.; Elmer, J.W.; Milewski, J.O.; Beese, A.M.; Wilson-Heid, A.; De, A.; Zhang, W. Additive manufacturing of metallic components—Process, structure and properties. *Prog. Mater. Sci.* **2018**, *92*, 112–224. [CrossRef]

29. Sames, W.J.; List, F.A.; Pannala, S.; Dehoff, R.R.; Babu, S.S. The metallurgy and processing science of metal additive manufacturing. *Int. Mater. Rev.* **2016**, *61*, 315–360. [CrossRef]

30. Santecchia, E.; Spigarelli, S.; Cabibbo, M. Material reuse in laser powder bed fusion: Side effects of the laser—Metal powder interaction. *Metals* **2020**, *10*, 341. [CrossRef]

31. Sames, W.J.; Medina, F.; Peter, W.H.; Babu, S.S.; Dehoff, R.R. Effect of process control and powder quality on IN 718 produced using Electron Beam Melting. In *8th International Symposium on Superalloy 718 and Derivatives*; Wiley: Hoboken, NJ, USA, 2014; pp. 409–423. [CrossRef]

32. Zhao, X.; Chen, J.; Lin, X.; Huang, W. Study on microstructure and mechanical properties of laser rapid forming Inconel 718. *Mater. Sci. Eng. A* **2008**, *478*, 119–124. [CrossRef]

33. Qi, H.; Azer, M.; Ritter, A. Studies of standard heat treatment effects on microstructure and mechanical properties of laser net shape manufactured INCONEL 718. *Metall. Mater. Trans. A Phys. Metall. Mater. Sci.* **2009**, *40*, 2410–2422. [CrossRef]

34. Ahmed, F.; Ali, U.; Sarker, D.; Marzbanrad, E.; Choi, K.; Mahmoodkhani, Y.; Toyserkani, E. Study of powder recycling and its effect on printed parts during laser powder-bed fusion of 17-4 PH stainless steel. *J. Mater. Process. Technol.* **2020**, *278*, 116522. [CrossRef]

35. Morgan, C.; Stokes, R.; Priddy, M.; Khanzadeh, M.K.; Bian, L.; Mazur, J.Y.T., II; Yadollahi, A.; Doude, H.; Hammond, V.H. Influence of recycling and process parameters on powder characteristics of direct energy deposition Ti-6Al-4V. In Proceedings of the Additive Manufacturing with Powder Metallurgy, Phoenix, AZ, USA, 23–26 June 2019; p. 538.

36. Cordova, L.; Campos, M.; Tinga, T. Revealing the effects of powder reuse for selective laser melting by powder characterization. *JOM J. Manag.* **2019**, *71*, 1062–1072. [CrossRef]

37. Hajnys, J.; Pagac, M.; Mesicek, J.; Petru, J.; Spalek, F. Research of 316L metallic powder for use in SLM 3D printing. *Adv. Mater. Sci.* **2020**, *20*, 5–15. [CrossRef]

38. Gorji, N.E.; O'Connor, R.; Brabazon, D. XPS, XRD, and SEM characterization of the virgin and recycled metallic powders for 3D printing applications. In Proceedings of the IOP Conference Series: Materials, Science and Engineering, Iasi, Romania, 19–22 June 2019; p. 591. [CrossRef]

39. Romero, C.; Yang, F.; Bolzoni, L. Fatigue and fracture properties of Ti alloys from powder-based processes—A review. *Int. J. Fatigue* **2018**, *117*, 407–419. [CrossRef]

40. Liu, B.; Wildman, R.; Tuck, C.; Ashcroft, I.; Hague, R. Investigation the effect of particle size distribution on processing parameters optimisation in selective laser melting process. In Proceedings of the 22nd Annual International Solid Freeform Fabrication Symposium—An Additive Manufacturing Conference SFF, Austin, TX, USA, 8–10 August 2011; pp. 227–238.

41. Benson, J.M.; Snyders, E. The need for powder characterisation in the additive manufacturing industry and the establishment of a national facility. *South Afr. J. Ind. Eng.* **2015**, *26*, 104–114. [CrossRef]

42. Sutton, A.T.; Kriewall, C.S.; Leu, M.C.; Newkirk, J.W. Powders for additive manufacturing processes: Characterization techniques and effects on part properties. *Solid Free. Fabr.* **2016**, *1*, 1004–1030.

43. Sutton, A.T.; Kriewall, C.S.; Leu, M.C.; Newkirk, J.W. Powder characterisation techniques and effects of powder characteristics on part properties in powder-bed fusion processes. *Virtual Phys. Prototyp.* **2017**, *12*, 3–29. [CrossRef]

44. Simmons, J.C.; Chen, X.; Azizi, A.; Daeumer, M.A.; Zavalij, P.Y.; Zhou, G.; Schiffres, S.N. Influence of processing and microstructure on the local and bulk thermal conductivity of selective laser melted 316L stainless steel. *Addit. Manuf.* **2020**, *32*, 100996. [CrossRef]

45. Beese, A.M.; Carroll, B.E. Review of mechanical properties of Ti-6Al-4V made by laser-based additive manufacturing using powder feedstock. *JOM J. Manag.* **2015**, *68*, 724–734. [CrossRef]

46. Quigley, E.; Luo, Z.; Murella, A.; Lee, W.L.; Adams, J.; Tasooji, A. *Effect of Powder Reuse on DMLS (Direct Metal Laser Sintering) Product Integrity: Why Honeywell Believes the Future is Additive Manufacturing*; Arizona State University: Tempe, AZ, USA, 2016.

47. Zhang, J.; Hu, B.; Zhang, Y.; Guo, X.; Wu, L.; Park, H.-Y.; Lee, J.-H.; Jung, Y.-G. Comparison of virgin and reused 15-5 PH stainless steel powders for laser powder bed fusion process. *Prog. Addit. Manuf.* **2018**, *3*, 11–14. [CrossRef]

48. Liu, P.; Huang, S.H.; Mokasdar, A.; Zhou, H.; Hou, L. The impact of additive manufacturing in the aircraft spare parts supply chain: Supply chain operation reference (scor) model based analysis. *Prod. Plan. Control.* **2013**, *25*, 1169–1181. [CrossRef]

49. Durach, C.F.; Kurpjuweit, S.; Wagner, S.M. The impact of additive manufacturing on supply chains. *Int. J. Phys. Distrib. Logist. Manag.* **2017**, *47*, 954–971. [CrossRef]

50. Yusuf, S.M.; Cutler, S.; Gao, N. Review: The impact of metal additive manufacturing on the aerospace industry. *Metals* **2019**, *9*, 1286. [CrossRef]

51. Khajavi, S.H.; Partanen, J.; Holmström, J. Additive manufacturing in the spare parts supply chain. *Comput. Ind.* **2014**, *65*, 50–63. [CrossRef]

52. Ghadge, A.; Karantoni, G.; Chaudhuri, A.; Srinivasan, A. Impact of additive manufacturing on aircraft supply chain performance. *J. Manuf. Technol. Manag.* **2018**, *29*, 846–865. [CrossRef]

53. Costa-Silva, B.; Aiello, N.M.; Ocean, A.J.; Singh, S.; Zhang, H.; Thakur, B.K.; Becker, A.; Hoshino, A.; Mark, M.T.; Molina, H.; et al. Pancreatic cancer exosomes initiate pre-metastatic niche formation in the liver. *Nat. Cell Biol.* **2015**, *17*, 816–826. [CrossRef] [PubMed]

54. Shen, Y.F.; Jia, N.; Wang, Y.D.; Sun, X.; Zuo, L.; Raabe, D. Suppression of twinning and phase transformation in an ultrafine grained 2 GPa strong metastable austenitic steel: Experiment and simulation. *Acta Mater.* **2015**, *97*, 305–315. [CrossRef]

55. Quintana, O.A.; Alvarez, J.; McMillan, R.; Tong, W.; Tomonto, C. Effects of reusing Ti-6Al-4V powder in a selective laser melting additive system operated in an industrial setting. *JOM J. Manag.* **2018**, *70*, 1863–1869. [CrossRef]

56. Galicki, D.; List, F.; Babu, S.S.; Plotkowski, A.; Meyer, H.M.; Seals, R.; Hayes, C. Localized changes of stainless steel powder characteristics during selective laser melting additive manufacturing. *Metall. Mater. Trans. A Phys. Metall. Mater. Sci.* **2019**, *50*, 1582–1605. [CrossRef]

57. Heiden, M.J.; Deibler, L.A.; Rodelas, J.M.; Koepke, J.R.; Tung, D.J.; Saiz, D.J.; Jared, B.H. Evolution of 316L stainless steel feedstock due to laser powder bed fusion process. *Addit. Manuf.* **2019**, *25*, 84–103. [CrossRef]

58. Spierings, A.B.; Voegtlin, M.; Bauer, T.; Wegener, K. Powder flowability characterisation methodology for powder-bed-based metal additive manufacturing. *Prog. Addit. Manuf.* **2015**, *1*, 9–20. [CrossRef]

59. Hausnerova, B.; Mukund, B.N.; Sanetrnik, D. Rheological properties of gas and water atomized 17-4PH stainless steel MIM feedstocks: Effect of powder shape and size. *Powder Technol.* **2017**, *312*, 152–158. [CrossRef]

60. Louvis, E.; Fox, P.; Sutcliffe, C.J. Selective laser melting of aluminium components. *J. Mater. Process. Technol.* **2011**, *211*, 275–284. [CrossRef]

61. Ardila, L.C.; Garciandia, F.; González-Díaz, J.B.; Álvarez, P.; Echeverria, A.; Petite, M.M.; Deffley, R.; Ochoa, J. Effect of IN718 recycled powder reuse on properties of parts manufactured by means of Selective Laser Melting. *Phys. Procedia* **2014**, *56*, 99–107. [CrossRef]

62. Mellin, P.; Shvab, R.; Strondl, A.; Randelius, M.; Brodin, H.; Hryha, E.; Nyborg, L. COPGLOW and XPS investigation of recycled metal powder for selective laser melting. *Powder Metall.* **2017**, *60*, 223–231. [CrossRef]

63. Lutter-Günther, M.; Gebbe, C.; Kamps, T.; Seidel, C.; Reinhart, G. Powder recycling in laser beam melting: Strategies, consumption modeling and influence on resource efficiency. *Prod. Eng.* **2018**, *12*, 377–389. [CrossRef]

64. Nezhadfar, P.D.; Soltani-Tehrani, A.; Shamsaei, N. Effect of preheating build platform on microstructure and mechanical properties of additively manufactured 316L stainless steel. In Proceedings of the 30th Annual International Solid Freeform Fabrication Symposium—An Additive Manufacturing Conference, Austin, TX, USA, 12–14 August 2019; pp. 415–425.

65. Jacob, G.; Brown, C.; Donmez, A.; Watson, S. *Effects of Powder Recycling on Stainless Steel Powder and Built Material Properties in Metal Powder Bed Fusion Processes*; NIST Advanced Manufacturing Series 100-6; NIST National Institute of Standard and Technology: Gaithersburg, MD, USA, 2016; pp. 1–51.

66. Seyda, V.; Kaufmann, N.; Emmelmann, C. Investigation of aging processes of Ti-6Al-4 v powder material in laser melting. *Phys. Procedia* **2012**, *39*, 425–431. [CrossRef]

67. Tang, H.P.; Qian, M.; Liu, N.; Zhang, X.Z.; Yang, G.Y.; Wang, J. Effect of powder reuse times on additive manufacturing of Ti-6Al-4V by selective electron beam melting. *JOM J. Manag.* **2015**, *67*, 555–563. [CrossRef]

68. Sun, Y.Y.; Gulizia, S.; Oh, C.H.; Doblin, C.; Yang, Y.F.; Qian, M. Manipulation and characterization of a novel titanium powder precursor for additive manufacturing applications. *JOM J. Manag.* **2015**, *67*, 564–572. [CrossRef]

69. Simonelli, M.; Tuck, C.; Aboulkhair, N.T.; Maskery, I.; Ashcroft, I.; Wildman, R.D.; Hague, R. A study on the laser spatter and the oxidation reactions during selective laser melting of 316L stainless steel, Al-Si10-Mg, and Ti-6Al-4V. *Metall. Mater. Trans. A Phys. Metall. Mater. Sci.* **2015**, *46*, 3842–3851. [CrossRef]

70. Terrassa, K.L.; Haley, J.C.; MacDonald, B.E.; Schoenung, J.M. Reuse of powder feedstock for directed energy deposition. *Powder Technol.* **2018**, *338*, 819–829. [CrossRef]

71. Jelis, E.; Clemente, M.; Kerwien, S.; Ravindra, N.M.; Hespos, M.R. Metallurgical and mechanical evaluation of 4340 steel produced by direct metal laser sintering. *JOM J. Manag.* **2015**, *67*, 582–589. [CrossRef]

72. Andani, M.T.; Dehghani, R.; Karamooz-Ravari, M.R.; Mirzaeifar, R.; Ni, J. A study on the effect of energy input on spatter particles creation during selective laser melting process. *Addit. Manuf.* **2018**, *20*, 33–43. [CrossRef]

73. Slotwinski, J.A.; Garboczi, E.J. Metrology needs for metal additive manufacturing powders. *JOM J. Manag.* **2015**, *67*, 538–543. [CrossRef]

74. Morrow, B.M.; Lienert, T.J.; Knapp, C.M.; Sutton, J.O.; Brand, M.J.; Pacheco, R.M.; Livescu, V.; Carpenter, J.S.; Gray, G.T., III. Impact of defects in powder feedstock materials on microstructure of 304L and 316L stainless steel produced by additive manufacturing. *Metall. Mater. Trans. A.* **2018**, *49*, 3637–3650. [CrossRef]

75. Harun, W.S.W.; Manam, N.S.; Kamariah, M.S.I.N.; Sharif, S.; Zulkifly, A.H.; Ahmad, I.; Miura, H. A review of powdered additive manufacturing techniques for Ti-6al-4v biomedical applications. *Powder Technol.* **2018**, *331*, 74–97. [CrossRef]

76. Acharya, R.; Sharon, J.A.; Staroselsky, A. Prediction of microstructure in laser powder bed fusion process. *Acta Mater.* **2017**, *124*, 360–371. [CrossRef]

77. Mower, T.M.; Long, M.J. Mechanical behavior of additive manufactured, powder-bed laser-fused materials. *Mater. Sci. Eng. A* **2016**, *651*, 198–213. [CrossRef]

78. Simchi, A. The role of particle size on the laser sintering of iron powder. *Metall. Mater. Trans. B* **2004**, *35*, 937–948. [CrossRef]

79. Thejane, K.; Chikosha, S.; Du Preez, W.B. Characterisation and monitoring of TI6AL4V (ELI) powder used in different selective laser melting systems. *S. Afr. J. Ind. Eng.* **2017**, *28*, 161–171. [CrossRef]

80. Lavery, N.P.; Cherry, J.; Mehmood, S.; Davies, H.; Girling, B.; Sackett, E.; Brown, S.G.R.; Sienz, J. Effects of hot isostatic pressing on the elastic modulus and tensile properties of 316L parts made by powder bed laser fusion. *Mater. Sci. Eng. A* **2017**, *693*, 186–213. [CrossRef]

81. Wilson, P.R.; Chen, Z. Characterisation of surface grain boundary precipitates formed during annealing of low carbon steel sheets. *Scr. Mater.* **2005**, *53*, 119–123. [CrossRef]

82. Wilson, P.R.; Chen, Z. The effect of manganese and chromium on surface oxidation products formed during batch annealing of low carbon steel strip. *Corros. Sci.* **2007**, *49*, 1305–1320. [CrossRef]

83. Hryha, E.; Gierl, C.; Nyborg, L.; Danninger, H.; Dudrova, E. Surface composition of the steel powders pre-alloyed with manganese. *Appl. Surf. Sci.* **2010**, *256*, 3946–3961. [CrossRef]

84. Brooks, J.A.; Thompson, A.W. Microstructural development and solidification cracking susceptibility of austenitic stainless steel welds. *Int. Mater. Rev.* **1991**, *36*, 16–44. [CrossRef]

85. Trevisan, F.; Calignano, F.; Lorusso, M.; Pakkanen, J.; Aversa, A.; Ambrosio, E.P.; Lombardi, M.; Fino, P.; Manfredi, D. On the selective laser melting (SLM) of the AlSi10Mg alloy: Process, microstructure, and mechanical properties. *Materials* **2017**, *10*, 76. [CrossRef]

86. Hryha, E.; Dudrova, E.; Nyborg, L. Critical aspects of alloying of sintered steels with manganese. *Metall. Mater. Trans. A Phys. Metall. Mater. Sci.* **2010**, *41*, 2880–2897. [CrossRef]

87. Bauereiß, A.; Scharowsky, T.; Körner, C. Defect generation and propagation mechanism during additive manufacturing by selective beam melting. *J. Mater. Process. Technol.* **2014**, *214*, 2522–2528. [CrossRef]

88. Pohlman, N.A.; Roberts, J.A.; Gonser, M.J. Characterization of titanium powder: Microscopic views and macroscopic flow. *Powder Technol.* **2012**, *228*, 141–148. [CrossRef]

89. Zekovic, S.; Dwivedi, R.; Kovacevic, R. Numerical simulation and experimental investigation of gas-powder flow from radially symmetrical nozzles in laser-based direct metal deposition. *Int. J. Mach. Tools Manuf.* **2007**, *47*, 112–123. [CrossRef]

90. Slotwinski, J.A.; Garboczi, E.; Stutzman, P.; Ferraris, C.F.; Watson, S.S.; Peltz, M.A. Characterization of metal powders used for additive manufacturing. *J. Res. Natl. Inst. Stand. Technol.* **2014**, *119*, 460–493. [CrossRef]

91. Strondl, A.; Lyckfeldt, O.; Brodin, H.; Ackelid, U. Characterization and control of powder properties for additive manufacturing. *JOM J. Manag.* **2015**, *67*, 549–554. [CrossRef]

92. Gaytan, S.M.; Murr, L.E.; Medina, F.; Martinez, E.; Lopez, M.I.; Wicker, R.B. Advanced metal powder based manufacturing of complex components by electron beam melting. *Mater. Technol.* **2009**, *24*, 180–190. [CrossRef]

93. Yusuf, S.M.; Chen, Y.; Boardman, R.; Yang, S.; Gao, N. Investigation on porosity and microhardness of 316L stainless steel fabricated by selective laser melting. *Metals* **2017**, *7*, 64. [CrossRef]

94. Cleary, P.W.; Sawley, M.L. DEM modelling of industrial granular flows: 3D case studies and the effect of particle shape on hopper discharge. *Appl. Math. Model.* **2002**, *26*, 89–111. [CrossRef]

Publisher's Note: MDPI stays neutral with regard to jurisdictional claims in published maps and institutional affiliations.

© 2020 by the authors. Licensee MDPI, Basel, Switzerland. This article is an open access article distributed under the terms and conditions of the Creative Commons Attribution (CC BY) license (http://creativecommons.org/licenses/by/4.0/).

Article

The Influence of Particle Shape, Powder Flowability, and Powder Layer Density on Part Density in Laser Powder Bed Fusion

Lukas Haferkamp [1,2,*], Livia Haudenschild [2], Adriaan Spierings [1], Konrad Wegener [2], Kirstin Riener [3,4], Stefan Ziegelmeier [3] and Gerhard J. Leichtfried [4]

[1] Inspire, Innovation Center for Additive Manufacturing Switzerland (ICAMS), Lerchenfeldstrasse 3, 9014 St. Gallen, Switzerland; spierings@inspire.ethz.ch

[2] ETH Zürich, Institute of Machine Tools and Manufacturing (IWF), Leonhardstrasse 21, 8092 Zürich, Switzerland; livia@haudenschild.com (L.H.); wegener@iwf.mavt.ethz.ch (K.W.)

[3] BMW Group FIZ, Knorrstraße 147, 80788 Munich, Germany; Kirstin.Riener@bmw.de (K.R.); Stefan.Ziegelmeier@bmw.de (S.Z.)

[4] Department of Mechatronics, Materials Science, Faculty of Engineering Science, University of Innsbruck, Technikerstrasse 13, 6020 Innsbruck, Austria; Gerhard.Leichtfried@uibk.ac.at

* Correspondence: haferkamp@inspire.ethz.ch

Abstract: The particle shape influences the part properties in laser powder bed fusion, and powder flowability and powder layer density (PLD) are considered the link between the powder and part properties. Therefore, this study investigates the relationship between these properties and their influence on final part density for six 1.4404 (316L) powders and eight AlSi10Mg powders. The results show a correlation of the powder properties with a Pearson correlation coefficient (PCC) of -0.89 for the PLD and the Hausner ratio, a PCC of -0.67 for the Hausner ratio and circularity, and a PCC of 0.72 for circularity and PLD. Furthermore, the results show that beyond a threshold, improvement of circularity, PLD, or Hausner ratio have no positive influence on the final part density. While the water-atomized, least-spherical powder yielded parts with high porosity, no improvement of part density was achieved by feedstock with higher circularities than gas-atomized powder.

Keywords: selective laser melting (SLM); laser powder bed fusion (LPBF); powder; particle size distribution; particle morphology; powder layer density; part density; flowability; Hausner ratio

Citation: Haferkamp, L.; Haudenschild, L.; Spierings, A.; Wegener, K.; Riener, K.; Ziegelmeier, S.; Leichtfried, G.J. The Influence of Particle Shape, Powder Flowability, and Powder Layer Density on Part Density in Laser Powder Bed Fusion. *Metals* **2021**, *11*, 418. https://doi.org/10.3390/met11030418

Academic Editor: Atila Ertas

Received: 25 January 2021
Accepted: 26 February 2021
Published: 4 March 2021

Publisher's Note: MDPI stays neutral with regard to jurisdictional claims in published maps and institutional affiliations.

Copyright: © 2021 by the authors. Licensee MDPI, Basel, Switzerland. This article is an open access article distributed under the terms and conditions of the Creative Commons Attribution (CC BY) license (https://creativecommons.org/licenses/by/4.0/).

1. Introduction

The quality of parts produced by laser powder bed fusion (LPBF) is strongly influenced by feedstock material, which is the powder. The most important characteristics of powder for LPBF are usually considered to be the size distribution in [1], particle shape in [1,2], chemical composition in [3,4], flowability in [5], powder layer density (PLD) in [6,7], and the amount of internal porosity in [1]. Despite the significance of the raw material, there is a lack of quantified and broadly accepted powder requirements, which are unknown or, according to Tan et al. [2], often undisclosed by the powder suppliers for competitive reasons. Furthermore, there are no quantitative standards for LPBF powder published to date, which necessitates the identification of relevant powder characteristics and to quantify their influence on the LPBF process.

One powder characteristic that lacks quantified requirements is the particle shape. There are several production techniques for metal powders, including water-atomization (WA), gas-atomization (GA), plasma-atomization (PA), and centrifugal-atomization techniques as the plasma rotating electrode process (PREP), and mechanical milling [8]. The particle shape, however, varies across the different production techniques [2]. Currently, GA powders are preferred in LPBF [2]. It is, however, unclear to what extent the powder particle shape affects processability and part quality in LPBF. Several studies have shown the negative impact of non-spherical WA powder on the process and part quality [3,9,10]. It is, however, not clear, whether more spherical, e.g., PA and PREP powders, further improve part quality.

Several studies investigated the influence of the particle shape on part quality. Most of them were focused on a comparison of WA and GA powders. Riener et al. [11] compared the processing of three GA powders and one plasma-atomized AlSi10Mg powder. They concluded that, in general, higher part densities could be achieved by using more spherical powders. Brika et al. [12] investigated two PA powders and one GA Ti-6Al-4V powder and also found a positive influence of particle sphericity on the powder flowability and part density. Engeli et al. [3] investigated seven GA powders and one WA IN738LC powder for the LPBF process. They found higher porosity in the parts made of WA powder, which they attribute to its more irregular morphology. They also observed an influence of the Hausner ratio on part porosity of GA powders. Baitimerov et al. [13] processed three gas atomized AlSi12 powders and concluded that non-spherical particles lead to a low apparent density of the powder and thus porosity in the parts. Li et al. [9] processed one GA and one WA 1.4404 (316L) powder and found higher porosity in parts made from WA powder, which they attributed to its high oxygen content and low packing density. Hoeges et al. [10] compared the density of parts made from a single WA powder and a GA powder. They were able to process both powders to a density of more than 99%; however, the morphology, flowability, and chemical composition of the WA powder posed a challenge during processing on a LPBF machine. Irrinki et al. [14] investigated the processability of three WA powders and one GA 17-4-PH stainless steel powder and found that the GA powder generally yielded parts with less porosity, but were able to process the WA powder into parts with comparable densities and mechanical properties at sufficiently high energy densities. Jeon et al. [15] processed four Fe-powders, of which three were mixtures of spherical and non-spherical powders. They found an increase in the part porosity of blended powders for increasing size misfit between the spherical and non-spherical powders.

The literature indicates that spherical PA and PREP powders might, compared to less spherical GA powders, further improve part quality, especially final part density, which is, according to Hoeges et al. [10], one of the major quality metrics of LPBF. This paper therefore investigates the influence of the particle shape on the process with a focus on final part density. The PLD [5,9] and powder flowability [5,16] are often considered as a link between the powder and final part properties and are therefore measured for all powders in this experiment, as they allow inferring the densification mechanisms during LPBF. In contrast to previous studies, mostly GA, or more spherical, powders were used in this experiment. Thus, the results are more relevant for industrial practice, where mostly spherical powders are used [2,17].

2. Materials and Methods

2.1. Powder Characterization

Six 1.4404 (316L) powders and eight AlSi10Mg powders were investigated. Their particle size distribution (PSD) was measured three times by laser diffraction with a LS230 laser diffractometer (Beckman Coulter, Inc., Brea, CA, USA) and the average PSD was used. The corresponding PSDs are displayed in Figure 1. The powder designation was chosen to reflect the alloy (Fe = 1.4404 (316L); Al = AlSi10Mg), the production process was as specified by the powder supplier (GA = gas-atomized; PREP = plasma rotating electrode process; WA = water-atomized; PA = plasma-atomized), and the volume-weighted D50 of a powder. The powders Fe-GA-38 and Fe-GA-47 as well as Fe-PREP-40 and Fe-PREP-47 were from the same powder batch and were separated into two powders by sieving. All other powders were from different batches. All powders were unused, as reuse of powder can change the powder properties, as discussed in [18]. A chemical influence can, however, not be entirely excluded, as the oxygen in the powder depends on the powder surface [19], which strongly depends on the PSD.

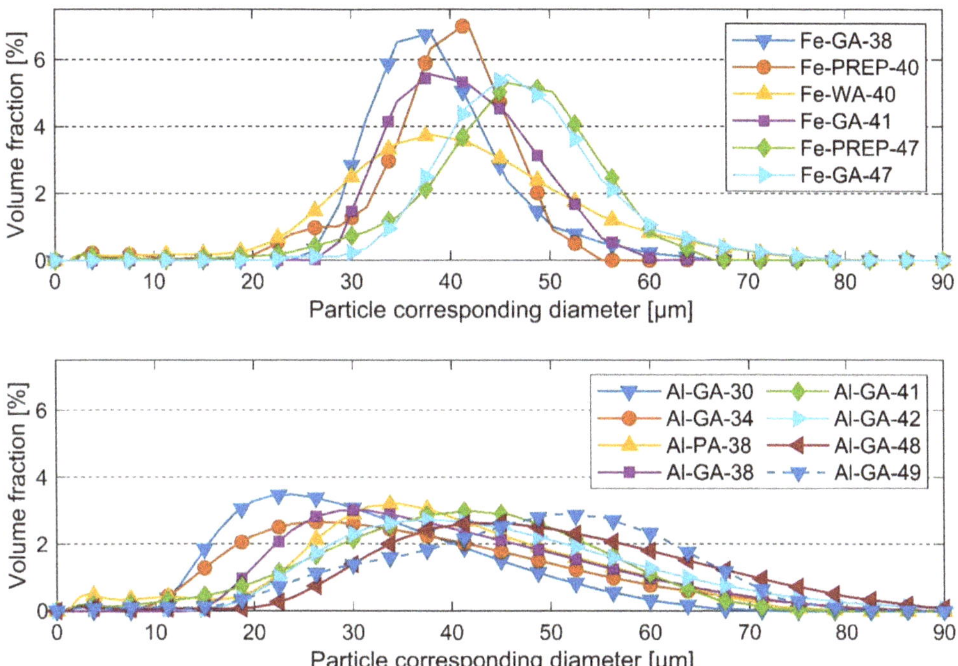

Figure 1. Particle size distribution (PSD) of the investigated powders: **top** 1.4404 (316L), **bottom** AlSi10Mg.

The particle shape factors were measured using a Leica DM6 light microscope (Leica Microsystems GmbH, Wetzlar, Germany). Images were taken by transmitted light microscopy and then processed in MATLAB (The MathWorks, Inc., Natick, MA, USA), where all agglomerates were manually removed until 3000 particles per powder were identified. The two commonly used shape factors, circularity and aspect ratio, were used in this study. The circularity f_{circ} was calculated according to the equation used by Bouwman et al. [20] as $f_{circ} = (4\pi A)/P^2$, where A is the area and P the perimeter of the projected particle. The aspect ratio f_{AR} was calculated as the ratio of the minimum Feret diameter d_{fmin} to the maximum Feret diameter d_{fmax} as $f_{AR} = d_{fmin}/d_{fmax}$. The apparent density ρ_a and tap density ρ_t of each powder were measured three times according to ASTM B417-18 [21] and ASTM B527-15 [22], respectively. The Hausner ratio, H, was then calculated as the ratio of the tap and apparent density as $H = \rho_t/\rho_a$.

2.2. Powder Layer Density Measurement

The PLD was measured by filling a cavity, as shown in Figure 2a, with a coating blade shown in Figure 2b. The PLD was then calculated from the powder mass and the known volume of the cavity. Two cavities were used, one with an average depth of 84.5 μm and the other with an average depth of 141.4 μm. Both were made from hardened 1.2379 steel with an outer cylinder diameter of 44.0 mm and a total height of 15.2 mm. The inner cavities with a diameter of 40 mm were produced by electrical discharge machining, and their depth was measured by a GelSight Benchtop system (GelSight, Inc., Waltham, MA, USA).

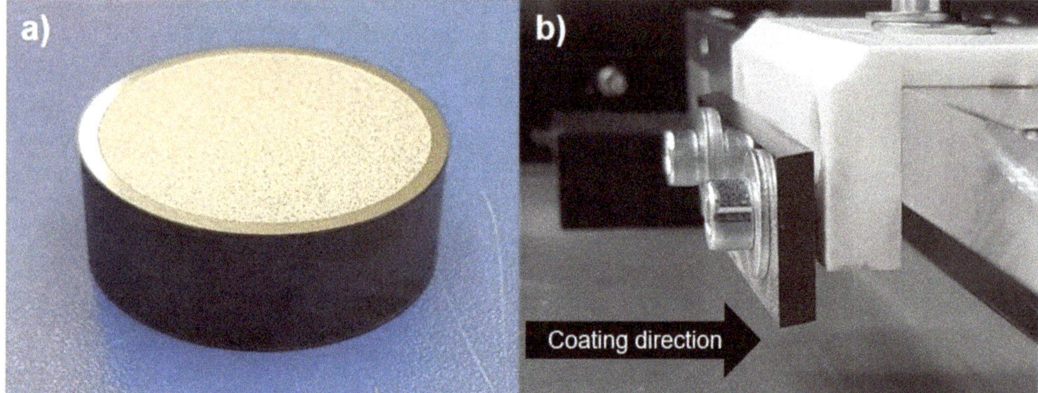

Figure 2. An (**a**) empty cavity and (**b**) coating blade from stainless steel.

The coating of the cavity was performed on a test bench, which was propelled with a toothed belt drive on the coater axis and a spindle screw drive on the z-axis. The cavity was placed on the z-platform and leveled such that the tip of the coating blade just touched the rim of the cavity along the entire movement on top of the cavity. The z-axis was then raised by 50 µm to create pressure between the coating blade and the rim of the cavity. This pressure ensured that no powder remained on the rim of the cavity after coating. The position of the cavity in the coating direction was secured with two mechanical stops on the z-platform.

The layer density of all powders was measured five times for each powder and cavity, thus moving the coating blade to its starting position in front of the z-platform. The cavity was then placed on the z-platform and the powder placement platform was placed around the cavity, as shown in Figure 3b. A powder with a volume of 12 cm^3 at apparent density was positioned in front of the cavity on the powder placement platform. The coater was then moved across the cavity at a speed of 100 mm/s. Eventually, the cavity was removed, cleaned from the outside, and weighed on an AX205 Delta Range scale by Mettler Toledo.

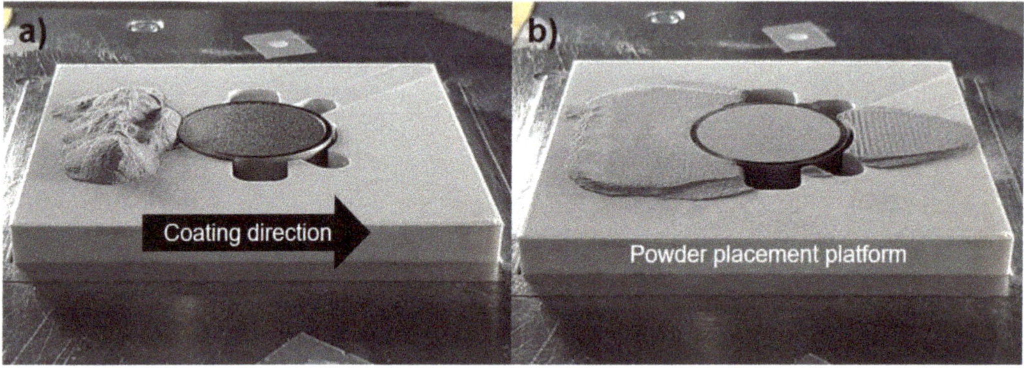

Figure 3. (**a**) Twelve cm^3 of powder placed in front of the cavity before the coating process, (**b**) powder in the cavity after the coating process.

2.3. Powder Processing

All powders were processed on a Concept Laser M2 (Concept Laser GmbH, Lichtenfels, Germany) using the parameters listed in Table 1. Five $10 \times 10 \times 10$ mm^3 cubes were built per scan speed, resulting in 25 cubes per build cycle. These cubes were evenly distributed over the build plate to compensate for possible location-dependent differences in part density. The parts were not placed on a regular build plate, but with a custom-made insert to reduce the build chamber volume. This reduced the build platform to 95 mm in diameter. A brush was used for coating, which was different from the blade used for measuring PLD. This was done as the layer density measurement is only possible with a blade, whereas a build cycle with a blade would cause collisions of the steel blade during the processing of some powders. The density of final parts was measured using the Archimedes principle in acetone according to the procedure described by Spierings et al. [23]. Density as a quality metric was chosen as it is easy to measure and does directly influence the mechanical properties, as has been discussed in [11,24].

Table 1. Parameters used for processing.

Parameter	Unit	Value
Laser power	(W)	180
Hatch	(µm)	1.4404 (316L): 75, AlSi10Mg: 100
Layer height	(µm)	30
Scan speed	(mm/s)	1000, 1250, 1500, 1750, 2000
Shielding gas	-	Nitrogen
Coating technology	-	Brush
Beam diameter (as measured)	(µm)	105
Coating speed	(mm/s)	100
Scanning strategy	-	90° alternating, parallel to the sides of the cube, crosswise

3. Results and Discussion

3.1. Circularity and Aspect Ratio

Figure 4 shows the circularity and aspect ratio of all investigated powders. The highest circularity was measured for the PREP powders, which had a circularity of up to 0.99, whereas the lowest circularity was measured for the WA powder, which had a circularity of 0.68. Most GA powders have a circularity of approximately 0.9. The aspect ratio was lower than the circularity for all powders. The absolute values of the circularity and aspect ratio were within the range of what has been reported in the literature for the respective production processes [11,25–29]. Furthermore, the results showed a strong correlation between powder circularity and aspect ratio with a Pearson correlation coefficient (PCC) of $r\,(12) = 0.91$, $p < 0.001$. Thus, subsequently, only circularity will be discussed as all dependencies are qualitatively interchangeable for circularity and aspect ratio.

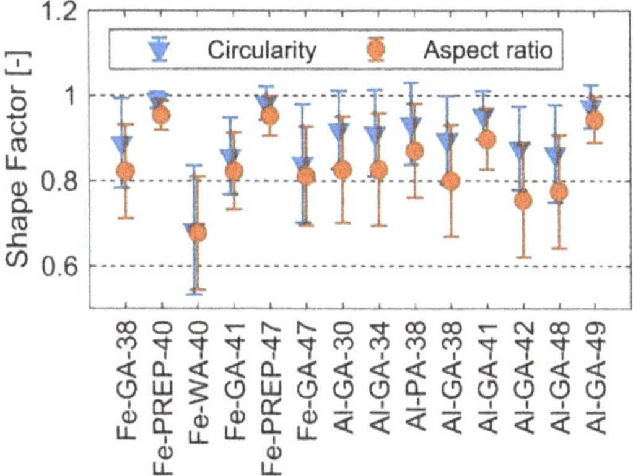

Figure 4. The mean circularity and mean aspect ratio of all powders with standard deviation.

3.2. Powder Layer Density, Tap Density, and Apparent Density

The PLD, tap density, and apparent density of all powders is displayed in Figure 5. For all powders, the PLD in the 84.5 µm cavity was lower than in the 141.4 µm cavity, and both PLDs were lower than the apparent density of the powders. This was due to the wall effect, which is the reduction of packing density through vacant sites in a packed powder in the presence of a wall, as discussed in [29,30]. For a solid material layer with a height of 30 µm, recent studies indicate that 141.1 µm is close to the realistic effective powder layer thickness for the LPBF process; Wischeropp et al. [6] measured it to be between 130 and 165 µm. These results are supported by the recent findings of Mahmoodkhani et al. [7], who measured the effective powder layer thickness to be greater than 100 µm in all of their experiments. Therefore, the subsequent discussion is mostly based on the PLD of the 141.4 µm cavity, which is more representative of the reality.

The measured PLDs for the 141.1 µm cavity are comparable to values in the literature, i.e., those of Wischeropp et al. [6], who measured the PLD to be between 44% and 53% for three LPBF powders, and Chen et al. [31], who measured a PLD for a single layer of 316L powder to be approx. 45% for a coating speed of 100 mm/s. The water-atomized powder Fe-WA-40, which was the least spherical, exhibited the lowest PLD, whereas the PREP powders, which were the most spherical ones, exhibited the highest layer densities. This trend will be discussed in more detail in the following sections.

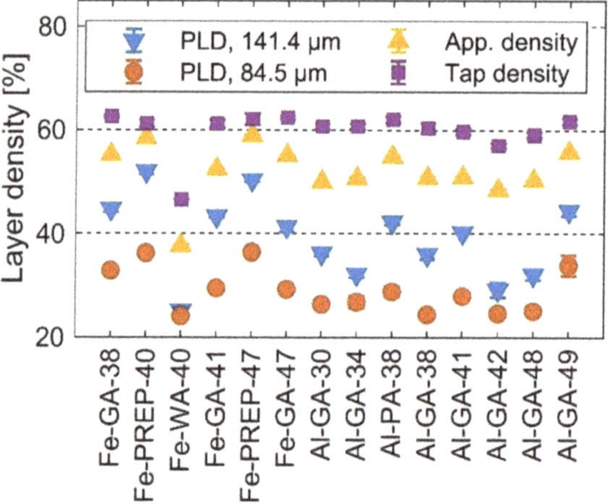

Figure 5. The mean powder layer density (PLD) for both cavities, apparent density, and tap density of all powders as a fraction of the full material density with standard deviation.

3.3. Hausner Ratio

The Hausner ratios of all powders are displayed in Figure 6, as calculated from the tap and apparent density in Figure 5. The values range from 1.05 for Fe-PREP-40 to 1.24 for Fe-WA-40. According to Sutton et al. [32], any powder with a Hausner ratio below 1.25 can be considered free-flowing, which is in accordance with the subjective perception of the powder flow of all powders. All powders had sufficient flowability to be processed on a LPBF machine, yet did not necessarily yield sufficient part quality, as it will be subsequently shown. The Hausner ratios were within the range of what has been reported in the literature for similar powders; i.e., approx. 1.17 for GA AlSi10Mg in [11], approx. 1.11 for PA AlSi10Mg in [11], a variety of Fe and Ni powders for LPBF with Hausner ratios mostly between 1.1 and 1.2 in [5], water-atomized 316L powder with 1.27 in [28], and others in [3,33].

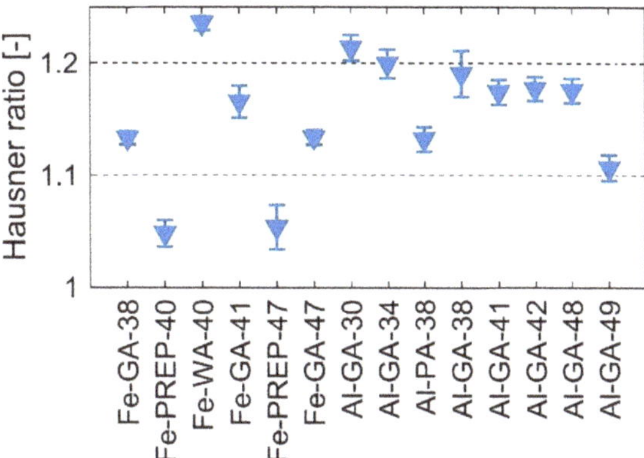

Figure 6. The mean Hausner ratio of all powders with standard deviation.

3.4. Relationship between PLD, Circularity, and Hausner Ratio

Figure 7 shows the relationship between PLD for both cavities, circularity, and Hausner ratio. The PLD for the 141.4 μm-deep cavity is plotted against the PLD of the 84.5 μm-deep cavity in Figure 7a. Besides the fact that the PLD was higher for the deeper cavity for every single powder, there was a linear relation with a PCC of r (12) = 0.92, $p < 0.001$ between the mean PLD for the 84.5 μm and 141.4 μm cavities. The PLD is plotted against the mean circularity in Figure 7b and shows a linear dependency between the PLD and the powder circularity with a PCC of r (12) = 0.72 $p < 0.01$. Furthermore, this trend seems dependent on the material, as the layer density values for AlSi10Mg powders were lower than those for 1.4404 (316L) powders for the same circularity.

The Hausner ratio is plotted against the mean circularity in Figure 7c. The data show a moderate linear correlation with a PCC of r (12) = $-0.67, p < 0.01$ for the combined 1.4404 (316L) and AlSi10Mg data. However, the dependence of Hausner ratios on circularity, just like that of PLD on circularity, seems to be material-dependent, as the Hausner ratio for AlSi10Mg powders was higher for the same circularity. The dependency of the Hausner ratio on the particle shape has been shown in [34], where the Hausner ratio increases for decreasing particle sphericity. There was also a linear relationship between the PLD for the 141.4 μm cavity and the Hausner ratio with a PCC of r (12) = $-0.89, p < 0.001$, as shown in Figure 7c. This trend, in contrast to the Hausner ratio and PLD, fits the data for both 1.4404 (316L) and AlSi10Mg powders independent of the material.

In summary, the PLD in both cavities correlated strongly. The PLD and Hausner ratio of a powder cannot be predicted only from its circularity, as they seem to be material-dependent. The Hausner ratio, however, is a good predictor for the PLD in LPBF and did not show a material dependency for the two alloys in this experiment. As the Hausner ratio depends on the friction between powder particles [35], caused by either particle shape, oxides, or other factors, it seems that low friction between the powder particles caused a higher PLD. This finding is also supported by the fact that the Al powders in Figure 7c exhibited a higher Hausner ratio than the Fe powders for the same circularity, which in turn caused the lower PLD of Al powders for the same circularity in Figure 7b.

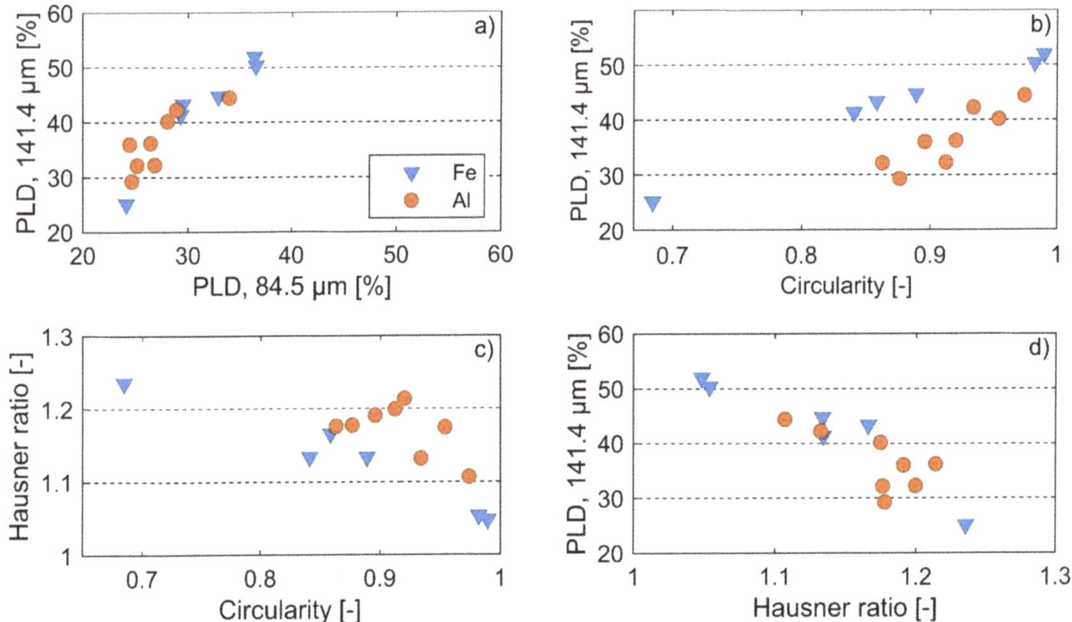

Figure 7. The correlation between: (**a**) mean PLD for the 141.4 µm and 84.5 µm cavity, (**b**) mean PLD for the 141.4 µm cavity and mean circularity, (**c**) Hausner ratio and mean circularity, and (**d**) PLD for the 141.4 µm cavity and Hausner ratio.

3.5. Part Density

All powders were processed on an LPBF machine with parameters reported in Section 2.3. The material density of parts made from every powder at five scan speeds is displayed in Figure 8 for (a) 1.4404 (316L) and (b) AlSi10Mg. Some of the cubes from the Fe-WA-40 powder showed delamination issues, as displayed in Figure 9. The density of these cubes was omitted from the data as it is affected by the porosity due to delamination.

The part density decreases, as typical for LPBF [36,37], with increasing scan speed for all powders. While the processing windows of Fe-GA-38, Fe-PREP-40, and Fe-GA-41 in Figure 8a were similar, the other three 14404 (316L) powders exhibited a different processing window. PREP-47 and Fe-GA-47 exhibited a similar processing window with a high decrease in part densities for scan speeds higher than 1250 mm/s. This can be explained by their coarser PSD, as Meiners [37] and Simchi [38,39] observed before and showed that coarser powders lead to increased part porosity at high scan speeds. The processing window of the water-atomized Fe-WA-40 differed from those of all other powders. For all scan speeds, the part density of parts made from WA powder was almost 2 percentage points lower than of parts made from gas-atomized and PREP powders with similar PSD. This high porosity in parts made from water-atomized powder in LPBF is widely known [3,9,14]; however, Hoeges et al. [10] demonstrated the processing of water-atomized 316L with densities of more than 99% by using adapted processing parameters.

Most density curves for the AlSi10Mg powders in Figure 8b only differed slightly. The exception is the processing window of AL-GA-48, which was approx. 1 percentage point lower than the processing window of the other powders. One contributor to the lower density is the coarser PSD, which has been shown to cause porosity in parts [37–39]. Surprisingly, the processing window of Al-GA-49 was similar to those of the finer powders. This might be explained by its circularity, which is the highest among the Al powders. The Al powder with the second-highest circularity, Al-GA-41, however, did not yield parts with a higher density than Al powders with a similar PSD. Therefore, the positive influence of circularity on part density cannot be concluded for all powders. These findings will be discussed in the next sections.

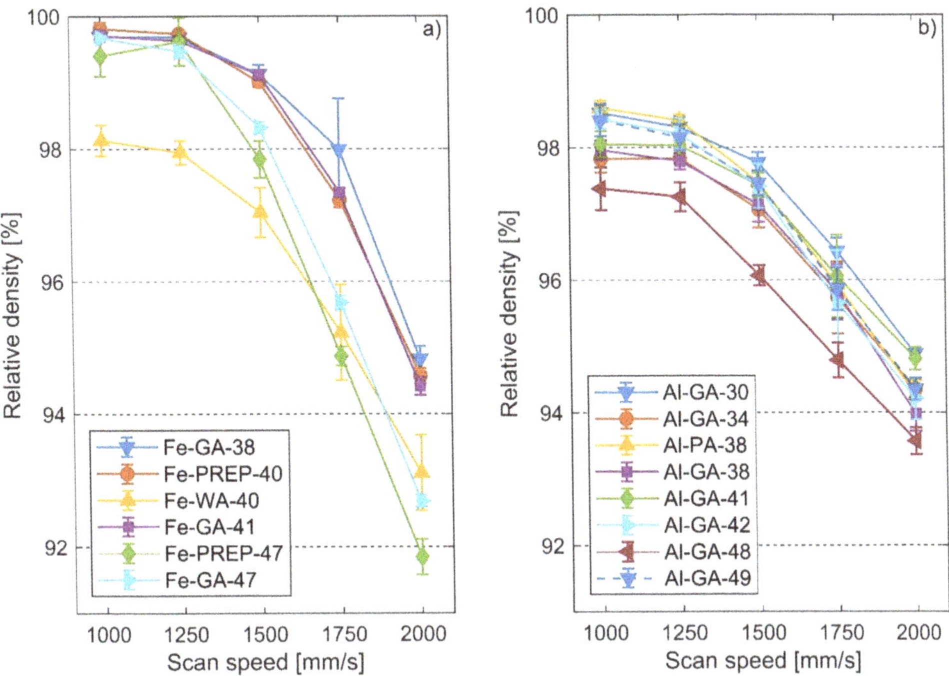

Figure 8. The dependency of the mean material density and its standard deviation on the scan speed of (**a**) 1.4404 (316L) and (**b**) AlSi10Mg for all powders.

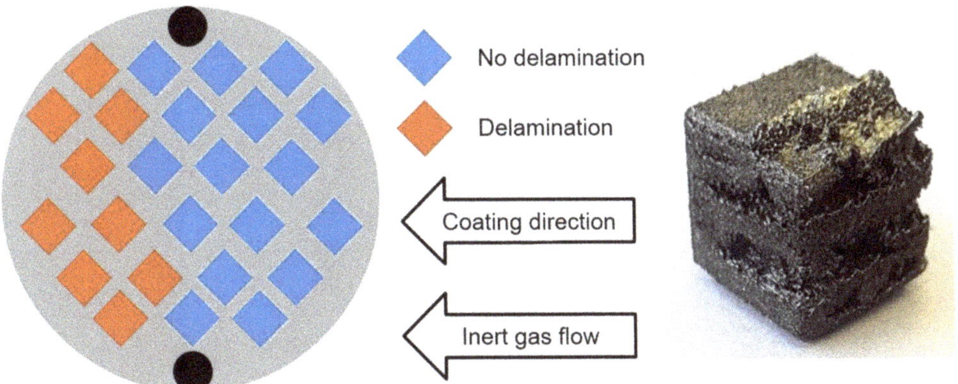

Figure 9. Delamination of the cubes built from the WA powder and a cube with delamination issues.

3.6. Dependence of the Part Density on the Particle Circularity

To identify causes for part porosity, the density data in Figure 8 is analyzed in more detail in this section. Figure 10 shows the mean part porosity plotted against powder circularity at scan speeds of 1000, 1500, and 2000 mm/s. All powders with D50 > 45 μm are shown in non-solid symbols. This is done to set them apart from the other powders, as their part porosity can be higher than the porosity of parts made from finer powders, as shown in [37–39]. The data show a difference in part densities between AlSi10Mg and 1.4404 (316L). At 1000 and 1500 mm/s, the porosity of the AlSi10Mg parts was approximately 1.5 percentage points higher for powders with D50 < 45 µm, the WA powder not considered. This was due to the material dependency of the processing windows for LPBF.

Furthermore, the data show that, except for Fe-WA-40, there was no influence of circularity on part density for both materials. Only the WA powder, which had the lowest circularity of all powders, showed a lower part density compared to powders made from the same alloy and with a similar PSD. The data for the four powders with a D50 > 45 µm also indicated no relationship between the part density and powder circularity.

The high part porosity of parts made from water-atomized powder has been shown before [3,9,12]. The finding for the other powders are in contrast to the results of Riener et al. [11], who compared three GA powders and one PA AlSi10Mg powder and Brika et al. [12], who compared one GA powders and two PA Ti-6Al-4V powders. Both found that more spherical particles lead to higher part density. Other authors report similar findings as in this experiment; i.e., little differences in part density for different particle shapes; e.g., Seyda, Herzog, and Emmelmann [40] found little difference in part density for three Ti-6Al-4V powders despite differences in particle morphology.

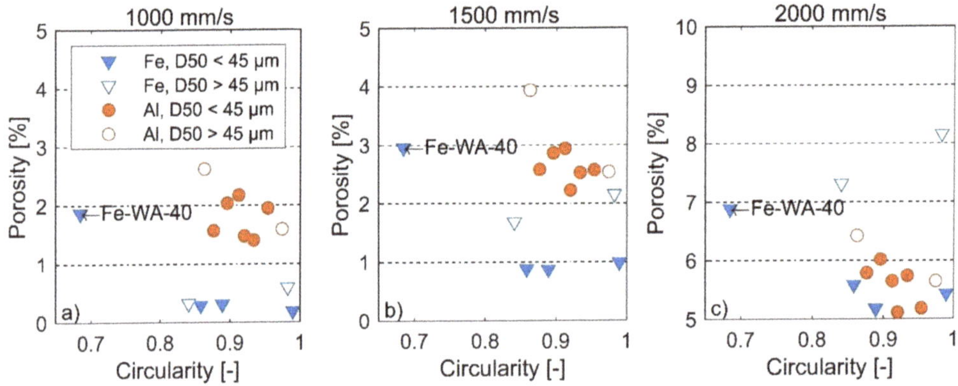

Figure 10. Mean part porosity plotted against the mean powder circularity for a scan speed of (**a**) 1000 mm/s, (**b**) 1500 mm/s, and (**c**) 2000 mm/s. The y-axis of graph (**c**) differs from that of graphs (**a**,**b**).

3.7. Dependence of the Part Density on the Hausner Ratio

Figure 11 shows the mean part porosity plotted against the Hausner ratio at scan speeds of 1000, 1500, and 2000 mm/s. The results once again show, except for Fe-Wa-40, that there was no clear influence of the Hausner ratio on part density. Since the Hausner ratio correlated with the particle circularity (compare Figure 7c) and the circularity of any but the water-atomized powder did not predict the part density, the Hausner ratio consequently also did not predict part porosity.

These findings are in contrast to the results of Engeli et al. [3], who found a relationship between the Hausner ratio and part density in LPBF for eight IN738LC powders. Brika et al. [12] also analyzed three powders, of which the one with the lowest Hausner ratio yielded parts with the lowest density, and Riener et al. [11] found in a comparison of three Ti-6Al-4V powders that the powder with the lowest Hausner ratio yielded parts with the highest density.

Powder with a too low flowability is undoubtedly unsuited for LPBF [5,41]. It is likely that the low flowability of the water-atomized Fe-WA-40 powder caused the low part porosity and also the delamination shown in Figure 9. It might, however, be that a decrease of the Hausner ratio of a powder beyond a certain threshold generated no further improvement of processability. This was also concluded by Pleass and Jothi [41], whose findings indicate that once adequate flowability is achieved, the processing parameters become much more relevant for part quality.

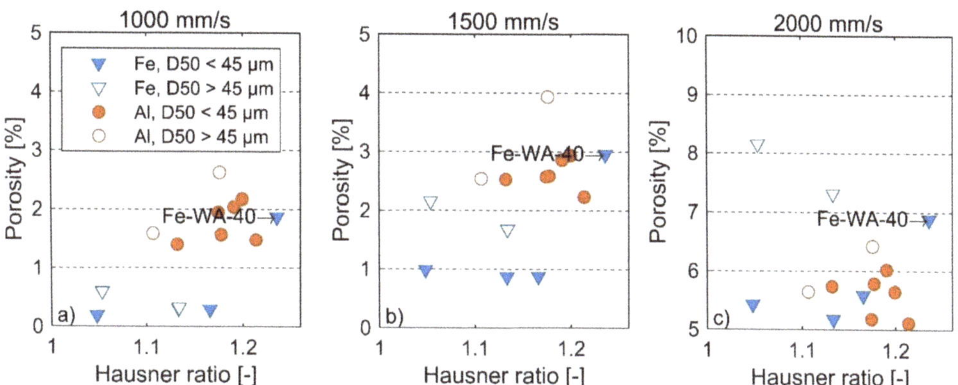

Figure 11. Mean part porosity against the Hausner ratio for a scan speed of (**a**) 1000 mm/s, (**b**) 1500 mm/s, and (**c**) 2000 mm/s. The y-axis of graph (**c**) differs from that of graphs (**a**,**b**).

3.8. Dependence of the Part Density on the Powder Layer Density

As the circularity and the Hausner ratio seemed to impact only the part density for WA powder, the influence of PLD on part density is evaluated in this section as the relevance of a high PLD for part quality is frequently stressed in literature [5,32,41,42]. Figure 12 shows the mean part porosity plotted against the PLD at scan speeds of 1000, 1500, and 2000 mm/s. The data show that, except for the Fe-Wa-40 powder and a material-dependent difference between Al and Fe powders, there was no influence of the PLD on the part density for either AlSi10Mg or 1.4404 (316L).

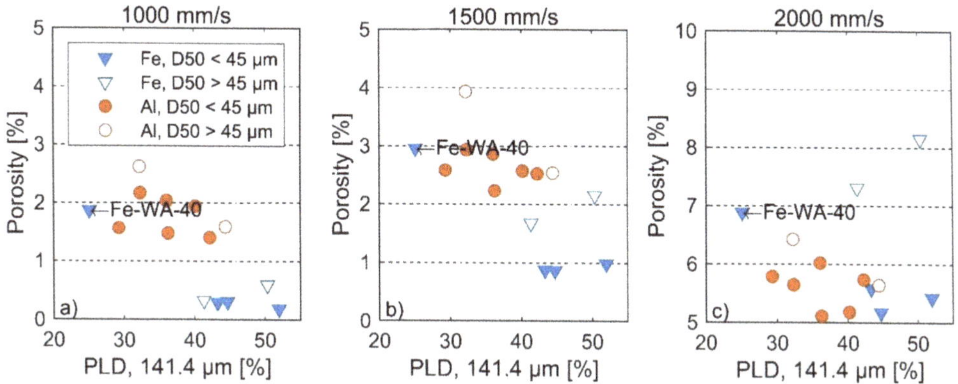

Figure 12. Mean part porosity against the PLD for the 141.4 μm cavity for a scan speed of (**a**) 1000 mm/s, (**b**) 1500 mm/s, and (**c**) 2000 mm/s. The y-axis of graph (**c**) differs from that of graphs (**a**,**b**).

An explanation for the limited impact of the PLD on the part density is given by the denudation of metal powder during LPBF, as has been described by Matthews et al. [43] and Bidare et al. [44]. Matthews et al. [43] showed that the metal vapor flux of the melt pool causes a shear gas flow, which entraps adjacent particles. The particles are then rearranged before the laser passes again and interacts with them. Consequently, the laser and melt pool do not interact with the powder bed as it has been deposited, but only with the rearranged particles. Furthermore, the law of conservation of mass causes the mass of the deposited powder to be the same as the mass of a solidified layer for the steady state, which is, as discussed in [45], reached after approximately seven layers. This makes the process, in contrast to other powder bed-based processes, less sensitive for variation in the PLD and powder layer homogeneity. Only when the flowability of a powder falls short of a minimal threshold does the powder bed become too inhomogeneous so that defects cannot be compensated for by the movement of particles in the shear gas flow, which then causes porosity, as is the case for the water-atomized Fe-WA-40 powder.

4. Conclusions

The results show a correlation between the particle shape, powder flowability, and PLD for the investigated powders. The more spherical the powders, the better their flowability, measured by the Hausner ratio. The flowability also directly affected the layer density, as better flowing powders generated layers with higher densities.

The WA powder, with a low circularity, exhibited an insufficient flowability, which caused low PLD, delamination, and part porosity. More spherical powders, such as the GA ones, showed better flowability and yielded parts with much lower porosity. Any improvement of the particle circularity beyond 0.8 did, however, have no positive influence on the part density. The more spherical, e.g., PREP, powders did exhibit better flowability, which resulted in a higher PLD. The increase in PLD did, however, not positively influence the part density. An explanation is given by the movement of powder particles in the shear gas flow. The laser and melt pool do not interact with the powder layer as it is deposited, but the particles are rearranged in the denudation zone of the previous, adjacent, scan track. This makes the LPBF less sensitive to inhomogeneity of the powder layer and particle circularity.

Future work should be done to measure the PLD in situ, as the results presented in this study might deviate from the true PLD. Furthermore, the influence of powder characteristics on other part quality metrics, such as the surface quality and mechanical properties, should be investigated in future experiments with a wider range of processing conditions.

Author Contributions: Conceptualization, L.H. (Lukas Haferkamp) and K.W.; methodology, L.H. (Lukas Haferkamp), A.S., K.R., S.Z., G.J.L.; formal analysis, L.H. (Lukas Haferkamp); investigation, L.H. (Lukas Haferkamp), L.H. (Livia Haudenschild); resources, K.W., K.R., S.Z., G.J.L.; writing—original draft preparation, L.H. (Lukas Haferkamp); writing—review and editing, L.H. (Lukas Haferkamp) and K.R.; visualization, L.H. (Lukas Haferkamp); supervision, K.W.; project administration, L.H. (Lukas Haferkamp). All authors have read and agreed to the published version of the manuscript.

Funding: This research received no external funding.

Institutional Review Board Statement: Not applicable.

Informed Consent Statement: Not applicable.

Data Availability Statement: The data presented in this study are available on request from the corresponding author.

Conflicts of Interest: The authors declare no conflict of interest.

References

1. Sames, W.J.; List, F.A.; Pannala, S.; Dehoff, R.R.; Babu, S.S. The metallurgy and processing science of metal additive manufacturing. *Int. Mater. Rev.* **2016**, *61*, 315–360. [CrossRef]
2. Tan, J.H.; Wong, W.L.E.; Dalgarno, K.W. An overview of powder granulometry on feedstock and part performance in the selective laser melting process. *Addit. Manuf.* **2017**, *18*, 228–255. [CrossRef]
3. Engeli, R.; Etter, T.; Hövel, S.; Wegener, K. Processability of different IN738LC powder batches by selective laser melting. *J. Mater. Process. Technol.* **2016**, *229*, 484–491. [CrossRef]
4. Slotwinski, J.A.; Garboczi, E.J. Metrology Needs for Metal Additive Manufacturing Powders. *J. Mater.* **2015**, *67*, 538–543. [CrossRef]
5. Spierings, A.B.; Voegtlin, M.; Bauer, T.; Wegener, K. Powder flowability characterisation methodology for powder-bed-based metal additive manufacturing. *Prog. Addit. Manuf.* **2016**, *1*, 9–20. [CrossRef]

6. Wischeropp, T.M.; Emmelmann, C.; Brandt, M.; Pateras, A. Measurement of actual powder layer height and packing density in a single layer in selective laser melting. *Addit. Manuf.* **2019**, *28*, 176–183. [CrossRef]

7. Mahmoodkhani, Y.; Ali, U.; Imani Shahabad, S.; Rani Kasinathan, A.; Esmaeilizadeh, R.; Keshavarzkermani, A.; Marzbanrad, E.; Toyserkani, E. On the measurement of effective powder layer thickness in laser powder-bed fusion additive manufacturing of metals. *Prog. Addit. Manuf.* **2019**, *4*, 109–116. [CrossRef]

8. Anatoliy, P.; Vadim, S. Metal Powder Additive Manufacturing. *New Trends 3D Print.* **2015**. [CrossRef]

9. Li, R.; Shi, Y.; Wang, Z.; Wang, L.; Liu, J.; Jiang, W. Densification behavior of gas and water atomized 316L stainless steel powder during selective laser melting. *Appl. Surf. Sci.* **2010**, *256*, 4350–4356. [CrossRef]

10. Hoeges, S.; Zwiren, A.; Schade, C. Additive manufacturing using water atomized steel powders. *Met. Powder Rep.* **2017**, *72*, 111–117. [CrossRef]

11. Riener, K.; Albrecht, N.; Ziegelmeier, S.; Ramakrishnan, R.; Haferkamp, L.; Spierings, A.B.; Leichtfried, G.J. Influence of particle size distribution and morphology on the properties of the powder feedstock as well as of AlSi10Mg parts produced by laser powder bed fusion (LPBF). *Addit. Manuf.* **2020**, *34*, 101286. [CrossRef]

12. Brika, S.E.; Letenneur, M.; Dion, C.A.; Brailovski, V. Influence of particle morphology and size distribution on the powder flowability and laser powder bed fusion manufacturability of Ti-6Al-4V alloy. *Addit. Manuf.* **2019**. [CrossRef]

13. Baitimerov, R.; Lykov, P.; Zherebtsov, D.; Radionova, L.; Shultc, A.; Prashanth, K.G. Influence of Powder Characteristics on Processability of AlSi12 Alloy Fabricated by Selective Laser Melting. *Materials* **2018**, *11*, 742. [CrossRef] [PubMed]

14. Irrinki, H.; Dexter, M.; Barmore, B.; Enneti, R.; Pasebani, S.; Badwe, S.; Stitzel, J.; Malhotra, R.; Atre, S.V. Effects of Powder Attributes and Laser Powder Bed Fusion (L-PBF) Process Conditions on the Densification and Mechanical Properties of 17-4 PH Stainless Steel. *J. Mater.* **2016**, *68*, 860–868. [CrossRef]

15. Jeon, T.J.; Hwang, T.W.; Yun, H.J.; VanTyne, C.J.; Moon, Y.H. Control of Porosity in Parts Produced by a Direct Laser Melting Process. *Appl. Sci.* **2018**, *8*, 2573. [CrossRef]

16. Seyda, V. *Werkstoff- und Prozessverhalten von Metallpulvern in der Laseradditiven Fertigung*; Springer: Berlin/Heidelberg, Germany, 2018. [CrossRef]

17. Cooke, A.; Slotwinski, J. *Properties of Metal Powders for Additive Manufacturing: A Review of the State of the Art of Metal Powder Property Testing*; US Department of Commerce, National Institute of Standards and Technology: Gaithersburg, MD, USA, 2012.

18. Gorji, N.E.; O'Connor, R.; Brabazon, D. X-ray Tomography, AFM and Nanoindentation Measurements for Recyclability Analysis of 316L Powders in 3D Printing Process. *Procedia Manuf.* **2020**, *47*, 1113–1116. [CrossRef]

19. Riener, K.; Oswald, S.; Winkler, M.; Leichtfried, G.J. Influence of storage conditions and reconditioning of AlSi10Mg powder on the quality of parts produced by laser powder bed fusion (LPBF). *Addit. Manuf.* **2021**, *39*, 101896. [CrossRef]

20. Bouwman, A.M.; Bosma, J.C.; Vonk, P.; Wesselingh, J.A.; Frijlink, H.W. Which shape factor(s) best describe granules? *Powder Technol.* **2004**, *146*, 66–72. [CrossRef]

21. ASTM. *ASTM B417—18, Standard Test Method for Apparent Density of Non-Free-Flowing Metal Powders Using the Carney Funnel*; ASTM: West Conshohocken, PA, USA, 2018. [CrossRef]

22. ASTM. *B527-15 Standard Test Method for Tap Density of Metal Powders and Compounds*; ASTM: West Conshohocken, PA, USA, 2015. [CrossRef]

23. Spierings, A.; Schneider, M.; Eggenberger, R. Comparison of Density Measurement Techniques for Additive Manufactured Metallic Parts. *Rapid Prototyp. J.* **2011**, *17*, 380–386. [CrossRef]

24. Mugica, G.W.; Tovio, D.O.; Cuyas, J.C.; González, A.C. Effect of porosity on the tensile properties of low ductility aluminum alloys. *Mater. Res.* **2004**, *7*, 221–229. [CrossRef]

25. Kiani, P.; Scipioni Bertoli, U.; Dupuy, A.D.; Ma, K.; Schoenung, J.M. A Statistical Analysis of Powder Flowability in Metal Additive Manufacturing. *Adv. Eng. Mater.* **2020**, *22*, 2000022. [CrossRef]

26. Uhlenwinkel, V.; Schwenck, D.; Ellendt, N.; Fischer-Bühner, J.; Hofmann, P. Gas recirculation affects the powder quality. In Proceedings of the Advances in Powder Metallurgy and Particulate Materials—2014, 2014 World Congress on Powder Metallurgy and Particulate Materials, PM, Orlando, FL, USA, 18–22 May 2014; pp. 246–253.

27. Snow, Z.; Martukanitz, R.; Joshi, S. On the development of powder spreadability metrics and feedstock requirements for powder bed fusion additive manufacturing. *Addit. Manuf.* **2019**, *28*, 78–86. [CrossRef]

28. Inaekyan, K.; Paserin, V.; Bailon-Poujol, I.; Brailovski, V. Binder-Jetting Additive Manufacturing with Water Atomized Iron Powders. In Proceedings of the AMPM 2016 Conference on Additive Manufacturing, Boston, MA, USA, 5–7 June 2016.

29. Haferkamp, L.; Spierings, A.; Rusch, M.; Jermann, D.; Spurek, M.A.; Wegener, K. Effect of Particle size of monomodal 316L powder on powder layer density in powder bed fusion. *Prog. Addit. Manuf.* **2020**. [CrossRef]

30. Karapatis, N.P.; Egger, G.; Gygax, P.E.; Glardon, R. Optimization of powder layer density in selective laser sintering. In Proceedings of the 10th Solid Freeform Fabrication Symposium (SFF), Austin, TX, USA, 9–11 August 1999.

31. Chen, H.; Chen, Y.; Liu, Y.; Wei, Q.; Shi, Y.; Yan, W. Packing quality of powder layer during counter-rolling-type powder spreading process in additive manufacturing. *Int. J. Mach. Tools Manuf.* **2020**, *153*, 103553. [CrossRef]

32. Sutton, A.T.; Kriewall, C.S.; Leu, M.C.; Newkirk, J.W. Powder characterisation techniques and effects of powder characteristics on part properties in powder-bed fusion processes. *Virtual Phys. Prototyp.* **2017**, *12*, 3–29. [CrossRef]

33. Coe, H.G.; Pasebani, S. Use of bimodal particle size distribution in selective laser melting of 316L stainless steel. *J. Manuf. Mater. Process.* **2020**, *4*, 8. [CrossRef]

34. Guo, A.; Beddow, J.K.; Vetter, A.F. A simple relationship between particle shape effects and density, flow rate and Hausner Ratio. *Powder Technol.* **1985**, *43*, 279–284. [CrossRef]

35. Abdullah, E.C.; Geldart, D. The use of bulk density measurements as flowability indicators. *Powder Technol.* **1999**, *102*, 151–165. [CrossRef]

36. Qiu, C.; Panwisawas, C.; Ward, M.; Basoalto, H.C.; Brooks, J.W.; Attallah, M.M. On the role of melt flow into the surface structure and porosity development during selective laser melting. *Acta Mater.* **2015**, *96*, 72–79. [CrossRef]

37. Meiners, W. Direktes Selektives Laser Sintern Einkomponentiger Metallischer Werkstoffe. Ph.D. Thesis, RWTH Aachen, Aachen, Germany, 1999.

38. Simchi, A. The role of particle size on the laser sintering of iron powder. *Metall. Mater. Trans. B* **2004**, *35*, 937–948. [CrossRef]

39. Simchi, A. Direct laser sintering of metal powders: Mechanism, kinetics and microstructural features. *Mater. Sci. Eng. A* **2006**, *428*, 148–158. [CrossRef]

40. Seyda, V.; Herzog, D.; Emmelmann, C. Relationship between powder characteristics and part properties in laser beam melting of Ti–6Al–4V, and implications on quality. *J. Laser Appl.* **2017**, *29*, 022311. [CrossRef]

41. Pleass, C.; Jothi, S. Influence of powder characteristics and additive manufacturing process parameters on the microstructure and mechanical behaviour of Inconel 625 fabricated by Selective Laser Melting. *Addit. Manuf.* **2018**, *24*, 419–431. [CrossRef]

42. Balbaa, M.A.; Ghasemi, A.; Fereiduni, E.; Elbestawi, M.A.; Jadhav, S.D.; Kruth, J.P. Role of powder particle size on laser powder bed fusion processability of AlSi10mg alloy. *Addit. Manuf.* **2020**. [CrossRef]

43. Matthews, M.J.; Guss, G.; Khairallah, S.A.; Rubenchik, A.M.; Depond, P.J.; King, W.E. Denudation of metal powder layers in laser powder bed fusion processes. *Acta Mater.* **2016**, *114*, 33–42. [CrossRef]

44. Bidare, P.; Bitharas, I.; Ward, R.M.; Attallah, M.M.; Moore, A.J. Fluid and particle dynamics in laser powder bed fusion. *Acta Mater.* **2018**, *142*, 107–120. [CrossRef]

45. Spierings, A.B.; Levy, G. Comparison of density of stainless steel 316L parts produced with selective laser melting using different powder grades. In Proceedings of the Solid freeform fabrication Symposium, Austin, TX, USA, 3–5 August 2009; pp. 342–353.

Article

Microstructure and Mechanical Properties of Nickel-Based Coatings Fabricated through Laser Additive Manufacturing

Shaoxiang Qian [1,2,*], Yongkang Zhang [1,3,*], Yibo Dai [4] and Yuhang Guo [4]

[1] College of Mechanical Engineering, Jiangsu University, Zhenjiang 212013, China
[2] School of Modern Equipment Manufacturing, Zhenjiang College, Zhenjiang 212028, China
[3] School of Electromechanical Engineering, Guangdong University of Technology, Guangzhou 510006, China
[4] School of Materials Science and Engineering, Jiangsu University of Science and Technology, Zhenjiang 212003, China; 199060030@stu.just.edu.cn (Y.D.); guoyuhang@just.edu.cn (Y.G.)
*   Correspondence: ujsqiansx@163.com or qiansx@zjc.edu.cn (S.Q.); zhangykujs@163.com or ykzhang@gdut.edu.cn (Y.Z.); Tel.: +86-511-88962013 (S.Q.)

Abstract: In this study, single-layer and three-layer nickel-based coatings were fabricated on 316L SS by laser additive manufacturing. The phase characterization, microstructure observation, and microhardness analysis of the coatings were carried out by X-ray diffraction (XRD), scanning electron microscope (SEM), and microhardness tester. And the wear resistance of the coatings was analyzed through dry sliding friction and wear test. The results show that the cross-section microstructure of the three-layer nickel-based coating is different from that of the single-layer one under the influence of heat accumulation; the dendrite structure in the central region of the former is equiaxial dendrite, while that of the latter still remains large columnar dendrites. The existence of solid solution phase γ-(Fe, Ni) and hard phases of $Ni_{17}Si_3$, Cr_5B_3, Ni_3B in the coating significantly improve the wear resistance of the coating, and the microhardness is nearly 2.5 times higher than that of the substrate. However, the average microhardness of multilayer cladding coating is about 48 $HV_{0.2}$ higher than that of the single-layer cladding coating. In addition, the fine surface structure of the three-layer nickel-based coating improves the wear resistance of the coating, making this coating with the best wear resistance.

Keywords: laser additive manufacturing; 316l ss; nickel alloy; microstructure; tribological behavior

Citation: Qian, S.; Zhang, Y.; Dai, Y.; Guo, Y. Microstructure and Mechanical Properties of Nickel-Based Coatings Fabricated through Laser Additive Manufacturing. *Metals* **2021**, *11*, 53. https://doi.org/met11010053

Received: 9 December 2020
Accepted: 24 December 2020
Published: 29 December 2020

Publisher's Note: MDPI stays neutral with regard to jurisdictional claims in published maps and institutional affiliations.

Copyright: © 2020 by the authors. Licensee MDPI, Basel, Switzerland. This article is an open access article distributed under the terms and conditions of the Creative Commons Attribution (CC BY) license (https://creativecommons.org/licenses/by/4.0/).

1. Introduction

Due to its excellent corrosion resistance and mechanical properties, 316L stainless steel is widely used in the chemical industry and aerospace field [1–3]. However, the low hardness and the poor wear resistance limit its applications to a certain extent. Therefore, surface modification is needed to further improve the properties of the material [4,5], including atmospheric plasma spraying (APS) [6], chemical vapor deposition (CVD) [7] and laser additive manufacturing (LAM) [8]. As a new manufacturing technology integrating laser, digitization, materials science, and other disciplines, LAM has obtained widespread attention in recent years since it can realize dimension reduction manufacturing, complex forming, and high material utilization [9–11]. According to material feed-in methods, LAM can be divided into powder spreading type selective laser melting and powder feeding type laser melting deposition [12,13]. Laser melting deposition technology is also called laser cladding (LC), which uses a high-energy laser beam to metallurgically bond the cladding material to the substrate surface, with a small heat-affected zone (HAZ), fast cooling speed, and many other advantages [14,15]. In addition, LC can inject powder with low dilution and specific quality into the substrate to improve material performance and repair material surface defects, which has been the focus of material researchers in recent years [16–18].

Ti-, Fe-, Co- and Ni-based alloy powders are widely used in laser additive manufacturing due to their excellent high-temperature resistance, high hardness, and excellent wear

133

resistance. Ertugrul et al. [5] studied the structure and hardness behavior of 15 vol% TiC reinforced 316L coating and found that the increase in material hardness was due to the refinement of TiC grains and the appearance of new carbides during solidification. Cheng et al. [19] found that the mechanical properties and wear resistance of the FeBSiNb coating prepared by broad-beam laser cladding (BLC) were significantly improved, because the Nb-B particles generated in situ were uniformly dispersed in each part of the coating. Yan et al. [14] applied laser cladding technology to prepare TaC/Stellite X-40 Co-based composite coating on a nickel aluminum bronze (NAB) substrate. The results showed that intermetallic reinforcement materials such as TaC, Cr_3C_2, and Co_3Ta were distributed in the substrate uniformly, improving the wear resistance and the electrochemical corrosion performance of the alloy. When the TaC content is 20 wt%, the material reached the best wear resistance and corrosion resistance.

Among the cemented carbide powders mentioned above, Ni-based alloy powders have received extensive attention due to their excellent surface properties. Liu et al. [20] deposited a nickel-based WC composite powder on the surface of the tunnel boring machine (TBM) tool ring by LC and found that there was a good metallurgical bond between the coating and the substrate. The widely distributed WC, W_2C, and other particles in the coating effectively improved the wear resistance of the surface of the TBM tool ring. Liu et al. [21] adopted a multilayer laser cladding method to prepare a nickel-based alloy coating on the surface of graphite cast iron. The results showed that with the increase in the number of cladding layers, the volume fraction of dendrites and the distance between secondary dendrite arms increased significantly. When 6 layers of the coating were deposited, the material received the best corrosion resistance at room temperature. SouSa et al. [22] tested the tribological properties of the Ni-Cr-Bo-Si coating prepared by the laser cladding method, and the results showed that the coating had good surface adhesion. With the increase of the CrC, the microhardness of the coating was increased by 10%.

From the above research, it can be found that the cladding of nickel-based alloy powder on the substrate surface can effectively improve the wear resistance of the material. However, most studies have focused on the influence of different nickel-based alloy powder compositions on coating properties, and few studies on process parameters. During multilayer cladding, the existence of heat accumulation is likely to affect the structure and performance of the coating. Therefore, in this paper, the single-layer nickel-based coating and the three-layer nickel-based coating were prepared. Physical characterization, microstructure observation, and microhardness analysis of the coatings were performed using XRD, SEM, and microhardness tester. The wear resistance and wear mechanism of the coating were analyzed by a dry sliding friction wear test.

2. Experimental Procedure

Commercially available Ni-based alloy (Ni-Cr-B-Si-Fe-C) powder was used as the cladding material. The SEM image of the powder is shown in Figure 1. It can be seen that its shape is almost equiaxed spherical the size is approximately 50~150 μm, and the d_{50} of powder is about 100 μm. 316L stainless steel cut into 80 mm × 50 mm × 10 mm (thickness) was applied as the substrate. Before cladding, sandblasting technique (Al_2O_3 ceramic particles) was used over the substrate to have a surface roughness (Ra) of 10 μm. Then, the samples were rinsed with alcohol and dried to keep the surface clean. The chemical compositions of 316L stainless steel and nickel-based alloy powder were tested by Tianjin Zhujin Technology Development Co., Ltd. (Tianjin, China) and shown in Table 1.

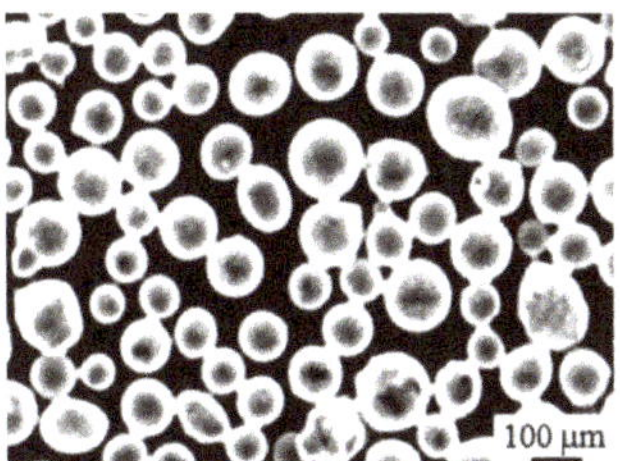

Figure 1. SEM image of Ni-based alloy powders.

Table 1. Chemical composition of powder and substrate metal.

Materials	Elements Composition (wt%)										
	C	P	Cr	S	Mn	Mo	Ni	Si	Fe	Co	B
316L steel	0.023	0.034	16.4	0.57	1.37	2.16	10.03	0.69	Bal.	-	-
Ni-based alloy	0.03	-	6	-	-	-	Bal.	1.5	0.38	-	3

RFL-C3300 (Raycus, Wuhan, China) high-power continuous fiber laser device was used in the laser cladding test. DPSF-2 powder feeder was used to feed the powder synchronously in an oblique direction, and argon was used as the power source to accurately send powder to the laser spot. During the cladding process, argon was used to protect the molten pool to avoid oxidation. In this experiment, single-layer (L1) and three-layer (L3) coatings were deposited respectively, as shown in Figure 2. The detailed parameters of laser cladding are listed in Table 2.

(a)   **(b)**

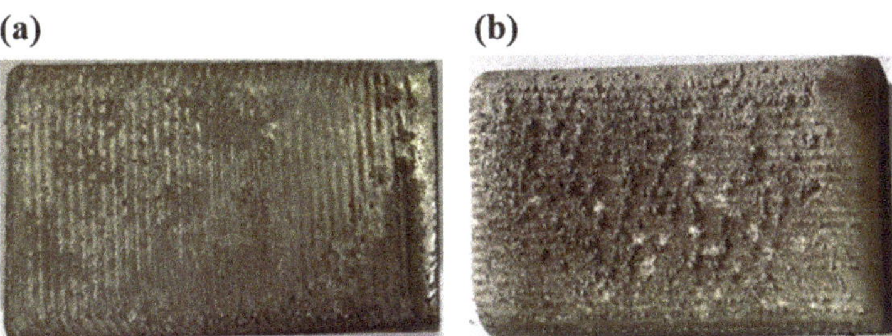

Figure 2. Macroscopic appearance of Ni-based laser-clad coating: (**a**) L1, (**b**) L3.

Table 2. Different parameters of laser cladding.

Laser Power (kW)	Powder Feeding Rate (g/min)	Scanning Velocity (mm/s)	Spot Diameter (mm)	Gas Flow Rate (L/min)	Overlap Rate
1.8	30	5	4	15	50%

When the laser cladding parameters are unchanged, the dilution rate of single-layer single-pass laser cladding is the same as single-layer multi-pass laser cladding and multi-layer multi-pass laser cladding. According to the laser cladding parameters in Table 2 (due to the single-layer single-pass cladding, the overlap rate parameter is not required here), the geometric dimensions of the single-layer single-pass laser cladding sample are shown in Figure 3. According to the literature, the dilution rate (λ) can be calculated by the area method [23], and the formula is simplified to:

$$\lambda = \frac{h}{H+h} \times 100\% \tag{1}$$

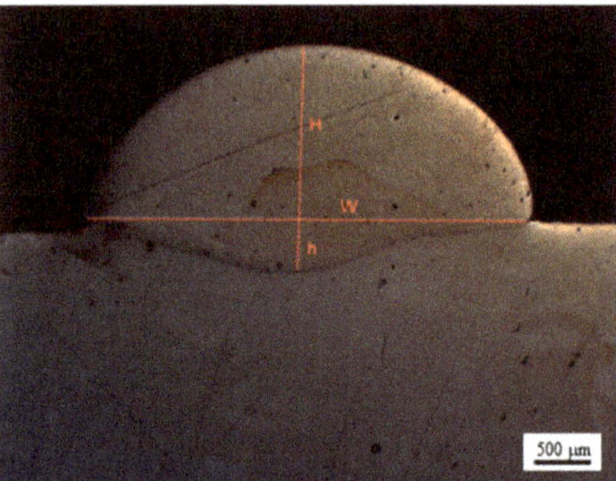

Figure 3. Feature size of single layer single pass laser cladding layer.

In the formula, h is the penetration depth of the base material, and H is the height of the cladding layer. According to the geometric dimensions shown in Figure 2, the dilution rate is 24.5%, indicating that the cladding layer and the substrate form a good bonding interface.

After cladding, the samples were cut, cleaned, and degreased with acetone alcohol, and then a standard metallographic sample preparation process (grinding, polishing, washing, and drying) was carried out to obtain mirror-surface metallographic and tribological samples. The phase identification of the cladding layer was performed by D8 AdvancE Multifunctional Powder Diffractometer (Bruker Corporation, Karlsruhe, Germany) with a Cu-k α radiation at a voltage of 40 kV. The micromorphology of the coating was observed by JSM-6480 tungsten filament scanning electron microscope (JEOL Corporation, Tokyo, Japan), and the element distribution in the micro area was measured by its equipped energy dispersive spectrometer (EDS). According to ASTM E92-2016 standard, KB-30S-FA Automatic Vickers Hardness Tester (KB Corporation, Assenheim, Germany) was used to test the hardness of the coating with the applied pressure of 0.2 kg and the holding time of 15 s. According to the ASTM G99-05 standard, the UMT-2 high temperature friction and wear tester (CETR Corporation, CA, USA) was used to study the dry sliding ball-on-plate wear of the nickel-based coating surface. For the dry sliding friction test, the sliding distance is 18 m, the relative speed between coating and counter-body is 10 mm/s, and the diameter of the counter-body ball is 10 mm. The surface roughness before wear test and the three-dimensional morphology of wear marks were observed by LEXTOLS4000 confocal laser scanning microscope (Leica Corporation, Wetzler, Germany). Before the wear test, the surface roughness (Sq) of the substrate, L1, and L3 were 0.012, 0.044, and 0.046 μm, respectively. Finally, the micro wear morphology was observed through SEM.

3. Results and Discussion

3.1. Microstructure and Composition

XRD patterns of coatings are shown in Figure 4. According to Table 1, the content of Fe in nickel-based alloy powder is very low, but face centered cubic structural γ-(Fe, Ni) phase has been detected in XRD. This is because the molten pool is formed on the substrate surface under the irradiation of laser beam, and interdiffusion occurs between the substrate and the nickel-based coating [24], making a large amount of Fe enter the coating. The radius of Fe atoms is very close to that of Ni atoms, and the rapid rise and decrease of temperature are similar to solid solution treatment. Thus γ-(Fe, Ni) solid solution appears on the coating surface of the single-layer cladding sample. However, during multilayer cladding, Fe is difficult to diffuse to the coating surface. Therefore, although the Fe-containing main dominant peaks are detected at 44° for both L1 and L3 samples, the phase content in L3 has changed, which can also be proved from the EDS results in Table 3. In addition, hard phases Cr_5B_3, $Ni_{17}Si_3$, and Ni_3B have also been detected on the coating surface, which is proved to be helpful to improve the hardness and wear resistance of the coating [25–29].

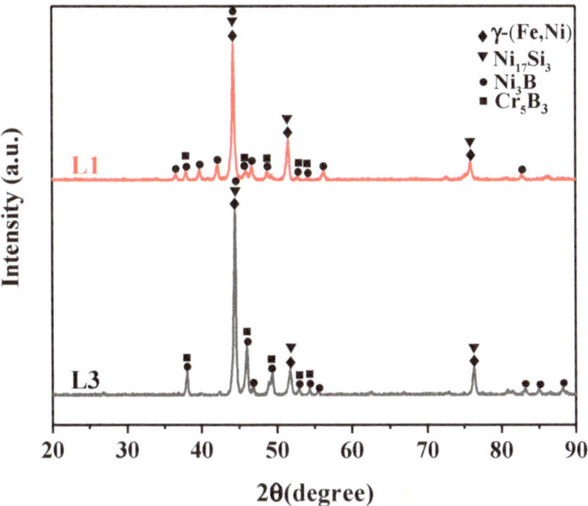

Figure 4. XRD patterns of the coatings.

Table 3. Chemical composition of phases in Figures 5 and 6.

Region	Element Composition (at.%)			
	Si	Cr	Fe	Ni
A	6.16	3.19	15.46	75.09
B_1	6.77	3.08	13.75	76.40
B_2	5.47	2.79	11.04	80.70
D_1	15.77	-	-	84.23
D_2	7.04	-	-	92.96
D_3	5.59	-	0.87	93.54

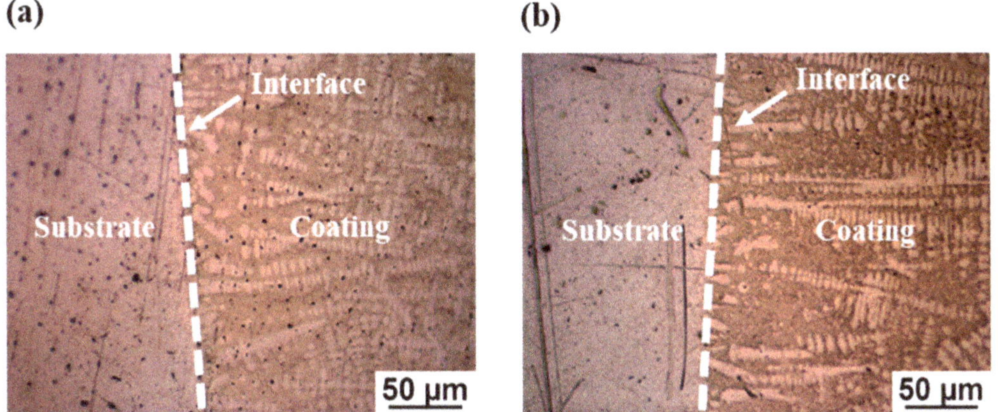

Figure 5. Cross-section OM images of (**a**) L1 and (**b**) L3.

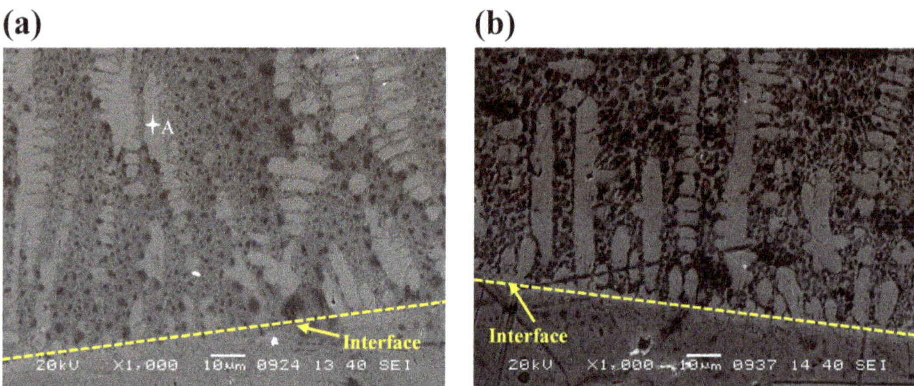

Figure 6. Cross-section morphology (SEM) near the coating-substrate interface (**a**) L1, (**b**) L3.

Figure 5 displays the metallographic diagram of the cross section of the nickel-based coating. It can be seen that the coating is well bonded with the substrate, and there are no obvious cracks. a lot of gas pores are observed in the coating of sample L1 (Figure 5a). During laser cladding, the temperature rises and decreases so rapidly that the gases have little time to escape, leaving them to remain in the coating and form pores. Moreover, these pores are randomly distributed and have nothing to do with the trajectory of the boundary, and the existence of pores will not affect the bonding quality between the substrate and the coating [30]. However, not many gas pores can be seen in the coating of sample L3 (Figure 5b), which is because the multilayer cladding will reheat the cladding area, greatly reducing the pores.

The SEM images of the cross section near the interface between coating and substrate are shown in Figure 6. According to Kou's study [31], the ratio of G to R can be used to describe the change of microstructure, where G is the temperature gradient and R is the growth rate. It is well known that the ratio of G to R is high due to the good heat dissipation of the substrate and the short solidification time of the solution near the interface during the solidification of the coating. Thus, some cellular crystals appear along the boundary line. As shown in Figure 6a,b, with the decrease of the temperature gradient, an obvious columnar dendrite structure can be observed near the boundary line, and the columnar dendrite grows along the temperature gradient direction. According to the EDS data in Table 3, the main component of point a in Figure 6a is Ni, with a small amount of Si, Cr, and Fe, indicating that the composition of dendrite is γ-Ni phase and the binary eutectic γ-Ni formed by γ-Ni phase and other metal elements (Si, Cr, and Fe). Due to the high cooling rate, a dense eutectic phase was both observed between the γ-Ni dendrites in Figure 6a,b, which shows a network structure in sample L1 while a honeycomb texture in sample L3 due to multiple heating. Many dark regions can be seen between the eutectic phases. Paul believed that the main component of this region is borides [32], and the XRD results also verify the existence of borides. Therefore, the dendrite structure of eutectic is boride and binary eutectic γ-Ni. The microstructure near the coating-substrate interface is a uniform combination of columnar dendrites and interdendritic crystals.

The SEM cross-section morphologies of the middle and surface areas of the coatings are shown in Figure 7. Due to the change of composition and the decrease of temperature gradient, the growth of columnar dendrites is limited. Thus, columnar dendrites are not observed in the middle of the coating, while uniformly arranged equiaxed dendrites appear instead. Meanwhile, the network eutectic structure between equiaxed dendrites becomes uniform due to the decrease of temperature gradient. Figure 7b is the SEM picture of the single-layer cladding coating surface and both columnar dendrites and steering dendrites can be observed. Due to the change of temperature gradient caused by the co-effect of the contact between the coating surface and air and the movement of laser beam [33], not only the columnar dendrites growth along the temperature gradient appear, but also the steering dendrites grow along the scanning direction on the coating surface occur. When the coating surface contacts the air, the cooling rate will be faster, and the grain growth will be limited. Thus it can be seen that the grain size in Figure 7b is significantly smaller than that in Figure 7a. In addition, more Fe elements have been detected in B_1 and B_2 in Figure 7b, indicating that single-layer cladding is more conducive to the diffusion of Fe elements to the coating surface so as to form γ-(Fe, Ni) lamellar eutectic with Ni [4]. Figure 7c is the SEM picture of the middle area of the multilayer cladding. When the second cladding layer solidifies, although the first cladding layer has solidified, it still has a certain amount of heat, resulting in heat accumulation. Due to the accumulation of heat, the value of G/R is not high enough to form planar crystal and cellular crystal, but columnar dendrite appears directly in Figure 7c. At the same time, the effect of heat accumulation increases the reheating area, and the solidification rate of grains decreases, resulting in larger columnar dendrite size [34]. On the other hand, heat accumulation reduces the effect of temperature gradient on the grain growth direction, making the columnar dendrites present different directions. From Figure 7d, it's hard to find obvious dendrites. This is because when it is near the top of the molten pool, the solidification rate of the molten pool increases rapidly due to the contact with air, so that the dendrites cannot grow sufficiently. Moreover, a large number of fine particles appear in Figure 7d and Mohamad et al. [35] believed that the particles not completely melted on the surface of the cladding layer would play the role of heterogeneous nucleation points. As the EDS data in Table 3 show, the main components of the points D_1, D_2, and D_3 are $Ni_{17}Si_3$, indicating that the large number of $Ni_{17}Si_3$ distributed on the coating surface during coating solidification help with the formation of a finer surface structure of the coating.

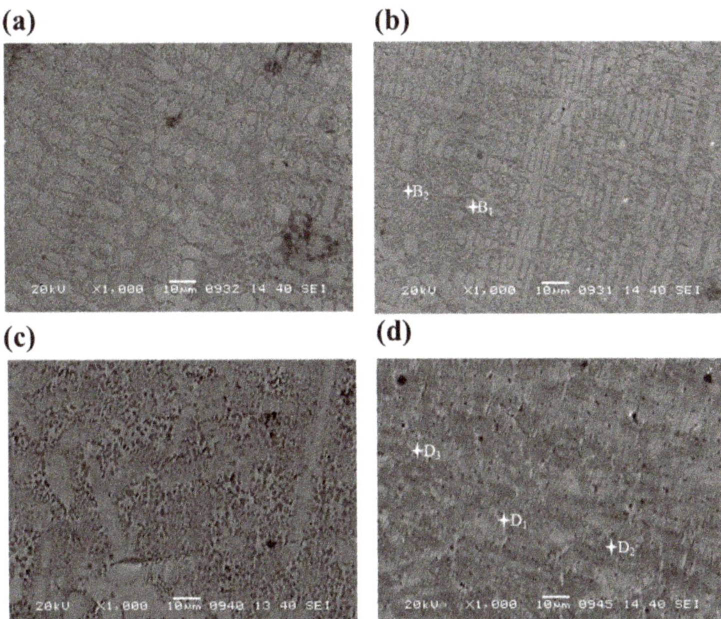

Figure 7. Cross-section SEM morphologies of the middle and surface of the coatings: (**a**) Middle region of L1, (**b**) Surface region of L1, (**c**) Middle region of L3, (**d**) Surface region of L3.

3.2. Wear Analysis

Figure 8 shows the cross-sectional microhardness distribution of different samples. It can be seen that the microhardness of both coatings is not less than 600 $HV_{0.2}$, which is nearly 2.5 times higher than that of the 316L SS substrate (about 250 $HV_{0.2}$). Due to the bonding of the coating and the substrate, the substrate is inevitably diluted [36]. Comparing Figure 8b to Figure 8a, the microhardness of the substrate close to the HAZ decreases slightly because of Fe dilution [4]. The HAZ of laser cladding is very small, Landowski [37] found that the HAZ of laser welded 316L SS joint is about 20 μm. Therefore, the presence of HAZ has a very limited effect on the microhardness of the substrate. During cladding, the microhardness distribution of the cross-section changes owing to the change of temperature. For the multilayer cladded L3 sample, the temperature changes much greater than the single-layer cladded L1 sample during laser cladding, resulting in a greater fluctuation of cross-sectional microhardness than L1. The significant improvement of microhardness of the coating compared with 316L SS substrate can be attributed to the addition of alloying elements and the γ-(Fe, Ni) solid solution phase and hard phases $Ni_{17}Si_3$, Cr_5B_3, and Ni_3B formed during the cladding process. In addition, the rapid rise and fall of temperature during laser cladding will produce residual stress in the coating, which also helps improve the microhardness of the coating [38]. The average microhardness of L1 coating is 593 (±6) $HV_{0.2}$, while that of L3 coating is 640 (±20)According to the analysis above, the improvement of microhardness is related to the fine surface structure of the coating.

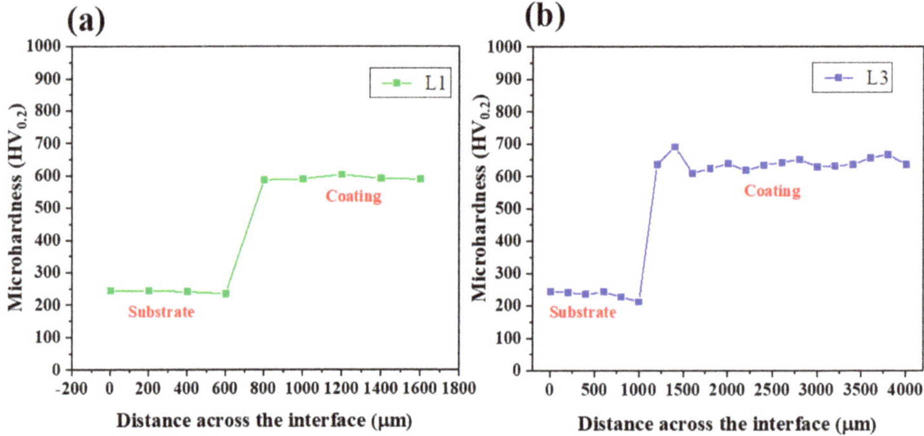

Figure 8. Microhardness profile of laser cladded cross section: (**a**) L1, (**b**) L3.

The dry sliding friction and wear test with a sliding friction load of 10 N and a wear time of 1800 s was conducted to evaluate the friction performance of the substrate and the coatings. The obtained friction coefficient curve of the substrate and the coatings and the relevant histogram of the average friction coefficient calculated from the friction coefficient curve are shown in Figure 9a,b, respectively. As shown in Figure 9a, all samples reach a high limited value of the friction coefficient before it stabilizes. This is because in the early stage of sliding friction, the material is continuously sheared, and the deformation layer is continuously accumulated. And the accumulated deformation layer is continuously washed to ensure the smooth progress of the experiment, so the friction coefficient is increased [39]. In addition, since the hardness of the substrate is relatively lower, the deformation layer accumulated in the sliding friction experiment is easier to be washed, so the friction coefficient of the substrate tends to be stable earlier than that of the coating. As the sliding friction experiment continues, the friction coefficient of the substrate shows a slight decrease. Kumar believes that this is because the transfer of materials during the sliding friction increases the carbon concentration on the surface of 316L SS, thus promoting the formation of the friction film [40]. It can be concluded from Figure 9b that the average friction coefficient of different samples meets the substrate (0.53 ± 0.02)

> sample L1 (0.47 $\pm$ 0.05) > sample L3 (0.44 $\pm$ 0.03), indicating that nickel-based alloy coating can effectively improve the wear resistance of 316L SS, and the improving effect of multilayer cladding is better than that of single-layer cladding.

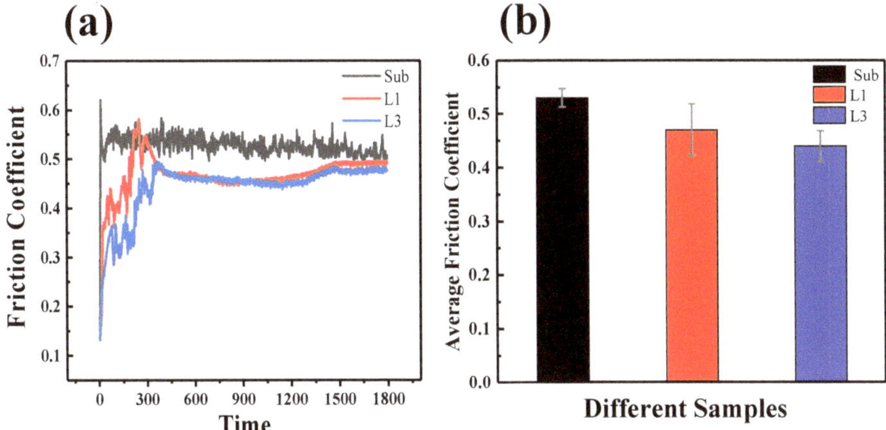

Figure 9. (**a**) Friction coefficient curves of the substrate (Sub), L1 and L3, (**b**) Histogram of average friction coefficient.

Figure 10 is the wear scar surface and the corresponding wear scar profile data obtained from the dry sliding friction and wear test. It is obvious that the wear resistance of 316L SS can be effectively improved by Ni-based alloy coating whether by single-layer cladding or multilayer cladding. Comparing Figure 10a–c, it can be found that the wear profile of the substrate surface shows more deflection. The difference in deflection depends on the difference in the contact area. Under the same load condition, the microhardness of the substrate surface is lower, which makes the contact area between the Al_2O_3 ceramic ball and the material surface larger. The higher the contact area, the higher the length of wear track and peak/valley heights, which in turn increases the wear rate of the material. At the same time, due to the lower microhardness of the substrate, greater deformation occurs during sliding friction, and obvious protrusions appear on both sides of the wear scar. In Figure 10b,c, samples L1 and L3 have similar 3D morphologies and the profiles of wear marks, indicating that the wear extent of L1 and L3 in dry sliding friction and wear test is similar. In addition, the relatively flat wear surface of samples L1 and L3 demonstrating that the higher microhardness of the coating surface makes the load applied to the coating be effectively distributed to the whole wear surface.

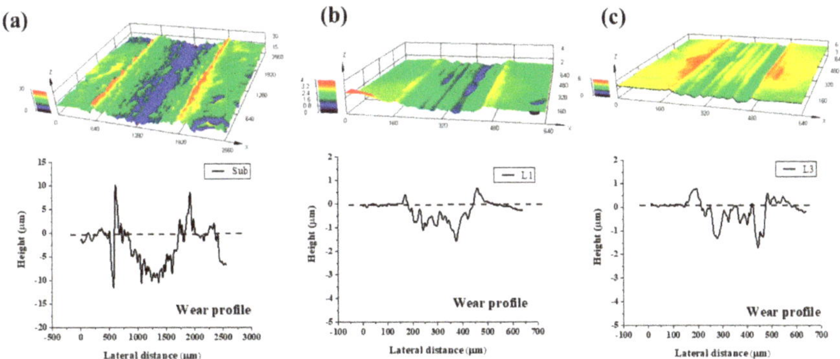

Figure 10. Surface wear of the substrate and the coating under the dry sliding friction and wear test conditions with a sliding friction load of 10N and a wear time of 1800 s: (**a**) Sub, (**b**) L1, (**c**) L3.

Figure 11 shows the SEM image of the morphologies of the substrate and the coatings after the dry sliding friction and wear test. Figure 11a,b are the wear SEM morphologies of the substrate. It can be seen that the wear surface is rough and uneven, and the particles peel off obviously, showing typical abrasive wear characteristics. As shown in Figure 11a, the substrate surface shows a high wear extent with some pores generated during sample preparation exposed. The surface in Figure 11b is even seriously worn, with large wear debris particles attached to it as well as many deep grooves in the area with severe particle spalling. Under a load of 10 N, particles frequently fall off from the surface, causing delamination on the wear surface, so as to greatly increase the wear rate. As seen in Figure 11c,d, the wear surfaces of the coatings are in good condition with few particles peeling off or delamination, and the coatings show excellent wear resistance. After an 1800 s sliding friction, some peeling-off appears on the surface of sample L1. At the same time, a small number of black patches appeared in Figure 11c, indicating that the wear mechanism of the L1 sample is a combination of abrasive wear and adhesive wear. In the figure, there are even obvious micro-cracks in the area where the spalling phenomenon is more serious. If the load or wear time continues to increase, the micro-cracks will expand and delamination will occur [41]. a large area of black patches can be seen in Figure 11d. During sliding friction, the particles will be transferred from the contact material. Because of the high hardness of the coating, the particles eventually appear in the form of patches, resulting in adhesive wear. Moreover, a large number of small pits in the middle of the patches indicate slight abrasive wear. Sometimes, the particles adhering to the contact surface will fill the gap (as shown in Figure 11d) during sliding, thus reducing the wear extent of the coating, but it won't affect the wear mechanism [42]. In the process of sliding friction, the positive pressure between the load and the tangential motion of Al_2O_3 ceramic ball causes plastic deformation of the material, resulting in shallow grooves along the sliding direction. In addition, the fine surface structure of the L3 coating can effectively improve the wear resistance of the coating, and there is no obvious particle spalling trace except slight wear marks. In general, no matter by single-layer cladding or by multi-layer cladding, nickel-based coatings change the wear mechanism of materials in the process of friction and wear, effectively reduce the abrasive wear extent and improve the wear resistance of materials; the three-layer nickel-based coating displays the best wear resistance.

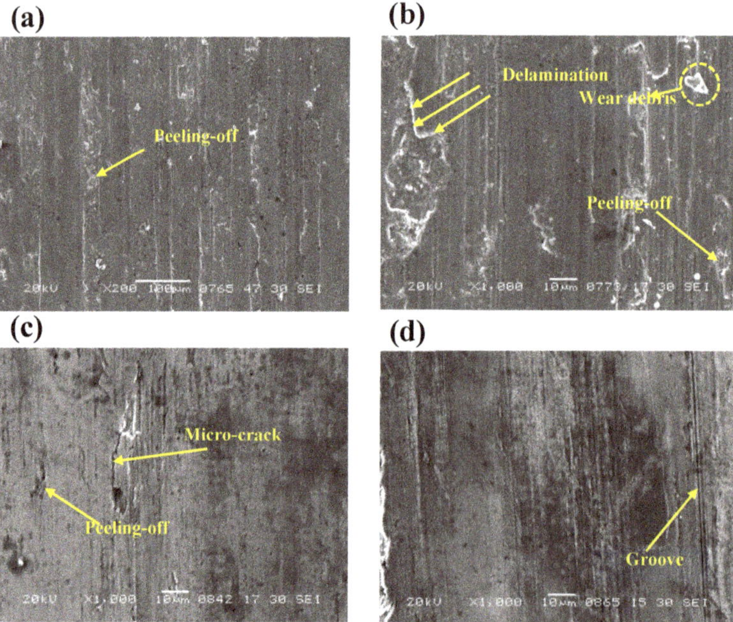

Figure 11. SEM morphologies of wear surfaces for (**a**,**b**) the substrate, (**c**) L1 and (**d**) L3.

4. Conclusions

The single-layer and three-layer nickel-based coatings have been fabricated on 316L SS substrate by laser additive manufacturing and are metallurgically bonded well with the substrate without obvious defects. The microstructure, surface precipitates and wear resistance of the coatings have been systematically studied and the following conclusions have been drawn:

(1) The XRD patterns of single-layer cladding and multilayer cladding coatings are similar, both including γ-(Fe, Ni) solid solution phase, Cr_5B_3, $Ni_{17}Si_3$, and Ni_3B phases. Due to the diffusion of Fe, the content of γ-(Fe, Ni) solid solution in the surface of single-layer coating is higher than that of the multilayer coating.

(2) Cellular crystals are observed in the coating-substrate interface of both single-layer cladding and multilayer cladding samples. With the change of G/R ratio, equiaxed dendrites appear in the middle of the single-layer coating, and columnar dendrites and steering crystals appear on the surface. However, due to heat accumulation, columnar dendrites with large size and different directions appear in the middle region of the multilayer cladded Ni-based coating. No obvious dendrite morphologies are observed in the surface area of the multilayer cladded coating, and the massively distributed $Ni_{17}Si_3$ phases help to obtain a finer surface structure.

(3) The surface microhardness of 316L SS has been significantly improved by Ni-based coating by about 2.5 times. The average microhardness of multilayer cladding coating is about 48 $HV_{0.2}$ higher than that of the single-layer cladding coating, and the average friction coefficient of three-layer coating is the lowest (0.44 $\pm$ 0.03) among the substrate and the single-layer coating. In addition, the fine surface structure of the three-layer nickel-based coating improves the wear resistance of the coating, making this coating with the best wear resistance.

(4) The Ni-based coating shows excellent wear resistance. After dry sliding friction and wear test, the substrate has been severely worn down, and the wear mechanism is mainly abrasive wear. Some peeling-off appear on the surface of single-layer nickel-based coating, and the main wear mechanism is a combination of abrasive wear and adhesive wear, while the three-layer coating is in good condition without serious abrasion due to its fine surface structure and the main wear mechanism is adhesive wear.

Author Contributions: Conceptualization, S.Q. and Y.Z.; methodology, S.Q.; software, Y.D.; validation, S.Q., Y.G. and Y.Z.; formal analysis, Y.G.; investigation, Y.G.; resources, S.Q.; data curation, Y.D.; writing—original draftpreparation, S.Q.; writing—review and editing, S.Q.; visualization, Y.D.; supervision, Y.Z.; project administration, Y.Z.; funding acquisition, Y.Z. All authors have read and agreed to the published version of the manuscript.

Funding: This research received no external funding.

Institutional Review Board Statement: "Not applicable" for studies not involving humans or animals.

Informed Consent Statement: Informed consent was obtained from all subjects involved in the study.

Data Availability Statement: The data that support the findings of this study are available on request from the corresponding author.

Conflicts of Interest: The authors declare no conflict of interest.

References

1. Al-Mamun, N.S.; Haider, W.; Shabib, I. Corrosion resistance of additively manufactured 316L stainless steel in chloride−thiosulfate environment. *Electrochim. Acta* **2020**, *362*, 137039. [CrossRef]
2. Canakci, A.; Ozkaya, S.; Erdemir, F.; Karabacak, A.H.; Celebi, M. Effects of Fe–Al intermetallic compounds on the wear and corrosion performances of AA2024/316L SS metal/metal composites. *J. Alloys Compd.* **2020**, *845*, 156236. [CrossRef]
3. Li, J.; Qin, W.; Peng, P.; Chen, M.; Mao, Q.; Yue, W.; Kang, J.; Meng, D.; She, D.; Zhu, X.; et al. Effects of geometric dimension and grain size on impact properties of 316L stainless steel. *Mater. Lett.* **2021**, *284*, 128908. [CrossRef]

4. Jeyaprakash, N.; Yang, C.-H.; Sivasankaran, S. Laser cladding process of Cobalt and Nickel based hard-micron-layers on 316L-stainless-steel-substrate. *Mater. Manuf. Process.* **2019**, *35*, 142–151. [CrossRef]
5. Ertugrul, O.; Enrici, T.M.; Paydas, H.; Saggionetto, E.; Boschini, F.; Mertens, A. Laser cladding of TiC reinforced 316L stainless steel composites: Feedstock powder preparation and microstructural evaluation. *Powder Technol.* **2020**, *375*, 384–396. [CrossRef]
6. Karaoglanli, A.C.; Ozgurluk, Y.; Doleker, K.M. Comparison of microstructure and oxidation behavior of CoNiCrAlY coatings produced by APS, SSAPS, D-gun, HVOF and CGDS techniques. *Vacuum* **2020**, *180*, 109609. [CrossRef]
7. Song, M.; Yang, Y.; Guo, J.; Xiang, M.; Zhu, Q.; Ge, Y. a new precursor for fabricating TiN coating on 316L stainless steel at low temperature without corrosive byproducts. *Ceram. Int.* **2019**, *45*, 18265–18272. [CrossRef]
8. Yu, Q.; Wang, C.; Wang, D.; Min, X. Microstructure and properties of Ti–Zr congruent alloy fabricated by laser additive manufacturing. *J. Alloys Compd.* **2020**, *834*, 155087. [CrossRef]
9. Liu, Y.; Zhang, Y. Microstructure and mechanical properties of TA15-Ti2AlNb bimetallic structures by laser additive manufacturing. *Mater. Sci. Eng. A* **2020**, *795*, 140019. [CrossRef]
10. Wang, H. Materials' fundamental issues of laser additive manufacturing for high-performance large metallic components. *Acta Aeronaut. Astronaut. Sin.* **2014**, *35*, 2690–2698.
11. Lu, Y.; Huang, Y.; Wu, J. Laser additive manufacturing of structural-graded bulk metallic glass. *J. Alloys Compd.* **2018**, *766*, 506–510. [CrossRef]
12. Fang, Z.-C.; Wu, Z.-L.; Huang, C.-G.; Wu, C.-W. Review on residual stress in selective laser melting additive manufacturing of alloy parts. *Opt. Laser Technol.* **2020**, *129*, 106283. [CrossRef]
13. Adeyemi, A.; Akinlabi, E.T.; Mahamood, R.M. Powder bed based laser additive manufacturing process of stainless steel: a review. *Mater. Today Proc.* **2018**, *5*, 18510–18517. [CrossRef]
14. Li, Z.; Yan, H.; Zhang, P.; Guo, J.; Yu, Z.; Ringsberg, J.W. Improving surface resistance to wear and corrosion of nickel-aluminum bronze by laser-clad TaC/Co-based alloy composite coatings. *Surf. Coat. Technol.* **2020**, 126592, in press. [CrossRef]
15. Javid, Y. Multi-response optimization in laser cladding process of WC powder on Inconel 718. *CIRP J. Manuf. Sci. Technol.* **2020**, *31*, 406–417. [CrossRef]
16. Zhu, L.; Liu, Y.; Li, Z.; Zhou, L.; Li, Y.; Xiong, A. Microstructure and properties of Cu-Ti-Ni composite coatings on gray cast iron fabricated by laser cladding. *Opt. Laser Technol.* **2020**, *122*, 105879. [CrossRef]
17. Wang, H.-H.; Li, H.; Li, H.; Liu, X.-S.; Kong, J.-A.; Zhou, H. Repair of SiC coating on carbon/carbon composites by laser cladding technique. *Ceram. Int.* **2020**, *46*, 19537–19544. [CrossRef]
18. Yuan, W.; Li, R.; Chen, Z.; Gu, J.; Tian, Y. a comparative study on microstructure and properties of traditional laser cladding and high-speed laser cladding of Ni45 alloy coatings. *Surf. Coat. Technol.* **2020**, 126582, in press. [CrossRef]
19. Sun, B.; Cheng, J.; Cai, Z.; Zhao, H.; Zhang, Z.; Qu, H.; Zhang, Q.; Hong, S.; Liang, X. Formation and wear property of broad-beam laser clad Fe-based coatings. *Surf. Coat. Technol.* **2020**, 126598. [CrossRef]
20. Hu, D.; Liu, Y.; Chen, H.; Wang, M. Microstructure and wear resistance of Ni-based tungsten carbide coating by laser cladding on tunnel boring machine cutter ring. *Surf. Coat. Technol.* **2020**, *404*, 126432. [CrossRef]
21. Liu, J.; Liu, H.; Tian, X.; Yang, H.; Hao, J. Microstructural evolution and corrosion properties of Ni-based alloy coatings fabricated by multi-layer laser cladding on cast iron. *J. Alloys Compd.* **2020**, *822*, 153708. [CrossRef]
22. De Sousa, J.M.S.; Ratusznei, F.; Pereira, M.; Castro, R.D.M.; Curi, E.I.M. Abrasion resistance of Ni-Cr-B-Si coating deposited by laser cladding process. *Tribol. Int.* **2020**, *143*, 106002. [CrossRef]
23. Tomków, J.; Czupryński, A.; Fydrych, D. The abrasive wear resistance of coatings manufactured on high-strength low-alloy (HSLA) offshore steel in wet welding conditions. *Coatings* **2020**, *10*, 219. [CrossRef]
24. Gargarella, P.; Almeida, A.; Vilar, R.; Afonso, C.R.M.; Peripolli, S.; Rios, C.T.; Bolfarini, C.; Botta, W.J.; Kiminami, C.S. Formation of Fe-based glassy matrix composite coatings by laser processing. *Surf. Coat. Technol.* **2014**, *240*, 336–343. [CrossRef]
25. Ramasubbu, V.; Chakraborty, G.; Albert, S.K.; Bhaduri, A.K. Effect of dilution on GTAW Colmonoy 6 (AWS NiCr–C) hardface deposit made on 316LN stainless steel. *Mater. Sci. Technol.* **2011**, *27*, 573–580. [CrossRef]
26. Zikin, A.; Hussainova, I.; Winkelmann, H.; Kulu, P.; Badisch, E. Plasma transferred arc hardfacings reinforced by chromium carbide based cermet particles. *Int. Heat Treat. Surf. Eng.* **2012**, *6*, 88–92. [CrossRef]
27. Hemmati, I.; Rao, J.C.; Ocelík, V.; De Hosson, J.T.M. Electron microscopy characterization of Ni-Cr-B-Si-C laser deposited coatings. *Microsc. Microanal.* **2013**, *19*, 120–131. [CrossRef]
28. Hemmati, I.; Ocelík, V.; De Hosson, J.T.M. Dilution effects in laser cladding of Ni–Cr–B–Si–C hardfacing alloys. *Mater. Lett.* **2012**, *84*, 69–72. [CrossRef]
29. Kumar, J.; Singh, D.; Kalsi, N.S.; Sharma, S.; Pruncu, C.I.; Pimenov, D.Y.; Rao, K.V.; Kapłonek, W. Comparative study on the mechanical, tribological, morphological and structural properties of vortex casting processed, Al–SiC–Cr hybrid metal matrix composites for high strength wear-resistant applications: Fabrication and characterizations. *J. Mater. Res. Technol.* **2020**, *9*, 13607–13615. [CrossRef]
30. Moskal, G.; Niemiec, D.; Chmiela, B.; Kałamarz, P.; Durejko, T.; Ziętala, M.; Czujko, T. Microstructural characterization of laser-cladded NiCrAlY coatings on Inconel 625 Ni-based superalloy and 316L stainless steel. *Surf. Coat. Technol.* **2020**, *387*, 125317. [CrossRef]
31. Kou, S. *Welding Metallurgy*, 2nd ed.; Wiley: Hoboken, NJ, USA, 2003; pp. 431–446.

32. Paul, C.P.; Jain, A.; Ganesh, P.; Negi, J.; Nath, A.K. Laser rapid manufacturing of Colmonoy-6 components. *Opt. Lasers Eng.* **2006**, *44*, 1096–1109. [CrossRef]
33. Hemmati, I.; Ocelík, V.; De Hosson, J.T.M. Microstructural characterization of AISI 431 martensitic stainless steel laser-deposited coatings. *J. Mater. Sci.* **2011**, *46*, 3405–3414. [CrossRef]
34. Luo, K.; Xu, X.; Zhao, Z.; Zhao, S.; Cheng, Z.; Lu, J. Microstructural evolution and characteristics of bonding zone in multilayer laser cladding of Fe-based coating. *J. Mater. Process. Technol.* **2019**, *263*, 50–58. [CrossRef]
35. Saeedi, R.; Shoja-Razavi, R.; Bakhshi, S.-R.; Erfanmanesh, M.; Ahmadi-Bani, A. Optimization and characterization of laser cladding of NiCr and NiCr–TiC composite coatings on AISI 420 stainless steel. *Ceram. Int.* **2020**. [CrossRef]
36. Conde, A.; Zubiri, F. Cladding of Ni–Cr–B–Si coatings with a high power diode laser. *Mater. Sci. Eng. A* **2002**, *334*, 233–238. [CrossRef]
37. Landowski, M.; Świerczyńska, A.; Rogalski, G.; Fydrych, D. Autogenous fiber laser welding of 316L austenitic and 2304 lean duplex stainless steels. *Materials* **2020**, *13*, 2930. [CrossRef] [PubMed]
38. Wang, D.; Hu, Q.; Zeng, X. Residual stress and cracking behaviors of Cr13Ni5Si2 based composite coatings prepared by laser-induction hybrid cladding. *Surf. Coat. Technol.* **2015**, *274*, 51–59. [CrossRef]
39. Wang, B.; Liu, B.; Zhang, X.; Gu, J. Enhancing heavy load wear resistance of AISI 4140 steel through the formation of a severely deformed compound-free nitrided surface layer. *Surf. Coat. Technol.* **2018**, *356*, 89–95. [CrossRef]
40. Upadhyay, R.; Kumar, A. Scratch and wear resistance of additive manufactured 316L stainless steel sample fabricated by laser powder bed fusion technique. *Wear* **2020**, *458*, 203437. [CrossRef]
41. Liu, H.; Sun, S.; Zhang, T.; Zhang, G.; Yang, H.; Hao, J. Effect of Si addition on microstructure and wear behavior of AlCoCrFeNi high-entropy alloy coatings prepared by laser cladding. *Surf. Coat. Technol.* **2020**, 126522. [CrossRef]
42. Upadhyay, R.; Kumar, A. Effect of humidity on the synergy of friction and wear properties in ternary epoxy-graphene-MoS2 composites. *Carbon* **2019**, *146*, 717–727. [CrossRef]

Article

Deep Learning-Based Ultrasonic Testing to Evaluate the Porosity of Additively Manufactured Parts with Rough Surfaces

Seong-Hyun Park [1], Jung-Yean Hong [1], Taeho Ha [2], Sungho Choi [3],* and Kyung-Young Jhang [4],*

[1] Department of Mechanical Convergence Engineering, Hanyang University, Seoul 04763, Korea;
seonghyun@hanyang.ac.kr (S.-H.P.); hongjy127@hanyang.ac.kr (J.-Y.H.)

[2] Department of 3D Printing, Korea Institute of Machinery & Materials, Daejeon 34103, Korea;
taehoha@kimm.re.kr

[3] Department of Flexible and Printable Electronics, LANL-JBNU Engineering Institute-Korea,
Jeonbuk National University, Jeonju-si 54896, Korea

[4] School of Mechanical Engineering, Hanyang University, Seoul 04763, Korea

* Correspondence: schoi@jbnu.ac.kr (S.C.); kyjhang@hanyang.ac.kr (K.-Y.J.); Tel.: +82-63-219-5437 (S.C.);
+82-2-2220-4220 (K.-Y.J.)

Abstract: Ultrasonic testing (UT) has been actively studied to evaluate the porosity of additively manufactured parts. Currently, ultrasonic measurements of as-deposited parts with a rough surface remain problematic because the surface lowers the signal-to-noise ratio (SNR) of ultrasonic signals, which degrades the UT performance. In this study, various deep learning (DL) techniques that can effectively extract the features of defects, even from signals with a low SNR, were applied to UT, and their performance in terms of the porosity evaluation of additively manufactured parts with rough surfaces was investigated. Experimentally, the effects of the processing conditions of additive manufacturing on the resulting porosity were first analyzed using both optical and scanning acoustic microscopy. Second, convolutional neural network (CNN), deep neural network, and multi-layer perceptron models were trained using time-domain ultrasonic signals obtained from additively manufactured specimens with various levels of porosity and surface roughness. The experimental results showed that all the models could evaluate porosity accurately, even that of the as-deposited specimens. In particular, the CNN delivered the best performance at 94.5%. However, conventional UT could not be applied because of the low SNR. The generalization performance when using newly manufactured as-deposited specimens was high at 90%.

Keywords: additive manufacturing; porosity; rough surface; ultrasonic testing; convolutional neural network; deep neural network; multi-layer perceptron

Citation: Park, S.-H.; Hong, J.-Y.; Ha, T.; Choi, S.; Jhang, K.-Y. Deep Learning-Based Ultrasonic Testing to Evaluate the Porosity of Additively Manufactured Parts with Rough Surfaces. *Metals* **2021**, *11*, 290. https://doi.org/10.3390/met11020290

Academic Editor: Atila Ertas

Received: 20 December 2020
Accepted: 10 January 2021
Published: 8 February 2021

Publisher's Note: MDPI stays neutral with regard to jurisdictional claims in published maps and institutional affiliations.

Copyright: © 2021 by the authors. Licensee MDPI, Basel, Switzerland. This article is an open access article distributed under the terms and conditions of the Creative Commons Attribution (CC BY) license (https://creativecommons.org/licenses/by/4.0/).

1. Introduction

Additive manufacturing (AM) is the process of depositing materials, layer upon layer, to create an object from a 3D computer-aided design [1–4]. The distinctive advantages of AM are that it can be used to produce innovative, complex designs and the fact that it is lightweight compared to conventional subtractive or casting manufacturing. Owing to these advantages, this manufacturing method has been actively studied in various fields [5–7]. A current major concern is manufacturing defects that can occur in the interior of AM parts and their effects on the integrity of these parts [8,9]. Porosity, defined as air-filled cavities inside a material, is a typical manufacturing defect found in AM parts, and is the result of deviation from the optimal AM processing conditions. Because porosity can severely aggravate the mechanical properties of AM parts, it has been of significant interest to researchers to evaluate the extent of porosity in manufactured parts [10–12].

Ultrasonic testing (UT) is a well-known non-destructive testing method and can be used to effectively evaluate the porosity [11–13] and properties of a material, including its strength, elastic modulus, and material density [14–16]. Previous studies have reported that an ultrasonic wave propagating through a porous medium is scattered and delayed so that the wave velocity decreases [17,18]. The relation between ultrasonic velocity and the amount of porosity has been studied [13]. Studies on UT for porosity evaluation in AM parts have also been reported. Slotwinski et al. [19] experimentally investigated the correlation between the porosity content and ultrasonic velocity in AM parts using a contact ultrasonic method. The relation between the ultrasonic velocity and attenuation, and the properties of microstructures, including the porosity and grain size, was studied by Karthik et al. [20], who used a water-immersion ultrasonic method. Javidrad et al. [21] correlated the elastic modulus affected by the porosity content with the ultrasonic velocity in varying the propagation directions of ultrasonic waves using a contact ultrasonic method.

Although the effectiveness of UT in evaluating the porosity of AM parts has previously been verified, as-deposited AM parts with rough surfaces still present a major problem in UT. Ultrasonic signals obtained from a rough surface have a low signal-to-noise ratio (SNR), which degrades the UT performance. To overcome this problem, previous studies commonly used various surface polishing methods as a post-processing step to prepare the surface of as-deposited AM parts whose roughness exceeded a certain level. However, this post-processing step often resulted in an increase in the overall time and cost of the AM process [22–25]. The aforementioned discussion highlights the need to develop an innovative UT that can evaluate porosity without surface polishing, that is, even under rough surface conditions.

Recently, deep learning (DL) methods have been actively employed in various fields, including speech recognition [26], computer vision [27], and signal processing [28,29]. Other approaches have involved the application of DL to UT for feature extraction in defect detection [30,31]. One advantage thereof is the excellent ability of DL to consistently and accurately interpret ultrasonic signals for the characterization and detection of defects, compared with conventional UT that are error-prone and often depend on the experience and skills of non-destructive inspectors [32,33]. In addition, DL is able to effectively extract and defect features, even from ultrasonic signals with a low SNR [34,35].

Although DL has been used in UT in many studies for the detection of cracks [36,37], corrosion [38], welding defects [39,40], and others [31,33–35,41], only a few studies on DL-based UT under low SNR conditions have been reported. For example, Munir et al. [34] used DL for the classification of welding defects at low SNRs. They measured ultrasonic signals from welding defects and then added various levels of artificial noise. They reported that their convolutional neural network (CNN) outperformed the fully connected deep neural network (DNN) and multi-layer perceptron (MLP) as the SNR decreased. Zhang et al. [42] investigated DL to diagnose bearing faults at low SNRs. They added various levels of artificial noise to bearing fault vibration signals measured by an acoustic emission method. Similar to the aforementioned study, the classification performance of CNN was more accurate than that of DNN, MLP, and a support vector machine at low SNRs. However, to the best of the authors' knowledge, studies on DL-based UT to inspect AM parts have not been reported. Apart from this, it would be difficult to directly apply the results of previous studies to DL-based UT at low SNRs for the inspection of AM parts. This is because the reduction of the SNR of ultrasonic signals as a result of the surface roughness of AM parts is more severe than that due to the artificial noise used in previous studies.

In this study, we investigated the performance of DL-based UT to evaluate the porosity of AM parts with a rough surface. Several surface polishing methods were used after the AM process to obtain AM specimens with different levels of surface roughness. The porosity content of AM specimens was controlled by varying the AM processing conditions. The experimental procedures were as follows. (1) The amount of porosity was quantitatively determined and the porosity generation mechanisms were evaluated using scanning acoustic microscopy (SAM) and optical microscopy (OM), respectively. (2) The training and testing datasets were composed of time-domain ultrasonic signals acquired by measuring AM specimens, which were divided into 10 classes and labeled according to their porosity content. CNN, DNN, and MLP models were used and trained using these datasets. The testing performance of the three models was evaluated and compared with respect to the surface roughness. Furthermore, the applicability of the conventional UT was also considered for performance comparison. (3) The generalization performance was evaluated using newly manufactured AM specimens that were not used to train the DL model in order to verify the generalizability of the pre-trained model.

2. Experiments

2.1. Fabrication of Porosity-Induced Specimens with Different Levels of Surface Roughness

Ten AM specimens with various levels of porosity were created using a commercial selective laser melting machine (SLM 280 2.0, SLM Solutions, Lübeck, Germany) and commercially pure Ti powder with a particle diameter range of 20–63 μm. The specimens were 20 mm × 10 mm × 3 mm in size and were numbered from #1 to #10. The porosity content was controlled by varying the AM processing parameters, such as the laser power and laser scanning speed, which are known as the main indices that affect the porosity content of AM specimens. The processing parameters were selected after several preliminary tests and are summarized in Table 1. Generally, increasing the laser power or decreasing the scanning speed causes over-melting porosity. The opposite situation gives rise to porosity with a lack of fusion (LOF) [43]. Recent studies have found that porosity may also develop differently as a result of the variation in other properties such as the conduction-keyhole mode conversion of the melt pool [44], laser absorptivity [45], energy dissipation rate, and interaction time [46]. A detailed analysis of porosity mechanisms is provided in the next section. After specimen fabrication, both sides of the surfaces were polished to obtain the "smooth condition" using wire electrical discharge machining (EDM), where the arithmetic mean roughness (R_a) measured by a general roughness tester was 0.65 μm.

Table 1. Laser power and laser scanning speed used for specimen creation.

Specimen Name	#1	#2	#3	#4	#5	#6	#7	#8	#9	#10
Laser power (W)	275	355	275	235	315	355	355	355	235	235
Scanning speed (mm/s)	1091	1315	849	1187	1458	1517	1644	1793	768	725

To consider different surface roughness conditions, we fabricated 10 additional AM specimens, numbered from #1′ to #10′, the surfaces of which on both sides had different degrees of roughness. The surface on one side was polished to attain the "medium condition" by a general hand grinder that was used to separate specimens from the baseplate of a 3D printer. The other side was "rough condition", which corresponded to the as-deposited raw surface. The R_a of each surface was 3.1 μm and 6.4 μm, respectively, which are also listed in Table 2 together with those of the #1–#10 specimens. Except for the roughness conditions, all the other properties were the same as #1–#10 specimens. A photograph of the AM specimens is shown in Figure 1. Only the porosity of the specimens with smooth surfaces was examined with SAM. The training/testing datasets were constructed by using all three surface conditions.

Table 2. Three different conditions of surface roughness used in the experiments.

Roughness Condition	Ra Value (μm)	Surface Polishing Method
Smooth condition	0.65	Wire electrical discharge machining
Medium condition	3.1	Hand grinder
Rough condition	6.4	X (As-deposited state)

Figure 1. Photographic image of the additive manufacturing (AM) specimens.

2.2. Porosity Examination

SAM was used to quantify the porosity content of the AM specimens. C-mode imaging using a scanning acoustic microscope (HS-1000, Sonix, Springfield, VA, USA) was conducted with a 75 MHz focusing-type transducer. Only AM specimens finished to obtain the "smooth condition" were tested because the scanning quality of this imaging is highly influenced by the surface condition of specimens [47]. The focal point and the C-mode window were positioned at the center of the specimen. The window size was set to 1.5 mm, which corresponded to half the thickness of the specimen, to obtain comprehensive results for porosity. After the SAM analysis, a general optical microscope (GX53, OLYMPUS, Inc., Olympus, Tokyo, Japan) was used to analyze the shape and type of porosity. For the OM analysis, the surfaces of AM specimens with the "smooth condition" were additionally polished to lower their surface roughness to 0.04 μm.

Figure 2 shows the C-mode images that were obtained. The images clearly show that the porosities, represented as bright spots, are distributed differently depending on the AM processing conditions. These C-mode images were used to quantify the amount of porosity by using the open-source software ImageJ [48]. Based on the 6 dB drop method, each image was subjected to the binarization process, and the calculated porosity contents, defined as the ratio of the pore area to the total area in two-dimensional space, are summarized in Table 3. Note that these porosity contents are relative values [48]. According to the amount of porosity, they were labeled from "Porosity level 1" to "Porosity level 10". The measured porosity content increases as the specimen number increases. Because specimen #1 was manufactured under the optimal processing conditions, its porosity content was the lowest at 0.7% as determined by SAM. Almost no porosity was observed in the OM image shown in Figure 3a. Under this condition, the volume laser energy input (LEI) in the melt pool was 70 J/mm^3. The porosity contents of specimens #2 and #3 were in the range of 2.5–7.5%. The LEI of these specimens was 75 and 90 J/mm^3, respectively, which was within the over-melting condition; this causes not only welded particles and wavy surfaces but also entrapped gas, resulting in small pits and gas porosity, as shown in Figure 3b. Specimens #4–#8 were manufactured under LEI conditions of 55–65 J/mm^3. The porosity content was within the range 7.5–27.5%, slightly higher than those of #2 and #3. Generally, more porosity is created under LOF conditions than under over-melting conditions [49]. The lack of LEI prevents the powder in the inter- and intra-layers from melting sufficiently, which results in LOF porosity with un-melted powder, as shown in Figure 3c. Although these specimens had similar LEI levels, more pores were observed in specimens #7 and #8 than in #4–#6, as shown in Figure 2. This may be due to the insufficient interaction time of the laser to melt the powder owing to the higher scanning speed [49]. The LEI of specimens #9 and #10 was 85–90 J/mm^3, within the over-melting condition. Despite the LEI being similar to that of #2 and #3, the porosity content was significantly higher (over 27.5%). The reason may be a combination of high LEI and low laser power, in which case a shallow melt pool is generated, which may not be able to penetrate the previously deposited layers. This result may yield a large number of pits with un-melted powder between the interlayers, with the result that these specimens have the highest porosity, as shown in Figure 3d.

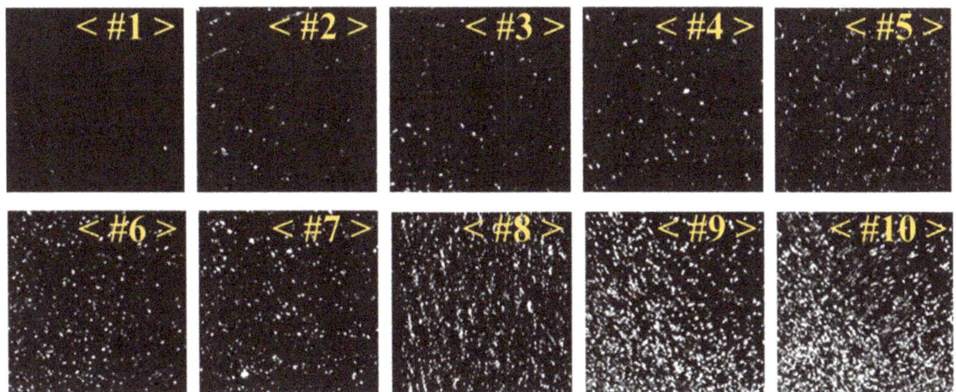

Figure 2. C-scan images obtained from specimens #1 to #10.

Table 3. Porosity content of specimens #1 to #10 as measured using scanning acoustic microscopy (SAM), and the corresponding porosity levels.

Specimen Name	#1	#2	#3	#4	#5	#6	#7	#8	#9	#10
Porosity (%)	0.7	3.2	5.2	7.6	11	13	23	26	29	39
Porosity Level	Lev. 1 (0–2.5)	Lev. 2 (2.5–5)	Lev. 3 (5–7.5)	Lev. 4 (7.5–10)	Lev. 5 (10–12.5)	Lev. 6 (12.5–15)	Lev. 7 (22.5–25)	Lev. 8 (25–27.5)	Lev. 9 (27.5–30)	Lev. 10 (30-)

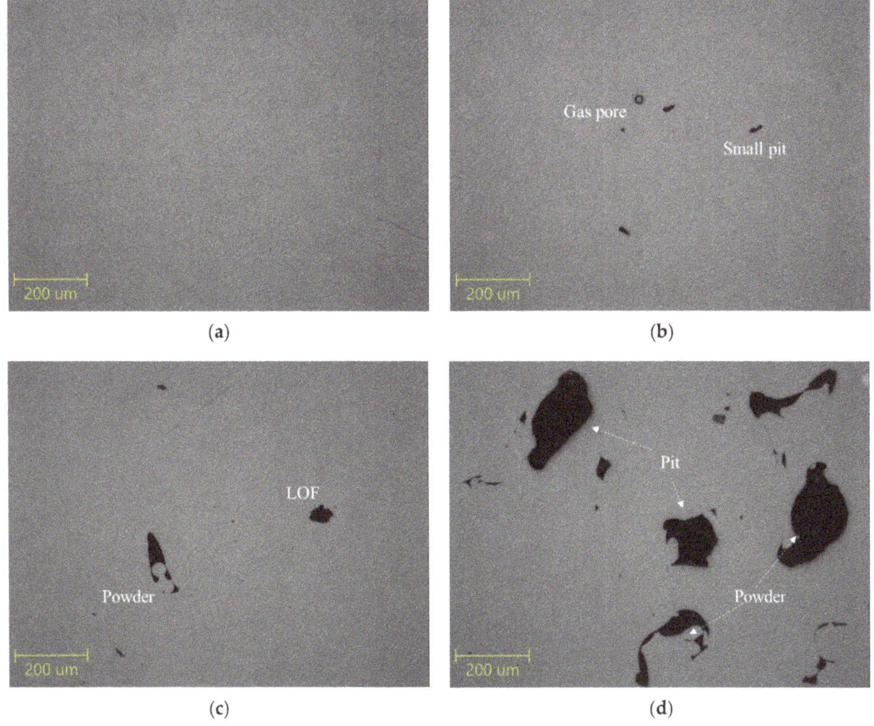

(a)

(b)

(c)

(d)

Figure 3. OM images of specimens (**a**) #1, (**b**) #3, (**c**) #6, and (**d**) #10.

2.3. Ultrasonic Measurements

Ultrasonic measurements were conducted by a pulse-echo mode using a contact transducer. This method is a well-known nondestructive testing technique that uses a pulsed signal with a broad bandwidth and several back-wall echo signals [50]. A schematic diagram and an image of the experimental setup are shown in Figure 4. A pulsed voltage signal, generated by a commercial pulser/receiver, was sent to a 5 MHz piezoelectric transducer. A longitudinal wave with a wavelength of approximately 1.2 mm was emitted by the transducer and was then incident on the AM specimen. At this wavelength, the ultrasonic diffraction effect is negligibly small because the ultrasonic wave propagation distance is in the range of the near field zone, which is obtained by $D^2/4\lambda = 19$ mm, where D is the transducer diameter and λ is the wavelength. The back-wall echo was received by the same transducer, and the ultrasonic signal was displayed and saved on a commercial oscilloscope. This echo signal reflects the effects of porosity in the ultrasonic propagation direction in the form of variations in the ultrasonic arrival time and ultrasonic attenuations [50]. To minimize the ultrasonic measurement errors, a pneumatic device that can apply a consistent pressure of 0.4 MPa was used, such that the contact condition between the transducer and the specimen was maintained consistently in each measurement [50]. Ultrasonic signals were obtained for the three different surface conditions: smooth, medium, and rough. Typical signals obtained for specimens #1 and #1′ prepared with three different surface conditions are plotted in Figure 5. The three measured signals overlap with each other. Although the porosity contents of specimens #1 and #1′ are the same, the difference in their ultrasonic amplitudes is clearly visible when the three signals are compared. In particular, the amplitude loss is very large on the surface of the "rough condition" specimen. The SNRs of these signals were 33 dB, 16 dB, and 10 dB, respectively, as shown in Figure 5. These SNRs are attributed to imperfect contact between the transducer and test specimens. The presence of air gaps owing to the imperfect contact results in impedance mismatch and also multi-reflections of the incident ultrasonic waves. Consequently, except for the "smooth condition", the additional changes in the properties of the incident ultrasonic waves make it difficult to evaluate the porosity of rough surfaces by using conventional UT such as ultrasonic velocity and ultrasonic attenuation coefficient measurements [22].

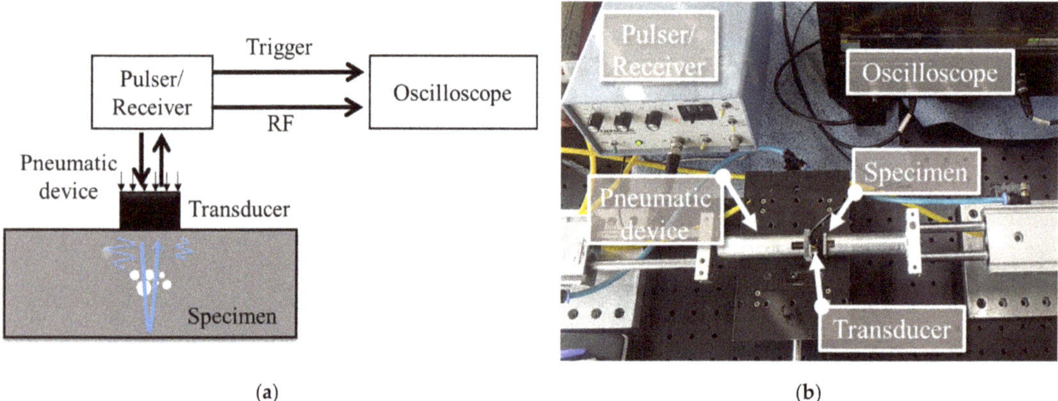

Figure 4. (**a**) Schematic diagram and (**b**) image of the experimental setup.

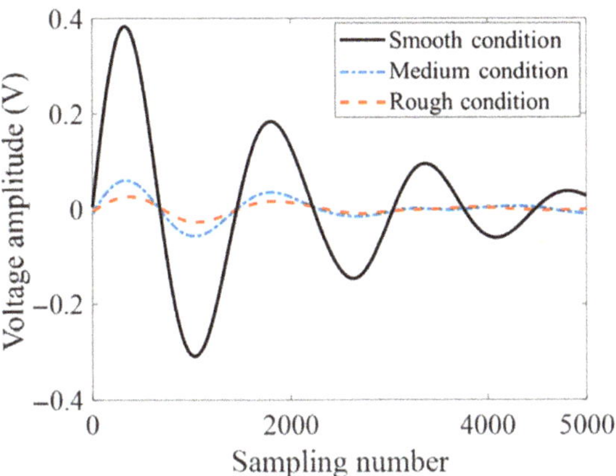

Figure 5. Ultrasonic signals obtained from the three different surface conditions. The signals were cropped to 5000 sampling numbers and were plotted to overlap with each other.

3. Porosity Evaluation in Rough Surface Conditions

3.1. Structures of Deep Learning Models

Artificial intelligence using neural network uses the group of correlated nodes motivated by biological neurons [51,52]. The simplest model is the MLP, which consists of an input layer, an output layer, and one hidden layer, where nodes are fully connected with each other [53]. DNN consists of more than two hidden layers with input and output layers, with the deeper structures being able to enhance the feature extraction capability. CNN is a type of MLP designed to use a feature extractor that requires minimal pre-processing with a fully connected neural network [51,54]. The feature extractor consists of one or more convolutional and pooling layers. A high-level feature map obtained from the feature extractor enables the network to be deeper with fewer parameters [55].

In this study, the CNN, DNN, and MLP models were used, and their performance was compared. Among several types of CNN, a one-dimensional (1D) CNN, which is effective not only to derive features from shorter segments of overall data but also accepts any type of signal as the input, was used. This network comprised a 1D array-type input layer, two convolutional layers, two max-pooling layers, a fully connected layer, and an output layer [56]. The input layer was restricted to (5000 × 1) nodes, which corresponded to the sampling numbers of the original ultrasonic signals. Note that the down-sampling of input nodes from original signals can reduce the computation time during model training. However, this down-sampling can also decrease the ability of the DL model to extract porosity features in the ultrasonic signals. In the first convolutional layer, the kernel size was set wide with (50 × 1) to restrain noise effectively, and the sizes of feature map and stride were 32 and (5 × 1), respectively. In the second convolutional layer, the kernel size was set to (4 × 1), considerably smaller than the first layer to extract a large number of features. The sizes of the feature map and stride in the second layer were 64 and (2 × 1), respectively. According to several simulation studies [34,35], these two convolutional layers with different kernel sizes showed good performance in noisy conditions. After each convolutional layer, one pooling layer was used, where both pooling and stride sizes were (2 × 1). The fully connected layer was set to (1000 × 1) nodes and connected to the output layer based on the softmax function $F(s_i)$ with cross-entropy (CE) loss for classification, derived as follows [34]:

$$F(s_i) = \frac{e^{s_i}}{\sum_{j=1}^{K} e^{s_j}},\tag{1}$$

$$CE = -\sum_{i}^{K} t_i log(s_i),\tag{2}$$

where s is the predicted output, subscript i indexes each output class, K is the total class numbers, and t_i is the real output. The activation function was a rectified linear unit (ReLU), presented as follows [57]:

$$ReLU(z) = \max(0, z). \tag{3}$$

To prevent an overfitting problem, dropout regularization [58] with a 70% training probability, which is a trick method to deactivate several nodes during training, was used before and after the fully connected layer. This dropout is also effective for the stronger robustness of the model. The learning rate was set to 0.001 after several trials. Note that too large a learning rate shows a corresponding effect for the down-sampling of the input nodes. Details of the CNN model are shown in Figure 6a. In the fully connected DNN model, instead of using the convolutional layer, the fully connected layers were set deeper than the used CNN model. Two hidden layers, i.e., (1000×1) and (1000×1) nodes, were used. The previous simulation studies [34,35] also reported that the deeper structures showed a better feature extraction ability of the DNN model. In the MLP model, only one hidden layer with (1000×1) nodes was used. The other parameters of the DNN and MLP models, such as the number of nodes of the input and output, and the dropout rate, were set to correspond to those of the CNN model, as shown in Figure 6b,c, and Table 4. The Relu and $F(s_i)$ with CE functions were also used. All the models were designed using TensorFlow and Keras.

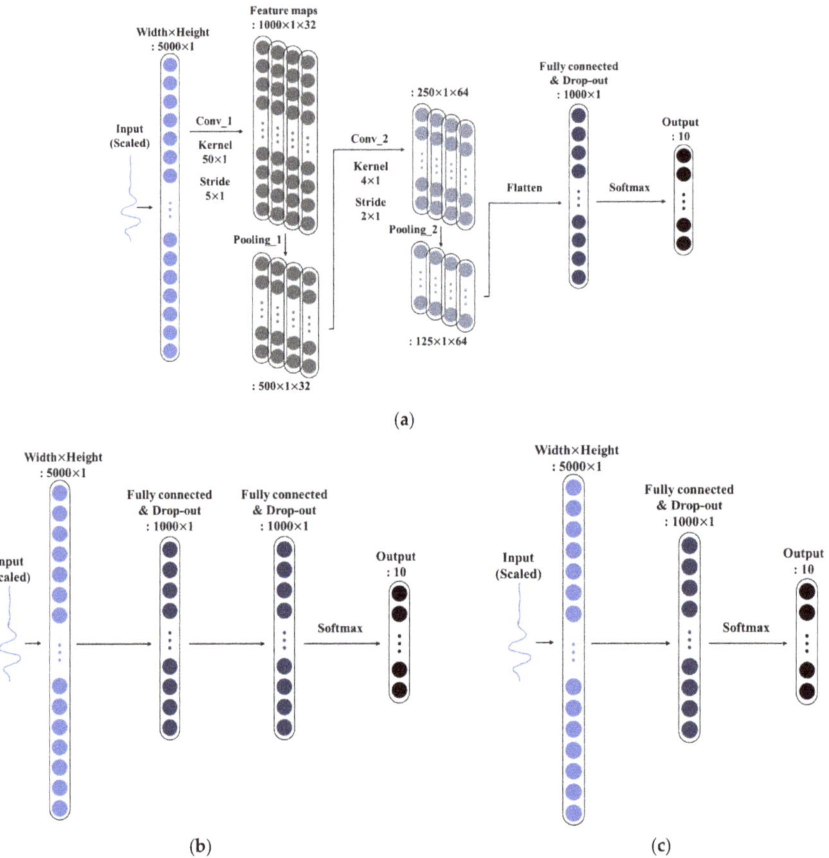

Figure 6. Architectures of the (**a**) convolutional neural network (CNN), (**b**) deep neural network (DNN), and (**c**) multi-layer perceptron (MLP) models.

Table 4. Parameters used to train each model.

Model	CNN	DNN	MLP
Input	5000	5000	5000
1st conv: Feature map/Kernel/Stride	32/50/5	-	-
1st max-pooling: Kernel/Stride	2/2	-	-
2nd conv: Feature map/Kernel/Stride	64/4/2	-	-
2nd max-pooling: Kernel/Stride	2/2	-	-
Fully connected (Wide)	1000	1000	1000
Fully connected (Deep)	1	2	1
Drop out	70%	70%	70%
Learning rate	0.001	0.001	0.001
Cost function	Softmax cross-entropy	Softmax cross-entropy	Softmax cross-entropy
Activation function	ReLU	ReLU	ReLU

3.2. Procedures to Train and Test the Models

The CNN, DNN, and MLP models were trained using the prepared training dataset to derive each specific function for porosity evaluation, after which the testing performance was compared. Training was conducted by using a classification method based on supervised learning for three models, with the respective training datasets generating the class label as the output. Ten labeled classes with porosity levels ranging from 1 to 10 were used with the levels based on the results of the porosity content measurements obtained with SAM. Ultrasonic signals were acquired from the surfaces with three different roughness levels. For each surface roughness level, 100 ultrasonic signals were measured, of which 80 ultrasonic signals were randomly extracted and used as the training dataset. The remaining 20 signals were used as the testing dataset. Considering previous research [36], the amount of data used to form the training and testing datasets is sufficient. The training and testing datasets for each of the 10 porosity levels and the three different roughness conditions are summarized in Table 5.

Table 5. Training and testing datasets according to roughness condition and porosity level.

Porosity Level	Smooth Condition			Medium Condition			Rough Condition		
	Specimen	No. of Signals		Specimen	No. of Signals		Specimen	No. of Signals	
		Train	Test		Train	Test		Train	Test
Lev. 1	#1	80	20	#1'	80	20	#1'	80	20
Lev. 2	#2	80	20	#2'	80	20	#2'	80	20
Lev. 3	#3	80	20	#3'	80	20	#3'	80	20
Lev. 4	#4	80	20	#4'	80	20	#4'	80	20
Lev. 5	#5	80	20	#5'	80	20	#5'	80	20
Lev. 6	#6	80	20	#6'	80	20	#6'	80	20
Lev. 7	#7	80	20	#7'	80	20	#7'	80	20
Lev. 8	#8	80	20	#8'	80	20	#8'	80	20
Lev. 9	#9	80	20	#9'	80	20	#9'	80	20
Lev. 10	#10	80	20	#10'	80	20	#10'	80	20

Figure 7 shows learning curves of the CNN, DNN, and MLP models for the "rough condition", which represents the testing accuracy and cost as a function of the number of epochs. The testing accuracy was defined as the classification performance at each epoch on the testing data, and was calculated as:

$$Testing\ accuracy\ (\%) = m_1/n_1 \cdot 100, \qquad (4)$$

where m_1 is the number of testing data points classified well, and n_1 is the overall number of testing data points. The performance of the model can be evaluated from the testing accuracy, which represents the classification performance at each epoch on the testing data. The cost represents the error between the real output and the predicted output of the tested model based on the testing dataset. Figure 7 shows that the testing accuracy of the respective CNN, DNN, and MLP model reached a global maximum and oscillated after approximately the 40th epoch. Therefore, we monitored the testing accuracy from the 40th epoch until the end, and the epoch that provided the highest testing accuracy was chosen for the respective CNN, DNN, and MLP model [59]. A commercial CPU device was used to train all models. The computation time for the respective CNN, DNN, and MLP model was approximately 760, 500, and 96 s, respectively.

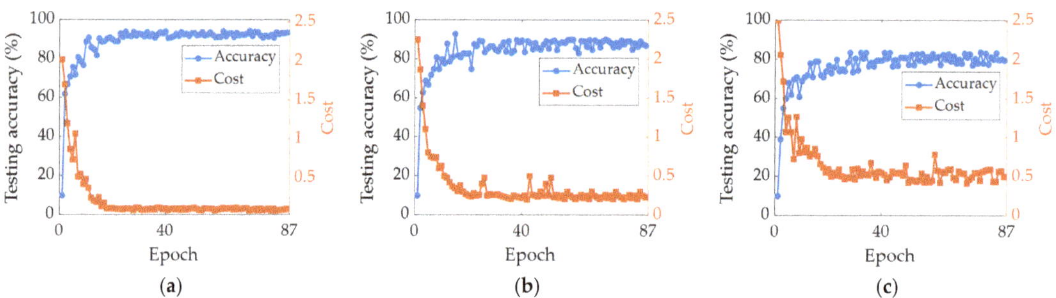

Figure 7. Learning curves of the (**a**) CNN, (**b**) DNN, and (**c**) MLP models for the "rough condition", showing the testing accuracy and cost with respect to the number of epochs.

Table 6 lists the testing performance of the three different models for the various surface roughness levels, and these results are compared in Figure 8. The testing performance results are also listed as confusion matrices in Appendix A. For the "smooth condition" surface, each CNN and DNN model delivered average testing performance of 98.5% and 98.0%, respectively. Although the performance of the MLP model was relatively lower than that of the others, it also performed well at 96.0%. Increased roughness levels caused the performance of all the models to decrease; however, the performance of all the models exceeded 80.5% in terms of their accuracy. The model that delivered the best performance for the "medium condition" and "rough condition" surfaces was CNN. The performance only decreased to 97.5% and 94.5% for surfaces with these two roughness levels, i.e., decreases of 1% and 4%, respectively, when compared with the "smooth condition" surface. Those of the DNN were 95.5% and 89.5% for the "medium condition" and "rough condition", respectively, a slightly larger decrease of 2.5% and 8.5%, respectively. This tendency is more pronounced for the MLP model, in which case the performance decreased to 92.0% and 80.5%, i.e., decreases of 4% and 16.5%, respectively. In comparison, the performance of all models decreased slightly at porosity levels 1 and 2 compared with the other levels regardless of the surface roughness conditions. Above porosity level 3, the average performance for all roughness conditions for each of the CNN, DNN, and MLP models was 99.4%, 96.3%, and 92.5%, respectively. However, below level 2, these values decreased to 86.7%, 86.7%, and 77.5%, respectively.

Conventional UT is based on ultrasonic velocity and ultrasonic attenuation coefficient measurements [13]. The use of these methods requires not only the first back-wall echo signal but also the second echo in pulse-echo mode to be measured to extract the ultrasonic velocity and attenuation coefficient parameters. A comparison of the extent to which the parameters vary enables the porosity to be evaluated. These parameters are calculated as follows [13]:

$$v = \frac{d}{\tau},\tag{5}$$

$$a = \frac{20}{d}log\left(\frac{A_1}{A_2}\right),\tag{6}$$

where v is the ultrasonic velocity, d is the ultrasonic wave propagation distance corresponding to twice the thickness of the specimen, τ is the time-of-flight difference between two consecutive echoes, a is the ultrasonic attenuation coefficient, and A_1 and A_2 are the amplitudes of two consecutive echoes, respectively. Generally, the amplitude of the second echo is smaller than that of the first echo because the second echo is propagated over a longer distance.

Table 6. Testing performance of the three different models for the three different surface roughness conditions.

Classification	Testing Performance (%)								
	Smooth Condition			Medium Condition			Rough Condition		
	CNN	**DNN**	**MLP**	**CNN**	**DNN**	**MLP**	**CNN**	**DNN**	**MLP**
Lev. 1	90	95	85	85	90	80	75	70	60
Lev. 2	95	95	90	90	90	80	85	80	70
Lev. 3	100	100	95	100	95	90	90	90	85
Lev. 4	100	100	95	100	95	95	100	85	75
Lev. 5	100	95	100	100	95	95	95	90	85
Lev. 6	100	100	100	100	100	100	100	95	85
Lev. 7	100	100	100	100	100	95	100	100	95
Lev. 8	100	95	100	100	95	95	100	95	90
Lev. 9	100	100	100	100	95	100	100	90	80
Lev. 10	100	100	95	100	100	90	100	100	80
Average	**98.5**	**98**	**96**	**97.5**	**95.5**	**92**	**94.5**	**89.5**	**80.5**

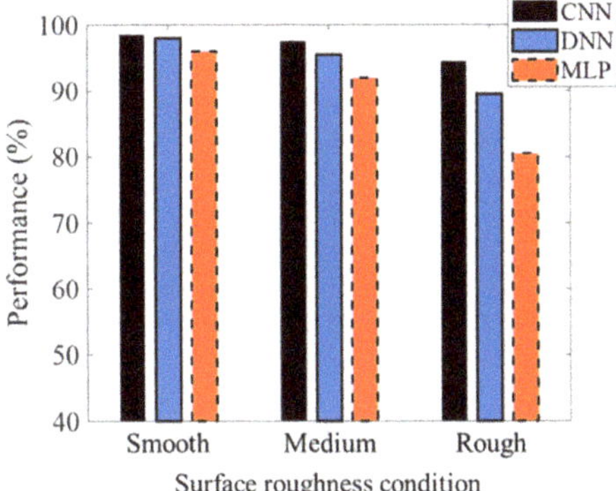

Figure 8. Performance of the CNN, DNN, and MLP models for the three different surface roughness conditions.

Figure 9 shows two consecutive echoes measured from specimens with the three surface roughness levels. For the "smooth condition" two echoes are clearly observed. However, for the other roughness levels, the levels of the second echo and background noise are almost similar owing to the amplitude loss from the rough surfaces. Consequently, rough surface conditions make it difficult to employ conventional UT for porosity evaluation.

Several reasons could exist for the high performance of the DL models in terms of their porosity evaluation of AM parts with rough surfaces. The first simple reason is their excellent ability to perform feature extraction. The use of DL models with deep and wide structures with hidden nodes is known to be more effective for extracting features than conventional UT [36]. The second reason is that the training dataset of the DL model consists of the raw ultrasonic signals, whereas conventional UT, which includes the ultrasonic velocity and attenuation measurements, only uses the velocity and attenuation coefficient parameters extracted from the ultrasonic signals. When the raw signal is used for training, various properties including not only the velocity and attenuation but also the ultrasonic backscattering and non-linearity can be used as features. Although not to the same extent as the velocity and attenuation, backscattering and non-linearity are also known to be related to the porosity content, which enhances the performance when DL models are used [50]. Our experimental results showed that the rougher the surface, that is, the lower the SNR, the more effective is the DL model. At the same time, the CNN model outperformed the DNN and MLP models because the CNN model, which uses a pre-processor, is beneficial for feature extraction from the waveform even for low SNRs. The waveform of the ultrasonic wave propagating through the porous medium varied locally. As mentioned above, the typical waveform variation is the delay in the arrival time and ultrasonic attenuation owing to local elastic inhomogeneity at the boundary of the pores. When the CNN model is used, both the convolutional and pooling layers in the pre-processor assign a greater weight to this variation in the waveform, thereby enabling the CNN to achieve more effective feature extraction than the other models.

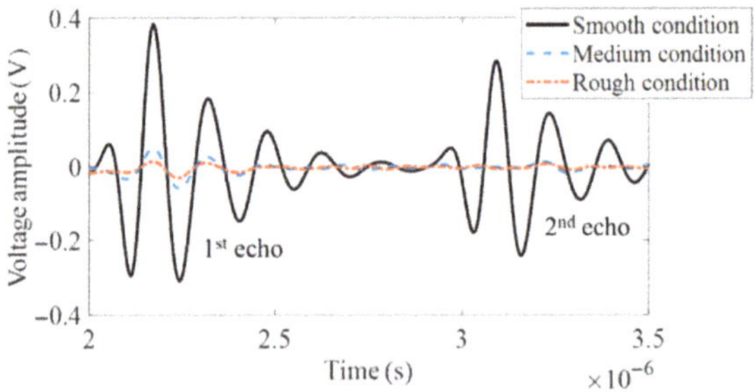

Figure 9. Ultrasonic 1st and 2nd echo signals measured from surfaces with the "smooth condition", "medium condition", and "rough condition".

Note that, in addition to the surface roughness issue, porosity with an irregular distribution pattern may affect the UT performance. For example, if the porosity is distributed non-uniformly in the direction parallel to the surface attached to the ultrasonic transducer, the UT performance may deteriorate depending on the positions at which measurements are conducted (where the surface in contact with the transducer is assumed to be constant). Generally, porosity originates from a lack of uniformity along the building direction because the cooling rate is varied during AM building. In contrast, the plane normal to the building direction is relatively uniform [21]. In our experiments, the ultrasonic measurement was conducted using a transducer attached to the surface in the direction normal to the building direction, as shown in Figure 4. In other words, an ultrasonic wave propagating in a direction parallel to the building direction reflects the effects of a non-uniform pore; however, the average porosity along this path is almost uniform in the direction parallel to the surface attached to the transducer. Therefore, there may be few errors in the UT performance owing to the irregular pattern in which the porosity is distributed. However, in the case of low levels of porosity, this assumption may be difficult to establish. In fact, our experimental results indicated that, below porosity level 2, the performance is slightly lower.

3.3. Evaluation of the Generalization Performance

To verify the applicability of the pre-trained model, a generalization performance test was carried out on newly fabricated AM specimens, which were not utilized to train the models. The generalization test was conducted on specimens in the as-deposited condition, i.e., the "rough condition". Only the pre-trained CNN model, which delivered the best performance for this roughness condition, was used. Two new specimens were manufactured by using the same AM process but different AM processing parameters. These parameters did not correspond to the processing conditions of the existing 10 AM specimens that were used to train the models. One-hundred ultrasonic signals were obtained for each specimen and were used as input to the pre-trained model. Figure 10 shows the results of the generalization performance of the two AM specimens using the pre-trained CNN model. This model assessed the Test#1 specimen as "Porosity level 2" with the highest probability of 89% and "Porosity level 1" with the second highest of 8%. This model also rated the Test-#2 specimen as "Porosity level 8" with 91% and "Porosity level 7" with 7%.

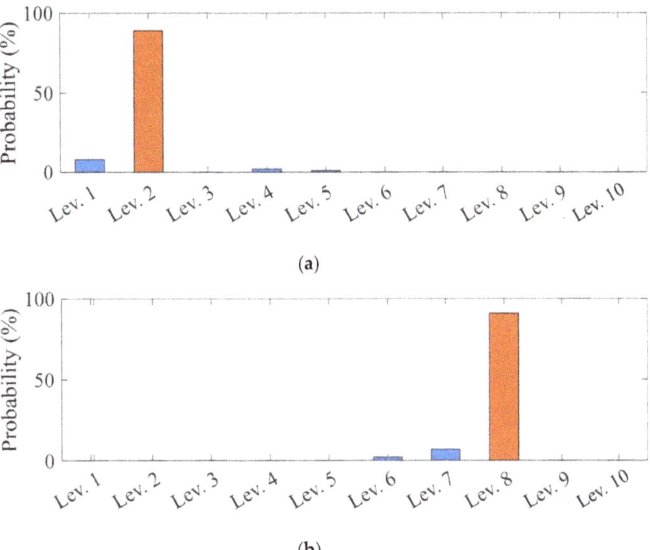

Figure 10. Generalization test results for the (**a**) Test-#1 and (**b**) Test-#2 specimens.

To validate the results obtained by the pre-trained CNN model, SAM was also used to assess the porosity content of the tested specimens. Because SAM cannot be employed to examine as-deposited specimens with rough surfaces, the test specimens were additionally polished using wire EDM. Figure 11 shows the obtained C-mode images. The porosity contents that were calculated from these images are presented in Table 7 alongside the assessment with the pre-trained CNN model. The calculated porosity contents of Test#1 and Test#2 were 4.3% and 27%, respectively, which were within the range of "Porosity level 2" and "Porosity level 8", respectively. In other words, the SAM results were in good correspondence with the results assessed as having the highest probability by the CNN model. In addition, the average generalization performance for the "rough condition" was 90%, which is slightly lower than the testing performance in Section 3.2. This might be due to differences in the AM processing conditions [1] and the experimental environment.

Table 7. Comparison of the results obtained with the pre-trained model and SAM.

| | Porosity Evaluation Results | | |
| AM Specimen | Pre-Trained Model | | SAM |
	Porosity Level	Porosity Content (%)	Porosity Content (%)
Test-#1	Lev. 2	2.5–5	4.3
Test-#2	Lev. 8	25–27.5	27

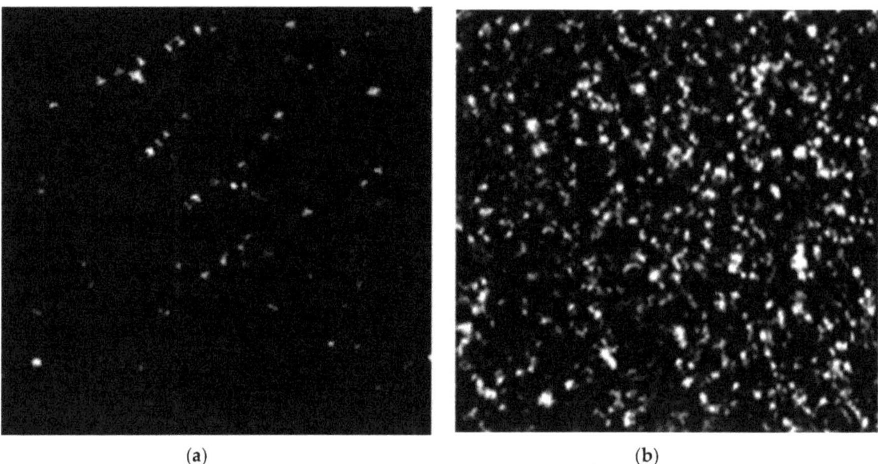

(a)   (b)

Figure 11. SAM results for the (**a**) Test-#1 and (**b**) Test-#2 specimens.

4. Conclusions

In this work, DL techniques were used in conjunction with UT to evaluate the porosity of AM parts with rough surfaces. Key research outcomes were as follows.

(1) Various porosity mechanisms were investigated through SAM and OM analysis. Porosity contents increased in the order of normal (the relative porosity content measured by SAM: 0.7%), over-melting (4.2%), LOF (16.1%), and over-melting with low laser power conditions (34%).

(2) A comparison of the performance results of the various DL models showed that all the models were highly accurate at over 80.5%, even for the as-deposited specimens with surfaces in the "rough condition". In particular, CNN was the most effective at 94.5%. Owing to the low SNR of the measured ultrasonic signal, conventional UT using ultrasonic velocity and ultrasonic attenuation coefficient measurements could not be used to assess "medium condition" and "rough condition" surfaces.

(3) A generalization test was also conducted using newly as-deposited AM specimens that were not used for training to evaluate the applicability of the pre-trained CNN model. The test results confirmed the model's high evaluation performance of 90.0%, which corresponded well with the results obtained with SAM.

These results suggest that the use of DL could be expected to enhance the UT performance with respect to the porosity evaluation of AM parts, even for as-deposited rough surfaces.

Author Contributions: Conceptualization, S.-H.P.; methodology, S.-H.P., S.C. and K.-Y.J.; validation, S.-H.P., J.-Y.H., T.H. and S.C.; formal analysis, S.-H.P., J.-Y.H., T.H. and S.C.; investigation, S.-H.P. and J.-Y.H.; resources, S.-H.P., J.-Y.H. and K.-Y.J.; writing—original draft preparation, S.-H.P.; writing—review and editing, S.-H.P., S.C. and K.-Y.J.; supervision, S.C. and K.-Y.J.; funding acquisition, T.H. and K.-Y.J. All authors have read and agreed to the published version of the manuscript.

Funding: This work was supported by a Korea Institute of Machinery and Materials grant funded by the Korea government (MSIT) (NK230l), and the Korea Institute of Energy Technology Evaluation and Planning (KETEP) and the Ministry of Trade, Industry and Energy (MOTIE) of the Republic of Korea (No.20181510102360).

Institutional Review Board Statement: Not applicable.

Informed Consent Statement: Not applicable.

Conflicts of Interest: The authors declare no conflict of interest.

Appendix A

Table A1. Confusion matrices for the CNN model in three surface conditions.

CNN									
Smooth condition									
(Unit: %)	Lev. 1	Lev. 2	Lev. 3	Lev. 4	Lev. 5	Lev. 6	Lev. 7	Lev. 8	Lev. 9	Lev. 10
Lev. 1	90	10	0	0	0	0	0	0	0	0
Lev. 2	5	95	0	0	0	0	0	0	0	0
Lev. 3	0	0	100	0	0	0	0	0	0	0
Lev. 4	0	0	0	100	0	0	0	0	0	0
Lev. 5	0	0	0	0	100	0	0	0	0	0
Lev. 6	0	0	0	0	0	100	0	0	0	0
Lev. 7	0	0	0	0	0	0	100	0	0	0
Lev. 8	0	0	0	0	0	0	0	100	0	0
Lev. 9	0	0	0	0	0	0	0	0	100	0
Lev. 10	0	0	0	0	0	0	0	0	0	100

Medium condition									
(Unit: %)	Lev. 1	Lev. 2	Lev. 3	Lev. 4	Lev. 5	Lev. 6	Lev. 7	Lev. 8	Lev. 9	Lev. 10
Lev. 1	85	10	0	5	0	0	0	0	0	0
Lev. 2	10	90	0	0	0	0	0	0	0	0
Lev. 3	0	0	100	0	0	0	0	0	0	0
Lev. 4	0	0	0	100	0	0	0	0	0	0
Lev. 5	0	0	0	0	100	0	0	0	0	0
Lev. 6	0	0	0	0	0	100	0	0	0	0
Lev. 7	0	0	0	0	0	0	100	0	0	0
Lev. 8	0	0	0	0	0	0	0	100	0	0
Lev. 9	0	0	0	0	0	0	0	0	100	0
Lev. 10	0	0	0	0	0	0	0	0	0	100

Rough condition									
(Unit: %)	Lev. 1	Lev. 2	Lev. 3	Lev. 4	Lev. 5	Lev. 6	Lev. 7	Lev. 8	Lev. 9	Lev. 10
Lev. 1	75	20	0	5	0	0	0	0	0	0
Lev. 2	10	85	5	0	0	0	0	0	0	0
Lev. 3	5	0	90	5	0	0	0	0	0	0
Lev. 4	0	0	0	100	0	0	0	0	0	0
Lev. 5	0	0	0	5	95	0	0	0	0	0
Lev. 6	0	0	0	0	0	100	0	0	0	0
Lev. 7	0	0	0	0	0	0	100	0	0	0
Lev. 8	0	0	0	0	0	0	0	100	0	0
Lev. 9	0	0	0	0	0	0	0	0	100	0
Lev. 10	0	0	0	0	0	0	0	0	0	100

Table A2. Confusion matrices for the DNN model in three surface conditions.

DNN									
Smooth condition									
(Unit: %)	Lev. 1	Lev. 2	Lev. 3	Lev. 4	Lev. 5	Lev. 6	Lev. 7	Lev. 8	Lev. 9	Lev. 10
Lev. 1	95	5	0	0	0	0	0	0	0	0
Lev. 2	5	95	0	0	0	0	0	0	0	0
Lev. 3	0	0	100	0	0	0	0	0	0	0
Lev. 4	0	0	0	100	0	0	0	0	0	0
Lev. 5	0	0	0	5	95	0	0	0	0	0
Lev. 6	0	0	0	0	0	100	0	0	0	0
Lev. 7	0	0	0	0	0	0	100	0	0	0
Lev. 8	0	0	0	0	0	0	0	95	5	0
Lev. 9	0	0	0	0	0	0	0	0	100	0
Lev. 10	0	0	0	0	0	0	0	0	0	100

Table A2. *Cont.*

	DNN									
	Medium condition									
(Unit: %)	Lev. 1	Lev. 2	Lev. 3	Lev. 4	Lev. 5	Lev. 6	Lev. 7	Lev. 8	Lev. 9	Lev. 10
Lev. 1	90	10	0	0	0	0	0	0	0	0
Lev. 2	10	90	0	0	0	0	0	0	0	0
Lev. 3	0	0	95	5	0	0	0	0	0	0
Lev. 4	5	0	0	95	0	0	0	0	0	0
Lev. 5	0	0	0	5	95	0	0	0	0	0
Lev. 6	0	0	0	0	0	100	0	0	0	0
Lev. 7	0	0	0	0	0	0	100	0	0	0
Lev. 8	0	0	0	0	0	0	0	95	0	5
Lev. 9	0	0	0	0	0	0	0	0	95	5
Lev. 10	0	0	0	0	0	0	0	0	0	100
	Rough condition									
(Unit: %)	Lev. 1	Lev. 2	Lev. 3	Lev. 4	Lev. 5	Lev. 6	Lev. 7	Lev. 8	Lev. 9	Lev. 10
Lev. 1	70	20	5	5	0	0	0	0	0	0
Lev. 2	15	80	5	0	0	0	0	0	0	0
Lev. 3	5	0	90	5	0	0	0	0	0	0
Lev. 4	10	0	5	85	0	0	0	0	0	0
Lev. 5	0	0	0	10	90	0	0	0	0	0
Lev. 6	0	0	0	0	5	95	0	0	0	0
Lev. 7	0	0	0	0	0	0	100	0	0	0
Lev. 8	0	0	0	0	0	0	0	95	0	5
Lev. 9	0	0	0	0	0	0	0	0	90	10
Lev. 10	0	0	0	0	0	0	0	0	0	100

Table A3. Confusion matrices for the MLP model in three surface conditions.

	MLP									
	Smooth condition									
(Unit: %)	Lev. 1	Lev. 2	Lev. 3	Lev. 4	Lev. 5	Lev. 6	Lev. 7	Lev. 8	Lev. 9	Lev. 10
Lev. 1	85	15	0	0	0	0	0	0	0	0
Lev. 2	5	90	0	5	0	0	0	0	0	0
Lev. 3	5	0	95	0	0	0	0	0	0	0
Lev. 4	0	0	0	95	0	0	5	0	0	0
Lev. 5	0	0	0	0	100	0	0	0	0	0
Lev. 6	0	0	0	0	0	100	0	0	0	0
Lev. 7	0	0	0	0	0	0	100	0	0	0
Lev. 8	0	0	0	0	0	0	0	100	0	0
Lev. 9	0	0	0	0	0	0	0	0	100	0
Lev. 10	0	0	0	0	0	0	0	0	5	95
	Medium condition									
(Unit: %)	Lev. 1	Lev. 2	Lev. 3	Lev. 4	Lev. 5	Lev. 6	Lev. 7	Lev. 8	Lev. 9	Lev. 10
Lev. 1	80	15	5	0	0	0	0	0	0	0
Lev. 2	15	80	0	5	0	0	0	0	0	0
Lev. 3	5	0	90	5	0	0	0	0	0	0
Lev. 4	0	0	0	95	0	0	5	0	0	0
Lev. 5	0	0	0	5	95	0	0	0	0	0
Lev. 6	0	0	0	0	0	100	0	0	0	0
Lev. 7	0	0	0	0	0	0	95	0	0	5
Lev. 8	0	0	0	0	0	0	0	95	0	5
Lev. 9	0	0	0	0	0	0	0	0	100	0
Lev. 10	0	0	0	0	0	0	5	0	5	90

Table A3. *Cont.*

	MLP									
	Rough condition									
(Unit: %)	Lev. 1	Lev. 2	Lev. 3	Lev. 4	Lev. 5	Lev. 6	Lev. 7	Lev. 8	Lev. 9	Lev. 10
Lev. 1	60	25	5	10	0	0	0	0	0	0
Lev. 2	15	70	5	5	0	0	5	0	0	0
Lev. 3	5	0	85	10	0	0	0	0	0	0
Lev. 4	10	0	10	75	0	0	5	0	0	0
Lev. 5	0	0	0	10	85	5	0	0	0	0
Lev. 6	0	0	0	5	5	85	5	0	0	0
Lev. 7	0	0	0	0	0	0	95	0	0	5
Lev. 8	0	0	0	0	0	0	0	90	0	10
Lev. 9	0	0	0	0	0	5	0	0	80	15
Lev. 10	0	0	0	0	0	0	5	0	15	80

References

1. Gorsse, S.; Hutchinson, C.; Gouné, M.; Banerjee, R. Additive manufacturing of metals: A brief review of the characteristic microstructures and properties of steels, Ti-6Al-4V and high-entropy alloys. *Sci. Technol. Adv. Mater.* **2017**, *18*, 584–610. [CrossRef] [PubMed]
2. Seifi, M.; Salem, A.; Beuth, J.; Harrysson, O.; Lewandowski, J.J. Overview of Materials Qualification Needs for Metal Additive Manufacturing. *JOM* **2016**, *68*, 747–764. [CrossRef]
3. Everton, S.K.; Hirsch, M.; Stravroulakis, P.; Leach, R.K.; Clare, A.T. Review of in-situ process monitoring and in-situ metrology for metal additive manufacturing. *Mater. Des.* **2016**, *95*, 431–445. [CrossRef]
4. Babu, B.; Lundbäck, A.; Lindgren, L.-E. Simulation of Ti-6Al-4V Additive Manufacturing Using Coupled Physically Based Flow Stress and Metallurgical Model. *Materials* **2019**, *12*, 3844. [CrossRef]
5. Wang, Z.; Palmer, T.A.; Beese, A.M. Effect of processing parameters on microstructure and tensile properties of austenitic stainless steel 304L made by directed energy deposition additive manufacturing. *Acta Mater.* **2016**, *110*, 226–235. [CrossRef]
6. Collins, P.C.; Brice, D.A.; Samimi, P.; Ghamarian, I.; Fraser, H.L. Microstructural Control of Additively Manufactured Metallic Materials. *Annu. Rev. Mater. Res.* **2016**, *46*, 63–91. [CrossRef]
7. Yang, K.; Xie, H.; Sun, C.; Zhao, X.; Li, F. Influence of Vanadium on the Microstructure of IN718 Alloy by Laser Cladding. *Materials* **2019**, *12*, 3839. [CrossRef]
8. Everton, S.; Dickens, P.; Tuck, C.; Dutton, B. Evaluation of laser ultrasonic testing for inspection of metal additive manufacturing. *Laser 3D Manuf. II Int. Soc. Opt. Photonics* **2015**, *9353*, 935316. [CrossRef]
9. Clark, D.; Sharples, S.D.; Wright, D.C. Development of online inspection for additive manufacturing products. *Insight* **2011**, *53*, 610–613. [CrossRef]
10. Eren, E.; Kurama, S.; Solodov, I. Characterization of porosity and defect imaging in ceramic tile using ultrasonic inspections. *Ceram. Int.* **2012**, *38*, 2145–2151. [CrossRef]
11. Honarvar, F.; Varvani-Farahani, A. A review of ultrasonic testing applications in additive manufacturing: Defect evaluation, material characterization, and process control. *Ultrasonics* **2020**, *108*, 106227. [CrossRef]
12. Chabot, A.; Laroche, N.; Carcreff, E.; Rauch, M.; Hascoët, J.-Y. Towards defect monitoring for metallic additive manufacturing components using phased array ultrasonic testing. *J. Intell. Manuf.* **2019**, *31*, 1191–1201. [CrossRef]
13. Jeong, H.; Hsu, D.K. Experimental analysis of porosity-induced ultrasonic attenuation and velocity change in carbon composites. *Ultrasonics* **1995**, *33*, 195–203. [CrossRef]
14. Park, S.-H.; Kim, J.; Jhang, K.-Y. Relative measurement of the acoustic nonlinearity parameter using laser detection of an ultrasonic wave. *Int. J. Precis. Eng. Manuf.* **2017**, *18*, 1347–1352. [CrossRef]
15. Jeong, H. Effects of Voids on the Mechanical Strength and Ultrasonic Attenuation of Laminated Composites. *J. Compos. Mater.* **1997**, *31*, 276–292. [CrossRef]
16. Kim, H.S.; Bush, M.B. The effects of grain size and porosity on the elastic modulus of nanocrystalline materials. *Nanostruct. Mater.* **1999**, *11*, 361–367. [CrossRef]
17. Birt, E.A.; Smith, R.A. A review of NDE methods for porosity measurement in fibre-reinforced polymer composites. *Insight* **2004**, *46*, 681–686. [CrossRef]
18. Hernandez, M.G.; Izquierdo, M.A.G.; Ibanez, A.; Anaya, J.J.; Ullate, L.G. Porosity estimation of concrete by ultrasonic NDE. *Ultrasonics* **2000**, *38*, 531–533. [CrossRef]
19. Slotwinski, J.A.; Garboczi, E.J.; Hebenstreit, K.M. Porosity Measurements and Analysis for Metal Additive Manufacturing Process Control. *J. Res. Natl. Inst. Stand. Technol.* **2014**, *119*, 494–528. [CrossRef]
20. Karthik, N.V.; Gu, H.; Pal, D.; Starr, T.; Stucker, B. High frequency ultrasonic non destructive evaluation of additively manufactured components. In Proceedings of the 24th International Solid Freeform Fabrication Symposium, Austin, TX, USA, 12–14 August 2013; pp. 311–325.

21. Javidrad, H.R.; Salemi, S. Determination of elastic constants of additive manufactured Inconel 625 specimens using an ultrasonic technique. *Int. J. Adv. Manuf. Technol.* **2020**, *107*, 4597–4607. [CrossRef]
22. Bakre, C.; Hassanian, M.; Lissenden, C. Influence of surface roughness from additive manufacturing on laser ultrasonics measurements. In *AIP Conference Proceedings*; AIP Publishing LLC: Melville, NY, USA, 2019; p. 020009.
23. Calignano, F.; Manfredi, D.; Ambrosio, E.P.; Iuliano, L.; Fino, P. Influence of process parameters on surface roughness of aluminum parts produced by DMLS. *Int. J. Adv. Manuf. Technol.* **2013**, *67*, 2743–2751. [CrossRef]
24. Tyagi, P.; Goulet, T.; Riso, C.; Stephenson, R.; Chuenprateep, N.; Schlitzer, J.; Benton, C.; Garcia-Moreno, F. Reducing the roughness of internal surface of an additive manufacturing produced 316 steel component by chempolishing and electropolishing. *Addit. Manuf.* **2019**, *25*, 32–38. [CrossRef]
25. Whip, B.; Sheridan, L.; Gockel, J. The effect of primary processing parameters on surface roughness in laser powder bed additive manufacturing. *Int. J. Adv. Manuf. Technol.* **2019**, *103*, 4411–4422. [CrossRef]
26. Richardson, F.; Reynolds, D.; Dehak, N. Deep Neural Network Approaches to Speaker and Language Recognition. *IEEE Signal Process. Lett.* **2015**, *22*, 1671–1675. [CrossRef]
27. Noda, K.; Yamaguchi, Y.; Nakadai, K.; Okuno, H.G.; Ogata, T. Audio-visual speech recognition using deep learning. *Appl. Intell.* **2015**, *42*, 722–737. [CrossRef]
28. Hamel, P.; Eck, D. Learning features from music audio with deep belief networks. *ISMIR* **2010**, *10*, 339–344.
29. Kim, Y.; Lee, H.; Provost, E.M. Deep learning for robust feature generation in audiovisual emotion recognition. In Proceedings of the IEEE International Conference on Acoustics, Speech and Signal Processing, Vancouver, BC, Canada, 26–31 May 2013; pp. 3687–3691.
30. Hou, W.; Wei, Y.; Guo, J.; Jin, Y.; Zhu, C. Automatic Detection of Welding Defects using Deep Neural Network. *J. Phys. Conf. Ser.* **2018**, *933*, 012006. [CrossRef]
31. Meng, M.; Chua, Y.J.; Wouterson, E.; Ong, C.P.K. Ultrasonic signal classification and imaging system for composite materials via deep convolutional neural networks. *Neurocomputing* **2017**, *257*, 128–135. [CrossRef]
32. Zhu, P.; Cheng, Y.; Banerjee, P.; Tamburrino, A.; Deng, Y. A novel machine learning model for eddy current testing with uncertainty. *NDT E Int.* **2019**, *101*, 104–112. [CrossRef]
33. Aldrin, J.C.; Forsyth, D.S. Demonstration of using signal feature extraction and deep learning neural networks with ultrasonic data for detecting challenging discontinuities in composite panels. *AIP Conf. Proc.* **2019**, *2102*, 020012.
34. Munir, N.; Kim, H.-J.; Park, J.; Song, S.-J.; Kang, S.-S. Convolutional neural network for ultrasonic weldment flaw classification in noisy conditions. *Ultrasonics* **2019**, *94*, 74–81. [CrossRef] [PubMed]
35. Gao, F.; Li, B.; Chen, L.; Wei, X.; Shang, Z.; He, C. Ultrasonic signal denoising based on autoencoder. *Rev. Sci. Instrum.* **2020**, *91*, 045104. [CrossRef] [PubMed]
36. Margrave, F.; Rigas, K.; Bradley, D.; Barrowcliffe, P. The use of neural networks in ultrasonic flaw detection. *Measurement* **1999**, *25*, 143–154. [CrossRef]
37. Yuan, S.; Wang, L.; Peng, G. Neural network method based on a new damage signature for structural health monitoring. *Thin-Walled Struct.* **2005**, *43*, 553–563. [CrossRef]
38. Fahad, M.; Kamal, K.; Zafar, T.; Qayyum, R.; Tariq, S.; Khan, K. Corrosion detection in industrial pipes using guided acoustics and radial basis function neural network. In Proceedings of the 2017 International Conference on Robotics and Automation Sciences (ICRAS), Hong Kong, China, 26–29 August 2017; pp. 129–133.
39. Cai, W.; Wang, J.; Zhou, Q.; Yang, Y.; Jiang, P. Equipment and Machine Learning in Welding Monitoring: A Short Review. In Proceedings of the 5th International Conference on Mechatronics and Robotics Engineering, Rome, Italy, 16–19 February 2019; pp. 9–15.
40. Wang, Y.; Shi, F.; Tong, X. A Welding Defect Identification Approach in X-ray Images Based on Deep Convolutional Neural Networks. In Proceedings of the International Conference on Intelligent Computing, Nanchang, China, 3–6 August 2019; pp. 953–964.
41. Isamail, L.; Maskuri, N.L.; Isip, N.J.; Lokman, S.F.; Abu Bakar, M.H. Deep Neural Network Modeling for Metallic Component Defects Using the Finite Element Model. In *Progress in Engineering Technology*; Springer: Cham, Switzerland, 2019; pp. 259–270.
42. Zhang, W.; Li, C.; Peng, G.; Chen, Y.; Zhang, Z. A deep convolutional neural network with new training methods for bearing fault diagnosis under noisy environment and different working load. *Mech. Syst. Signal Process.* **2018**, *100*, 439–453. [CrossRef]
43. Grasso, M.; Colosimo, B.M. Process defects and in situ monitoring methods in metal powder bed fusion: A review. *Meas. Sci. Technol.* **2017**, *28*, 044005. [CrossRef]
44. Qi, T.; Zhu, H.; Zhang, H.; Yin, J.; Ke, L.; Zeng, X. Selective laser melting of Al7050 powder: Melting mode transition and comparison of the characteristics between the keyhole and conduction mode. *Mater. Des.* **2017**, *135*, 257–266. [CrossRef]
45. Guo, Q.; Zhao, C.; Qu, M.; Xiong, L.; Escano, L.I.; Hojjatzadeh, S.M.H.; Parab, N.D.; Fezzaa, K.; Everhart, W.; Sun, T.; et al. In-situ characterization and quantification of melt pool variation under constant input energy density in laser powder bed fusion additive manufacturing process. *Addit. Manuf.* **2019**, *28*, 600–609. [CrossRef]
46. Leung, C.L.A.; Marussi, S.; Atwood, R.C.; Towrie, M.; Withers, P.J.; Lee, P.D. In situ X-ray imaging of defect and molten pool dynamics in laser additive manufacturing. *Nat. Commun.* **2018**, *9*, 1355. [CrossRef]

47. Jung, K.-H.; Kim, D.-H.; Kim, H.-J.; Park, S.-H.; Jhang, K.-Y.; Kim, H.-S. Finite element analysis of a low-velocity impact test for glass fiber-reinforced polypropylene composites considering mixed-mode interlaminar fracture toughness. *Compos. Struct.* **2017**, *160*, 446–456. [CrossRef]

48. Ziółkowski, G.; Chlebus, E.; Szymczyk, P.; Kurzac, J. Application of X-ray CT method for discontinuity and porosity detection in 316L stainless steel parts produced with SLM technology. *Arch. Civ. Mech. Eng.* **2014**, *14*, 608–614. [CrossRef]

49. Koester, L.W.; Taheri, H.; Bigelow, T.A.; Collins, P.C.; Bonds, L.J. Nondestructive Testing for Metal Parts Fabricated Using Powder-Based Additive Manufacturing. *Mater. Eval.* **2018**, *76*, 514–524.

50. Jhang, K.-Y.; Choi, S.; Kim, J. Measurement of Nonlinear Ultrasonic Parameters from Higher Harmonics. In *Measurement of Nonlinear Ultrasonic Characteristics*; Springer: Singapore, 2020; pp. 9–60.

51. LeCun, Y.; Bengio, Y.; Hinton, G. Deep learning. *Nature* **2015**, *521*, 436–444. [CrossRef]

52. Hinton, G.E.; Osindero, S.; Teh, Y.-W. A Fast Learning Algorithm for Deep Belief Nets. *Neural Comput.* **2006**, *18*, 1527–1554. [CrossRef] [PubMed]

53. Sambath, S.; Nagaraj, P.; Selvakumar, N.; Arunachalam, S.; Page, T. Automatic detection of defects in ultrasonic testing using artificial neural network. *Int. J. Microstruct. Mater. Prop.* **2010**, *5*, 561. [CrossRef]

54. LeCun, Y.; Bengio, Y. Convolutional Networks for Images, Speech, and Time-Series. *Handb. Brain Theory Neural Netw.* **1995**, *3361*, 1995.

55. Aghdam, H.H.; Heravi, E.J. *Guide to Convolutional Neural Networks*; Springer: New York, NY, USA, 2017; Volume 10, pp. 973–978.

56. Wang, W.; Zhu, M.; Wang, J.; Zeng, X.; Yang, Z. End-to-end encrypted traffic classification with one-dimensional convolution neural networks. In Proceedings of the IEEE International Conference on Intelligence and Security Informatics (ISI), Beijing, China, 22–24 July 2017; pp. 43–48. [CrossRef]

57. Dahl, G.E.; Sainath, T.N.; Hinton, G.E. Improving deep neural networks for LVCSR using rectified linear units and dropout. In Proceedings of the 2013 IEEE International Conference on Acoustics, Speech and Signal Processing, Vancouver, BC, Canada, 26–31 May 2013; pp. 8609–8613.

58. Srivastava, N.; Hinton, G.; Krizhevsky, A.; Sutskever, I.; Salakhutdinov, R. Dropout: A simple way to prevent neural networks from overfitting. *J. Mach. Learn. Res.* **2014**, *15*, 1929–1958.

59. Zhang, C.; Bengio, S.; Hardt, M.; Recht, B.; Vinyals, O. Understanding deep learning requires rethinking generalization. *arXiv* **2016**, arXiv:1611.03530.

Article

Innovative Methodology for the Identification of the Most Suitable Additive Technology Based on Product Characteristics

Antonio Del Prete and Teresa Primo *

Department of Engineering for Innovation, University of Salento, Via per Arnesano, Building "O", 73100 Lecce, Italy; antonio.delprete@unisalento.it
* Correspondence: teresa.primo@unisalento.it

Abstract: This paper reports the study and development case of an innovative application of the Cloud Manufacturing paradigm. Based on the definition of an appropriate web-based application, the infrastructure is able to connect the possible client requests and the relative supply chain product/process development capabilities and then attempt to find the best available solutions. In particular, the main goal of the developed system, called AMSA (Additive Manufacturing Spare parts market Application), is the definition of a common platform to supply different kinds of services that have the following common reference points in the Additive Manufacturing Technologies (DFAM, Design For Additive Manufacturing): product development, prototypes, or small series production and reverse engineering activities to obtain Computer-Aided Design (CAD) models starting from a physical object. The definition of different kinds of services allows satisfying several client needs such as innovative product definition characterized by high performance in terms of stiffness/weight ratio, the possibility of manufacturing small series, such as in the motorsport field, and the possibility of defining CAD models for the obsolete parts for which the geometrical information is missed. The AMSA platform relies on the reconfigurable supply chain that is dynamic, and it depends on the client needs. For example, when the client requires the manufacture of a small series of a component, AMSA allows the technicians to choose the best solutions in terms of delivery time, price, and logistics. Therefore, the suppliers that contribute to the definition of the dynamic supply chain have an important role. For these reasons, the AMSA platform represents an important and innovative tool that is able to link the suppliers to the customers in the best manner in order to obtain services that are characterized by a high-performance level. Therefore, a provisional model has been implemented that allows filtering the technologies according to suitable performance indexes. A specific aspect for which AMSA can be considered unique is related with the given possibility to access Design for Additive Manufacturing Services through the Web in accordance with the possible additive manufacturing technologies.

Keywords: additive manufacturing; key performance indicators; topology optimization; design for additive manufacturing; design for additive manufacturing services

Citation: Del Prete, A.; Primo, T. Innovative Methodology for the Identification of the Most Suitable Additive Technology Based on Product Characteristics. *Metals* **2021**, *11*, 409. https://doi.org/10.3390/met11030409

Academic Editors: Eric Hug and Atila Ertas

Received: 14 December 2020
Accepted: 21 February 2021
Published: 2 March 2021

Publisher's Note: MDPI stays neutral with regard to jurisdictional claims in published maps and institutional affiliations.

Copyright: © 2021 by the authors. Licensee MDPI, Basel, Switzerland. This article is an open access article distributed under the terms and conditions of the Creative Commons Attribution (CC BY) license (https://creativecommons.org/licenses/by/4.0/).

1. Introduction

Recently, the manufacturing sector has seen several changes in its own reference markets. These changes are linked to different factors such as the advent of new technologies and customer needs. For the first aspect, the advent of new technologies, it has introduced new ways to manufacture products that allow obtaining high flexibility and new possibilities to manufacture components with innovative shapes. In addition to the technologies, the customer needs have assumed a strategic importance because they are going toward high quality levels, high customization, and high complexity of the products, leading from the "Mass Production" model to the "Mass Customization" one. Another important aspect that supports these market changes is the new capabilities that the IoT (Internet of Things) offers. This term is a neologism that refers to an Internet extension toward the world of objects and places, which was introduced by Kevin Ashton in 1999 and further developed by the research agency Gartner [1].

Additive Manufacturing (AM) is defined as "a process of joining materials to make objects from 3D model data, usually layer upon layer, as opposed to subtractive manufacturing methodologies" [2], and it is considered one of the most promising manufacturing technologies, owing mainly to the geometrical flexibility it provides, but also to assets such as production flexibility, lead time reduction for short series production, reduced cradle-to-grave energy usage, and CO_2 footprint [3]. With the rise of AM in the last decade, a multitude of processes has become available to engineers, at a much lower cost than previously [4]. This has led to considering AM for manufacturing anything from appliances to automobiles and aircrafts. However, the applicability of AM to such areas is questionable. Moreover, although AM processes share a common approach and thus a few common characteristics, each one of the market-available AM technologies presents its own unique opportunities and limitations [4].

Zaman et al. introduce a generic decision methodology, based on multi-criteria decision-making tools, that will not only provide a set of compromised AM materials, processes, and machines but will also act as a guideline for designers to achieve a strong foothold in the AM industry by providing practical solutions containing design-oriented and feasible material-machine combinations from a current database of 38 renowned AM vendors in the world. An industrial case study, related to aerospace, has also been tested in detail via the proposed methodology [5].

AM has the potential to simultaneously build an object's material and geometry but considering unlimited potential does not guarantee having unlimited capability. The designers working in the AM industry have to not only concentrate on the types of constraints involved in procedures such as Computer-Aided Design (CAD) and the digitization of its ideas [6], discretization (digital and physical) of the parts to be produced, assessing capabilities of AM machines, and processing of materials to gauge the impact on properties, but they also have to cater for new challenges and requirements associated with metrology and quality control, maintenance, repair and recycling, lack of generic interdependency between materials and processes, limitation in material selection, longer design cycle than manufacturing cycle, surface finishing issues, and post-processing requirements [7,8]. Since the stakeholders in AM industry related to part manufacture are not altering the design completely in the "design phase", thereby resulting in an increase in the costs incurred both due to manufacturability and production time, it is highly important to address the relationship between manufacturing constraints, customer requirements, and design guidelines so that the overall cost including assembly and logistics is minimized [9].

Colosimo et al. present a cost model to evaluate the economic impact of defects and process instability in metal Additive Manufacturing (AM) [10].

The introduction of new technologies could cause deep changes to the entire supply chain configuration. A new approach designed to understand the importance of supply chain considerations for a suitable technology assessment has been studied. It is aimed at proposing a quantitative model for the evaluation of different structures of supply chains based on different production technologies. In particular, by defining a set of Key Performance Indicators and applying a multi-criteria decision method, a final score is computed, giving important information about both environmental and economic aspects [11,12].

Bikas et al. define a framework to assist non-expert potential users of AM technology with evaluating their specific use cases [13].

Di et al. establish a mathematical cost model to quantify the different cost components in the direct metal laser sintering process, and it is applicable for evaluating the cost performance when adopting dynamic process planning with different layer-wise process parameters. The case study results indicate that 12.73% of the total production cost could be potentially reduced when applying the proposed dynamic process planning algorithm based on the complexity level of geometries. In addition, the sensitivity analysis results suggest that the raw material price and the overhead cost are the two key cost drivers in the current additive manufacturing market [14].

Sustainability aspects and cost efficiency in the product lifecycle have great potential for improvement, so cost models are necessary to assist managers in selecting part-specific allocation strategies for metal spare parts realized in AM.

Ott et al. propose a two-stage model as a basis for decision support in spare part allocation. The first stage introduces a multi-criteria part classification regarding classical criteria as well as criteria referring to AM. The impacts on different spare part allocation strategies such as final stockpiling, conventional spare part production, or AM on demand will be the focus. Based on the first stage, a conceptual model for a comprehensive activity-based cost assessment will be adopted to assess the arising costs that occur for each of the compared allocation strategies [15].

Sabiston et al. expand upon existing mathematical constructs by providing an algorithm to minimize the cost and time associated with additively manufactured parts within a three-dimensional topology optimization framework. The formulation has been constructed in such a manner to accommodate large-scale topology optimization problems, including a filtering scheme requiring minimal storage of additional mesh information and an iterative finite element analysis solver. A rigorous trade-off analysis is conducted to determine the optimal contribution of additive manufacturing factors to minimize build time [16].

Liu et al. present a manufacturing cost constrained topology optimization algorithm considering the laser powder bed additive manufacturing process. The proposed algorithm would provide an opportunity to balance the manufacturing cost while pursuing the superior structural performance through topology optimization [17].

Palanisamy et al. adopt a multi-criteria decision-making (MCDM) technique, namely the best-worst method (BWM), to select the suitable material for the product. This is along with the end user expectations in AM. In the initial phase, the suitable machine to be selected from the available machines is based on the parameters such as cost, accuracy, variety of materials, and material wastage. From the variety of materials, the suitable material was selected based on the respondent requirement. The criteria that had a greater influence on the overall cost of the product manufacture through AM are identified and used. According to the BWM, the criteria to be selected by the decision maker based on the respondent expectations are identified. In the BWM method, pairwise comparisons are carried out between the best and worst criterion suggested by the decision makers, as that leads to the selection of the suitable material [18].

Within a research project, the authors have developed a web-based innovative solution that is able to supply services thanks to innovative technologies such as internet capabilities, additive manufacturing processes, product development know-how, and reverse engineering systems.

This paper aims to report the obtained results thanks to the application of AMSA (Additive Manufacturing Spare parts market Application) methodology to one of the selected case studies in order to validate it also for complex geometries. The aim of the defined methodology is to provide support to the AMSA operator to properly evaluate the selection of suppliers based on the characteristics of the geometry to be created. Therefore, a provisional model has been implemented that allows filtering the technologies according to suitable performance indexes. The AMSA platform is a web-based service developed on the basis of the main guidelines defined by the modern paradigm of Cloud Manufacturing (https://www.amsacloudmanufacturing.it/, accessed on 1 December 2020).

The potential advantages that distinguish this paradigm are the following:

- Flexibility;
- Tool to support the decision-making process;
- Intelligent production "on demand";
- Production capacities intended as services.

After a brief description of the case studies, the appropriate Key Performance Indicators (KPIs) have been defined and then calculated for a selected case study in order to simulate the working condition of the AMSA operator, in the attempt to identify the most suitable technologies to adopt for the component production. Starting from the Compatibility Index (CBA, provides a relative compatibility index between the product and the AM machine, which is better described in the Appendix A) results, the components production was therefore started and, once realized, the requirements validation was carried out.

At present, the procedure has been developed and tested with the AMSA test cases, so the different indices values need to be further refined with the print of additional components. The implemented relationships are the result of an analytical interpretation of empirical evaluations matured on the field.

2. AMSA Methodology Description

The AMSA methodology was created with the aim to provide a functional tool, that is integrable within a web platform to favor the selection of the most appropriate additive technologies and/or machinery depending on the product characteristics.

In order to facilitate the inclusion of this platform within the additive market, it was decided to start from market research to evaluate the most commercially interesting context. Following this study, it was decided to concentrate activities on additive technologies aimed at the production of metal components:

- Selective Laser Melting (SLM)
- Electron Beam Melting (EBM)
- Directed Energy Deposition (DED).

In addition, it is worth mentioning that they represent almost 100% of the production capacity present on the market, and the three technologies also ensure the possibility of managing a wide range of product requirements, from the large components proposed by DED processes to the dimensional accuracy obtainable with SLM processes, up to the management of particularly reactive materials thanks to the EBM ones. For each of the above-mentioned technologies, an analytical–parametric model has been developed in order to correlate the main process parameters with product performance in terms of costs quality and mechanical characteristics. The developed models, supported by suitable experimental tests that are able to calibrate and optimize the simulation parameters, have allowed us to identify the most relevant process variables responsible for the additive production success for each technology.

Figure 1 shows the configured supply chains for new products and spare parts. As reported, the supply chain is not fixed, and it varies, taking into account the typology of services. One of the straight points of the AMSA platform is the possibility to adapt the supply chain and the resources to the market changes. AMSA represents also a good opportunity for the suppliers to improve the action range and then their business.

Starting from these results, the AMSA methodology has been developed, with the aim to extrapolate and to create appropriate relationships among different data sources able to provide, for each component to be produced by additive manufacturing, a series of key information:

- The most suitable technology;
- The suggested machine type;
- The production time;
- The production cost.

Considering the large amount of variables to be managed, it was decided to consider an approach based on the selection of a series of technical KPIs, whose objective is to provide an index of compatibility of each machine type (non-technology) with respect to the requirements of the component to be produced (Table 1).

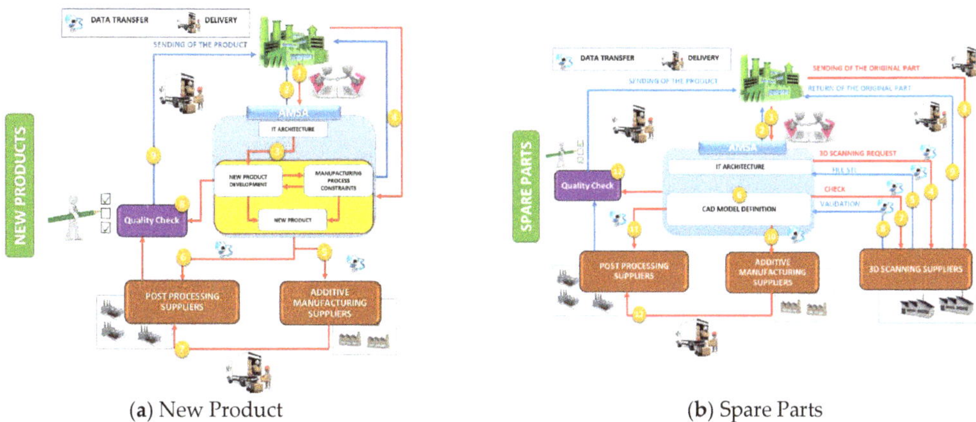

(a) New Product   (b) Spare Parts

Figure 1. AMSA (Additive Manufacturing Spare parts market Application) supply chain: (**a**) new product; (**b**) spare part.

Table 1. Matrix example that associates the Key Performance Indicator (KPIs) with the various machine types available on the platform (Powder Bed, PB or Directed Energy Deposition, DED).

		AVAILABLE MACHINES					
PRODUCT A		Machine PB_1	Machine PB_2	Machine PB_3	Machine DED_1	Machine DED_2	Machine DED_3
	KPI1	29%	94%	83%	79%	45%	61%
	KPI2	25%	75%	51%	92%	7%	97%
	KPI3	13%	99%	25%	11%	71%	90%
	KPI4	26%	53%	89%	41%	39%	8%
KPI	KPI5	76%	97%	44%	66%	18%	50%
	KPI6	78%	91%	26%	79%	57%	44%
	KPI7	25%	61%	40%	33%	12%	29%
	KPI8	52%	76%	54%	83%	3%	10%
	KPI9	58%	16%	61%	48%	100%	97%
	KPI10	30%	98%	93%	25%	82%	97%
	KPI11	16%	28%	11%	77%	76%	2%
COMPATIBILITY		39%	72%	53%	58%	46%	53%

The technical KPIs have a percentage values from 0 to 100% and different calculation methods depending on the considered case.

Once the KPIs calculation is completed, the platform provides a list of solutions ordered based on a compatibility KPI (CBA), which were obtained as an average (properly weighed) of the other available KPIs. With the compatibility KPI, the platform provides two other fundamental KPIs, which provide a complete summary that is able to facilitate the choice of the most appropriate machinery:

- Cost KPI (CST): provides an evaluation of the component price to be produced associated with the various machines available in AMSA;
- Time KPI (TMP): provides an estimate of production times according to the considered technology.

Identification of the technical KPIs, which define the CBA compatibility KPI, has been defined from a comprehensive analysis of the following aspects:

- The physical and mechanical characteristics of the materials selected for industrial applications;
- Analytical models selected from what is available in the technical and scientific literature;
- Analytical–empirical models developed by the research group during previous R&D (Research and Development) activities;
- Finite Element (FE) models where useful to make the correlation between process parameters and obtainable final results objective;
- The performance specifications description to be achieved for the selected industrial applications.

Therefore, the final list of the defined technical KPIs is the following:

- Material evaluation (MAT);
- Overall dimensions study (ING);
- Precision analysis (PRE);
- Resolution analysis (RIS);
- Roughness analysis (RGS);
- Study of the undercuts (STQ).

To formulate each one of the KPIs, the research team has tried to take into consideration all the aspects that are evaluated by the specialists during the process design phase for a specific product. These aspects have been correlated with each other in specific flow diagrams and then translated into analytical expressions, which are appropriately calibrated according to a series of experimental results. Figure 2 shows, as an example, the flow chart of the MAT KPI.

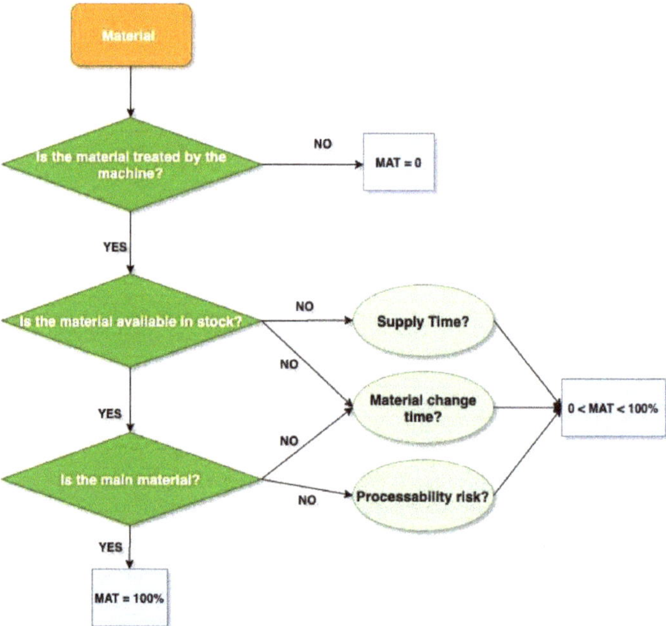

Figure 2. Flow chart of the material evaluation (MAT) Key Performance Indicator (KPI).

In the specific case of the material to be used for the production of the component, it is a direct request from the customer, which must be crossed with the available technologies to verify that it is actually obtainable with the available machines and evaluate any problems related to the timing or processability. MAT KPI is an index between 0 and 1, and has as its objective the expression of the process described in the Figure 1: once the request for an estimate has been received, the material indicated by the customer must be compared with the databases of the machines made available by the AMSA platform to identify the list of those who are actually capable of handling the material. Therefore, the list of machines is ordered according to three criteria:

- Powder supply time;
- Time to change material in the machine;
- Risk of material processability.

Evaluating the aspects highlighted in Figure 1, the following expression of MAT was proposed; the details are provided in the Appendix A.

$$\text{MAT} = \theta_{\text{mat}} \cdot \left(\alpha_{\text{appr}} \cdot \frac{1}{\sqrt{1 + \tau_{\text{apprw[g]}}}} + \alpha_{\text{mt}} \cdot \frac{1}{\sqrt{1 + \tau_{\text{AMmt[g]}}}} + \alpha_{\text{fr}} \cdot (2 - f_r) \right) \tag{1}$$

- θ_{mat}: compatibility between material and machine;
- α_{appr}: metal powder supply coefficient;
- $\tau_{\text{apprw[g]}}$: supply time for metallic powders (days);
- α_{mt}: material change coefficient;
- $\tau_{\text{AMmt[g]}}$: material change time (days);
- α_{fr}: material processability risk coefficient;
- f_r: risk factor of material processability, which represents a factor for increasing the mass of material required by the process, and it is requested to the supplier.

A clearly different approach was instead reserved for the KPIs of Cost (CST) and Time (TMP), which present particularly complex formulations that are able to take into consideration all the production phases of a product from design to finishing (Figure 3).

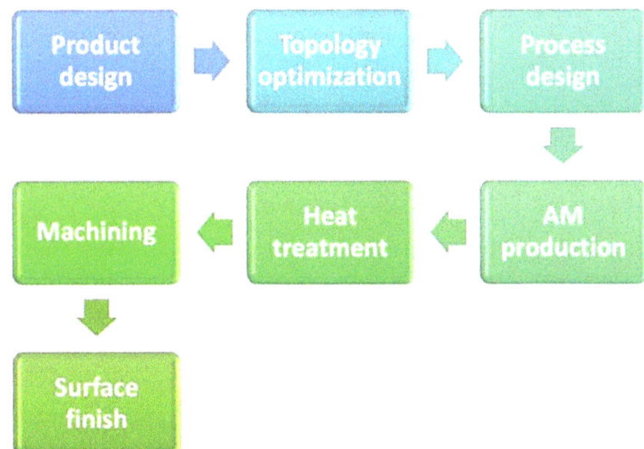

Figure 3. List of production phases considered for the Cost KPI (CST) and Time KPI (TMP) calculation.

In addition, in this case, the complete formulations are provided in Appendix A. With the application of the methodology, appropriately integrated within the AMSA platform, users are therefore able to evaluate the main process aspects and compare them, as described below.

3. Test Cases

- Automotive manifold (Figure 4): The use case representative of the possible industrial application for the spare part is a manifold that is used in the automotive sector for the discharge of exhausted gas. The product is currently made by welding three different components: a flat base suitably drilled, a truncated cone of connection, and a final neck on which the exhaust pipe is inserted. The welds are clearly visible in the image shown in Figure 4. The material is 316L stainless steel, which is able to withstand the high temperatures of the gas generated by combustion.

Figure 4. Real automotive manifold.

- Fused Deposition Modeling (FDM) head (Figure 5): The use case identified to represent the classes of components indicated as "new product" and "small series" is an assembly of metal parts welded together, which was designed by the research group to connect an FDM nozzle to the Z-axis of a Cartesian machine carried out within another research project. The prototype works in two different configurations:
  - FDM for traditional plastics (PLA, ABS, PP);
  - FDM for PEEK in a controlled temperature atmosphere.

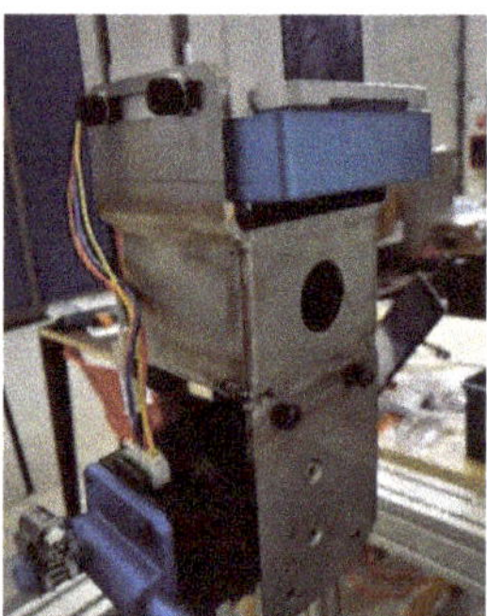

Figure 5. Fused Deposition Modeling (FDM) head of PEEK extrusion machine.

The difference between the two configurations consists in the assembly of a heated chamber conveniently sized to guarantee a constant temperature of 120 °C and thus reduce thermal stress due to the additive process to a minimum that can generate important tensions and distortions, such as compromising the correct growth of the component. In the case of PEEK, this minimum requires process temperatures over 350 °C.

4. Methodology Application to the Case Studies

In this section, the KPIs have been calculated based on the three machine types available: DED, SLM, and EBM. As an example, the application of the methodology for the FDM head for traditional plastics is reported.

4.1. KPIs Calculation for the FDM Head for Traditional Plastics

The FDM head for traditional plastics has been made in A357 and must respect defined functional requirements.

4.1.1. MAT KPI

This index evaluates the effects of compatibility between the material requested by the customer and the analyzed AM machine.

The A357 material is compatible with DED and SLM technologies. For this last machine, it represents a "main material"; for this reason, in this case, the supply time and the material change time are equal to zero. The EBM does not manage aluminum alloys because due to working conditions very close to the vacuum, there have been detected cases of material vaporization subjected to electron beam bombardment. This problem has no consequences only from the point of view of the deposition process, but it represents a critical condition also from the point of view of safety, because the vaporized particles generate scattering in the electrons, which are therefore spread uncontrollably inside the chamber. Regarding DED, a rather high-risk coefficient has been inserted, since the aluminum alloys processability has always been particularly problematic with this technology (Table 2).

Table 2. MAT calculation for the FDM head for traditional plastics.

MAT	Allowed Values	Description	DED	SLM	EBM
		Material	A357	A357	A357
θ_{mat}	[0]; [1]	0 = incompatible material 1 = compatible material	1	1	0
α_{appr}	[0:1]	Weight related to the powder supply time	0.5	0.5	0.5
$\tau_{apprw[g]}$	[0:∞)	Time expressed in days equal to the powder supply time	7	0	14
α_{mt}	[0:1]	Weight related to material change time	0.3	0.3	0.3
$\tau_{AMmt[g]}$	[0:∞)	Time expressed in days equal to the powder change time in the machine for a given material	2	0	5
α_{fr}	[0:1]	Weight related to the risk of material processability in the machine	0.2	0.2	0.2
f_r	[1: ∞)	Material processability risk in the machine - Increase factor of material mass required by the process	1.5	1	1
		MAT	45%	100%	0%

4.1.2. ING KPI

ING KPI evaluates the compatibility between the dimensions of the component(s) to be produced with respect to the AM machine. The aspects considered by this KPI are as follows:

- Compatibility between the component overall dimensions and the machine working volume;
- Convenience of work volume saturation in the case of powder bed processes.

The considered component has dimensions such that it can be processed by all the considered machinery. The lowest value of ING for powder bed technologies derives from the fact that the production of a single piece does not saturate the working volume, which instead happens (for obvious reasons) in the case of DED (Table 3).

Table 3. Overall dimensions study (ING) calculation for the FDM head for traditional plastics.

ING	Allowed Values	Description	DED	SLM	EBM
θ_{ing}	[0]; [1]	Bigger component of the machine working volume Compatible component with the machine working volume	1	1	1
x_{AM}	$[0:\infty)$	Machine working volume expressed as X-Y-Z in m	1.000	0.245	0.200
y_{AM}	$[0:\infty)$		1.500	0.245	0.200
z_{AM}	$[0:\infty)$		2.000	0.300	0.300
x_d	$[0:\infty)$	Volume of the parallelepiped containing the component to be produced expressed as X-Y-Z in m	0.100	0.100	0.100
y_d	$[0:\infty)$		0.100	0100	0.100
z_d	$[0:\infty)$		0.185	0.185	0.185
m	$[0:\infty)$	Maximum number of components that can be produced within a job	1	4	8
n	$[0:\infty)$	Number of components of the lot	1	1	1
n_{cycle}	$[0:\infty)$	Cycles number required by the machine to produce the entire lot	1	1	1
		ING	96%	39%	29%

4.1.3. PRE KPI

PRE KPI evaluates the compatibility between the precision required by the component and the precision guaranteed by the machine. The aspects considered by the KPI are as follows:

- Reference precision of the component;
- Accuracy guaranteed by the machine.

The precision required by the component is compatible only with SLM and EBM technologies; this means that in the case of DED, it is necessary to provide subsequent mechanical processes to ensure that the customer's requirements are respected (Table 4).

Table 4. Precision analysis (PRE) calculation for the FDM head for traditional plastics.

PRE	Allowed Values	Description	DED	SLM	EBM
θ_{pre}	[0]; [1]	$\theta_{pre} = 0 \rightarrow \zeta_d < \zeta_{AM}$ $\theta_{pre} = 1 \rightarrow \zeta_d > \zeta_{AM}$	0	1	1
ζ_d	$[0:\infty)$	Reference precision of the component	0.200	0.200	0.200
ζ_{AM}	$[0:\infty)$	Machine precision	0.500	0.050	0.100
		PRE	0%	75%	50%

4.1.4. RIS KPI

RIS KPI evaluates the compatibility between the smallest feature of the component and the machine's capabilities in making it. The aspects considered by the KPI are as follows:

- The smallest feature dimension in the component;
- Machine resolution.

The resolution required by the component is compatible with all technologies; therefore, it is not necessary to provide subsequent mechanical processes to remove excess material (Table 5).

Table 5. Resolution analysis (RIS) calculation for the FDM head for traditional plastics.

RIS	Allowed Values	Description	DED	SLM	EBM
θ_{ris}	[0]; [1]	$\theta_{ris} = 0 \rightarrow \xi_d < \xi_{AM}$ $\theta_{ris} = 1 \rightarrow \xi_d > \xi_{AM}$	1	1	1
ξ_d	[0:∞)	Reference resolution of the component	2.000	2.000	2.000
ξ_{AM}	[0:∞)	Machine resolution	1.000	0.300	0.400
		RIS	50%	85%	80%

4.1.5. RGS KPI

RGS KPI evaluates the compatibility between the roughness required by the component and the roughness obtainable by the AM machine. The aspects considered by the KPI are as follows:

- Component reference roughness;
- Roughness guaranteed by the machine.

The roughness required by the component is not compatible with the available machinery for which, for the interested areas by that type of roughness, it is necessary to provide subsequent mechanical processing (Table 6).

Table 6. Roughness analysis (RGS) calculation for the FDM head for traditional plastics.

RGS	Allowed Values	Description	DED	SLM	EBM
θ_{rgs}	[0]; [1]	$\theta_{rgs} = 0 \rightarrow \chi_d < \chi_{AM}$ $\theta_{rgs} = 1 \rightarrow \chi_d > \chi_{AM}$	0	0	0
χ_d	[0:∞)	Component reference roughness	3.2	3.2	3.2
χ_{AM}	[0:∞)	Roughness obtainable with the machine	15.0	5.0	10.0
		RGS	0%	0%	0%

4.1.6. STQ KPI

STQ KPI evaluates the compatibility between the undercut surfaces and the capacity of the considered AM machine. The aspects considered by the KPI are as follows:

- Surface of the undercut areas;
- Total component surface.

The amount of surfaces in undercut conditions is rather important compared to the total; therefore, the KPI suggests the powder bed technologies that are more easily able to manage it (Table 7).

Table 7. Study of the undercuts (STQ) calculation for the FDM head for traditional plastics.

STQ	Allowed Values	Description	DED	SLM	EBM
S_d	[0:∞)	Total surface of the component	0.028	0.028	0.028
S_t	[0:∞)	Total surface in undercut	0.003	0.003	0.003
ψ	[0:1]	Support management coefficient	0.1	0.6	1.0
		STQ	9%	78%	91%

4.1.7. CST KPI

The objective of this KPI is to provide an indicative economic quotation of the product according to the indicated technology. The aspects considered by the KPI are costs due to the following:

- Material;
- Printer usage time;
- Operator;
- Geometry complexity.

The CST KPI presents results only for SLM and DED, because the EBM does not support aluminum and the methodology also sets the result of CST to zero. In this case, the value of the DED machine is slightly lower, but the SLM technology provides a valid alternative due to the fact that DED has several disadvantages on aluminum management. In this case, the quantity of lot numbers has its weight: if the lot number had been higher and next to 4 (work volume saturation for SLM), the CST value would clearly favor the SLM machine (Table 8).

Table 8. CST calculation for the FDM head for traditional plastics.

CST	Allowed Values	Description	DED	SLM	EBM
θ_{mat}	[0]; [1]	0 = incompatible material 1 = compatible material	1	1	0
θ_{ing}	[0]; [1]	0 = bigger component of the machine working volume 1 = component compatible with the working volume of the machine	1	1	1
n	[0:∞)	Number of components of the lot	1	1	1
δ_d	[0:∞)	Density of the material indicated by the customer (kg/m^3)	2670	2670	2670
V_d	[0:∞)	Volume of the component to be produced (m^3)	5.03×10^{-8}	5.03×10^{-8}	$5.03 \; 10^{-8}$
S_d	[0:∞)	Total surface of the component (m^2)	0.0282	0.0282	0.0282
γ_{rt}	[0:∞)	Average machining allowances thickness identified by technology (m)	0.002	0.0002	0.0005
φ_w	[0:1]	Ratio between the surface subject to machining allowance and the total surface of the component (%)	8%	8%	8%
ε_{wr}	[0:1]	Powder management efficiency in the machining allowance production (%)	70%	85%	85%
ε_{wd}	[0:1]	Powder management efficiency in the component production (%)	70%	85%	85%
C_w	[0:∞)	Unit cost of metal powders (€/kg)	70	130	0
δ_t	[0:∞)	Material density chosen for the supports (kg/m^3)	2670	2670	2670
V_{boudt}	[0:∞)	Volume of the parallelepiped containing the support structures (m^3)	0.0003	0.0003	0.0003
φ_t	[0:1]	Technological coefficient for the supports material mass (%)	0.1	0.6	0.3
C_t	[0:∞)	Unit material cost required for the supports (€/kg)	70	130	0
δ_g	[0:∞)	Density of the anchor plate material (kg/m^3)	2670	2670	2670
μ_g	[0:∞)	Anchor plate thickness (m)	0.025	0.025	0.025
S_{bound}	[0:∞)	Surface of the component bounding box (m^2)	0.01	0.01	0.01
φ_g	[0:∞)	Coefficient of increase of the bounding box surface (%)	10%	5%	5%

Table 8. *Cont.*

CST	Allowed Values	Description	DED	SLM	EBM
f_r	[0:1]	Material risk coefficient	1.5	1	1
C_g	[0:∞)	Unit cost of anchor plates (€/kg)	5	5	0
K_d	[0:∞)	Complexity coefficient for process design	1.3	1	1
c_{dop}	[0:∞)	Cost operator time for process design €/h	30	30	30
τ_{dor}	[0:∞)	Operator time to identify component orientation (h)	0.5	0.5	0.5
τ_{dsl}	[0:∞)	Operator time to identify the optimal slicing strategy (h)	2	2	2
τ_{dps}	[0:∞)	Operator time to identify process parameters (h)	0.5	0.5	0.5
τ_{dcm}	[0:∞)	Operator time to set the tool path (h)	0.5	0.5	0.5
K_{AM}	[0:1]	Complexity coefficient for additive production	1.3	1	1
c_{AMeq}	[0:∞)	Hourly cost for Additive Production due to amortization	50	46.5	81.4
c_{AMmh}	[0:∞)	Hourly cost for Additive Production due to maintenance	10	7.1	15
c_{AMen}	[0:∞)	Hourly cost for Additive Production due to energy consumption	1.93	1.68	2.57
c_{AMop}	[0:∞)	Hourly cost for additive production due to the operator	0	0	0
τ_{AMmt}	[0:∞)	Material change time (h)	0.5	0	5
$\tau_{AMstart}$	[0:∞)	Machine start-up time (h)	1	3	3
τ_{AMrisc}	[0:∞)	Machine preheating time (h)	0	1	1
m	[0:∞)	Maximum components number that can be produced within a job	1	4	8
τ_{AMcc}	[0:∞)	Cycle change time (h)	2	4	1
V_d	[0:∞)	Total volume to be melted (m3)	0.000105976	0.000223014	0.000142268
P_{AM}	[0:∞)	Machine productivity (m³/h)	0.000100	0.000015	0.000080
$\tau_{AMraffr}$	[0:∞)	Room cooling time (h)	2	2	8
$\tau_{AMclean}$	[0:∞)	Excess powder removal time (h)	0.5	4	4
		CST (€)	633.39 €	1,541.08 €	

4.1.8. TMP KPI

TMP KPI provides an estimation of component delivery time. The aspects considered by the KPI are as follows:

- Powder supply time;
- Material change time in the machine;
- Time for process design;
- Accessory times for the machine heating and cooling phases;
- Production time;
- Machine cleaning time.

This KPI provides an estimate of the lot production times. In the case of EBM, it remains equal to zero because, as already mentioned, it does not support the used material. In the case of the SLM, it is the main material for which TMP only represents the production time equal to about one day. For the DED, on the other hand, it also includes delivery times for the powder that is not in stock (Table 9).

Table 9. TMP calculation for the FDM head for traditional plastics.

TMP	Allowed Values	Description	DED	SLM	EBM
θ_{mat}	[0]; [1]	0 = incompatible material 1 = compatible material	1	1	0
θ_{ing}	[0]; [1]	0 = Bigger component of the machine working volume 1 = component compatible with the working volume of the machine	1	1	1
n	[0:∞)	Number of components of the lot	1	1	1
V_d	[0:∞)	Volume of the component to be produced (m^3)	5.03×10^{-8}	5.03×10^{-8}	5.03×10^{-8}
$\tau_{apprw[h]}$	[0:∞)	Powder supply time (h)	168	0	336
τ_{dor}	[0:∞)	Operator time to identify component orientation (h)	0.5	0.5	0.5
τ_{dsl}	[0:∞)	Operator time to identify the optimal slicing strategy (h)	2	2	2
τ_{dps}	[0:∞)	Operator time to identify process parameters (h)	0.5	0.5	0.5
τ_{dcm}	[0:∞)	Operator time to set the tool path (h)	0.5	0.5	0.5
τ_{AMmt}	[0:∞)	Material change time (h)	0.5	0	5
$\tau_{AMstart}$	[0:∞)	Machine start-up time (h)	1	3	3
τ_{AMrisc}	[0:∞)	Machine preheating time (h)	0	1	1
m	[0:∞)	Maximum components number that can be produced within a job	1	4	8
τ_{AMcc}	[0:∞)	Cycle change time (h)	2	4	1
V_d	[0:∞)	Total volume to be melted (m^3)	0.00010598	0.00022301	0.00014227
P_{AM}	[0:∞)	Machine productivity (m^3/h)	0.000100	0.000015	0.000080
$\tau_{AMraffr}$	[0:∞)	Room cooling time (h)	2	2	8
$\tau_{AMclean}$	[0:∞)	Excess powder removal time (h)	0.5	4	4
		TMP (gg)	7.5	2.6	0.0

4.1.9. CBA KPI

The compatibility coefficient is represented by the appropriately weighted sum of a series of contributions given by specific KPIs. This last can be analyzed separately by the AMSA technician in the analysis of the estimate quotation, but the values will flow into the CBA calculation, which will be used to order the list of suppliers to be evaluated (Table 10).

Table 10. Compatibility Index (CBA) calculation for the FDM head for traditional plastic.

CBA	Allowed Values	Description	DED	SLM	EBM
θ_{mat}	[0]; [1]	0 = incompatible material 1 = compatible material	1	1	0
θ_{ing}	[0]; [1]	0 = Bigger component of the machine working volume 1 = component compatible with the machine working volume	1	1	1
PRE	[0:1]	Precision KPI	0%	75%	50%
RIS	[0:1]	Resolution KPI	50%	85%	80%
RGS	[0:1]	Roughness KPI	0%	0%	0%
STQ	[0:1]	Undercuts KPI	9%	78%	91%
CST (€)	[0:∞)	Cost KPI	633.39 €	1,541.08 €	
TMP (gg)	[0:∞)	Time KPI	7.5	2.6	0
		CBA (%)	40%	53%	0%

It provides a relative Compatibility Index between the product and the considered AM machine.

Indeed, CBA provides an overview of the various KPIs:

- Both the material and the component overall dimensions are supported by SLM and DED; the material is not supported by the EBM, which is therefore automatically excluded from the CBA calculation.
- The required precision is manageable only by SLM; therefore, in the case of DED, it is necessary to provide a specific post-processing step.
- The required resolution is largely managed by all the machines.
- The required roughness requires post-processing for all machines.
- There are undercuts; therefore, they should be more easily managed by powder bed technologies.
- The cost slightly favors the DED, while the production time clearly favors the SLM.

Therefore, the final result of CBA suggests the geometry production by SLM.

5. Methodology Validation

Component production has allowed us to obtain the necessary information for the KPI compilation and therefore the methodology validation.

5.1. FDM Head for Traditional Plastics

According to the KPI results, the head for traditional plastics should be produced by SLM (A357). In the present paper, only the geometry and material analysis are presented, since both the process parameters identification and the CAM development are not the object of the present work.

5.2. Geometry and Material Analysis

For the FDM head, it has been necessary to proceed with a specific topology optimization phase. The geometry is reported in Figure 6, where two slots are visible in the upper part for the Cartesian Z-axis fixing (1), the holes (2) for the extruder mounting, and the holes (3) for the folding assembly, which protects the aluminum structure from the heat produced by the process chamber. Altair's INSPIRE® software [19] has been used for topology optimization. The four screw fixings have been constrained to the Z-axis of the Cartesian. As load, we considered the component and extruder weight and the force due to the head movement, which has been applied on the lower plate as an acceleration equal to 0.1 g in the X and Y-directions (Figure 7).

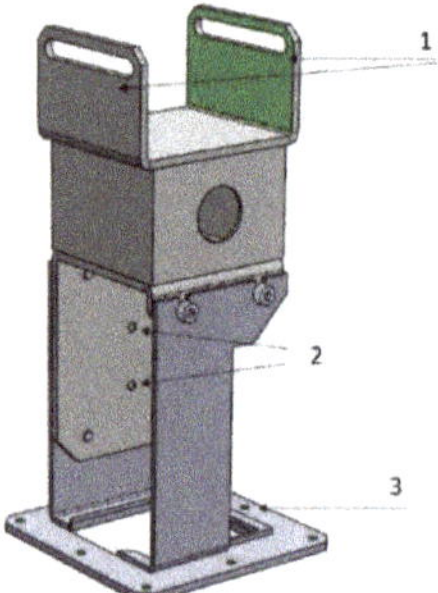

Figure 6. Component coupling surface.

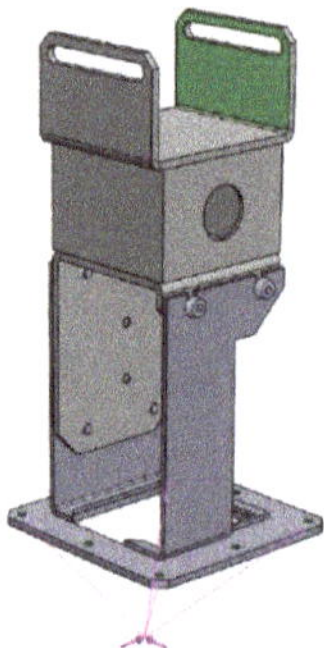

Figure 7. Accelerations application point.

To simplify the component redesign after the optimization, the non-design zones have been remodeled, and the slots for fixing to the Cartesian have become bushings, as well as the holes for extruder fixing (Figure 8). The base plate has been cut by a laser, and to ensure assembly, other bushings have been modeled for coupling screws insertion. Tetra elements have been modeled to maintain a flush between this component and the Z-axis.

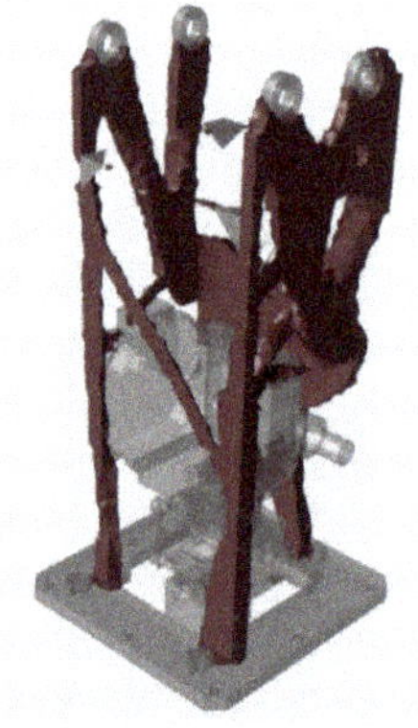

Figure 8. Material density distribution after topology optimization.

Since the base plate must also allow the capacitive sensor to be mounted in the case of PEEK production, and it must necessarily be removable, it was decided to remove it from the "design space" and provide, on the optimized geometry, four holes to ensure this assembly option.

The following optimizations have been realized:

- Optimization for A357 Aluminum to obtain the head for traditional plastics (SLM—small series);
- Optimization for Ti6Al4V Titanium to obtain the head for traditional plastics (SLM—new product).

The stiffness maximization has been defined as an objective function. The two results have provided very similar geometries between them, and then, they have been remodeled with Inspire's PolyNurbs feature, which allows us to "coat" imperfect geometries deriving from topology optimization with more regular shapes, in order to obtain components almost ready for printing (Figure 9). Subsequently, the model has been engineered in the SolidWorks® environment (Figure 10).

Figure 9. Remodeled component with PolyNurbs.

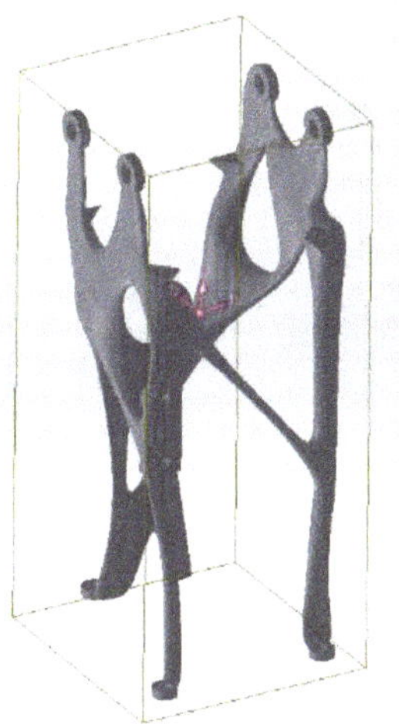

Figure 10. A357 component for Selective Laser Melting (SLM).

Particular attention has been given to the interface areas between the PolyNurbs and the non-design areas: the transition between the two portions of the CAD model must be gradual to avoid the stress concentration in these areas during the printing process.

After the CAD modeling, the components were verified again by static analysis to detect any stress concentrations. Figure 11 shows the real component created that was detached from the sacrificial platforms and analyzed to verify the success of the production process.

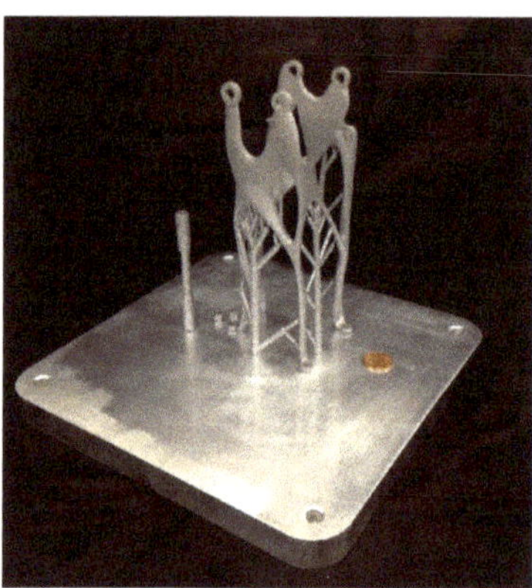

Figure 11. Real component made by SLM (small series).

6. Functional Requirements Verification and KPIs Validation for Small Series (SLM)

The created component has been removed from the sacrificial platforms and analyzed to verify the obtained quality. As can be seen in Table 11, all the positioning tolerances as well as the dimensional tolerances have been respected. The planarity of the reference plans for the head mounting was verified.

Table 11. Functional requirements verification of the FDM head for traditional plastics.

FDM Head for Traditional Plastic (Small Series)	Characteristic	Functional Requirement	
Head fixing component Z-axis	Positioning tolerances of the fixing holes	mk 0.1	✔
	Reference plan planarity	mk 0.1	✔
	Dimensional tolerance of through fixing holes	mk 0.1	✔
	Reference plan planarity roughness	Ra 3.2	✔
Nozzle head fixing component	Positioning tolerances of the fixing holes	mk 0.1	✔
	Dimensional tolerance of through fixing holes	mk 0.1	✔
	Internal roughness of the holes	Ra 1.6	✔
Folding head fixing component	Positioning tolerances of the fixing holes	mk 0.1	✔
	Reference plan planarity	mk 0.1	✔
	Dimensional tolerance of through fixing holes	mk 0.1	✔
Structure and material	Vibration resistance	Able to support acceleration of 0.1g	✔

However, the holes' internal surfaces roughness did not respect the reference values, and they have required further processing in order to consider the requirement satisfied. The roughness has been measured and has been equal to 5.2 Ra on the reference plans and 7.7 Ra in the nozzle fixing holes.

Furthermore, the vibrations resistance was verified only from a numerical point of view; in fact, the topological optimization was carried out based on this constraint.

From KPIs validation point of view, it is possible to compare the real data with those predicted by the implemented methodology. Table 10 shows how precision, and undercuts, as expected, were easily managed by technology. It is also confirmed that the roughness was not obtainable with the additive process, and in fact, the process was able to provide a maximum of 5.2 Ra.

Finally, as regards the production cost, a comparison was made with the calculated real results, and a deviation of approximately 3.8% was found. Specifically, considering the number of productive hours of an industrial machine equal to 2800, a cost of €1600 was obtained, compared to the calculated value of €1541 (Table 10).

7. Conclusions

The target of this study has been to evaluate the capabilities of the AMSA methodology considering a defined test case representative of a typical application in order to validate the complex geometries that AMSA service could face. For this reason, the main operations that the AMSA technician could manage for a possible production order request have been simulated. The KPIs have been calculated considering the formulation described in Appendix A.

Starting from the CBA results, the component has been manufactured. After the production, in order to evaluate the AMSA methodology performance, a verification of the functional requirements and the KPI values have been carried out. The obtained results highlight that the AMSA methodology is robust enough to manage complex geometries, and it represents a valid tool for manufacturing engineers during the decision process related to the handling of possible production orders. The procedure has been developed for AMSA test cases, and the different indices values need to be further refined with the printing of additional components.

On the basis of the obtained results, the methodology seems to respond, above all qualitatively, quite well to the variables introduced, even for complex components such as the considered use cases.

From a quantitative point of view, especially as regards the production cost, it would be good to carry out other tests with different geometries in order to have a statistical basis capable of further improving the formulation of the technological coefficients that have a significant weight in the CST calculation. What has been demonstrated certainly represents an important result that can be the starting point for the drafting of real algorithms able to assist the AMSA operator in deciding which technology to use for production, highlighting all the critical aspects that should be kept into account.

Author Contributions: Conceptualization, A.D.P. and T.P.; methodology, A.D.P. and T.P.; Validation, A.D.P. and T.P. All authors have read and agreed to the published version of the manuscript.

Funding: This research was funded by Ministry of Economic Development, Bando Horizon 2020-PON 2014/2020.

Informed Consent Statement: Not applicable.

Data Availability Statement: Data sharing not applicable.

Conflicts of Interest: The authors declare no conflict of interest.

Appendix A

To better understand the details, the complete formulation of the indexes listed above is shown below.

Appendix A.1. Material Index "MAT"

The following is the complete formulation of the MAT index with the relative control variables.

$$\text{MAT} = \theta_{\text{mat}} \cdot \left(\alpha_{\text{appr}} \cdot \frac{1}{\sqrt{1 + \tau_{\text{apprw}[g]}}} + \alpha_{\text{mt}} \cdot \frac{1}{\sqrt{1 + \tau_{\text{AMmt}[g]}}} + \alpha_{\text{fr}} \cdot (2 - f_{\text{r}}) \right) \tag{A1}$$

- θ_{mat}: compatibility between material and machine
- α_{appr}: metal powder supply coefficient
- $\tau_{\text{apprw}[g]}$: supply time for metallic powders [days]
- α_{mt}: material change coefficient
- $\tau_{\text{AMmt}[g]}$: material change time [days]
- α_{fr}: material processability risk coefficient
- f_{r}: risk factor of material processability, which represents a factor for increasing the mass of material required by the process, and it is requested to the supplier.

Appendix A.2. Production Cost Index "CST"

Below, we describe the complete formulation of the CST index with the list of related control variables.

$$\begin{aligned}
\text{CST} = \theta_{\text{mat}} \cdot \theta_{\text{ing}} \\
\cdot \Big\{ \Big[(n \cdot \delta_d) \cdot \Big(\frac{V_d}{\varepsilon_{W_d}} + \frac{S_d \cdot \gamma_{\text{rt}} \cdot \varphi_w}{\varepsilon_{W_r}} \Big) \Big] \cdot C_w + [n \cdot \delta_t \cdot V_{\text{boundt}} \cdot \varphi_t] \cdot C_t \\
+ \Big[n \cdot \mu_g \cdot S_{\text{bound}} \cdot (1 + \varphi_g) \cdot \delta_g \Big] \cdot C_g \Big\} \cdot f_r \\
+ \Big\{ K_d \cdot \Big(C_{d_{\text{op}}} \cdot \Big(\tau_{d_{\text{or}}} + \tau_{d_{\text{sl}}} + \tau_{d_{\text{ps}}} + \tau_{d_{\text{cm}}} \Big) \Big) + K_{\text{am}} \\
\cdot \Big[\Big(C_{\text{AM}_{\text{eq}}} + C_{\text{AM}_{\text{mh}}} + C_{\text{AM}_{\text{en}}} + C_{\text{AM}_{\text{op}}} \Big) \\
\cdot \Big(\tau_{\text{AM}_{\text{mt}}} + (\tau_{\text{AM}_{\text{start}}} + \tau_{\text{AM}_{\text{risc}}}) \cdot \left[\tfrac{n}{m} \right] + \tfrac{V_d}{V_{\text{AM}}} \cdot n + \tau_{\text{AM}_{\text{cc}}} \cdot \left(\left[\tfrac{n}{m} \right] - 1 \right) \\
+ (\tau_{\text{AM}_{\text{raffr}}} + \tau_{\text{AM}_{\text{clean}}}) \cdot \left[\tfrac{n}{m} \right] \Big) \Big] \Big\}
\end{aligned} \tag{A2}$$

- θ_{mat}: compatibility between material and machine
- θ_{ing} : compatibility between component and the machine working volume
- δ_d: material density indicated by the customer (kg/m^3);
- V_d: volume of the component to be produced (m^3);
- S_d: total area of the component (m^2);
- γ_{rt}: average machining allowance thickness identified by technology (m);
- φ_w: ratio between the surface subject to machining allowance and the total surface area of the component (%);
- ε_{W_r}: powders management efficiency in the production of machining allowance (%);
- ε_{W_d}: powders management efficiency in the production of the component (%);
- C_w: unit cost of metal powders (€/kg);
- δ_t: material density chosen for the supports (kg/m^3);
- V_{bound_t}: parallelepiped volume of the containment of support structures (m^3);
- φ_t: technological coefficient for the mass of material for the supports (%);
- C_t: unit cost of the material required for the supports (€/kg);
- δ_g: material density of the anchor plate (kg/m^3);
- μ_g: thickness of the anchor plate (kg/m^3);
- S_{bound}: surface of the component's bounding box (kg/m^3);
- φ_g: increase coefficient of the bounding box surface (%);
- f_r: material risk coefficient;
- C_g: unit cost of the anchor plates (€/kg);
- K_d: complexity coefficient for process design;

- K_{AM}: complexity coefficient for additive production;
- $c_{d_{op}}$: operator hourly cost for process design;
- $\tau_{d_{or}}$: operator time to identify component orientation (h)
- $\tau_{d_{sl}}$: operator time to identify the optimal slicing strategy (h)
- $\tau_{d_{ps}}$: operator time to identify process parameters (h)
- $\tau_{d_{cm}}$: operator time to set the tool path (h)
- $c_{AM_{eq}}$: hourly cost for Additive Production due to amortization;
- $c_{AM_{mh}}$: hourly cost for Additive Production due to maintenance;
- $c_{AM_{en}}$: hourly cost for Additive Production due to energy consumption;
- $c_{AM_{op}}$: hourly cost for Additive Production due to the operator;
- $\tau_{AM_{mt}}$: material change time (h);
- $\tau_{AM_{start}}$: machine start-up time (h)
- $\tau_{AM_{risc}}$: machine preheating time (h);
- n: components of the lot number;
- m: maximum number of components that can be produced within a job;
- $\left[\frac{n}{m}\right]$: number of jobs required by a machine with a maximum capacity of m components to make n components;
- $\tau_{AM_{cc}}$: cycle change time (h);
- P_{AM}: machine productivity (m^3/h);
- $\tau_{AM_{raffr}}$: chamber cooling time (h);
- $\tau_{AM_{clean}}$: excess powders removal time (h).

Appendix A.3. Production Time Index "TMP"

The following is the complete formulation of the TMP index with the relative control variables.

$$
\begin{aligned}
\text{TMP} = \theta_{mat} \cdot \theta_{ing} \\
\cdot \Big\{ \tau_{appr_{w[h]}} + \left(\tau_{d_{or}} + \tau_{d_{sl}} + \tau_{d_{ps}} + \tau_{d_{cm}} \right) \\
+ \left[\tau_{AM_{mt}} + \left(\tau_{AM_{start}} + \tau_{AM_{risc}} \right) \cdot \left(\frac{n}{m} \right) \right] \\
+ \left[\frac{V_d}{P_{AM}} \cdot n + \tau_{AM_{cc}} \cdot \left(\left(\frac{n}{m} \right) - 1 \right) \right] + \left[\left(\tau_{AM_{raffr}} + \tau_{AM_{clean}} \right) \cdot \left(\frac{n}{m} \right) \right] \Big\}
\end{aligned}
\tag{A3}
$$

- θ_{mat}: compatibility between material and machine
- θ_{ing} : compatibility between component and the machine working volume
- $\tau_{appr_{w[h]}}$: powder supply time (h)
- $\tau_{d_{or}}$: operator time to identify component orientation (h)
- $\tau_{d_{sl}}$: operator time to identify the optimal slicing strategy (h)
- $\tau_{d_{ps}}$: operator time to identify process parameters (h)
- $\tau_{d_{cm}}$: operator time to set the tool path (h)
- $\tau_{AM_{mt}}$: material change time (h)
- $\tau_{AM_{start}}$: machine start-up time (h)
- $\tau_{AM_{risc}}$: machine preheating time (h)
- m: maximum components number that can be produced within a job
- $\tau_{AM_{raffr}}$: room cooling time (h)
- $\tau_{AM_{clean}}$: excess powder removal time (h)
- $\tau_{AM_{cc}}$: cycle change time (h)
- n : number of components of the lot
- P_{AM}: machine productivity (m^3/h)
- V_d : total volume to be melted (m^3)

Appendix A.4. Overall Dimensions of the Component in the Machine "ING"

The following is the complete formulation of the ING index with the relative control variables.

$$\text{ING} = \theta_{\text{ing}} \cdot \left\{ \frac{\left[\left(\frac{[(x_{AM} - x_d)]}{x_{AM}} + \frac{(y_{AM} - y_d)}{y_{AM}} + \frac{(z_{AM} - z_d)}{z_{AM}} \right) / 3 \right] + \left[\frac{n}{m} - \left(n_{\text{cycle}} - 1 \right) \right]}{2} \right\} \tag{A4}$$

- θ_{ing}: compatibility between component and the machine working volume
- x_{AM}
- y_{AM}: machine working volume
- z_{AM}
- x_d
- y_d: volume of the parallelepiped containing the component
- z_d
- n: number of components of the lot
- n_{cycle}: job numbers
- m: maximum number of components that can be produced within a job

Appendix A.5. Technology Precision "PRE"

The following is the complete formulation of the PRE index with the relative control variables.

$$\text{PRE} = \theta_{\text{pre}} \cdot \frac{(\zeta_d - \zeta_{AM})}{\zeta_d} \tag{A5}$$

- θ_{pre}: product—machine coefficient, $\theta_{\text{pre}} = 0 \rightarrow \zeta_d < \zeta_{AM}$ $\theta_{\text{pre}} = 1 \rightarrow \zeta_d > \zeta_{AM}$;
- ζ_d: reference precision of the component;
- ζ_{AM}: machine precision.

Appendix A.6. Technology Resolution "RIS"

The following is the complete formulation of the RIS index with the relative control variables.

$$\text{RIS} = \theta_{\text{ris}} \cdot \frac{(\xi_d - \xi_{AM})}{\xi_d} \tag{A6}$$

- θ_{ris}: product—machine coefficient, $\theta_{\text{ris}} = 0 \rightarrow \zeta_d < \zeta_{AM}$ $\theta_{\text{ris}} = 1 \rightarrow \zeta_d > \zeta_{AM}$;
- ζ_d: reference resolution of the component;
- ζ_{AM}: machine resolution.

Appendix A.7. Undercuts Management "STQ"

The following is the complete formulation of the STQ index with the relative control variables.

$$\text{STQ} = \frac{S_d}{\left(\frac{S_t}{2} + S_d \right)} \tag{A7}$$

- S_d : total component surface;
- S_t : total undercut surface;
- ψ : support management coefficient.

Appendix A.8. Technology Roughness "RGS"

The following is the complete formulation of the RGS index with the relative control variables.

$$\text{RGS} = \theta_{\text{rgs}} \cdot \frac{(\chi_d - \chi_{AM})}{\chi_d} \tag{A8}$$

- θ_{rgs}: product—machine coefficient, $\theta_{\text{rgs}} = 0 \rightarrow \chi_d < \chi_{AM}$ $\theta_{\text{rgs}} = 1 \rightarrow \chi_d > \chi_{AM}$;

- χ_d: component reference roughness;
- χ_{AM}: roughness obtainable with the machine.

Appendix A.9. Compatibility Index "CBA"

The following is the complete formulation of the CBA index with the relative control variables.

$$
\begin{aligned}
CBA = \theta_{mat} \cdot \theta_{ing} \\
\cdot \left\{ \alpha_{gmt} \cdot \left(\frac{PRE + RIS + RGS + STQ}{4} \right) + \alpha_{cst} \right. \\
\left. \cdot \left(\frac{1 - CST}{CST_{max} \cdot (1 + \alpha_{cst})} \right) + \alpha_{tmp} \cdot \left(\frac{1 - TMP}{TMP_{max} \cdot (1 + \alpha_{tmp})} \right) \right\}
\end{aligned}
\tag{A9}
$$

The KPI in equal to zero when θ_{mat} or θ_{ing} are null, which are conditions corresponding respectively to an incompatibility in terms of material (the material is not supported by the machine) or overall dimensions (the component is too large compared to the machine's working volume).

If PRE, RIS, RGS, or STQ are equal to zero, CBA does not become null because the component is actually feasible; it requires only additional treatments or workings that cannot be ignored if the customer's request is to be satisfied.

- θ_{mat}: $\theta_{mat} = 0 \rightarrow$ incompatible material $\theta_{mat} = 1 \rightarrow$ compatible material;
- θ_{ing} : compatibility between component and the machine working volume
- α_{gmt}: weight associated with geometric KPIs;
- α_{cst}: weight associated with cost KPIs;
- α_{tmp}: weight associated with time KPIs;
- PRE: technology precision;
- RIS: technology resolution;
- RGS: technology roughness;
- STQ: undercuts management;
- CST: production cost;
- TMP: production time;
- CST_{max}: maximum CST value calculated among all available machines;
- TMP_{max}: maximum TMP value calculated among all available machines.

References

1. Ashton, K. That 'internet of things' thing. *RFID J.* **2009**, *22*, 97–114.
2. ASTM F2792-12a. *Standard Terminology for Additive Manufacturing Technologies, (Withdrawn 2015)*; ASTM International: West Conshohocken, PA, USA, 2012.
3. Bikas, H.; Stavropoulos, P.; Chryssolouris, G. Additive manufacturing methods and modelling approaches: A critical review. *Int. J. Adv. Manuf. Technol.* **2016**, *83*, 389–405. [CrossRef]
4. Wohlers, T.T.; Campbell, I.; Diegel, O.; Kowen, J.; Caffrey, T. *Wohlers Report. 3D Printing and Additive Manufacturing State of the Industry*; Wohlers Associates: Fort Collins, CO, USA, 2007.
5. Uz Zaman, U.K.; Rivette, M.; Siadat, A.; Mousavi, S.M. Integrated product-process design: Material and manufacturing process selection for additive manufacturing using multi-criteria decision making. *Robot. Comput. -Integr. Manuf.* **2018**, *51*, 169–180. [CrossRef]
6. Huang, Y.; Leu, M.C.; Mazumder, J.; Donmez, A. Additive manufacturing: Current state, future potential, gaps and needs, and recommendations. *J. Manuf. Sci. Eng.* **2015**, *137*, 014001. [CrossRef]
7. Vaezi, M.; Chianrabutra, S.; Mellor, B. Multiple material additive manufacturing—part1: A review. *J. Virtual Phys. Prototyp.* **2013**, *8*, 19–50. [CrossRef]
8. Cozmei, C.; Caloian, F. Additive manufacturing flickering at the beginning of existence. *Procedia Econ. Financ.* **2012**, *3*, 457–462. [CrossRef]
9. Gibson, I.; Rosen, D.; Stucker, B. *Additive Manufacturing Technologies: 3D Printing, Rapid Prototyping, and Direct Digital Manufacturing*, 2nd ed.; Springer: New York, NY, USA; Heidelberg, Germany; Dordrecht, The Netherlands; London, UK, 2015; ISBN 978-1-4939-2112-6.

10. Colosimo, B.M.; Cavalli, S.; Grasso, M. A cost model for the economic evaluation of in-situ monitoring tools in metal additive manufacturing. *Int. J. Prod. Econ.* **2020**, *223*, 107532. [CrossRef]
11. Rinaldi, M.; Caterino, M.; Fera, M.; Manco, P.; Macchiaroli, R. Technology selection in green supply chains—The effects of additive and traditional manufacturing. *J. Clean. Prod.* **2021**, *282*, 124554. [CrossRef]
12. Afshari, H.; Searcy, C.; Jaber, M.Y. The role of eco-innovation drivers in promoting additive manufacturing in supply chains. *Int. J. Prod. Econ.* **2019**, *223*, 107538. [CrossRef]
13. Bikas, H.; Koutsoukos, S.; Stavropoulos, P. A decision support method for evaluation and process selection of Additive Manufacturing. *Procedia CIRP* **2019**, *81*, 1107–1112. [CrossRef]
14. Di, L.; Yang, Y. Cost Modeling and Evaluation of Direct Metal Laser Sintering with Integrated Dynamic Process Planning. *Sustainability* **2021**, *13*, 319. [CrossRef]
15. Ott, K.; Pascher, H.; Sihn, W. Improving sustainability and cost efficiency for spare part allocation strategies by utilisation of additive manufacturing technologies. *Procedia Manuf.* **2019**, *33*, 123–130. [CrossRef]
16. Sabiston, G.; Kim, I.Y. 3D topology optimization for cost and time minimization in additive manufacturing. *Struct. Multidiscip. Optim.* **2020**, *61*, 731–748. [CrossRef]
17. Liu, J.; Chen, Q.; Liang, X.; To, A.C. Manufacturing cost constrained topology optimization for additive manufacturing. *Front. Mech. Eng.* **2019**, *14*, 213–221. [CrossRef]
18. Palanisamy, M.; Pugalendhi, A.; Ranganathan, R. Selection of suitable additive manufacturing machine and materials through best–worst method (BWM). *Int. J. Adv. Manuf. Technol.* **2020**, *107*, 2345–2362. [CrossRef]
19. Altair's INSPIRE®User Guide. Available online: https://www.altair.com/inspire/ (accessed on 2 March 2021).

Article

Feasibility Study of the Cranial Implant Fabricated without Supports in Electron Beam Melting

Khaja Moiduddin [1,*], Syed Hammad Mian [1], Wadea Ameen [1], Hisham Alkhalefah [1] and Abdul Sayeed [2]

[1]   Advanced Manufacturing Institute, King Saud University, Riyadh 11421, Saudi Arabia;
      smien@ksu.edu.sa (S.H.M.); wqaid@ksu.edu.sa (W.A.); halkhalefah@ksu.edu.sa (H.A.)
[2]   Department of Mechanical Engineering, College of Engineering, King Saud University,
      Riyadh 11421, Saudi Arabia; 439106628@student.ksu.edu.sa
*   Correspondence: khussain1@ksu.edu.sa; Tel.: +966-11-469-7372

Abstract: Additive manufacturing (AM), particularly electron beam melting (EBM), is becoming increasingly common in the medical industry because of its remarkable benefits. The application of personalized titanium alloy implants produced using EBM has received considerable attention in recent times due to their simplicity and efficacy. However, these tailored implants are not cost-effective, placing a tremendous strain on the patient. The use of additional materials as support during the manufacturing process is one of the key causes of its high cost. A lot of research has been done to lessen the use of supports through various types of support designs. There is indeed a noticeable paucity of studies in the literature that have examined customized implants produced without or minimal supports. This research, therefore, reports on the investigation of cranial implants fabricated with and without supports. The two personalized implants are evaluated in terms of their cost, fabrication time, and accuracy. The study showed impressive results for cranial implants manufactured without supports that cost 39% less than the implants with supports. Similarly, the implant's (without supports) build time was 18% less than its equivalent with supports. The two implants also demonstrated similar fitting accuracy with 0.2613 mm error in the instance of implant built without supports and 0.2544 mm for the implant with supports. The results indicate that cranial implants can be produced without EBM supports, which can minimize both production time and cost substantially. However, the manufacture of other complex implants without supports needs further study. The future study also requires a detailed review of the mechanical and structural characteristics of cranial implants built without supports.

Keywords: electron beam melting; customized implant; cost analysis; fitting accuracy; cranial reconstruction

Citation: Moiduddin, K.; Mian, S.H.; Ameen, W.; Alkhalefah, H.; Sayeed, A. Feasibility Study of the Cranial Implant Fabricated without Supports in Electron Beam Melting. *Metals* **2021**, *11*, 496. https://doi.org/10.3390/met11030496

Academic Editor: Atila Ertas

Received: 22 February 2021
Accepted: 11 March 2021
Published: 17 March 2021

Publisher's Note: MDPI stays neutral with regard to jurisdictional claims in published maps and institutional affiliations.

Copyright: © 2021 by the authors. Licensee MDPI, Basel, Switzerland. This article is an open access article distributed under the terms and conditions of the Creative Commons Attribution (CC BY) license (https://creativecommons.org/licenses/by/4.0/).

1. Introduction

Craniofacial skeleton injuries that are often induced from tumor, trauma, and external neurosurgical decompression mandate appropriate restoration measures. The purpose of cranial restoration is to provide cerebral protection, preserve esthetics, and relieve neurological symptoms [1]. The performance of cranial or any reconstruction implant relies on the defect's preoperative assessment, the design of the implant, manufacturing approach, and eventually the surgeon's skills and execution. Earlier studies in cranial reconstruction addressed the utilization of computer-aided cranial implant design, although without the usage of medical modeling software and state-of-the-art Additive Manufacturing (AM) technology, resulting in a great deal of manual activities [2,3]. Injection molding [4], extrusion [5], and casting [6] are among the conventional techniques for the fabrication of cranial reconstruction, but they are time-consuming as well as expensive in terms of accuracy and precision.

With the advent of AM techniques in the medical sector and surgical performance over the last decade, greater focus has been given to the manufacture of personalized implants. AM allows direct digital manufacturing of patient-specific models with high precision and productivity in surgical preparation for different biomedical applications. AM also known as three-dimensional (3D) printing, is a processing technique capable of manufacturing complex 3D components by incorporating material layers, beginning with a digital computer-aided design (CAD) model [7]. Metals, polymers, ceramics, composites, and various other materials can be processed using AM technologies. Amongst several metallic biomaterials used in AM, including stainless steel, cobalt-chromium, and titanium alloys. Ti6Al4V ELI (extra-low interstitial) offers enhanced corrosion resistance, high specific strength and young's modulus closer to the bone. Ti6AL4V ELI has a range of advantages over pure titanium (commercial titanium), including greater strength and resistance to fatigue [8]. In addition, Ti6Al4V ELI though being a paramagnetic metal, its magnetic susceptibilies is much lower than that of other biocompatible metals, thus decreasing the influence of magnetic resonance imaging (MRI) and avoiding the hindrance in the diagnosis under MRI.

The integration of clinical imaging, CAD software, and AM has contributed to substantial improvements in surgical and biomedical applications. Electron beam melting (EBM) technology is the primary option for the manufacture of titanium surgical implants among the many AM technologies. EBM has been approved by the FDA (Food and Drug Administration) and Conformité Européene (CE) for the manufacture of medical implants in America and Europe [9,10]. EBM produced Ti6Al4V ELI implants exhibits promising results in the fabrication of mandible, hip, orthopedic and cranial implants clinically [11–14]. EBM produces completely dense metallic components, touching 99.9 percent, compared to other metallic AM technologies, which is excellent for biomedical implants [15].

However, the build parts need additional structures (supports) during the EBM fabrication process to assist the overhang surfaces, decrease the geometric distortion and to minimize the warping effect. The sintered powder covering the building component can also act as support structures, thus reducing the cost and time during the construction process. The usage of support structures is appropriate only if the optimum orientation is unable to support the overhang surfaces.

The utilization of support structures in overhang surfaces created using EBM technology was examined by Cheng and Chou [16]. Research suggests that supports increases heat flow and restores thermal behavior as well as decrease overhang deformation [17]. They also serve as the heat dissipaters in high-temperature builds, as in the case of metal based AM [18]. Support structures serve as anchors, assisting in the protection of the part's shape from delamination or deformation during the building process. The anchors are scrapped once the construction is completed. A very thin layer (40–60 μm) is melted or sintered by an electron or a laser beam in the powder bed fusion process according to the specified geometry. During the solidification process, considerable thermal gradients and internal stresses are produced, which contributes to distortions and delamination [19]. As a consequence, temporary support material that serves as an anchor dissipates the heat and provides structural resistance to distortions and sagging. The use of support structures in the powder bed fusion process has a number of advantages, but it also has a number of shortcomings [20]. Support removal is usually more difficult in metal-based AM than in the case of polymer-based AM. As of now, there is no established automated procedure to remove the anchors, and are usually detached by the manual procedure. The requirement to produce and then eliminate the attachments leads to a substantial increase in material consumption, costs, energy usage, and total production time.

The existence of support structures certainly presents serious challenges in addition to increasing production time, unnecessary material use, post-processing time in the removal of supports, as well as the danger and challenges of separating the supports without compromising the building portion [21–23]. The presence of support structures has many downsides. After fabrication, the removal of support structures also involves a considerable amount of manual labor. The support structures result in additional expense and time in the milling, cutting, and grinding of the supports [24]. In addition, various types of support generate different surface roughness, requiring post-processing. The support structures are of no use once built, because they cannot be recycled when removed and lead to waste materials, thus generating higher material costs [25]. Supports also incur additional energy costs and time of production as well as they are often difficult to remove, particularly for smaller, complicated, and complex geometries [26]. Another problem with support structures is that they are less ideal and inactive relative to usable components. Parts with support structures also influence the surface finish of the build, leaving a mark on the surface after removal [27].

Preferably, the sintered powder serves as a support in the powder bed fusion process, thereby discarding the use of external supports, but depending on the geometry, it is often not enough to support the melted component. By default, the 3D printing software designs the supports in most of the overhang areas. Therefore, reducing the use of unnecessary support structures in overhang sections is extremely necessary. In polymer-based 3D printing, several methods has been used to eliminate them, such as water solvable supports, simple breakaway supports, but in the case of 3D metal printing, it presents a major conundrum [16]. Instead of a continuous mode, Jhabvala et al. [28] reported a revolutionary pulse laser method for the manufacture of support structures and stated that the fabricated supports had similar mechanical properties and were much easier to remove. Similarly, several researchers have studied cellular lattice structures with different support unit cells and investigated their effects on their geometric properties [29,30]. The removal of supports through a feasible orientation in lightweight builds had also been suggested by few researchers [31,32]. Similarly, an experimental analysis involving optimal self-supporting structures for overhang surfaces was proposed by Calignano [33]. It is of significant relevance, based on literature research, to further study and minimizes the use of support structures in AM wherever applicable. It is evident that a considerable amount of work has been conducted to minimize the use of supports especially by employing different kinds of support designs. Nevertheless, there is a notable lack of literature studies investigating personalized implant build without or limited supports. Consequently, this study focuses on the examination and comparison of cranial implants produced with and without supports. The two personalized implants are evaluated in terms of their cost, fabrication time, and accuracy. The fitting accuracy is assessed to ensure that there is no major deformation in the implant manufactured in the absence of the supports. This is one of the few works in the scientific literature, to the author's knowledge, on the workflow for implant fitting accuracy study and manufacturing of metallic AM cranial implants without supports.

2. Methodology

The methodology adopted in this study as shown in Figure 1, consists of four major steps: (1) The creation of an artificial defect in the clean skull, (2) the design of a personalized cranial implant, (3) the manufacture of implants with and without supports, and (4) the evaluation of the implants manufactured.

2.1. Creation of Artificial Defect

A clean skull model (Figure 2a) is imported into Materialise Mimics 18.0 (Materialise, NV, Leuven, Belgium) in this study and an unnatural defect with a tumor presumption is produced in the left skull region. Rather than the actual defect, the artificial defect is used to prevent the hassle of obtaining permission from the patient to use their data in the publication. Moreover, as a result of artificial defect, a reference model in the form of a clean skull is available for accuracy assessment. The cutting operation is conducted to split the clean skull (Figure 2a) into two halves (Figure 2b,c), and a region is marked on the left side of the skull to ascertain the tumor outline (Figure 2d). Finally, a void (Figure 2e,f) is generated on the left side of the skull that mimic the defective or affected region.

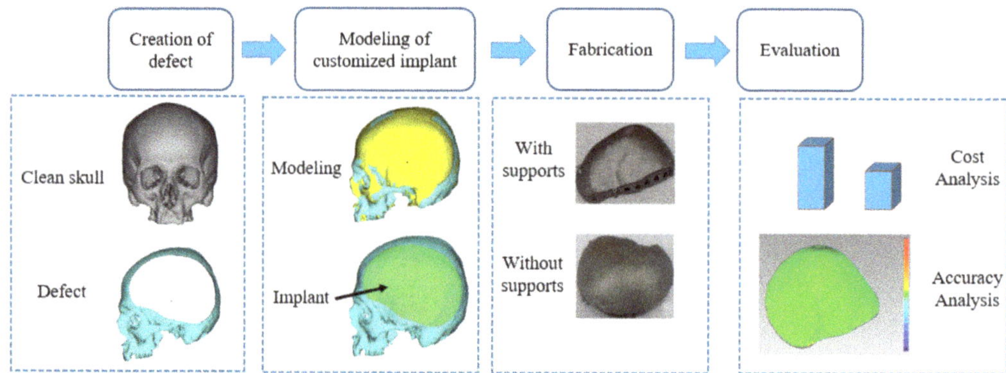

Figure 1. Process flow of adopted methodology.

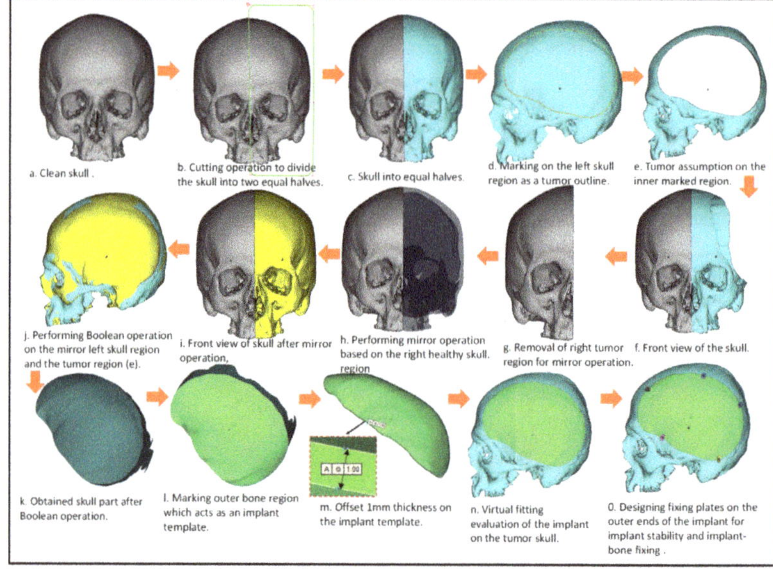

Figure 2. Workflow for the reconstruction of customized cranial implant.

2.2. Design of Customized Implant

The mirroring operation is performed based on the healthy right side of skull region. For this purpose, the entire left affected region across the symmetric plane is removed as depicted in Figure 2g. Subsequently, the mirroring operation is implemented to imitate the right healthier side on the left side of the skull (Figure 2h,i). Then, the Boolean subtraction operation is performed on the mirror and tumor assumed model in order to get the implant template (Figure 2j,k). The outer region of the implant template is marked to get the implant outline pattern (Figure 2l). An offset thickness of 1mm is provided on the pattern to create the final implant design (Figure 2m). Finally, the implant design is evaluated through a virtual fitting on the polymer model (Figure 2n) and fixing plates are designed onto the outer ends of the implant to attain implant stability and implant–bone fixing (Figure 2o).

2.3. Fabrication

Upon successful virtual fitting and rehearsal operation, the customized cranial implant and the skull are manufactured using AM technologies. The skull (Figure 3b) is fabricated using Acryl butadiene Styrene (ABS) material in Zortrax M200 Layer plastic deposition printer (Zortrax, Olsztyn, Poland) (Figure 3a).

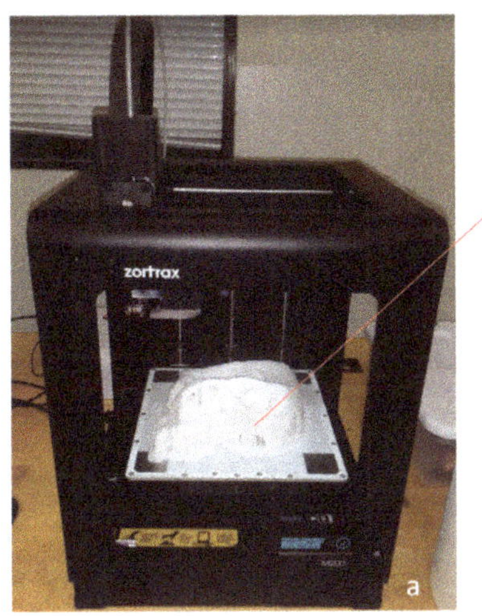
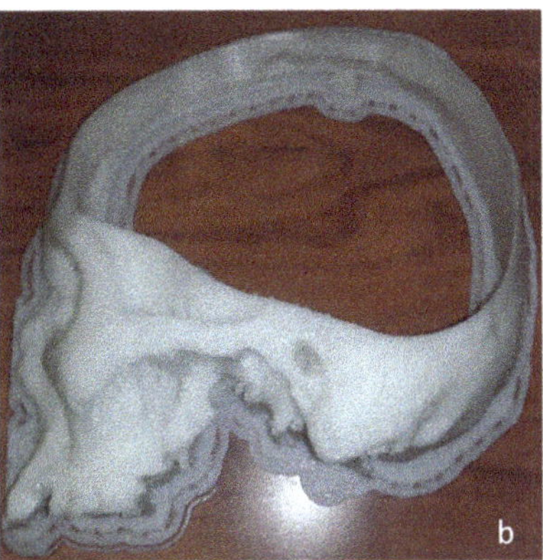

Figure 3. (**a**) Zortrax additive manufacturing (AM) machine with the fabricated (**b**) polymer skull.

To generate the support structures, the customized cranial implant design is imported into Magics software (Materialise, NV, Leuven, Belgium). By default, based on the geometry, the software itself generate self-supporting structures onto the overhang parts to prevent deformation and to increase the heat dissipation as shown in the Figure 4.

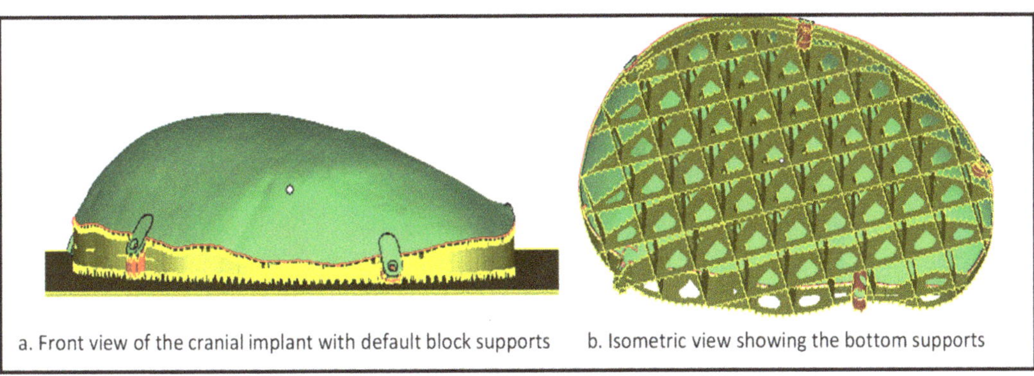

Figure 4. Virtual simulation of customized cranial implant with default block support structures.

For the fabrication of customized titanium cranial implant one with supports and one without supports, the sliced standard tessellation language (STL) file is loaded into ARCAM EBM following the same orientation. The schematic diagram as illustrated in Figure 5 demonstrates the fabrication process. The metal powder (Ti6Al4V ELI) is loaded and a vacuum is created inside the close build chamber to ensure a clean and controlled environment. The EBM powder layering system equally distributes the powder over the powder bed using rake. A powerful 6 KW of high energy electron beam heats and melts the powder bed as per the geometry of the cranial implant. The focus and deflection coils guide the electron beam onto the path for part accuracy. Throughout the build process, the build temperature is kept high at approximately 1000 °C, thus resulting in avoiding residual stresses. After completion of each layer, a new layer of fresh powder is raked over the previous build layer, this process is repeated until the final build is obtained.

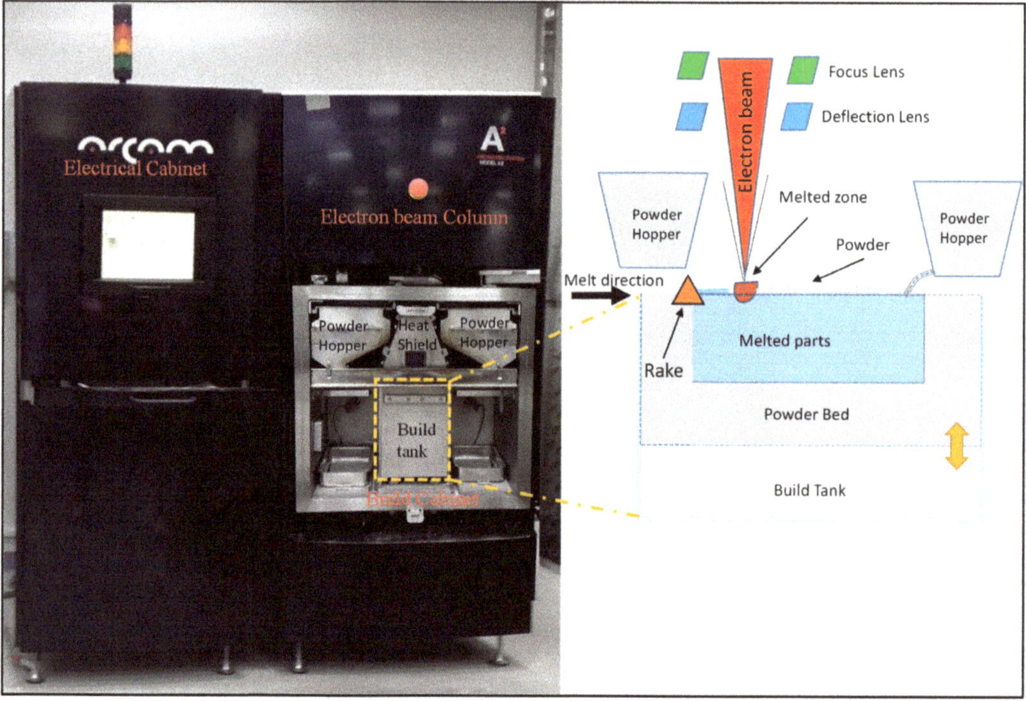

Figure 5. ARCAM's electron beam melting (EBM) machine with schematic diagram illustrating the fabrication process.

Two customized cranial implants are fabricated using EBM, one with default support structures (Figure 6d–f) and other without supports (Figure 6a–c). Upon completion of final build, the parts are moved to Powder recovery system (PRS) for the removal of semi-sintered and excess powder attached to the build part. The blasted powder is filtered using sewing machines and recycled for the next job.

The time taken to remove the semi-sintered powder from the implant without support is approximately 5 min, whereas it is around 10 min for the implant with supports. This may be due to the presence of semi-sintered powder around the supports, which is difficult to blast. The blasted implant (with supports) is then subjected to post processing where the support structures are manually removed using pliers as shown in Figure 7a. Even after supports removal, support protrusions could still be noticed on the cranial implant as illustrated in Figure 7b,c. The time taken to remove the supports, including the cleaning of surface protrusion using sandpaper, is almost 100 to 110 min. The post-processed cranial implant is fixed onto the polymer skull model for fitting evaluation as demonstrated in Figure 7d.

Finally, the cost analysis and fitting evaluation of the two implants are carried out to investigate the implications of support structures.

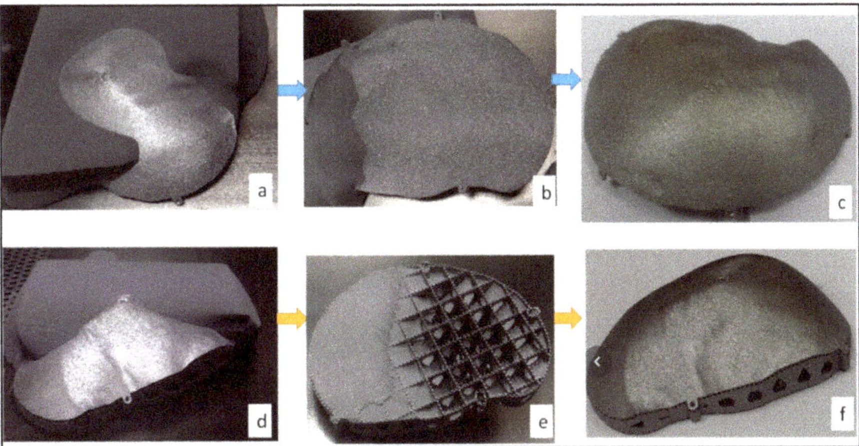

Figure 6. The EBM produced customized cranial implant without supports (**a**–**c**) and implant with support structures (**d**–**f**).

Figure 7. (**a**) Removal of support structures using pliers, (**b**,**c**) illustration of protrusion onto the surface of the cranial implant after the removal of supports, and (**d**) implant fitted onto the polymer skull model.

2.4. Evaluation

The scanning electron microscopy (SEM) analysis is conducted in order to study the powder size morphology. Figure 8 shows the SEM images of Ti-6Al-4V powder: (a) Low magnification showing the population of particle sizes and (b) high magnification is of an individual powder. This analysis confirms that the formation of the powder particles is primarily spherical in shape with slight variation in the geometry.

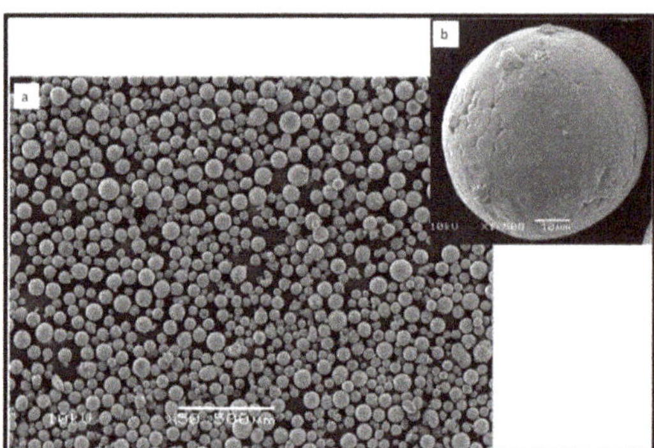

Figure 8. (**a**) SEM image of Ti-6Al-4V powder particles of different sizes and (**b**) an individual particle.

A fine powder particle of 50–100 μm is used in the EBM process. The EBM build platform and its sintering kinetics between the powder particles are influenced by the size and shape of the powder. Thus, a laser diffraction analysis as shown in Figure 9 is performed to measure and confirm the geometrical dimension of the feedstock powder.

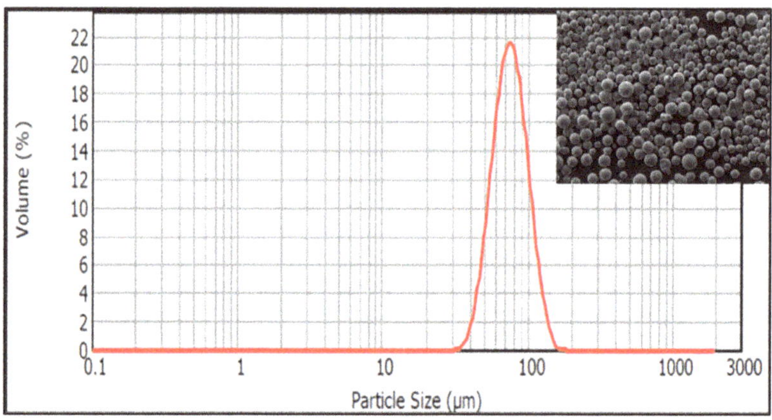

Figure 9. Laser diffraction technique used to study the powder size distribution.

The chemical composition of Ti6Al4V ELI as revealed in Figure 10 is 6.04% Aluminum (Al), 4.05% Vanadium (V), 0.013% Carbon (C), 0.0107% Iron (Fe), and 0.13% Oxygen (O), with the remaining constituent as Titanium (Ti) in weight percent. Based on the results, the chemical composition of the EBM build part of Ti6Al4V ELI material do not diverge much from the initial powder feedstock.

2.4.1. Implant Cost and Time Analysis

Reliable cost estimation of AM parts is of utmost importance, especially in the medical industry because of the high investment costs related to the product development phases. Moreover, the wrong estimates result in expensive consequences and may result in the production loss. Previous researchers have investigated and developed numerous models for the calculation of AM cost. For example, Ruffo et al. [34] performed a comprehensive cost model study of direct and indirect cost for AM, in which material related cost was considered as a direct cost and the fabrication cost, machine and administration cost were counted as an indirect cost. Similarly, Hopkins and Dickens [35] proposed a comparative cost analysis study of three AM processes involving Stereolithography, fused deposition modeling and laser sintering with injection molding. Syam et al. [36] also performed the EBM cost analysis and proved that EBM cost was significantly dependent on the number of units produced in a single fabrication cycle when compared to traditional manufacturing. Lindermann et al. [37] in his study developed a methodology that helped to better understand and analyze the largest cost drivers in the fabrication of metal AM products. Baumers et al. [38] also performed a comparative metal AM study of EBM and selective laser melting based on their cost performance. It is quiet noticeable that previous research done in the AM cost model, mainly focused on the cost structure of the AM product and few on the comparative studies with conventional machining, but there is hardly any study that has considered AM support structures.

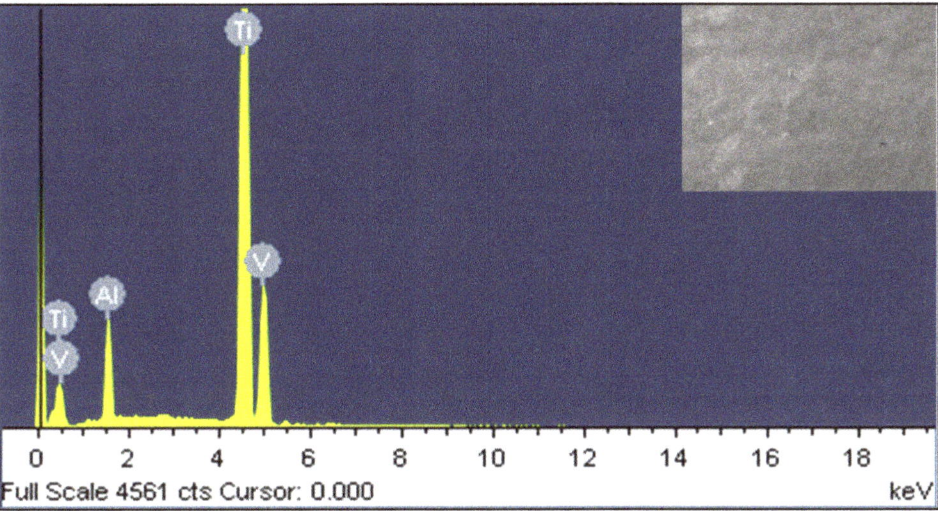

Figure 10. Chemical composition of Ti6Al4V powder used in the implant fabrication.

Henceforth, an inclusive cost and time model developed by Priarone et al. [39] is adopted in this study to investigate the economics of EBM fabricated cranial implants (with supports and without supports). The main driving factors considered in this study are the material consumption, build time and the total cost involved in building the implants. The implemented cost model to estimate the building cost (C_{Build}) involves the material cost as well as the energy consumption cost as presented in Equation (1). As shown in Equation (2), the material cost is made up of the price and the consumption of material for each build whereas the energy consumption cost is associated with the cost of running the EBM machine while fabricating cranial implant. Other cost factors such as EBM ownership cost and EBM maintenance cost are not considered, as they remain constant for both cranial implants with and without supports.

$$C_{Build} = \text{Material cost} + \text{Energy consumption cost} \tag{1}$$

$$C_{Build} = [(M_{Consumption} \times C_{Raw}) + (T_{Fabrication} \times E_{Build} \times P_{Energy})] \tag{2}$$

where,
$T_{Fabrication}$: EBM implant fabrication time.
$M_{Consumption}$: Material consumption for the cranial implant in grams.
C_{Raw}: Cost of the raw material (Ti6Al4V ELI) measured in $/gram.
E_{Build}: Energy consumption for the fabrication of cranial implant.

P_{Energy}: Price of energy consumption (Electricity cost for EBM process), measured in $/KWh.

B_{Hours}: Build time for the fabrication of cranial implant measured in hh:mm and simplified to hours for multiplication.

PP_{Time}: Time taken for post-processing the cranial implant in hh:mm and simplified to hours for multiplication.

$$T_{Analysis} = [\, T_{Fabrication} + PP_{Time}\,] \tag{3}$$

Time Analysis ($T_{Analysis}$) for the fabrication of cranial implant with and without supports includes the time taken to build the cranial implant as well as the post-processing time as explained in Equation (3). The post processing time involves the time taken to remove the semi-sintered powder attached to the implant after build and the removal of supports. Moreover, the weights (in grams) required for the cost model are obtained through the weighing scale (Ohaus Corporation, Parsippany, NJ, USA) as shown in Figure 11.

Figure 11. Weighing scale reading of cranial implants (**a**) with supports and (**b**) without supports.

2.4.2. Accuracy Analysis

The two cranial implants which include the implant manufactured without supports and the implant produced with supports are also inspected for precise fitting. The accuracy assessment is carried out to estimate the gap between the implant and the skull as well as to gauge the consistency of the aesthetics and the external profile of the revived cranium.

In the application of AM technology, model accuracy has a significant impact on surgical planning [40]. Higher accuracy in implant fitting is very important from a medical standpoint, particularly in large cranial defects. Elkatatny and Eldabaa [41] reported that a large proportion of their tumor patients or post-traumatic patients with minor mutilating defects required surgery to improve their aesthetics or cosmetic appearance in addition to providing cerebral protection. If the implant is correctly fitted on the skull, it will have a pleasant cosmetic appearance and the patient will not need to undergo re-surgery. According to Hohne et al. [42], good biocompatibility, adequate defect closure with the precise fitting of the implant reconstruction to the osseous rims, and particularly a pleasing cosmetic outcome are all important elements in cranioplasty. The patient is exposed to all of the risks associated with the repeated surgical operation, including those associated with anesthesia [43]. An accurately fabricated implant, according to Toth et al. [44], fits into the defect properly and reduces the likelihood of subsequent movement, dislodgement, and extrusion. Numerous instances of titanium cranioplasty were examined, and it was discovered that the majority of the ill-fitting or aesthetically poor implants resulted in increased intraoperative time and asymmetrical reconstruction [45–47]. The implant possessing an exact match to the patient's cranial defect leads to a very symmetric skull profile and the implant lies passively on the body surface all around the defect. By ensuring a precise fit of the AM-produced implant, Maravelakis et al. [48] reduced roughly 30% of the overall operation time, thus minimizing intraoperative manipulations to achieve implant fixation.

Figure 12 illustrates the procedure employed to compute the 3D deviation. In this analysis, the defect or tumor is purposely produced in the normal skull and then the healthier right half is replicated on the left faulty portion. The unreal defect is incorporated so that we have the patient's real skull as a guide for comparison thereafter. In the case of the genuine defect, it is not feasible to get a patient's scan (and hence the actual reference) when he/she was healthy. The mirrored model is therefore treated as the reference model in most instances [49]. The utilization of the mirrored model as a reference is not a very reliable approach as it does not incorporate the mirroring bias. In this work, the defect is therefore assumed in the healthier patient to accurately measure the possible error produced in the customized implants. First, the mirrored model is evaluated with the healthy skull of the patient and the overall variance (in the outer direction) is measured. This characterizes the error of mirroring. Next the mirrored model is assessed relative to the virtual cranial implant (because the mirrored model is used as a template to create the implant). This measures the design error of the virtual implant. Finally, the fabricated implants (with and without supports) are examined relative to virtual implants. This gives us the fabrication error. The overall difference in the produced implants can therefore be determined by summing up mirroring, design and manufacturing errors, as shown in Figure 12. In this analysis, d2 symbolizes the manufacturing error of the cranial implant with support, while d2$'$ denotes the fabrication error of the cranial implant without support. The d3 is the cumulative error (or the fitting accuracy) of the cranial implant with supports and the d4 implies the total error of the implant without supports.

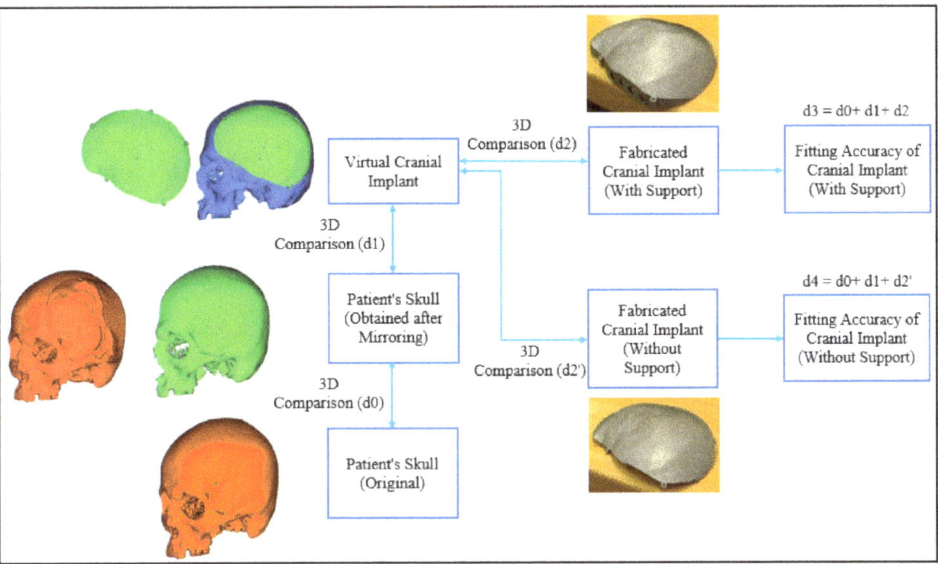

Figure 12. Process flow to assess the accuracy of the cranial implant fitting.

As shown in Figure 13, the outer surfaces of the fabricated implants are digitized using a laser line scanner mounted on the FARO platinum arm (FARO, Lake Mary, FL, USA). The FARO arm (with laser scanner) is preferred due to its availability, a reasonable precision of 20 μm and the ability to acquire a large number of points in short time.

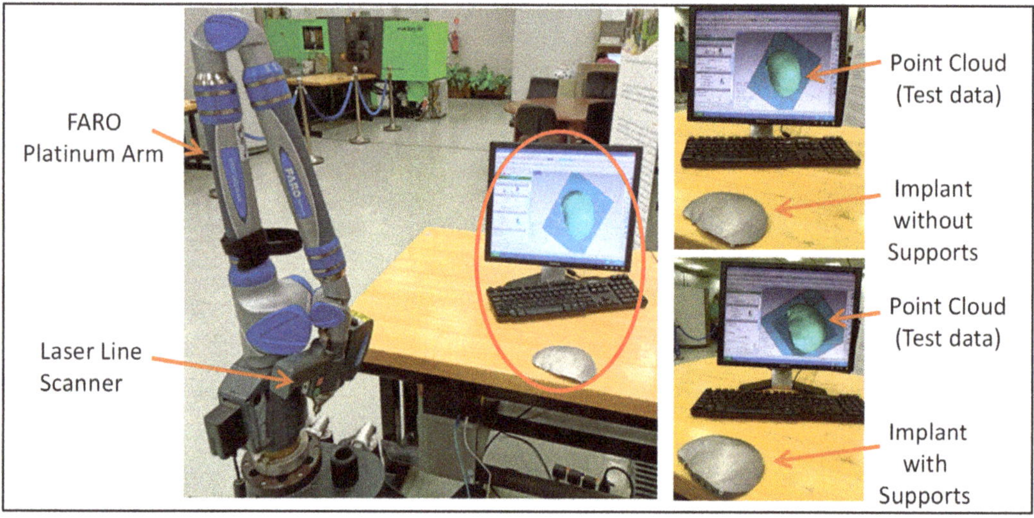

Figure 13. Set up to capture point cloud data for the fabricated models.

The 3D data captured by the FARO scanner is a point cloud that is transformed into triangulation model and then saved in STL after post-processing, using a reverse engineering program (Geomagics Studio 2014, 3D System, Valencia, CA, USA). The "Alignment" feature of the 3D evaluation software (Geomagics Control 2014, 3D System, Valencia, CA, USA) is used with each implant model to superimpose the two datasets (test and reference). In this study, the acquired dataset (of the fabricated implants) is specified as the test model, while the reference is assigned to the virtual (or designed) implant. In Geomagics Control software, the "deviation analysis" algorithm is implemented to conduct a 3D comparison between the reference and test models. To demonstrate the variations in the test models, a graphic of color scales is generated. The results are determined on the basis of almost 100,000 points on the 3D scanned model. The "average error" between each pair, described as the average of all the distances between the closest point pairs on the reference and the test model, is estimated. The closest point pairs are searched and matched automatically by the software algorithm.

3. Results and Discussion

The cost and time analysis results are illustrated in Tables 1 and 2 respectively. Figure 14a indicates, that the material consumption decreases significantly by 47% from 125.92 g to 66.75 g in the case of the implant without supports. Thus, the material consumed in the manufacture of cranial implants with supports is almost double that of the cranial implant without supports. The higher material consumption in turn raises the final cost of the cranial implant with support in comparison to the implant without supports. The cranial implant without supports costs roughly $21 in comparison to $34 for the implant with supports, thereby resulting in a 39% reduction in overall cost (Figure 14b).

Table 1. Cost estimation for the EBM build cranial implants.

Influencing Factors	Description	Values	Estimation
Material consumption (grams) $M_{Consumption}$	Weight of implant with supports	114.47 g	
	Weight of Implant without supports	60.69 g	
	Material wastage including support structures [50]	10% of built mass	
	$M_{consumption}$ (Material consumed for cranial implant with supports)	125.91 g	Total mass of EBM built implant = (Implant mass + 10% material wastage of built mass) = 114.47 + 11.44
	$M_{consumption}$ (Material consumed for cranial implant without supports)	66.75 g	60.69 + 6.06
Material cost (per gram)	C_{Raw} (Ti6Al4V ELI cost price)	$0.22/g	=$220/kg
$(M_{consumption} \times C_{Raw})$ Supports	Implant with supports	$27.70	(125.91 g $\times$ $0.22/g)
$((M_{consumption} \times C_{Raw})$ Without supports	Implant without supports	$14.68	(66.75 g $\times$ $0.22/g)
$T_{Fabrication}$ (cranial implants with supports)	Time to obtain desired vacuum level	0:3 hh:mm	
	Time to heat start plate	0:45 hh:mm	
	EBM cool down time	4:00 hh:mm	
	Build time for cranial implant with supports	5:39 hh:mm	
	$T_{Fabrication}$ Time for completion of cranial build with supports	10.9 h	Time for desired vacuum level + heating start plate + EBM cool down time + part build time (0:30 + 0:45 + 4:00 + 5:39) = 10:54 hh:mm = 10.9 h
$T_{Fabrication}$ (cranial implants without supports)	Build time for cranial implant without supports	5.11 hh:mm	
	$T_{Fabrication}$ Time for completion of cranial build without supports	10.43 h	Time for desired vacuum level + heating start plate + EBM cool down time + part build time (0:30 + 0:45 + 4:00 + 5:11) = 10:26 hh:mm = 10.43 h
$T_{Fabrication}$ (with supports)	Implant with support	10.90 h	
$T_{Fabrication}$ (without supports)	Implant without support	10.43 h	
EBM energy consumption (KW) for Implant fabrication E_{Build}	EBM Power supply	7 KW [51]	
EBM Electricity cost (Per hour) P_{Energy}	P_{Energy} (EBM energy consumption cost)	$0.085/KWh	Electricity tariff = SAR 0.32/KWh(https://www.se.com.sa/en-us/customers/Pages/TariffRates.aspx (accessed on 18 August 2020)) Conversion of SAR to $ = $0.085/KWh

Table 1. *Cont.*

Influencing Factors	Description	Values	Estimation
$E_{Build} \times T_{Fabrication}$ (with supports) $\times P_{Energy}$	Cranial implant with supports	$6.48	=(EBM power consumption x EBM build time for cranial implant with supports x EBM energy consumption cost) = 7 KW $\times$ 10.9 hours' $\times$ $0.085/KWh
$E_{Build} \times T$ Fabrication (without supports) $\times P_{Energy}$	Cranial implant without supports	$6.20	=7 KW $\times$ 10.43 h $\times$ $0.085/KWh
Total Cost for building implant	Implant with supports	$34.18	(material cost + Energy consumption cost) = $27.70 + $6.48
	Implant without support	$20.88	=$14.68 + $6.20
Percentage Difference in cost		39%	=(13.3/34.18) $\times$ 100

Table 2. Time analysis for fabrication of cranial implant with and without supports.

Factors	Description	Values	Estimation
Build Time (Hours)	B_{Hours} (Build Hours for Cranial Implant with Supports)	10:54 hh:mm	Time Taken for the Fabrication of Cranial Implant with Supports- (0:3 + 0:45 + 4:00 + 5:39) = 10:54 hh:mm
	B_{Hours} (Build Hours for Cranial Implant without Supports)	10:26 hh:mm	=(0:3 + 0:45 + 4:00 + 5:11) =10:26 hh:mm
EBM Post-Processing Time (Hours) PP_{Time}	Post processing time for implant with supports	1:55 hh:mm	Post processing time includes the removal of supports + removal of semi-centered powder (1:45 + 0:10) hh:mm
	Post processing time for implant without supports	0:05 hh:mm	Removal of semi-centered powder.
Total Build time for Cranial implant	Total build time for cranial implant with support after post-processing	12:49 hh:mm	Total Build time = (fabrication time + post-processing time) (10:54 + 1:55) hh:mm =12:49 hh:mm = 12.81 h
	Total build time for cranial implant without support after post-processing	10:31 hh:mm	(10:26 + 0:05) hh:mm
Percentage difference in build time		18%	=(12:49–10:31) =2:18 hh:mm (2.30 h) =(2.30/12.81) $\times$ 100

As indicated in Table 2, the cranial implant with support structures take approximately 10 min for sandblasting the semi-sintered powder and 1 h 45 min for the support's removal. Removing support structures is a tedious and time-consuming process and additional precautions should be taken to prevent implant damage. The time taken for post-processing of unsupported cranial implant is just 5 min, which is the removal of semi-sintered powder. The difference in time between the implant with and without supports in the removal of semi-sintered powder is primarily due to the extra time required to remove the concealed powder within the supports. The building time for implants without supports is 18% shorter than for implants with supports (Figure 14c).

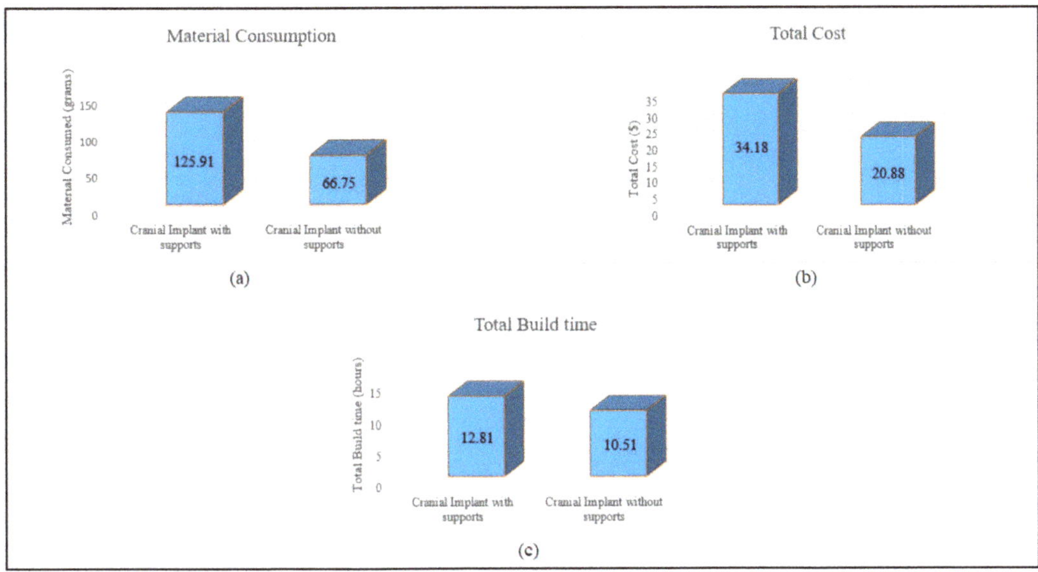

Figure 14. Comparative analysis of cranial implants with and without supports (**a**) material consumption; (**b**) total cost; and (**c**) total build time.

It is apparent from the cost and time analysis that, if the support structures are not used, a considerable amount of time and cost could be saved. The user often implements support mechanisms due to normal protocol, or default settings. Certainly, there is always the likelihood that some sections can be constructed effectively in EBM without the supports. The current investigation, for instance, has demonstrated that the cranial implant is successfully built without support structures. Significant costs, time, and effort needed to withdraw supports (during post-processing) are saved. There are, however, many components that cannot be fabricated without supports, either because of their complexity, size, or shape. In these cases, users must at least try to reduce support by defining the critical support locations or through applying various types of support design. There is always a risk that due to the lack of support, the fitting accuracy of the implant could be compromised. Henceforth, a detailed fitting accuracy analysis for the two implants is also carried out in this investigation.

Table 3 presents the findings of 3D deviation analysis (d0, d1, d2, d2′, d3, and d4) in terms of average deviations. The visual deviation is also shown in Figure 15, including a color-coded map to display the variations between each test model and the reference. It is noted that the greener the color, the nearer the reference model is to it.

Table 3. Outcome of 3D deviation analysis.

Models Combination	Notations	Deviation (mm)
Original (Reference) and Mirroring (Test)	d0	0.1458
Mirroring (Reference) and Virtual Implant (Test)	d1	0.0182
Virtual Implant (Reference) and Fabricated Implant with Supports (Test)	d2	0.0904
	d3	0.2544
Virtual Implant (Reference) and Fabricated Implant without Supports (Test)	d2′	0.0973
	d4	0.2613

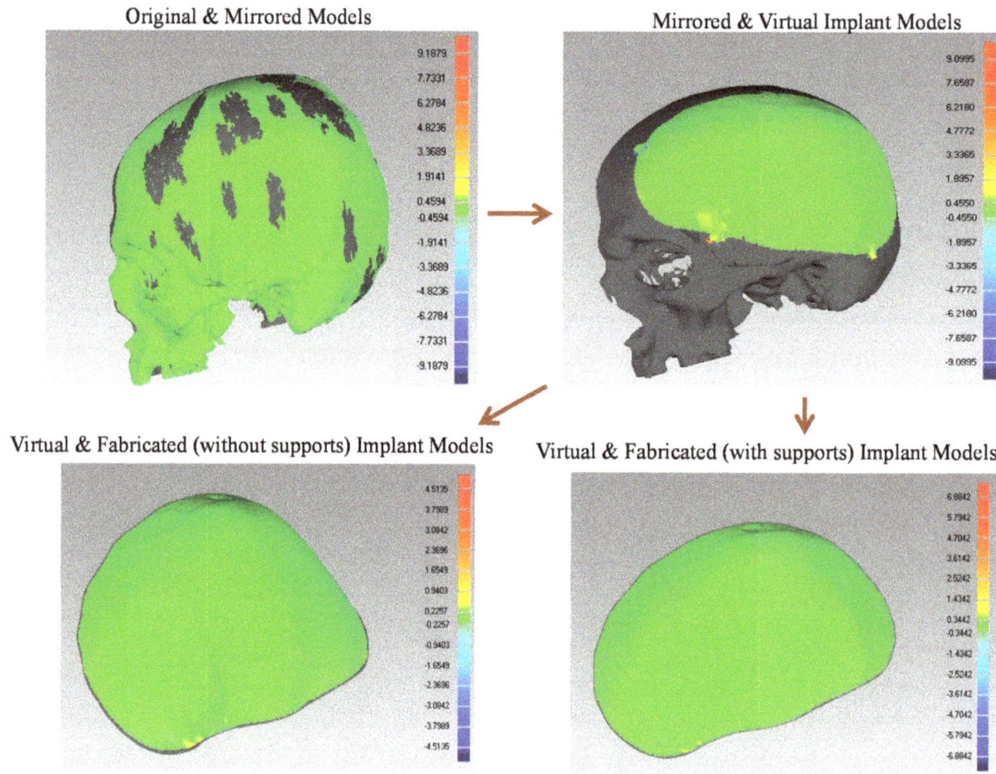

Figure 15. Graphical display of 3D deviation analysis.

The specimens display similar variations from the reference, according to Figure 15, and the implant with support presents a slightly better result, with a lower variance from the reference. The divergence is, however, negligible between implants with support and without support. This is a difference of less than 3%, which is quite marginal. The average cumulative implant error (without support) with respect to the original skull of the patient is 0.2613 mm, whereas that with support is 0.2544 mm.

4. Conclusions

This has been a recognized fact that conventional cranial reconstruction techniques are ineffective because they cannot handle customization effectively. Therefore, the EBM is progressively being employed in the domain of implant reconstruction due to its efficacy and performance. Nevertheless, the manufacture of EBM entails higher costs and greater production time, which makes it expensive for the general populace. It can be stated from the cost analysis that the material cost and the machine running cost are the one that contributes substantially to the total implant cost. It has also been noted that a larger amount of material used as supports results in higher production time, making the manufacture of implants using EBM very costly. In addition, the EBM produced implant without supports must be examined for fitting accuracy to preserve the outer appearance and attain the desired aesthetics. The cost analysis illustrates significant improvements for non-support cranial implants as they cost 39% less than the implant with supports. Likewise, the implant's production time (without support) is found to be 18% shorter than its equivalent implant with supports. The two implants also display similar fitting accuracy with 0.2613 mm error in the instance of implant fabricated without supports and 0.2544 mm for the implant with supports. The study thus found that the cranial implants can be manufactured without EBM supports, which can dramatically reduce both manufacturing time and cost expenses. However, more study is needed to produce other complex implants in the absence of supports. The future research also aims to include a thorough examination of the mechanical and structural aspects of cranial implants without supports.

Author Contributions: Conceptualization, K.M., S.H.M., and W.A.; methodology, K.M., S.H.M., and W.A.; validation, K.M., S.H.M., and H.A.; formal analysis, K.M., S.H.M., and A.S.; investigation, K.M., S.H.M., H.A., and A.S.; data curation, W.A., and H.A.; writing—original draft preparation, K.M. and S.H.M.; writing—review and editing, K.M., S.H.M., W.A., H.A., and A.S.; Project administration, K.M. All authors have read and agreed to the published version of the manuscript.

Funding: This research was financially supported by Deanship of Scientific Research, King Saud University: Research group No. RG-1440-034.

Institutional Review Board Statement: Not applicable.

Informed Consent Statement: Written informed consent has been obtained from the subject to publish this paper.

Acknowledgments: The authors extend their appreciation to the Deanship of Scientific Research at King Saud University for funding this work through Research group no RG-1440-034.

Conflicts of Interest: The authors declare no conflict of interest.

References

1. van de Vijfeijken, S.E.C.M.; Münker, T.J.A.G.; Spijker, R.; Karssemakers, L.H.E.; Vandertop, W.P.; Becking, A.G.; Ubbink, D.T. CranioSafe group autologous bone is inferior to alloplastic cranioplasties: Safety of autograft and allograft materials for cranioplasties, a systematic review. *World Neurosurg.* **2018**, *117*, 443–452. [CrossRef]
2. Kim, B.; Hong, K.-S.; Park, K.-J.; Park, D.-H.; Chung, Y.-G.; Kang, S.-H. Customized cranioplasty implants using three-dimensional printers and polymethyl-methacrylate casting. *J. Korean Neurosurg. Soc.* **2012**, *52*, 541–546. [CrossRef]
3. Unterhofer, C.; Wipplinger, C.; Verius, M.; Recheis, W.; Thomé, C.; Ortler, M. Reconstruction of large cranial defects with Poly-Methyl-Methacrylate (PMMA) using a rapid prototyping model and a new technique for intraoperative implant modeling. *Neurol. Neurochir. Pol.* **2017**, *51*, 214–220. [CrossRef]
4. Chang, S.C.N.; Tobias, G.; Roy, A.K.; Vacanti, C.A.; Bonassar, L.J. Tissue engineering of autologous cartilage for craniofacial reconstruction by injection molding. *Plast. Reconstr. Surg.* **2003**, *112*, 793–799, discussion 800–801. [CrossRef]
5. Brown, A.E.; Banks, P. Late extrusion of alloplastic orbital floor implants. *Br. J. Oral Maxillofac. Surg.* **1993**, *31*, 154–157. [CrossRef]
6. Wolfaardt, J.F.; Coss, P. An impression and cast construction technique for implant-retained auricular prostheses. *J. Prosthet. Dent.* **1996**, *75*, 45–49. [CrossRef]
7. Gibson, I.; Rosen, D.; Stucker, B. *Additive Manufacturing Technologies: 3D Printing, Rapid Prototyping, and Direct Digital Manufacturing*, 2nd ed.; Springer: New York, NY, USA, 2015; ISBN 978-1-4939-2112-6.
8. Vandenbroucke, B.; Kruth, J.-P. Selective laser melting of biocompatible metals for rapid manufacturing of medical parts. *Rapid Prototyp. J.* **2007**, *13*, 196–203. [CrossRef]
9. 3D Printing Industry (3DPI). Arcam Announces FDA Clearance of Implants Produced with Additive Manufacturing. Available online: http://www.arcam.com/arcam-announces-fda-clearance-of-implants-produced-with-additive-manufacturing (accessed on 31 March 2015).
10. Chua, C.K.; Wong, C.H.; Yeong, W.Y. *Standards, Quality Control, and Measurement Sciences in 3D Printing and Additive Manufacturing*; Academic Press: Cambridge, MA, USA, 2017; ISBN 978-0-12-813490-0.
11. Ameen, W.; Al-Ahmari, A.; Mohammed, M.K.; Abdulhameed, O.; Umer, U.; Moiduddin, K. Design, Finite Element Analysis (FEA), and fabrication of custom titanium alloy cranial implant using electron beam melting additive manufacturing. *Adv. Produc. Engineer. Manag.* **2018**, *13*, 267–278. [CrossRef]
12. Al-Ahmari, A.; Nasr, E.A.; Moiduddin, K.; Anwar, S.; Kindi, M.A.; Kamrani, A. A comparative study on the customized design of mandibular reconstruction plates using finite element method. *Adv. Mech. Eng.* **2015**, *7*. [CrossRef]
13. Ginestra, P.; Ferraro, R.M.; Zohar-Hauber, K.; Abeni, A.; Giliani, S.; Ceretti, E. Selective laser melting and electron beam melting of Ti6Al4V for orthopedic applications: A comparative study on the applied building direction. *Materials* **2020**, *13*, 5584. [CrossRef]
14. Murr, L.E.; Gaytan, S.M.; Martinez, E.; Medina, F.; Wicker, R.B. Next Generation Orthopaedic Implants by Additive Manufacturing Using Electron Beam Melting. Available online: https://www.hindawi.com/journals/ijbm/2012/245727 (accessed on 3 February 2021).
15. Kotzem, D.; Ohlmeyer, H.; Walther, F. Damage tolerance evaluation of a unit cell plane based on Electron Beam Powder Bed Fusion (E-PBF) manufactured Ti6Al4V alloy. *Procedia Struct. Integr.* **2020**, *28*, 11–18. [CrossRef]

16. Cheng, B.; Chou, K. Geometric consideration of support structures in part overhang fabrications by electron beam additive manufacturing. *Comput. Aided Des.* **2015**, *69*. [CrossRef]

17. Umer, U.; Ameen, W.; Abidi, M.H.; Moiduddin, K.; Alkhalefah, H.; Alkahtani, M.; Al-Ahmari, A. Modeling the effect of different support structures in electron beam melting of titanium alloy using finite element models. *Metals* **2019**, *9*, 806. [CrossRef]

18. Tran, T.Q.; Chinnappan, A.; Lee, J.K.Y.; Loc, N.H.; Tran, L.T.; Wang, G.; Kumar, V.V.; Jayathilaka, W.A.D.M.; Ji, D.; Doddamani, M.; et al. 3D printing of highly pure copper. *Metals* **2019**, *9*, 756. [CrossRef]

19. Shiomi, M.; Osakada, K.; Nakamura, K.; Yamashita, T.; Abe, F. Residual stress within metallic model made by selective laser melting process. *CIRP Ann.* **2004**, *53*, 195–198. [CrossRef]

20. Hussein, A.; Hao, L.; Yan, C.; Everson, R.; Young, P. Advanced lattice support structures for metal additive manufacturing. *J. Mater. Process. Technol.* **2013**, *213*, 1019–1026. [CrossRef]

21. Ameen, W.; Al-Ahmari, A.; Mohammed, M.; Mian, S. Manufacturability of overhanging holes using electron beam melting. *Metals* **2018**, *8*, 397. [CrossRef]

22. Wang, D.; Mai, S.; Xiao, D.; Yang, Y. Surface quality of the curved overhanging structure manufactured from 316-L stainless steel by SLM. *Int. J. Adv. Manuf. Technol.* **2016**, *86*, 781–792. [CrossRef]

23. Wang, D.; Yang, Y.; Zhang, M.; Lu, J.; Liu, R.; Xiao, D. Study on SLM fabrication of precision metal parts with overhanging structures. In Proceedings of the 2013 IEEE International Symposium on Assembly and Manufacturing (ISAM), Xi'an, China, 30 July–2 August 2013; pp. 222–225.

24. Ford, S.; Despeisse, M. Additive manufacturing and sustainability: An exploratory study of the advantages and challenges. *J. Clean. Prod.* **2016**, *137*, 1573–1587. [CrossRef]

25. Jiang, J.; Xu, X.; Stringer, J. Support structures for additive manufacturing: A review. *J. Manuf. Mater. Process.* **2018**, *2*, 64. [CrossRef]

26. Morgan, D.; Agba, E.; Hill, C. Support structure development and initial results for metal powder bed fusion additive manufacturing. *Procedia Manuf.* **2017**, *10*, 819–830. [CrossRef]

27. Samant, R.; Ranjan, R.; Mhapsekar, K.; Anand, S. Octree data structure for support accessibility and removal analysis in additive manufacturing. *Addit. Manuf.* **2018**, *22*, 618–633. [CrossRef]

28. Jhabvala, J.; Boillat, E.; André, C.; Glardon, R. An innovative method to build support structures with a pulsed laser in the selective laser melting process. *Int. J. Adv. Manuf. Technol.* **2012**, *59*, 137–142. [CrossRef]

29. Yan, C.; Hao, L.; Hussein, A.; Raymont, D. Evaluations of cellular lattice structures manufactured using selective laser melting. *Int. J. Mach. Tools Manuf.* **2012**, *62*, 32–38. [CrossRef]

30. Park, S.; Rosen, D.; Duty, C. Comparing mechanical and geometrical properties of lattice structure fabricated using electron beam melting. In Proceedings of the 2014 Annual International Solid Freeform Fabrication Symposium—An Additive Manufacturing Conference, Austin, TX, USA, 6 August 2014; Volume 1, pp. 1359–1370.

31. Li, Z.; Zhang, D.Z.; Dong, P.; Kucukkoc, I. A lightweight and support-free design method for selective laser melting. *Int. J. Adv. Manuf. Technol.* **2017**, *90*, 2943–2953. [CrossRef]

32. Langelaar, M. Topology optimization of 3D self-supporting structures for additive manufacturing. *Addit. Manuf.* **2016**, *12*, 60–70. [CrossRef]

33. Calignano, F. Design optimization of supports for overhanging structures in aluminum and titanium alloys by selective laser melting. *Mater. Des.* **2014**, *64*, 203–213. [CrossRef]

34. Ruffo, M.; Tuck, C.; Hague, R. Cost estimation for rapid manufacturing-laser sintering production for low to medium volumes. *Proc. Inst. Mech. Eng. Part B* **2006**, *220*, 1417–1427. [CrossRef]

35. Hopkinson, N.; Dicknes, P. Analysis of rapid manufacturing—Using layer manufacturing processes for production. *Proc. Inst. Mech. Eng. Part C* **2003**, *217*, 31–39. [CrossRef]

36. Syam, W.; Al-Ahmari, A.; Mannan, M.; Al-Shehri, H.; Al-Wazzan, K. Metallurgical, accuracy and cost analysis of Ti6Al4V dental coping fabricated by electron beam melting process. In Proceedings of the 5th International Conference on Advanced Research in Virtual and Rapid Prototyping, Leiria, Portugal, 28 September–1 October 2011; ISBN 978-0-415-68418-7.

37. Lindemann, C.; Jahnke, U.; Habdank, M.; Koch, R. Analyzing product lifecycle costs for a better understanding of cost drivers in additive manufacturing. In Proceedings of the 23th Annual International Solid Freeform Fabrication Symposium—An Additive Manufacturing Conference, Austin, TX, USA, 6–8 August 2012.

38. Baumers, M.; Dickens, P.; Tuck, C.; Hague, R. The cost of additive manufacturing: Machine productivity, economies of scale and technology-push. *Technol. Forecast. Soc. Chang.* **2016**, *102*, 193–201. [CrossRef]

39. Priarone, P.C.; Robiglio, M.; Ingarao, G.; Settineri, L. Assessment of cost and energy requirements of Electron Beam Melting (EBM) and machining processes. In *International Conference on Sustainable Design and Manufacturing, Bologna, Italy, 26–28 April 2017*; Campana, G., Howlett, R.J., Setchi, R., Cimatti, B., Eds.; Smart Innovation, Systems and Technologies; Springer: Cham, Switzerland, 2017; Volume 68, pp. 723–735. ISBN 978-3-319-57077-8.

40. Choi, J.-Y.; Choi, J.-H.; Kim, N.-K.; Kim, Y.; Lee, J.-K.; Kim, M.-K.; Lee, J.-H.; Kim, M.-J. Analysis of errors in medical rapid prototyping models. *Int. J. Oral Maxillofac. Surg.* **2002**, *31*, 23–32. [CrossRef]

41. Mostafa Elkatatny, A.A.A.; Eldabaa, K.A. Cranioplasty: A new perspective. *Open Access Maced. J. Med. Sci.* **2019**, *7*, 2093–2101. [CrossRef] [PubMed]

42. Höhne, J.; Brawanski, A.; Gassner, H.; Schebesch, K.-M. Feasibility of the custom-made titanium cranioplasty CRANIOTOP®. *Surg. Neurol. Int.* **2013**, *4*, 88. [CrossRef] [PubMed]
43. Brown, D.A.; Wijdicks, E.F.M. Decompressive craniectomy in acute brain injury. *Handb. Clin. Neurol.* **2017**, *140*, 299–318. [CrossRef] [PubMed]
44. Toth, B.A.; Ellis, D.S.; Stewart, W.B.; Jeffrey, L.; Marsh, D.; Vannier, M.W. Computer-designed prostheses for orbitocranial reconstruction. *Plast. Reconstr. Surg.* **1988**, *81*, 323–3241. [CrossRef]
45. Joffe, J.M.; McDermott, P.J.; Linney, A.D.; Mosse, C.A.; Harris, M. Computer-generated titanium cranioplasty: Report of a new technique for repairing skull defects. *Br. J. Neurosurg.* **1992**, *6*, 343–350. [CrossRef]
46. Webb, P.A. A Review of Rapid Prototyping (RP) techniques in the medical and biomedical sector. *J. Med. Eng. Technol.* **2000**, *24*, 149–153. [CrossRef]
47. Moreira-Gonzalez, A.; Jackson, I.T.; Miyawaki, T.; Barakat, K.; DiNick, V. Clinical outcome in cranioplasty: Critical review in long-term follow-up. *J. Craniofac. Surg.* **2003**, *14*, 144–153. [CrossRef]
48. Maravelakis, E.; David, K.; Antoniadis, A.; Manios, A.; Bilalis, N.; Papaharilaou, Y. Reverse engineering techniques for cranioplasty: A case study. *J. Med. Eng. Technol.* **2008**, *32*, 115–121. [CrossRef]
49. Moiduddin, K.; Hammad Mian, S.; Umer, U.; Ahmed, N.; Alkhalefah, H.; Ameen, W. Reconstruction of complex zygomatic bone defects using mirroring coupled with EBM fabrication of titanium implant. *Metals* **2019**, *9*, 1250. [CrossRef]
50. Alsing, L.; Storm, S.J. Sustainability of Additive Manufacturing—Electron Beam Melting of IN718. Ph.D. Thesis, University West, Trollhättan, Sweden, 2019.
51. Arcam A2 Setting the Standard for Additive Manufacturing. Available online: http://www.arcam.com/wp-content/uploads/Arcam-A2.pdf (accessed on 12 July 2019).

metals

Article

Model for the Prediction of Deformations in the Manufacture of Thin-Walled Parts by Wire Arc Additive Manufacturing Technology

Mikel Casuso [1], Fernando Veiga [1,*], Alfredo Suárez [1], Trunal Bhujangrao [1], Eider Aldalur [1], Teresa Artaza [1], Jaime Amondarain [2] and Aitzol Lamikiz [3]

[1] TECNALIA, Basque Research and Technology Alliance (BRTA), Parque Científico y Tecnológico de Gipuzkoa, E20009 Donostia-San Sebastián, Spain; mikel.casuso@tecnalia.com (M.C.); alfredo.suarez@tecnalia.com (A.S.); trunal.bhujangrao@tecnalia.com (T.B.); eider.aldalur@tecnalia.com (E.A.); teresa.artaza@tecnalia.com (T.A.)

[2] TALLERES AMONDARAIN I, Barrio Akezkoa, S/N, 20150 Zizurkil, Spain; jaime@keytech.tech

[3] Department of Mechanical Engineering, University of the Basque Country (UPV/EHU), E48013 Bilbao, Spain; aitzol.lamikiz@ehu.eus

* Correspondence: fernando.veiga@tecnalia.com; Tel.: +34-902-760-000

Abstract: Gas Metal Arc Welding (GMAW) is a manufacturing technology included within the different Wire Arc Additive Manufacturing alternatives. These technologies have been generating great attention among scientists in recent decades. Its main qualities that make it highly productive with a large use of material with relatively inexpensive machine solutions make it a very advantageous technology. This paper covers the application of this technology for the manufacture of thin-walled parts. A finite element model is presented for estimating the deformations in this type of parts. This paper presents a simulation model that predicts temperatures with less than 5% error and deformations of the final part that, although quantitatively has errors of 20%, qualitatively allows to know the deformation modes of the part. Knowing the part areas subject to greater deformation may allow the future adaptation of deposition strategies or redesigns for their adaptation. These models are very useful both at a scientific and industrial level since when we find ourselves with a technology oriented to Near Net Shape (NNS) manufacturing where deformations are critical for obtaining the final part in a quality regime.

Keywords: thin wall manufacturing; additive manufacturing; process modelling

Citation: Casuso, M.; Veiga, F.; Suárez, A.; Bhujangrao, T.; Aldalur, E.; Artaza, T.; Amondarain, J.; Lamikiz, A. Model for the Prediction of Deformations in the Manufacture of Thin-Walled Parts by Wire Arc Additive Manufacturing Technology. *Metals* **2021**, *11*, 678. https://doi.org/10.3390/met11050678

Academic Editor: Atila Ertas

Received: 6 April 2021
Accepted: 19 April 2021
Published: 21 April 2021

Publisher's Note: MDPI stays neutral with regard to jurisdictional claims in published maps and institutional affiliations.

Copyright: © 2021 by the authors. Licensee MDPI, Basel, Switzerland. This article is an open access article distributed under the terms and conditions of the Creative Commons Attribution (CC BY) license (https://creativecommons.org/licenses/by/4.0/).

1. Introduction

Additive Manufacturing (AM) is an emerging and promising field that is gaining increasing research attention and its applications in different industrial sectors are spreading day by day. Wire Arc Additive Manufacturing (WAAM) is a type of AM technique classified into Directed Energy Deposition (DED) technologies, where a metal wire acts as material feedstock and is melted by a heat source [1]. According to the nature of heat source, there are three types of WAAM technologies: Gas Metal Arc Welding (GMAW)-based WAAM, Gas Tungsten Arc Welding (GTAW)-based WAAM and Plasma Arc Welding (PAW)-based WAAM, each of them with different features and advantages [2,3].

It is inaccurate to see WAAM as opposed to traditional subtractive manufacturing technologies, but as complementary to them, since WAAM is still far from achieving final parts without dimensional inaccuracies and good surface quality. Because of this issue, subtractive manufacturing, like machining, is nowadays the usual finishing step for parts coming from WAAM. Therefore, WAAM should meet Near Net Shape (NNS) manufacturing paradigm, thus reducing machining operations aiming to save material, energy and time. In order to achieve this, distortions and residual stresses should be predicted and minimized to the extent possible, since they are two of the main problems of

WAAM that mostly affect subsequent machining and that hinder its deployment [4]. They are inherent to WAAM, so they can be predicted and reduced, but not suppressed [5].

On the one hand, distortions in parts are caused by thermal cycles of melting and cooling that happened in layer-by-layer deposition, with their corresponding volumetric expansion and shrinkage. They are a concerning issue in structures such as thin walls [5].

On the other hand, residual stresses are the mechanical stresses that remain in a part once their cause (external forces or thermal gradients) disappears and the part is at equilibrium with its surroundings [6]. Unclamping or machining the part may disrupt this equilibrium and distort the part [7]. Poor mechanical properties and cracking may also appear [5]. There are different kind of residual stresses, and the ones caused in WAAM can be classified as inherent, bulk residual stresses [8].

Trial and error experiments are the most usual way to gain knowledge of WAAM processes so as to optimize process and operation parameters [9,10] and reduce distortion and residual stress, but they are costly and time consuming [3], so modeling and simulation are key tools to overcome this problem.

Due to the rapid and spatially variable cooling and heating of products manufactured by AM, they are often deformed and subject to residual stresses. Deformations due to welding appear to have a negative effect on exterior appearance, dimensional accuracy and other conditions such as various structural strengths. Finite element modeling (FEM) can be used to analyze deformation issues such as distortion and temperature field, allowing for more advanced planning early in the WAAM planning process to avoid expensive rework.

Srivastava et al. [11] describes in detail the features such models present. Most of them are based on previous welding models, precedent of current WAAM. To date, these models are mesh-based numeric models, both 2D and 3D Finite Element (FE) models. There are not analytical models yet. Likewise, Rodrigues et al. [3] only mentions FE analysis to avoid trial and error experiments.

As the mechanical properties of the part coming from WAAM have thermal origin, the modeling should include thermal modeling and mechanical modeling. The interaction between two models can be coupled or weakly coupled [11]. Coupled models simultaneously carry out both analyses, as the distortion induced heat affects thermal properties, whereas weakly coupled ones are performed sequentially, because they consider that the energy input from the heat source is much higher than the heat induced by distortion, so they neglect the latter [4]. Temperature history of the part is the main result from the thermal modelling, and it is an input for the mechanical modelling, which has distortions and residual stress fields as results [11].

Ding et al. [12] compared two thermo-mechanical weakly coupled 3D FE models for WAAM: a steady state model with Eulerian reference frame, and a transient model with Lagrangian reference frame. The "element birth technique" is used to simulate the addition of material. The software employed was ABAQUS, (Dassault Systèmes, Vélizy-Villacoublay, Ile-De-France, France) and the material was mild steel. The main outputs were the temperature, distortion and residual stress distributions. It was validated for a thin wall. In a more recent work, Ding et al. [13] developed a weakly coupled transient thermo-mechanical 3D FE model, optimizing it to save computational time, based on the finding that the maximum temperature a point reaches during the WAAM process determines the residual stress of that point.

Zhao et al. [14] simulated a thin wall manufactured by GMAW, by means of a coupled thermo-mechanical 3D FE model. It concluded that the deposition direction, especially on the deposition of the last layer, is a very influencing factor in the residual stress field of the whole component. The employed software was MSC Marc, and the deposited material was H08Mn2Si.

Cadiou et al. [15] developed a transient 2D axysimetric numerical model for PAW. In addition to thermal and mechanical laws, it also considered electromagnetic forces. It focused on droplet generation, deposition and dynamics in the melt pool, taking into account only operating parameters and basing on level-set method by Osher and Sethian [16] to distinguish between gas and metal areas and define their properties. The software employed was COMSOL Multiphysics® and the material was 304 stainless steel. The main outputs were the evolution of the melt pool, and the height and the weight of the rod, as well as its temperature. It was validated for the first deposited two rods.

Oyama et al. [17] compared residual stresses, distortion and manufacturing duration for different deposition and heating strategies in WAAM, using 3D FE simulation. Simufact welding software was employed, and the materials were Al-5Mg and Al-3Si aluminum alloys. It was validated for a complete thin wall.

As it has been shown, the great majority of numeric models have been validated for thin walls, but FEM has not been successfully applied to a complex part of reduced thickness yet. Therefore, broadly, it can be concluded that there is still a gap regarding the prediction of the complete real geometry of final complex parts by means of modelling and simulation. This gap is a concerning issue, since most of the real industrial parts are complex, and they present a combination of different geometries as curve and plane surfaces. Therefore, in order to allow the complete industrial deployment of parts manufactured by WAAM, to determine if FEM analysis is suitable for predicting their temperature and distortion profiles is a key step.

Consequently, in this research, a method development on predicting both the temperature each layer reaches and the deformed state of WAAM will be investigated by utilizing FEM analysis in a form of multi-layered process. The manufactured part is a geometrically complex mold of ER70S-6 steel.

This steel is usually used in applications such as construction works, automotive industry, pipes, shafts and tanks, so it is an industrially relevant material. Up to now, researchers are mainly focused on selecting optimal operation parameters regarding mechanical properties and microstructure, by means of experimental tests [18,19], but there is not yet an FE analysis of a complex part, aiming to predict final distortion.

Additionally, Dttmann et al. [20] studied the machining of a part of this material manufactured by GMAW, so a tool as FEM, which can predict distortion and temperatures prior to this machining operation, is highly valuable.

The good agreement between the results obtained from FEM simulation and a methodology for the prediction of temperatures during the process and deformations in the final part has been validated with the measurement of the part using a laser tacker (FARO Technologies Inc., Lake Mary, FL, USA).

2. Materials and Methods

2.1. Set-Up for the Wire Arc Additive Manufacturing of the Part

This document details the manufacturing process of an ER70 steel mold using WAAM technology based on GMAW. Figure 1 shows the CAD design of the mold to be manufactured. Both the design of the mold and the materials, equipment, methodology and parameters used are determined. In addition, the problems encountered during manufacturing and the solution that has been taken for each of them are also shown.

This demonstrator was made of ER70S-6 mild steel, Table 1 showed the composition of the material according to the provider. This ER70S-6 steel was provided in the form of commercial wire with a diameter of 1.2 mm, first 3 coils (Praxair, Danbury, CT, USA) and the last coil (Bohler, Düsseldorf, Germany), and 8 mm flat S235JR steel plates were used as a substrate.

Table 1. Chemical composition of ER70S-6 steel wire (% of the weight).

Mn	Si	C	Cr	Cu	Ni	S	P	Mo	Ti	Zr
1.64	0.94	0.06	0.02	0.02	0.02	0.016	0.013	0.005	0.004	0.002

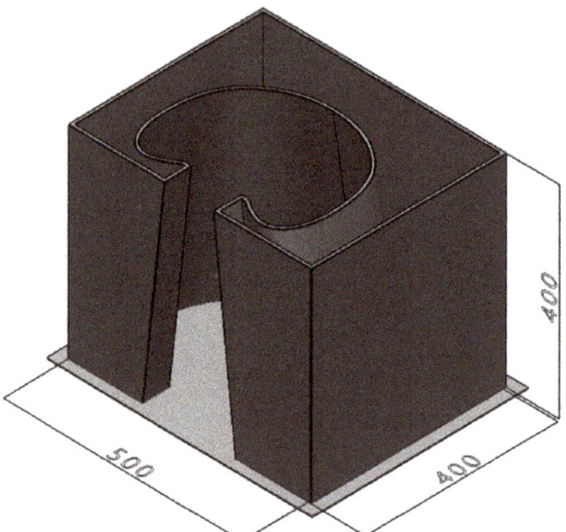

Figure 1. Mold design: main dimensions of the part.

A robotic fixed table welding system was used to manufacture this mold, as can be seen in Figure 2. In this system, the Alpha Q 552 puls (EWM, Mündersbach, Germany) welding equipment feeds the GMAW welding torch. The torch was placed on a Fanuc Arc Mate 100-iC (Fanuc, Oshino, Japan) robotic arm to be able to manufacture parts layer by layer. This set was also equipped with the M drive 4 Rob5 XR RE (EWM, Mündersbach, Germany) wire feeding equipment and the shielding gas system. The welding torch had been equipped with a compact Optris pyrometer (Optris GmbH, Berlin, Germany); the measurement was made on the surface of the bead once deposited with a delay of 10 ms (estimated based on the travel speed used). The emissivity of the surface was considered 0.9 for an incandescent body made of ER70 steel.

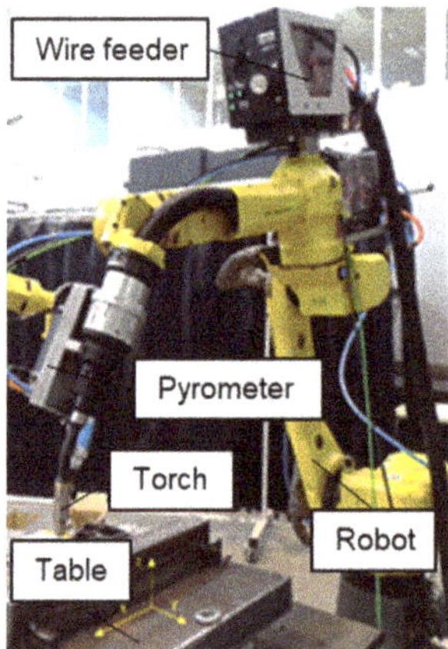

Figure 2. Set-up of the robotic system for the manufacture of parts with GMAW-based WAAM technology.

Taking into account the design of the part, the material was provided continuously in a single cord. The trajectories were designed using PowerMill (Autodesk, San Rafael, CA, USA) In this program the layer thickness and the type of path to follow could be defined and by means of internal algorithms the program calculated the desired paths. In order to be able to make the continuous deposition on a part of the straight back wall, the height of a layer was raised in z. These trajectories, once defined, were post-processed directly from the PowerMill. Each cell (robot or CNC machine) had its own post-processor, which allowed the part to be directly placed in the desired position, taking into account the characteristics of each equipment. In this case, the part was post-processed for the Fanuc robot post-processor on a fixed to the table as can be seen in Figure 3.

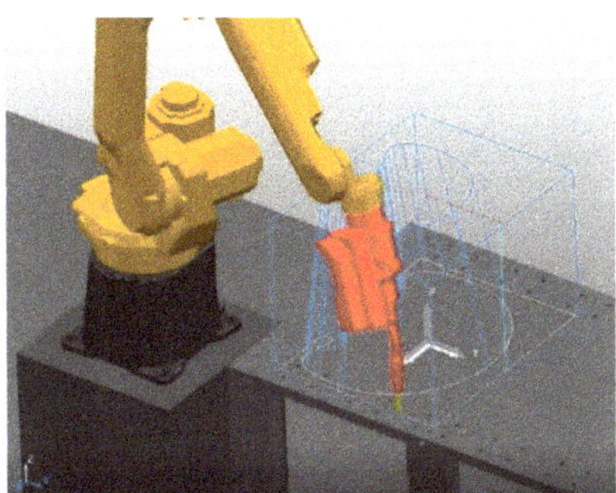

Figure 3. Path planning in PowerMill software.

In the present work, modes Cold Arc and Pulsed-GMAW (Job 194 and 9) were mainly used, which allowed possible wire feed speeds of between 4 and 16 m/min for the used material. Furthermore, the gas mixture % 20 CO_2-% 80 Ar with a flow rate of 18 L/min was chosen as the shielding gas. Regarding the inclination of the torch with respect to the substrate, it had worked with angles of 90° and a Stick-out of 17 mm. The diameter of the nozzle used was 20 mm and the parameters chosen were the following:

First, the initial programs were produced with the Cold Arc working mode with 5 m/min wire feed speed and 50 cm/min travel speed. But the number of splashes had been high, and the nozzle filled quickly so it was decided to change the mode to Pulsed-GMAW and program 3 is manufactured. Finally, it was decided to increase the travel speed to 65 cm/min and reduce the layer height to 1.3 mm. Manufacturing conditions are summarized on Table 2.

Table 2. Parameters used to Wire-Arc Additive Manufacture the mold.

Program Number	Material	Density (kg/m³)	Material Sustrate	Wire Diameter (mm)	Mode	Job Number	Wire Feed (m/min)	Deposition Rate (kg/h)	Stick Out (mm)	Travel Speed (cm/min)	Bead Width (mm)	Stimated Height (mm)
N# 1	ER70S-6	7850	S235JR	1.2	Cold Arc	194	5	2.66	17	50	7.2	1.6
N# 2	ER70S-6	7850	S235JR	1.2	Pulsed-GMAW	9	5	2.66	17	50	7.2	1.6
N# 3	ER70S-6	7850	S235JR	1.2	Pulsed-GMAW	9	5	2.66	17	65	7.2	1.3

The manufacture of the piece was therefore carried out with a combination of two transfer modes. A Cold-Arc transfer mode was used in the first ten passes, then another ten were made in Pulsed-GMAW at the same traverse speed and ended with the same conditions in the Pulsed-GMAW mode at higher travel speed. This strategy was considered adequate since in the first layers the Cold-Arc mode introduced less heat to the substrate, and it transitioned to a Pulsed-GMAW mode where fewer splashes were produced and finally the speed was increased to introduce less energy so that it prevented the collapse of the wall.

From the control of the machine and the welding source, the relevant data of the process were extracted that allowed the relevant information to be obtained. This information gave an overview of both the stability of the process and the energy introduced in the process for the simulation of the manufacture of the part. These signals were sampled with a sample rate of two hertz. The energy was calculated as the product of the intensity (I) by the voltage (V) and the travel speed (TS), expressed in units [kJ/cm].

$$Energy\left[\frac{kJ}{cm}\right] = \frac{I \times V}{TS} \tag{1}$$

Figure 4 shows the volumetric representation of the energy calculated following Equation (1). Analyzing the energy results, it was observed that the measured values of the machine control were not constant. This value depended both on the stick-out, the distance the wire goes out from the nozzle to the upper surface of the part and on fluctuations in travel speed, this type of energy behavior was already reported previously by Wang et al. [21].

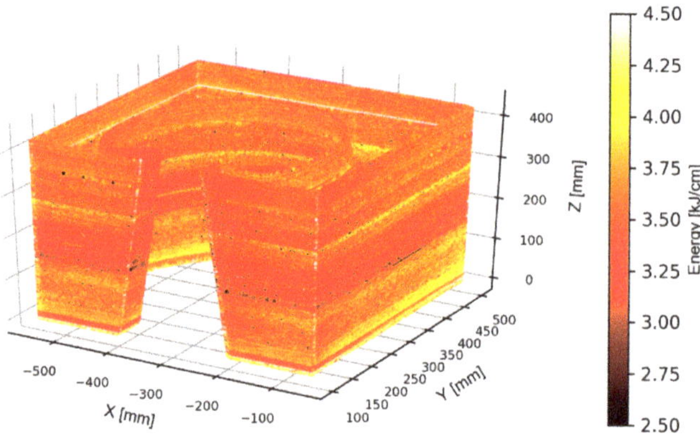

Figure 4. Evolution of energy during the manufacture of the mold.

2.2. Methodology for Material Characterization and Part Verification

Measuring molds for large parts in industrial environments used to be a difficult and time-consuming with high precision. In this paper, a Portable FARO (FARO Technologies Inc., Lake Mary, FL, USA) Vantage Laser Trackers made on-site measurements in 3D coordinates by tracking a target that the user moved from one point to another of the wire-arc manufactured part, as it can be seen on Figure 5a. The measurements were then compared to nominal CAD data, Figure 5b.

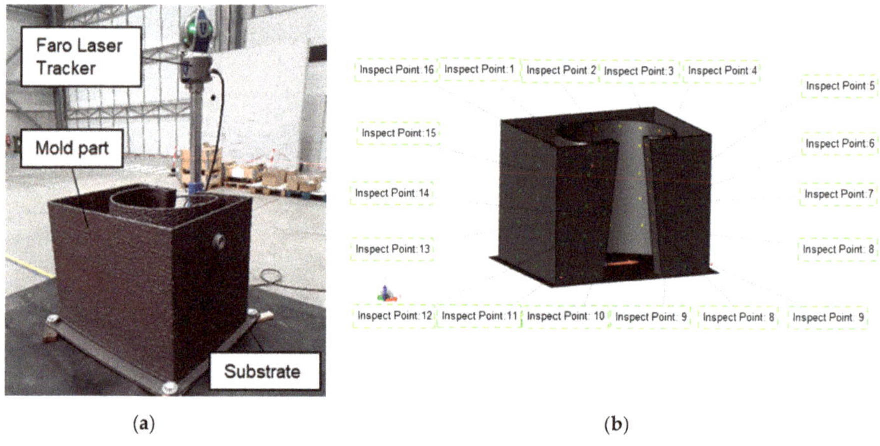

(**a**)　　　　　　　　　　(**b**)

Figure 5. Set-up for the geometric verification of the part: (**a**) laser tacker and (**b**) reference points on the CAD part.

3. Results

3.1. Mechanical Characterization of Additive Manufacturing Material

A sample wall had been manufactured using same additive conditions that were in the mold part. The wall was manufactured to take out specimens with threaded terminals to perform the tensile tests according to the ISO 6892-1 standard [22]. The tensile tests in uniaxial direction were carried out at room temperature in an Instron (Instron, Norwood, MA, USA) 5585H bench test that gave 100 kN of maximum load equipped with an Instron EX2620-602 (contact extensometer). In addition, the same samples were performed at room temperature using Vickers hardness tests at different heights on a Struers Duramin A-300 machine (Struers, Copenhagen, Denmark).

The following Table 3 summarizes the results obtained from the tensile tests carried out on the specimens. Six tensile tests were carried out in each of the horizontal and vertical directions to determine the anisotropy of the mechanical properties. ER70 steel is an easily weldable mild steel that exhibits similar mechanical properties in both directions. In this case, the tests do not show large differences, being comparable to the data of the material provided by the supplier. Additionally, a Vickers hardness test was carried out at different heights, which placed its hardness around 150 HV.

Table 3. Summary of mechanical characterization on the sample walls.

Material	Direction	Tensile Test			Vickers Test
		UTS (MPa)	YS 0.2% (MPa)	Elong. (%)	Hardness [HV]
WAAM sample wall	Horizontal	498 ± 9	368 ± 12	36 ± 4	151 ± 9
	Vertical	501 ± 3	368 ± 4	32 ± 1	
ER70 as welded	-	500–640	>420	28	

As part of the analysis of the properties of the additive material, the cross-sectional macrographs of the bead are presented in the following Figure 6. The dissimilarity inherent in the formation of the layers of the beads is observed. Nevertheless, no macrostructural flaws such as lack of filling or pores are seen. To reveal the grain and to know the crystallography of the ER70 additive material, polished and finally etched with a solution 2% Nital-Nitric and Ethanol acid- to reveal the grain structure were used. Ferrite areas and ferrite / bainite acicular areas are generally found. In the upper part, as there have been thermal cooling-heating cycles, there are acicular parts of ferrite. In the central part in every layer limit also can be found acicular ferrite/bainite small areas but the microstructure mainly is composed by polygonal ferrite zones. A more detailed analysis of the microstructure can be found in the articles by Eider et al. [23,24].

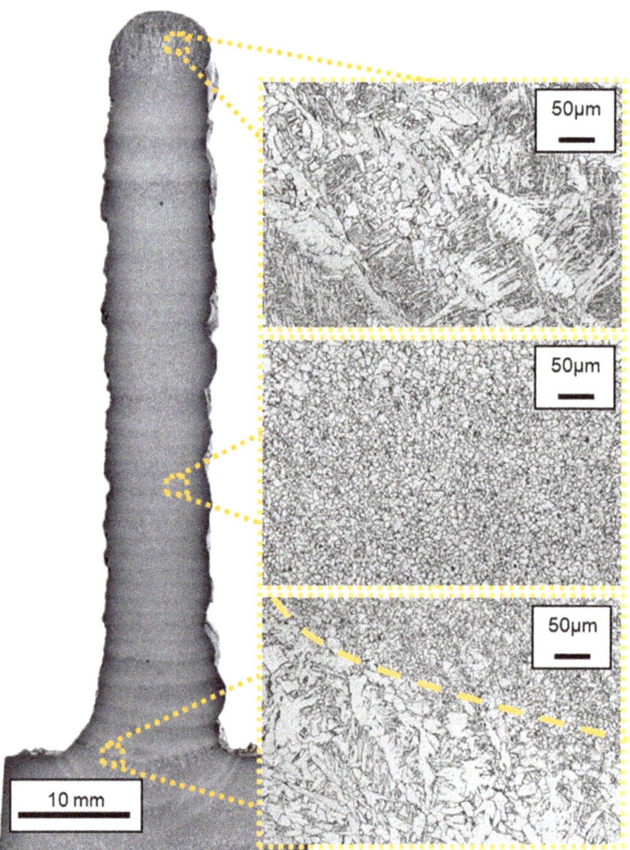

Figure 6. Macrograph and micros of a wall made of ER70 steel, at different heights.

3.2. Model for the Simulation of Thin-Wall Deformations

In this section, the model simulation of the additive manufacturing process by material addition of the mold ER70 steel part is presented. The analyses performed aimed to predict the thermo-mechanical behavior of the part, by analyzing the temperature field and permanent displacements during the process. One of the most frequent problems that occurs when parts are manufactured with this technology is the appearance of permanent deformations, the stored heat begins to transfer and the different parts of the model begin to expand depending on the temperature they reach and their properties. Since there are usually important thermal gradients in the part, different dilatations are produced. From the point of view of heat transfer, conduction, convection and radiation mechanisms are considered. The phase change of the material used in the addition is also taken into account. These type of simulations are very nonlinear, due to the nonlinearity of the materials and to the fact that the addition of material implies, in numerical terms, to frequently modify the stiffness matrix, which makes the resolution of the problem costly or even unfeasible from a resolution time point of view. Regarding the software, two programs are used for the simulation of the additive manufacturing process. Thus, the mesh is built with Siemens' NX software (Siemens, Munich, Germany), while the rest of the pre- and post-processing tasks are completed through MSC's Mentat. The solver used to solve the thermo-mechanical problem in MSC-Marc. According to Zhang et al. [25], the temperature-dependent material parameters were in this case the thermal expansion coefficient, thermal conductivity, elastic modulus, yield stress and specific heat. These temperature-dependent materials were obtained from the MSC Marc (MSC Software, Irvine, CA, USA) software database [26]. In this research, the double ellipsoid heat source according to Goldak [27] was considered, which is a proposed heat source model that promotes heat input as a function to generate heat while being able to control the overall amount of power delivered into the substrate and filler. This is a typical approach for common arc welding simulations, which reproduces the thermal energy input well [28,29]. The power density distribution in the heat source is as following. In the front half of the heat source:

$$q_f(x, y, z) = \frac{6\sqrt{3}f_f Q}{abc_f \pi \sqrt{\pi}} \exp\left(-\frac{3x^2}{a^2} - \frac{3y^2}{b^2} - \frac{3z^2}{c_f^2} \right) \tag{2}$$

In the rear half of the heat source:

$$q_r(x, y, z) = \frac{6\sqrt{3}f_r Q}{abc_r \pi \sqrt{\pi}} \exp\left(-\frac{3x^2}{a^2} - \frac{3y^2}{b^2} - \frac{3z^2}{c_r^2} \right) \tag{3}$$

where q_f is the power density distribution in the front half of the heat source and q_r is that in the rear half of the heat source; a is the width of the heat source and b is the depth; c_f and c_r are the forward and rear lengths of the heat source along the length of the layers; Q is the power of the heat source and Q = ηVI. η is the heat source efficiency and V and I are the voltage and the current; f_f and f_r are the fractions of the heat deposited in the front and rear halves of the heat source and $f_f + f_r = 2$.

The heat source efficiency of GMAW is assumed to be 85% [30] and the parameters defining the dimensions of molten pool used for the volume weld flux are chosen according to actual dimensions of the molten pool. The distribution parameters of double ellipsoidal heat source are assumed to be the same in the numerical simulation of all layer depositions. Considering the high temperature gradient of deposition metal, the meshes in the deposition area and the heat-affected zone are dense. The finite element mesh consists of hexahedron elements, and the element size in the deposition area is about 1 mm. Simulation has been performed using several element sizes close to 1.3 mm by keeping the total height constant. It is overserved that there is not any difference in the temperature variation in layers and in total deformation of the mold. Hence, in this paper the average element size is considered as 1 mm. The material is modeled as elastic perfectly plastic. The overall geometry, including both filler and base metal, was discretized using 15,398, 8-nodes brick elements. Free convection boundary conditions were set up on the base plate top and bottom surfaces and on the wall vertical surfaces. Convection coefficients values, set according to literature correlations, were: 8.5 W/m^{2}K for the base plate top surface, 4.0 W/m^{2}K for the bottom surface and 12.0 W/m^{2}K for the wall vertical surface. A boundary condition of general radiation to environment was included, setting material emissivity was set to 0.3. Environment and material initial temperatures were set to 20 °C. At the room temperature (25 °C), the Poisson's ratio is 0.29 and the elastic modulus is 190 GPa. The yield stress is 448 MPa and the thermal expansion coefficient is 13 µm/(m·K). The specific heat is 470 J/(kg·K) and the thermal conductivity is 50 W/(m·K) Figure 7a shows the finite element mesh multi-layer deposition of the original model and in Figure 7b shows the temperature field during the manufacturing processes. The simulated temperatures and the measured temperatures are around 1000 °C, although the melting temperature of the material is over 1400 °C. The simulation is focus on the quasi-stationary state of the process, where an abrupt drop in temperature is observed in the first moments of the deposition. It has been chosen to show the quasi-stationary state of the process, being more comparable to the measurements made on the material moments after deposition.

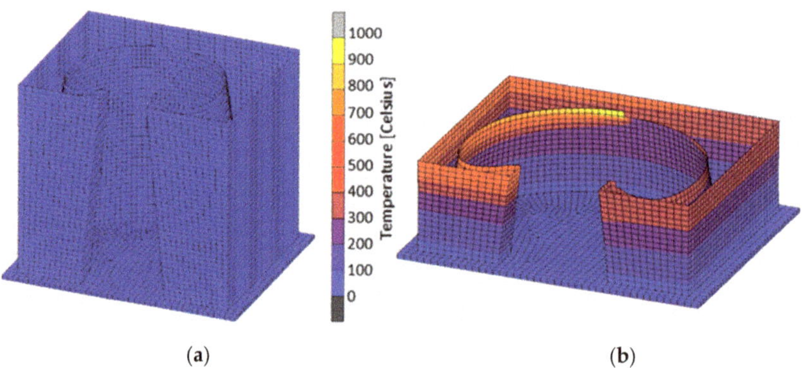

Figure 7. Simulation of temperatures in the manufacturing process: (**a**) meshing of the original CAD model and (**b**) result of the prediction of temperatures in the i layer.

Similarly, Figure 8 shows the temperatures measured by the pyrometer that continued to manufacture the part. A temperature of just over 900 °C has been reached, a little behind the simulated ones. It should be noted that the measurement of the pyrometer on the surface is collected with a slight delay, despite that the fit between the simulation and the real measurements presents correct results with a difference of less than 100 °C.

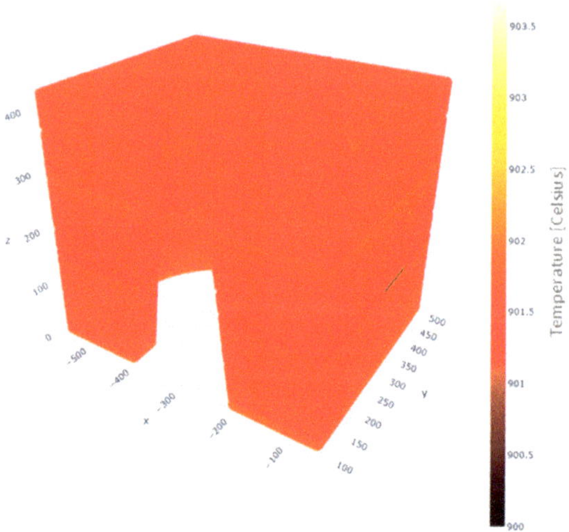

Figure 8. Temperature measured by the pyrometer installed in the torch during the additive process.

From Finite element simulation, the distribution of displacements on the thin-walled part examined and gathered information of the distortion of nodes in the FEM simulation. The estimated deformations for the manufacture of the part are shown in the Figure 9. The largest deformations are in the lower part of the curved zone and in the largest straight surface. This results then compare with the real time measuring techniques discussed in next section.

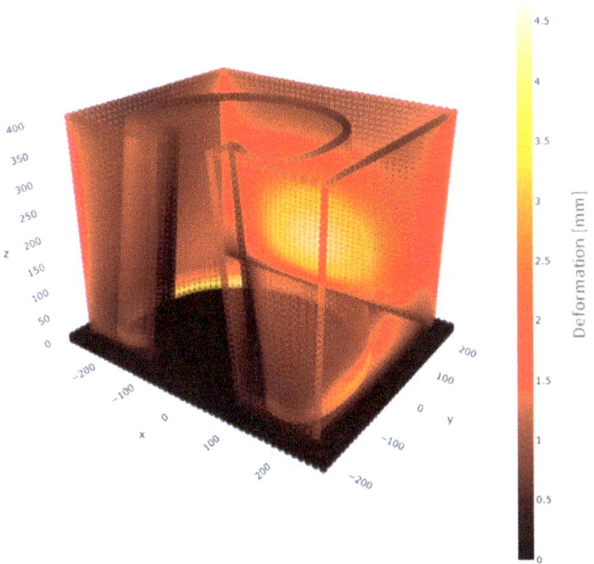

Figure 9. Deformations predicted by the finite element model for additive manufacturing of thin-walled part.

3.3. Measurement of Final Part Deformation

Once the GMAW manufacturing process of the molded part was completed, a measurement of its geometry was performed by an external company. As can be seen in the Figure 10, where the measured reference points are shown as black points against the predicted surface, the final part is within tolerances. Once the machining process (finishing) has been carried out, a fully functional final part would be obtained. The main sources of divergence between the measurement of the real part and the predictions of the simulated part are: (i) deformations and relaxations of the substrate that affect the geometry, (ii) transport and handling of the part and finally, (iii) wire-arc additive manufactured walls have a wavy surface, as it can be seen in Figures 5 and 6, that infers large variations in the measurements.

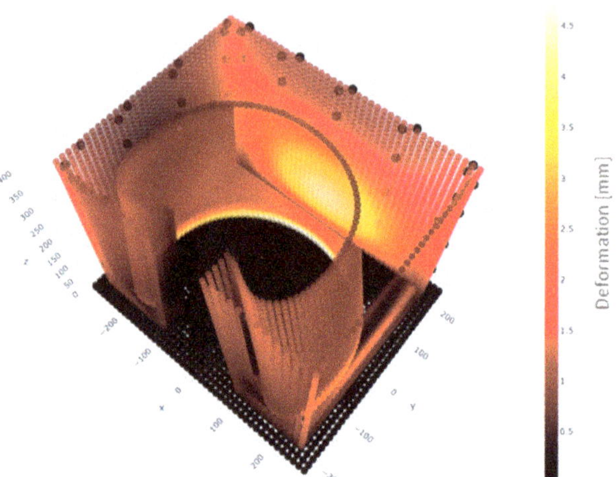

Figure 10. Measured outer surface points (black points) vs. predicted deformed surface from the model.

4. Discussion

Comparison of the Results Predicted by the Model and the Final Part Measurements

For a more detailed comparison of the measurements made with the results of the model, the part has been divided into different reference surfaces. With divided surfaces, as it can be seen on Figure 11, it is needed to focus on surface to surface. By calculating the distances between points according to Equation (4), the four points of the original surface closest to the measurement points are found.

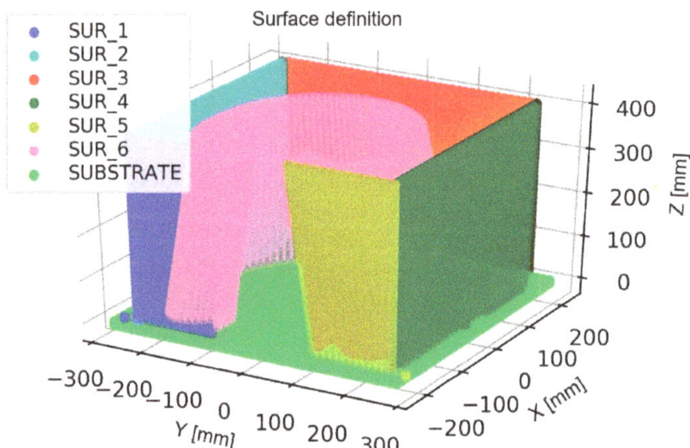

Figure 11. Volume division on reference surfaces.

These points will be the same as those taken on the deformed surface to compare with the measured points. Being P_1 with coordinates (x_1, y_1, z_1) and P_2 with coordinates (x_2, y_2, z_2), Equation (4) measures the modulus of the distance between the points.

$$d(P_1, P_2) = \sqrt{(x_2 - x_1)^2 + (y_2 - y_1)^2 + (z_2 - z_1)^2} \tag{4}$$

With the selected points of both the original surface and the simulated deformed surface the equation of the plane that contain the point can be calculated. Then, the distance from the measurement point to the plane is calculated according to Equation (3). given the point P (x_0, y_0, z_0) and the plane π: $Ax + By + Cz + D = 0$.

$$d(P, \pi) = \frac{|Ax_0 + By_0 + Cz_0 + D|}{\sqrt{A^2 + B^2 + C^2}} \tag{5}$$

This produces a csv file that includes the distances between the measured points and the original surface, the deformed surface and the original and the measured points and the deformed surface.

In addition, the YZ or XZ plane has been plotted according to the case of the flat surfaces of the part in the case of the measured points and the deformed surface, mapping the color as a function of the value of the remaining coordinate.

Among the analyzed surfaces, what happened on surface 1 and 3 is highlighted because it reflects what happens on all other surfaces (see Figure 11).

Going in further detail into surface 1 (Figure 12a), something similar is observed; quantitatively the simulation does not exactly predict the displacement, but the way in which this wall deforms. In the real case, the lower part of the wall deforms outwards, and the upper part moves slightly towards the inside of the piece. In the simulation, the displacement of the lower part is not appreciated, but the deformation of the upper part towards the interior of the part is.

While the simulated and the real surface 3 quantitatively differs. The way in which this wall has been deformed is similar in the real case and the simulated one, as seen in Figure 12b, so qualitatively could be accepted. It is observed that the lower central part enters towards inside the part and the upper part deforms outwards. The simulation shows good agreement on distortion tendency compared to the results of experiment with relative percentage error up to only 20%.

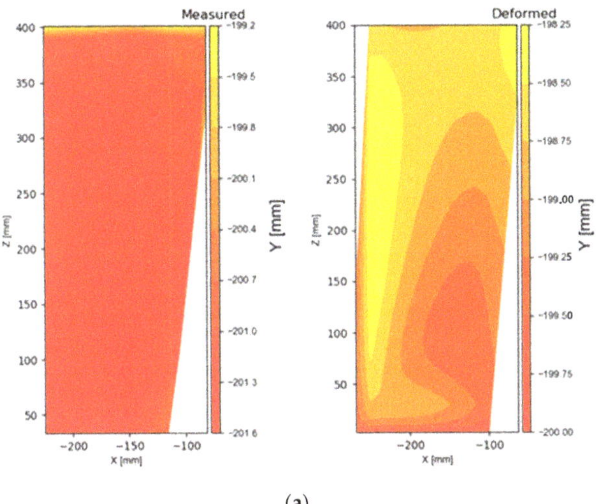

(a)

Figure 12. *Cont.*

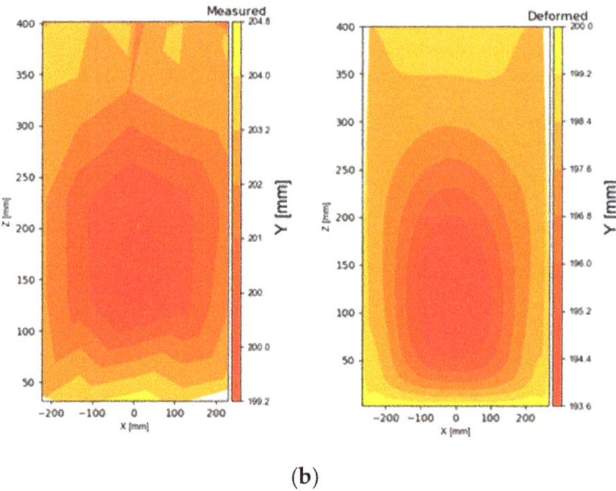

(b)

Figure 12. Comparison of the surfaces between the measurements and the predictions of the model in (**a**) surface 1 and in (**b**) surface 3.

5. Conclusions

This paper has presented a model for the prediction of deformations in a thin-walled mold part manufactured by means of WAAM technology. The main novelty presented by this article is the holistic vision of additive manufacturing by WAAM, with the use of finite element models to assist in the prediction of temperatures and deformations in the final part. The conclusions that can be obtained from this work are the following:

- The methodology for the manufacture of a mold piece by means of GMAW-WAAM technology with the combination of two transfer methods Cold-Arc and pulsed-GMAW has been presented.
- The deposition conditions ensure good metallographic quality and mechanical properties like those given by the supplier of the wire material.
- The finite element model predicts the process temperature. The adjustment of the predicted values with the temperature measured by means of pyrometric techniques shows accurate fitting. Results show deviation of less than a 5% in the temperature prediction.
- The deformation prediction model allows to know qualitatively the deformation mode of the part. Qualitatively the results are correct but can be improved, due to the external sources of error described. The surface waviness of the part manufactured, which reaches 2 mm from peak to valley, is one of most important.
- Finally, this paper has covered the purpose of prospecting the application of finite element simulation models for the prediction of deformations in thin-walled parts manufactured by WAAM technology.

Author Contributions: Conceptualization, M.C.; Data curation, F.V., T.B., T.A. and J.A.; Formal analysis, M.C., F.V. and E.A.; Funding acquisition, A.S. and J.A.; Investigation, E.A.; Methodology, F.V.; Project administration, A.S.; Resources, A.S.; Supervision, A.S.; Validation, A.L.; Visualization, T.B. and T.A.; Writing—original draft, M.C., F.V. and T.B.; Writing—review & editing, A.S., E.A., J.A. and A.L. All authors have read and agreed to the published version of the manuscript.

Funding: This research was funded by the vice-counseling of technology, innovation and competitiveness of the Basque Government grant agreement kk-2019/00004 (PROCODA project) and the QUALYFAM project, through the ELKARTEK 2020 (KK-2020/00042) and the ADIFIX project funded by HAZITEK 2019 and 2020 (ZL-2019/00738, ZL-2020/00073) programs and the Spanish Government CDTI-Red Cervera Programme (EXP 00123730/IDI-20191162).

Institutional Review Board Statement: Not applicable.

Informed Consent Statement: Not applicable.

Data Availability Statement: Data available on request.

Acknowledgments: The authors also thank the company KEYTECH S.L. for its support when performing the tests.

Conflicts of Interest: The funders had no role in the design of the study; in the collection, analyses or interpretation of data; in the writing of the manuscript or in the decision to publish the results.

Nomenclature

E	Energy
I	Intensity
V	Voltage
TS	Travel speed
UTS	Ultimate tensile strength
YS	Yield stress
q_f	Power density distribution in the front half of the heat source
q_r	Power density distribution in the rear half of the heat source
f_f	Fraction of the heat deposited in the front half of the heat source
f_r	Fraction of the heat deposited in the rear half of the heat source
c_f	Forward length of the heat source along the length of the layers
c_r	Rear length of the heat source along the length of the layers
a	Width of the heat source
b	Depth of the heat source
Q	Power of the heat source
η	Heat source efficiency
$d(P_1, P_2)$	Distance between two points
$d(P, \pi)$	Distance between a point and a plane

References

1. Teresa, A.; Suárez, A.; Veiga, F.; Braceras, I.; Tabernero, I.; Larrañaga, O.; Lamikiz, A. Wire arc additive manufacturing Ti6Al4V aeronautical parts using plasma arc welding: Analysis of heat-treatment processes in different atmospheres. *J. Mater. Res. Technol.* **2020**, *9*, 15454–15466.
2. Cunningham, C.; Flynn, J.; Shokrani, A.; Dhokia, V.; Newman, S. Invited review article: Strategies and processes for high quality wire arc additive manufacturing. *Addit. Manuf.* **2018**, *22*, 672–686. [CrossRef]
3. Rodrigues, T.A.; Duarte, V.; Miranda, R.M.; Santos, T.G.; Oliveira, J.P. Current status and perspectives on wire and arc additive manufacturing (WAAM). *Materials* **2019**, *12*, 1121. [CrossRef] [PubMed]
4. Lu, X.; Lin, X.; Chiumenti, M.; Cervera, M.; Hu, Y.; Ji, X.; Ma, L.; Yang, H.; Huang, W. Residual stress and distortion of rectangular and S-shaped Ti-6Al-4V parts by directed energy deposition: Modelling and experimental calibration. *Addit. Manuf.* **2019**, *26*, 166–179. [CrossRef]
5. Wu, B.; Pan, Z.; Ding, D.; Cuiuri, D.; Li, H.; Xu, J.; Norrish, J. A review of the wire arc additive manufacturing of metals: Properties, defects and quality improvement. *J. Manuf. Process.* **2018**, *35*, 127–139. [CrossRef]
6. D'Alvise, L.; Chantzis, D.; Schoinochoritis, B.; Salonitis, K. Modelling of part distortion due to residual stresses relaxation: An aerOnautical case study. *Procedia CIRP* **2015**, *31*, 447–452. [CrossRef]
7. Li, J.; Wang, S. Distortion caused by residual stresses in machining aeronautical aluminum alloy parts: Recent advances. *Int. J. Adv. Manuf. Technol.* **2017**, *89*, 997–1012. [CrossRef]
8. Casuso, M.; Polvorosa, R.; Veiga, F.; Suárez, A.; Lamikiz, A. Residual stress and distortion modeling on aeronautical aluminum alloy parts for machining sequence optimization. *Int. J. Adv. Manuf. Technol.* **2020**, *110*, 1219–1232. [CrossRef]
9. Tomaz, L.; Colaço, F.H.G.; Sarfraz, S.; Pimenov, D.Y. Investigations on quality characteristics in gas tungsten arc welding process using artificial neural network integrated with genetic algorithm. *Int. J. Adv. Manuf. Technol.* **2021**, *113*, 3569–3583. [CrossRef]
10. Arora, H.; Kumar, V.; Prakash, C.; Pimenov, D.; Singh, M.; Vasudev, H.; Singh, V. Analysis of Sensitization in Austenitic Stainless Steel-Welded Joint. In *Advances in Metrology and Measurement of Engineering Surfaces, Lecture Notes in Mechanical Engineering*; Prakash, C., Krolczyk, G., Singh, S., Pramanik, A., Eds.; Springer: Singapore, 2021; pp. 13–24.
11. Srivastava, S.; Garg, R.K.; Sharma, V.S.; Sachdeva, A. Measurement and mitigation of residual stress in wire-arc additive manufacturing: A review of macro-scale continuum modelling approach. *Arch. Comput. Methods Eng.* **2020**, 1–25. [CrossRef]
12. Ding, J.; Colegrove, P.; Mehnen, J.; Ganguly, S.; Almeida, P.S.; Wang, F.; Williams, S. Thermo-mechanical analysis of wire and arc additive layer manufacturing process on large multi-layer parts. *Comput. Mater. Sci.* **2011**, *50*, 3315–3322. [CrossRef]

13. Ding, J.; Colegrove, P.; Mehnen, J.; Williams, S.; Wang, F.; Almeida, P.S. A computationally efficient finite element model of wire and arc additive manufacture. *Int. J. Adv. Manuf. Technol.* **2014**, *70*, 227–236. [CrossRef]
14. Zhao, H.; Zhang, G.; Yin, Z.; Wu, L. Three-dimensional finite element analysis of thermal stress in single-pass multi-layer weld-based rapid prototyping. *J. Mater. Process. Technol.* **2012**, *212*, 276–285. [CrossRef]
15. Cadiou, S.; Courtois, M.; Carin, M.; Berckmans, W.; Masson, P.L. Heat transfer, fluid flow and electromagnetic model of droplets generation and melt pool behaviour for wire arc additive manufacturing. *Int. J. Heat Mass Transf.* **2020**, *148*, 119102. [CrossRef]
16. Osher, S.; Sethian, J.A. Fronts propagating with curvature-dependent speed: Algorithms based on Hamilton-Jacobi formulations. *J. Comput. Phys.* **1988**, *79*, 12–49. [CrossRef]
17. Oyama, K.; Diplas, S.; M'Hamdi, M.; Gunnæs, A.E.; Azar, A.S. Heat source management in wire-arc additive manufacturing process for Al-Mg and Al-Si alloys. *Addit. Manuf.* **2019**, *26*, 180–192. [CrossRef]
18. Rafieazad, M.; Ghaffari, M.; Nemani, A.V.; Nasiri, A. Microstructural evolution and mechanical properties of a low-carbon low-alloy steel produced by wire arc additive manufacturing. *Int. J. Adv. Manuf. Technol.* **2019**, *105*, 2121–2134. [CrossRef]
19. Prado-Cerqueira, J.L.; Camacho, A.M.; Diéguez, J.L.; Rodríguez-Prieto, Á.; Aragón, A.M.; Lorenzo-Martín, C.; Yanguas-Gil, Á. Analysis of favorable process conditions for the manufacturing of thin-wall pieces of mild steel obtained by wire and arc additive manufacturing (WAAM). *Materials* **2018**, *11*, 1449. [CrossRef]
20. Dttmann, A.; Gomes, J.D.O. Evaluation of additive manufacturing parts machinability using automated GMAW ER70S-6 with nodular cast iron. *U. Porto J. Eng.* **2021**, *7*, 88–97. [CrossRef]
21. Wang, Q.; Qi, B.; Cong, B.; Yang, M. Output characteristic and arc length control of pulsed gas metal arc welding process. *J. Manuf. Process.* **2017**, *29*, 427–437. [CrossRef]
22. Metallic materials—Tensile testing—Part 1: Method of test at room temperature. *Met. Mater.* **2020**, 1–78, ISO 6892-1:2019. Available online: https://www.iso.org/standard/78322.html (accessed on 6 April 2021).
23. Aldalur, E.; Veiga, F.; Suárez, A.; Bilbao, J.; Lamikiz, A. Analysis of the wall geometry with different strategies for high deposition wire arc additive manufacturing of mild steel. *Metals* **2020**, *10*, 892. [CrossRef]
24. Aldalur, E.; Veiga, F.; Suárez, A.; Bilbao, J.; Lamikiz, A. High deposition wire arc additive manufacturing of mild steel: Strategies and heat input effect on microstructure and mechanical properties. *J. Manuf. Process.* **2020**, *58*, 615–626. [CrossRef]
25. Zhang, H.; Zhang, G.; Cai, C.; Gao, H.; Wu, L. Fundamental studies on in process controlling angular distortion in asym-metrical double-sided double arc welding. *J. Mater. Process. Technol.* **2008**, *205*, 214–223. [CrossRef]
26. MSC Mentat Marc. *MARC User Guide*; Version 2010; MSC Software Corporation: Santa Ana, CA, USA, 2010.
27. Goldak, J.; Chakravarti, A.; Bibby, M. A new finite element model for welding heat sources. *Met. Mater. Trans. A* **1984**, *15*, 299–305. [CrossRef]
28. Ferro, P.; Berto, F.; James, N. Asymptotic residual stress distribution induced by multipass welding processes. *Int. J. Fatigue* **2017**, *101*, 421–429. [CrossRef]
29. Graf, M.; Pradjadhiana, K.P.; Hälsig, A.; Manurung, Y.H.P.; Awiszus, B. Numerical simulation of metallic wire arc additive manufacturing (WAAM). *AIP Conf. Proc.* **2018**, *1960*, 140010. [CrossRef]
30. Abid, M.; Siddique, M. Numerical simulation to study the effect of tack welds and root gap on welding deformations and residual stresses of a pipe-flange joint. *Int. J. Press. Vessel. Pip.* **2005**, *82*, 860–871. [CrossRef]

metals

MDPI

Article

Fabrication of a TiC-Ti Matrix Composite Coating Using Ultrasonic Vibration-Assisted Laser Directed Energy Deposition: The Effects of Ultrasonic Vibration and TiC Content

Yunze Li [1], Dongzhe Zhang [1], Hui Wang [2] and Weilong Cong [1,*]

[1] Department of Industrial, Manufacturing, and Systems Engineering, Texas Tech University, Lubbock, TX 79409, USA; yunze.li@ttu.edu (Y.L.); dongzhe.zhang@ttu.edu (D.Z.)

[2] Department of Industrial & Systems Engineering, Texas A&M University, College Station, TX 77843, USA; huiwang@tamu.edu

* Correspondence: weilong.cong@ttu.edu; Tel.: +1-806-834-6178

Citation: Li, Y.; Zhang, D.; Wang, H.; Cong, W. Fabrication of a TiC-Ti Matrix Composite Coating Using Ultrasonic Vibration-Assisted Laser Directed Energy Deposition: The Effects of Ultrasonic Vibration and TiC Content. *Metals* **2021**, *11*, 693. https://doi.org/10.3390/met11050693

Academic Editors: Andreas Chrysanthou and Ayrat Nazarov

Received: 28 March 2021
Accepted: 20 April 2021
Published: 23 April 2021

Publisher's Note: MDPI stays neutral with regard to jurisdictional claims in published maps and institutional affiliations.

Copyright: © 2021 by the authors. Licensee MDPI, Basel, Switzerland. This article is an open access article distributed under the terms and conditions of the Creative Commons Attribution (CC BY) license (https://creativecommons.org/licenses/by/4.0/).

Abstract: Titanium and its alloys exhibit superior properties of high corrosion resistance, an excellent strength to weight ratio and outstanding stiffness among other things. However, their relatively low hardness and wear resistance limit their service life in high-performance applications of structure parts, gears and bearings, for example. The fabrication of a ceramic reinforced titanium matrix composite (TMC) coating could be one of the solutions to enhance the microhardness and wear resistance. Titanium carbide (TiC) is a preferable candidate due to the advantages of self-lubrication, low cost and a similar density and thermal expansion coefficient with titanium. The fabrication of TiC-TMC coatings onto titanium using a laser directed energy deposition (LDED) process has been conducted. The problems of TiC aggregation, low bonding quality and the generation of fabrication defects still exist. Considering ultrasonic vibration could generate acoustic steaming and transient cavitation actions in melted materials, which could homogenize the distribution of reinforcement materials and promote the dissolution of TiC into liquid titanium. In this study, for the first time, we investigate the ultrasonic vibration-assisted LDED of TiC-TMC coatings. The effects of ultrasonic vibration and reinforcement content on the phase compositions, reinforcement aggregation, bonding quality, fabrication defects and mechanical properties (including microhardness and wear resistance) of LDED deposited TiC-TMC coatings have been investigated. With the assistance of ultrasonic vibration, the aggregation of TiC was reduced, the porosity was decreased, the defects in the bonding interface were reduced and the mechanical properties including microhardness and wear resistance were increased. However, the excessive TiC content could significantly increase the TiC aggregation and manufacturing defects, resulting in the reduction of the mechanical properties.

Keywords: ultrasonic vibration; laser directed energy deposition; coating; TiC-TMC

1. Introduction

Titanium and its alloys have been widely used in many industries (including the automotive industry, aerospace industry and medical industry) due to their superior properties of strong corrosion resistance, a high strength to weight ratio and outstanding stiffness among other things [1,2]. However, their relatively low surface mechanical properties (hardness and wear resistance) limit the service life in high-performance applications of structure parts, gears, bearings and jet engine compressors, for example. [3,4]. Ceramic reinforced titanium matrix composites (TMCs) were coated onto titanium to improve the mechanical properties [5,6]. Compared with other ceramic reinforcements (such as Al_2O_3, SiC, TiN and TiB [7–10]), TiC exhibits the unique property of self-lubrication with enhanced the wear resistance, a relativity low material cost and a similar thermal expansion coefficient with titanium [11]. TiC-TMC coatings have been successfully fabricated by laser

Metals **2021**, *11*, 693. https://doi.org/10.3390/met11050693   https://www.mdpi.com/journal/metals

additive manufacturing processes [12–15]. The mechanism of the dissolution of TiC into titanium and the phase transformations during the fabrication have been investigated. The results show that the precipitated TiC with a refined microstructure could significantly improve the surface microhardness and wear resistance.

Laser directed energy deposition (LDED) has been widely used in the fabrications of TMC coatings onto a metallic workpiece. In LDED, the melted material can be solidified at a high cooling rate (10^3–10^6 K/s) [16], which contributes to the development of coating layers with a relatively fine-grained microstructure [17,18]. LDED also has the capability of the functionally gradient coatings fabrication, which can reduce the discontinuity of properties between base materials and reinforced layers [19]. TiC-TMC, as a kind of preferable coating material, has been deposited onto titanium and its alloys by an LDED process [12,13,20–24]. The results showed that the LDED fabricated TiC-TMC coatings could significantly improve the surface properties of microhardness and wear resistance. By optimizing the laser density, the microhardness and wear resistance could be further increased [21,22]. In addition, by utilizing the functionally gradient coating layers, the bonding quality, density and surface properties were improved [12,24].

However, there are still a few problems that exist in the LDED fabricated TiC-TMC coatings such as the reinforcement aggregation and the generation of fabrication defects. Due to the high cooling rate of LDED, TiC powders were always partially melted and dissolved into liquid titanium during the fabrication [22]. These solid TiC particles resulted in the problem of the inhomogeneous distribution of unmelted reinforcement particles. In addition, the existence of solid TiC particles in the molten pool caused the lack of fusion, which was the major reason of the generation of fabrication defects. Both problems could reduce the microhardness and wear resistance of TiC-TMC coatings fabricated by the LDED process [13]. Utilizing ultrasonic vibration-assisted LDED to fabricated TiC-TMC coatings could be a possible solution to solve the existing problems. Ultrasonic vibration had been widely utilized in the LDED process to fabricate alloys and metal matrix composites. Cong et al. pointed out that the assistance of ultrasonic vibration could reduce the grain size and porosity, increase the size of the molten pool and improve the Rockwell hardness [25]. Wang et al. found the assistance of ultrasonic vibration could refine the Laves phase in LDED fabricated Inconel 718 parts, which improved the microhardness and wear resistance [26]. Li et al. investigated the effects of ultrasonic vibration on LDED fabricated Ni/WC/La$_2$O$_3$ coatings [27]. The assistance of ultrasonic vibration could disrupt the dendrites, refine the grain size and improve the hardness and wear resistance. The reinforcement size and mechanical properties of TiB reinforced Ti matrix composites were improved by the assistance of ultrasonic vibration, as stated by Ning et al. [28]. Ultrasonic vibration induced two direct actions on liquid materials including acoustic streaming and transient cavitation. The acoustic streaming was a steady flow in the fluid materials driven by the absorption of acoustic oscillations. Such actions could mix and stir the liquid materials in the molten pool, which could homogenize the distribution of reinforcement [25,26]. The transient cavitation was the dynamic process of growth and collapse of microbubbles in liquid materials, which promoted the dissolution of solid particles [29]. In addition, ultrasonic vibration could provide extra energy to the molten pool and promote the melting of powder materials [25].

In this study, the TiC-TMC coatings were successfully coated onto titanium by the ultrasonic vibration-assisted LDED process. The effects of ultrasonic vibration and TiC content on the phase compositions, reinforcement aggregation, bonding quality, fabrication defects, microhardness and wear resistance of TiC-TMC coatings were investigated.

2. Materials and Methods

2.1. Powder Materials and Treatment

The powder materials used in this study were TiC powder (99.7% purity) and Ti powder (99.9% purity) (Atlantic Equipment Engineers Inc., Upper Saddle River, NJ, USA). A pure Ti plate with a thickness of 6.65 mm was used to coat the substrate.

According to the results of the preliminary experiments, three feedstock material powders (Ti, Ti + 5 wt.% TiC and Ti + 10 wt.% TiC) were adopted to study the effects of TiC content. As shown in Figure 1, before the LDED process TiC powder and Ti powder were mixed by a ball milling machine (ND2L, Torrey Hills Technologies LLC., San Diego, CA, USA). The weight ratio of the milling balls to powders was 1:1. The milling time was 4 h with a consistent rotation speed of 200 rpm. The TiC powders were partially embedded on the surface of the Ti powders after the ball milling process.

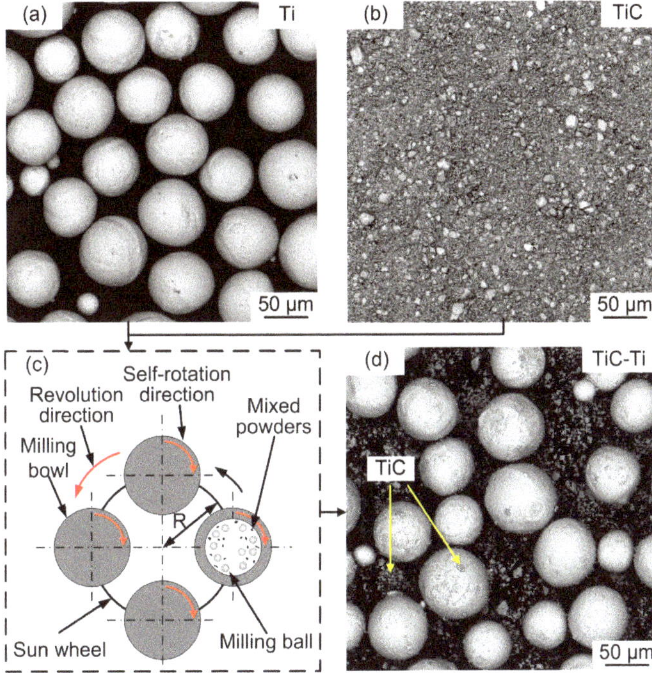

Figure 1. Powder pretreatment: (**a**) Ti powder; (**b**) TiC powder; (**c**) planetary ball milling process; (**d**) prepared TiC-Ti powder.

2.2. Experimental Setup

Experiments were conducted on an LDED system (LENS 450, Optomec Inc., Albuquerque, NM, USA). Figure 2 shows the experimental setup of the ultrasonic vibration-assisted LDED system. To avoid the oxidation of Ti at a high temperature, the chamber system was purged by argon gas until the oxygen level was lower than 50 ppm. Inside the chamber, a ceramic vibrator with a frequency of 29 kHz was fixed under the Ti substrate to provide ultrasonic vibration. A laser system equipped with a 400 W fiber laser source (YLM-1070, IPG Photonics, Oxford, MA, USA) was used to generate the laser beam. The movement of the substrate and the cladding head were controlled by the control system to build the designed 3D structures. During the fabrication, the laser beam was transmitted to the surface of the substrate to generate a molten pool, which caught and melted the material powders. When the laser beam moved away, the molten pool was solidified to fabricate the first layer. After the fabrication of the first layer, the cladding head moved up the distance of the layer thickness. The second layer was fabricated on top of the first layer. The designed coatings were deposited layer by layer. The laser coating parameters in this study are listed in Table 1.

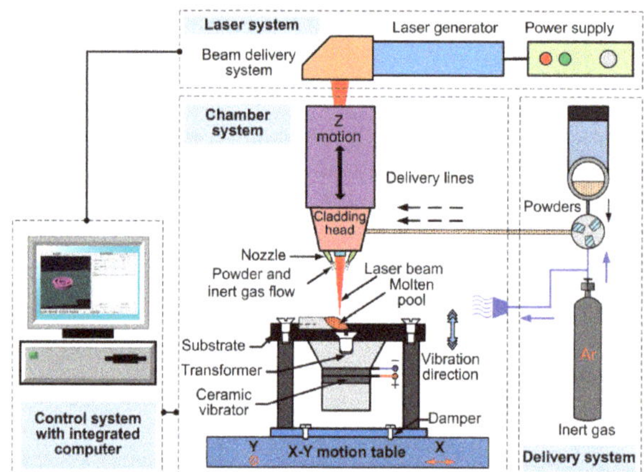

Figure 2. Experimental setup.

Table 1. Laser coating parameters.

Input Fabrication Variables	Value
Laser power (W)	375
Beam diameter of laser (μm)	400
Wavelength of laser (μm)	1.07
Deposit head scanning speed (mm/min)	11
Hatch distance (μm)	380
Layer thickness (μm)	432
Powder feeding rate (g/min)	2.5
Number of layers	3
Oxygen level (ppm)	<50
Argon gas flow rate (L/min)	6
Scanning orientation (°)	45, alternate 90 per layer

2.3. Measurement Procedures

After fabrication, the deposited coating layers were ground and polished (perpendicular to the deposition direction) by a grinder/polisher machine (MetaServ 250, Buehler, Lake Bluff, IL, USA). The whole cross-sectional surfaces and bonding quality were observed by an optical microscope (OM) (DSX-510, OLYMPUS, Tokyo, Japan). ImageJ software (1.8.0_172, LOCI, University of Wisconsin, Madison, WI, USA) was used to analyze the observed images under the mode of black and white [30]. The morphologies of the powders and the microstructure of the fabricated coatings were observed by scanning electron microscopy (SEM) (Phenom Pharos, Nanoscience, Phoenix, AZ, USA), which was equipped with a backscatter electron detector (BSD) system and an energy dispersive X-ray spectroscopy (EDS) system. The element compositions and phases were analyzed by an EDS and X-ray diffraction (XRD) machine (Ultima III, Rigaku Corp., Woodlands, TX, USA), respectively. In the XRD, the samples were scanned from 20 to 80 degrees (2θ) with a scanning step of 0.02 degrees (2θ), a wavelength of 0.154 nm, a voltage of 40 kV and a current of 44 mA. The weight percentages of each phase were calculated by MDI/JADE software (Version 2020, Materials Data, Livermore, CA, USA).

The microhardness of the deposited coating layers was tested by a Vickers microhardness tester (Phase II, Upper Saddle River, Bergen, NJ, USA) with a 10 N normal load and a 10 s dwell time. For each combination of inputs, two samples fabricated by the LDED process were tested. For each sample, the microhardness values were measured on ten random positions of the cross-sectional surface. The wear rate was tested and measured by dry sliding tests with a 1 mm radium silicon carbide (SiC) ball at room temperature using a mechanical testing system (PB1000, Nanovea, Irvine, CA, USA). During the dry sliding test, the SiC ball slid on the surface of the coating for 0.25 h with a load of 0.2 N, a constant sliding speed of 3 mm/s and a sliding distance of 3 mm. The wear volume lost, V, was calculated by Equation (1) [26].

$$V = L \times [(\pi R^2)/180 \times \arcsin(W/2R) - W/2 \times (R^2 - (W/2)^2)^{0.5}] \tag{1}$$

where L was the sliding distance, mm; R was the radius of the sliding ball, mm and W was the scratching width, mm. The wear rate W_r was calculated by Equation (2).

$$W_r = V/(F(vT)) \tag{2}$$

where F was the normal load used in the dry sliding test, N; v was the sliding speed, mm/s and T was the time of the dry sliding test, s.

3. Results and Discussion

3.1. Phase Compositions

The XRD results on the phase compositions are shown in Figure 3. The peaks were fitted and identified according to the information in the powder diffraction file (PDF) cards. Both TiC(O) and non-stoichiometric Ti_xC_y(P) had a cubic lattice structure [9,21]. The lattice parameter a could be calculated by Bragg's law, as shown in Equation (3) [31]:

$$1/(d_{hkl})^2 = (h^2 + k^2 + l^2)/a^2 \tag{3}$$

where d_{hkl} was the lattice spacing, which could be calculated by the location of the peak (Degree 2-theta) and h, k and l were the Miller indices of the Bragg plane, which could be found in PDF cards. The lattice parameter of TiC(O) was 4.337 Å. As a comparison, the lattice parameter of Ti_xC_y(P) was 4.272 Å, which was lower than that of TiC(O).

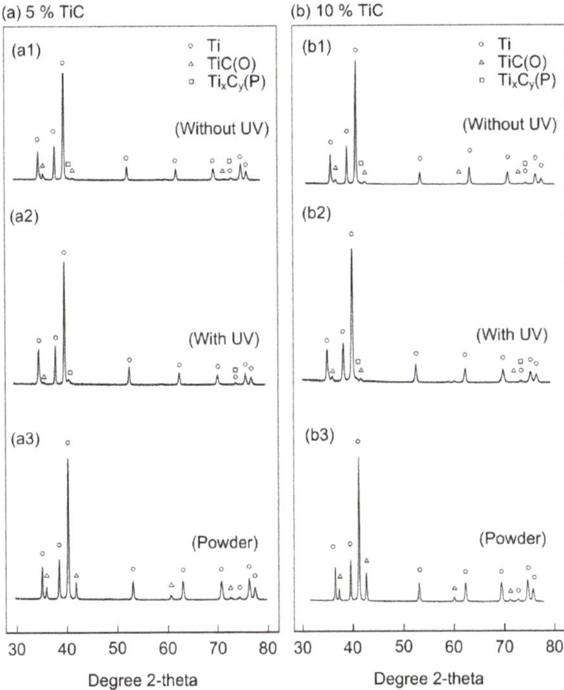

Figure 3. The effects of ultrasonic vibration and TiC content on the phase compositions of TiC-TMC coatings: (**a**) 5% TiC; (**b**) 10% TiC.

The TiC(O) was the phase of original feedstock TiC powders in which the atomic ratio of Ti and C was 1:1. The $Ti_xC_y(P)$ phase was non-stoichiometric TiC, which was precipitated from the TiC-Ti solutions during the fabrication [9,32]. Its atomic ratio of Ti and C could be changed from 1:0.55 to 1:1. The phase compositions of the feedstock powders and the LDED fabricated TiC-TMC coatings with 5% TiC and 10% TiC are shown in Tables 2 and 3, respectively. As shown in Figure 3a, at the low level of TiC content with the assistance of ultrasonic vibration, the area of TiC(O) peaks was significantly decreased. The detailed phase compositions results showed that the content of the TiC(O) phase in the TiC-TMC coatings decreased from 2.94 wt.% to 0.72 wt.%. In addition, the content of the $Ti_xC_y(P)$ phase increased from 4.36 wt.% to 6.98 wt.%. The reason for the phase composition changes was that ultrasonic vibration could promote the dissolution and precipitation process during the fabrication. On one hand, the acoustic streaming could mix and stir the liquid materials inside the molten pool, which could enhance the movement of both solvent and solute [22,33]. In addition, the transient cavitation induced by ultrasonic vibration could increase the diffusion rate between the TiC(O) particles and the liquid titanium [34]. Due to these two actions, more solid TiC(O) particles could be directly dissolved in liquid Ti. On the other hand, the additional energy was induced to the molten pool by ultrasonic vibration, which increased the temperature of the liquid materials [15]. More TiC(O) particles could be melted at a high temperature and then dissolved into the liquid titanium. During the solidification, the solubility of the TiC in the titanium was reduced. There were more $Ti_xC_y(P)$ phases precipitated from the TiC-Ti solution.

Table 2. Phase compositions of the feedstock powder and the TiC-TMC coatings with 5% TiC.

Conditions	TiC(O) (wt.%)	$Ti_xC_y(P)$ (wt.%)	Ti (wt.%)
Feedstock powder	5.24	0	balance
TiC-TMC coatings without UV	2.94	4.36	balance
TiC-TMC coatings with UV	0.72	6.98	balance

Table 3. Phase compositions of the feedstock powder and the TiC-TMC coatings with 10% TiC.

Conditions	TiC(O) (wt.%)	$Ti_xC_y(P)$ (wt.%)	Ti (wt.%)
Feedstock powder	10.71	0	balance
TiC-TMC coatings without UV	6.89	5.87	balance
TiC-TMC coatings with UV	5.12	8.48	balance

The XRD results of the feedstock powders and the fabricated TiC-TMC coatings with a high level of TiC content are shown in Figure 3b. It could be observed that the peaks pf of TiC(O) and $Ti_xC_y(P)$ had slight changes. The detailed phase compositions are shown in Table 3. Similar to the conditions with a lower TiC content, with the assistance of ultrasonic vibration, the content of TiC(O) decreased and the content of $Ti_xC_y(P)$ increased. However, the TiC(O) content in the TiC-TMC coatings was slightly reduced in comparison with that in the coatings with a lower TiC content. It meant that the effects of ultrasonic vibration on the phase transformation (from the TiC(O) phase to the $Ti_xC_y(P)$ phase) were suppressed. The higher content of TiC had two major effects. First, feedstock powders need more energy to melt due to the high melting point of TiC. At a relatively high TiC content of 10 wt.%, the TiC(O) particles were harder to melt and then be dissolved into liquid Ti, which prevented the precipitation of $Ti_xC_y(P)$ particles. Second, the increase of solid TiC(O) particles reduced the fluidity of the molten pool. The actions of ultrasonic vibration on the liquid materials were suppressed, which further prevented the dissolution and precipitation process.

3.2. Microstructure

Figure 4 shows the element compositions analyzed by EDS on the cross-section of the TiC-TMC coatings fabricated by the LDED process. There were three different kinds of regions (black, grey and light). In these three different regions, the element compositions were analyzed (as shown in Figure 4b–d). It could be seen that the irregular-shaped black regions had an atomic ratio of titanium to carbon of 1:1 (the same as the atomic ratio of titanium to carbon in the TiC(O) phase). Associated with the XRD analysis, it could be considered that the black regions were the TiC(O) phase. The grey regions had 36 at.% of the C element, which could be considered as $Ti_xC_y(P)$. The major reason was that the composition range of $Ti_xC_y(P)$ was extraordinarily wide according to the Ti-C phase diagram and its C element content was lower than that in the TiC(O) (50 at.%) [12]. Figure 4a shows that the size of the individual spherical-shaped grey regions was smaller than the size of the feedstock TiC powders, indicating the grey regions should be generated during the solidification. Similar phenomena have also been reported in other investigations on the fabrication of TiC-TMC parts [13,33]. Beside the independently distributed grey regions, there were also grey regions surrounded by the boundary of black regions. Due to the high cooling rate in the LDED process, the TiC particles with larger sizes could not be fully dissolved into the liquid titanium. The dissolution of TiC(O) into titanium took place at the boundary of the solid TiC(O) particles, which formed the TiC-Ti solution around the undissolved TiC(O) cores. During the solidification, the solubility of TiC in the TiC-Ti solution decreased, resulting in the interfacial reaction product of $Ti_xC_y(P)$. A similar phenomenon was also reported in the investigations of sintered TiC-TMC materials [35–37]. The light regions had 95.5 at.% of the Ti element, which indicated that these regions were the titanium matrix.

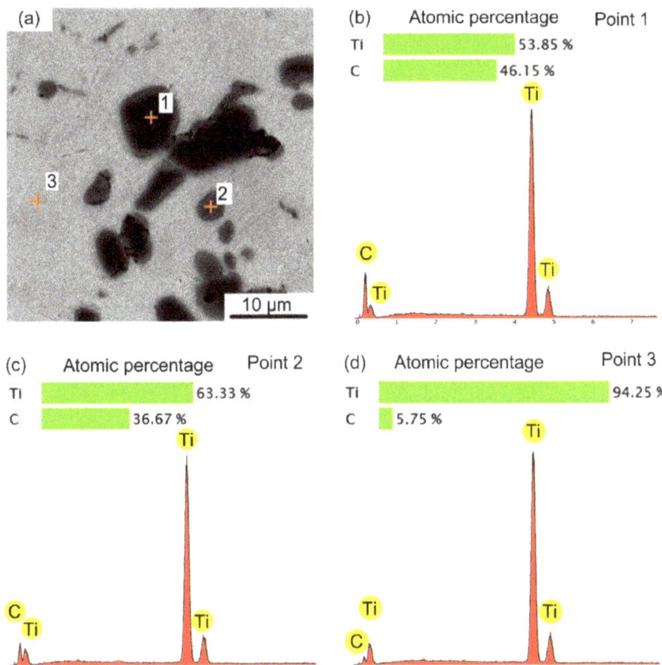

Figure 4. The EDS results of a laser DED fabricated TiC-TMC coating: (**a**) the microstructure morphology of the TiC-TMC coatings; (**b–d**) the element compositions of different regions.

The effects of ultrasonic vibration and TiC content on the microstructure are shown in Figure 5. Figure 5b,c,f,g show the cross-sectional OM images of the coatings fabricated by LDED. It could be observed that the reinforcements aggregated with the laser direction. Figure 5a,d,e,h show the enlarged view of SEM images of the aggregated regions of the reinforcements. It could be seen that the grey particles with a smaller size were relatively evenly distributed in the Ti matrix. As a comparison, the black particles mainly caused the aggregation. As discussed in Section 3.2, the small-sized grey particles were in the Ti$_x$C$_y$(P) phase and the large-sized black particles were in the TiC(O) phase. According to the SEM images, it could be confirmed that the reinforcement aggregation was mainly caused by the TiC(O) particles. It can be seen in Figure 5e–h that with the assistance of ultrasonic vibration, there were less TiC(O) particles aggregated. As discussed in Section 3.1, with the assistance of ultrasonic vibration, the amount and size of the TiC(O) particles in the TiC-TMC coatings could be significantly decreased. The reduction of unmelted and undissolved particles could increase the fluidity of the liquid materials in the molten pool. Similar phenomena were also reported in the laser melting of TiC-Al composites [38]. These solid TiC(O) particles recirculated in the molten pool faster, which improved the distribution of the undissolved TiC(O). Both the reduction of the TiC(O) particles and the better distribution of TiC(O) could significantly release the reinforcement aggregation.

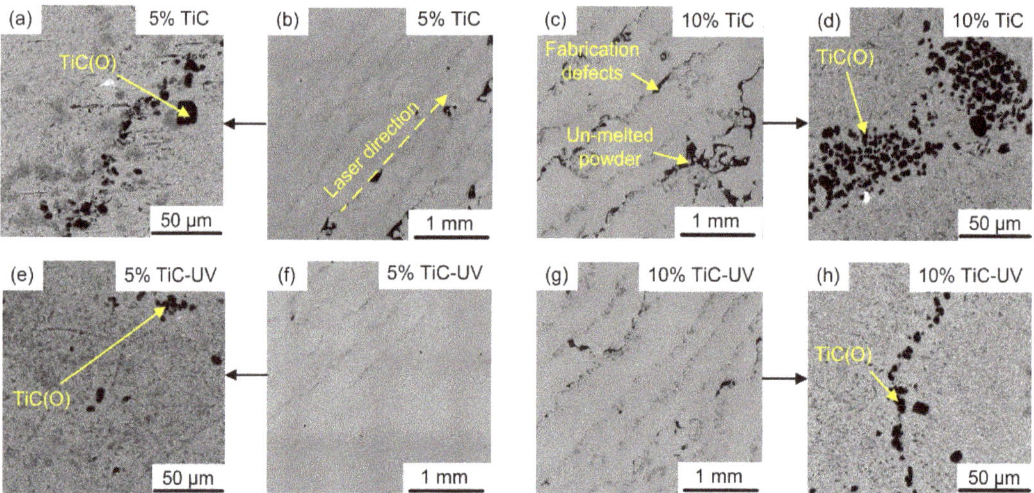

Figure 5. The microstructure of a laser DED fabricated TiC-TMC coating: (**a,d,e,h**) the SEM images; (**b,c,f,g**) the OM images.

As shown in Figure 5c,g, with the increase of TiC content, more TiC(O) particles aggregated in the LDED fabricated coatings. As discussed in Section 3.1, the TiC-TMC coatings with a higher content of TiC had more TiC(O) particles. The larger amount of solid TiC(O) particles could reduce the fluidity of the molten pool, which suppressed the movement of solid TiC(O) particles [39]. In addition, the specific heat capacity and laser absorptivity of TiC and Ti were different. The increase of the TiC content could enlarge the difference of the temperature and solidification rates inside the molten pool [40]. In the regions with a low temperature, liquid materials solidified faster. It could also suppress the movement of solid TiC(O) particles and resulted in the variation of the distribution of TiC(O) particles.

3.3. Bonding Quality

The molten pool of TiC-TMC coatings generated in the laser DED process with and without ultrasonic vibration are shown in Figure 6a. With the assistance of ultrasonic vibration, the width and depth of the molten pool became larger. The major reason was that the actions of acoustic streaming could stir the liquid material in the molten pool. For the Gaussian laser used in this study, the energy at the center of the laser beam was much higher than that at the boundary, leading to uneven heat density in the molten pool [41,42]. The actions of mixing and stirring promoted the dispersion of the high-temperature liquid from the center to the boundary of the molten pool. The temperature at the boundary then increased, which promoted the melting of the substrate materials [43]. In addition, ultrasonic vibration provided extra energy to the molten pool and increased the temperature of the liquid materials. A greater number of substrate materials could be melted at a higher temperature.

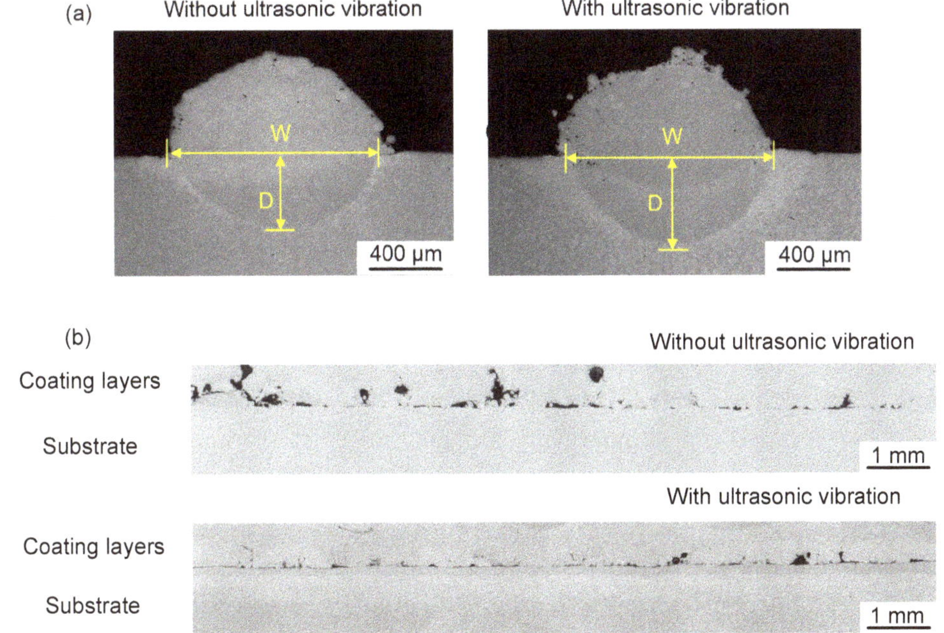

Figure 6. The effects of ultrasonic vibration on bonding quality: (**a**) the molten pool of the TiC-TMC coatings; (**b**) the bonding interface of the TiC-TMC coatings.

Figure 6b shows the bonding interface of the TiC-TMC coatings. The defects in the bonding regions were mainly caused by the insufficient overlap between adjacent layers, which could be significantly reduced by the assistance of ultrasonic vibration. The major reason was that ultrasonic vibration increased the depth of the molten pool on the substrate. The metallic bonding between the substrate materials and coating layers could be significantly improved. In addition, as discussed in Section 3.3, the fluidity of the liquid materials in the molten pool was improved by ultrasonic vibration. The higher fluidity increased the powder absorbability of the molten pool, which was also helpful for generating sufficient overlaps. A similar result was reported in the fabrication of zirconia-alumina ceramics using an LDED process [44].

3.4. Fabrication Defects

The fabrication defects of the LDED fabricated TiC-TMC coatings on a cross-sectional surface are shown in Figure 7. It could be seen that most fabrication defects were in irregular shapes. In the LDED process, the irregular-shaped fabrication defects on the cross-sectional surface were usually caused by lack of fusion, as demonstrated by Zhang et al. [45]. As the molten pool was enlarged by the assistance of ultrasonic vibration, the lack of fusion at the boundary of the molten pool was reduced. In addition, more powder could be caught by the molten pool during the fabrication, promoting the formation of sufficient overlaps.

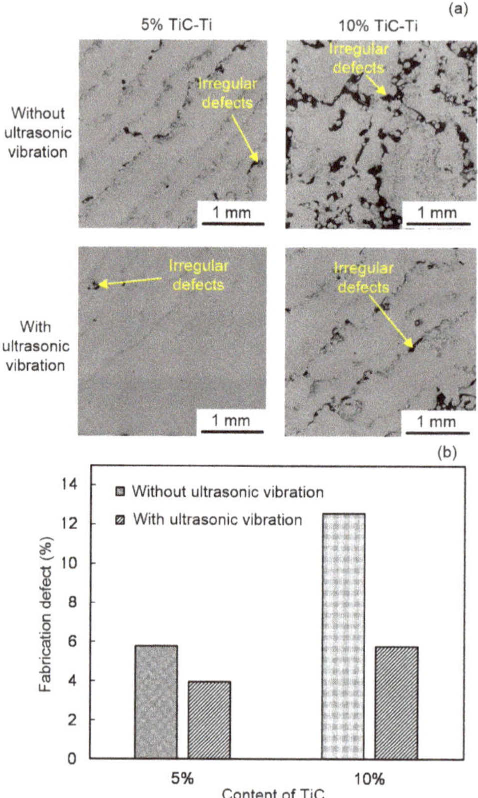

Figure 7. The effects of ultrasonic vibration on the fabrication defects: (**a**) the molten pool of the TiC-TMC coatings; (**b**) the bonding interface of the TiC-TMC coatings.

With the increased TiC content, the amount of irregular fabrication defects increased due to the lack of fusion. In this study, the laser input energy was constant. The TiC needed to absorb more energy to be melted than titanium. The TiC-TMC coatings with a higher TiC content had more unmelted TiC(O) particles in the molten pool during the fabrication, which aggravated the lack of fusion. Moreover, the large number of solid TiC(O) particles reduced the fluidity of the liquid materials in the molten pool resulting in the aggregation of unmelted TiC(O) particles. In these unmelted TiC(O)-rich regions, a lack of fusion was more likely to happen, as reported in the LDED fabricated Ti_6Al_4V with trace boron and the selected laser melting of a TiB_2 coating on Ti_6Al_4V [46,47].

3.5. Mechanical Properties

3.5.1. Microhardness

Figure 8 shows the effects of ultrasonic vibration and TiC content on microhardness. With the assistance of ultrasonic vibration, the microhardness value increased. As discussed in Section 3.1, more refined $Ti_xC_y(P)$ particles were precipitated in the titanium matrix with the assistance of ultrasonic vibration. These refined reinforcements could evenly bear the load and increase the resistance of plastic deformation during the microhardness tests. A similar result was demonstrated by Shen et al. through numerical methods [48]. Moreover, as discussed in Section 3.4, with the assistance of ultrasonic vibration, the fabrication defects decreased significantly. The higher density increased the ability to support the load, which could also increase the microhardness. Compared with commercial pure titanium coating layers, TiC-TMC coatings had a larger microhardness. The major reason was that TiC could support the load and reduce the deformation. However, the excessive TiC content led to the reduction of microhardness. As discussed before, the major reason was that the increase of fabrication defects reduced the microhardness.

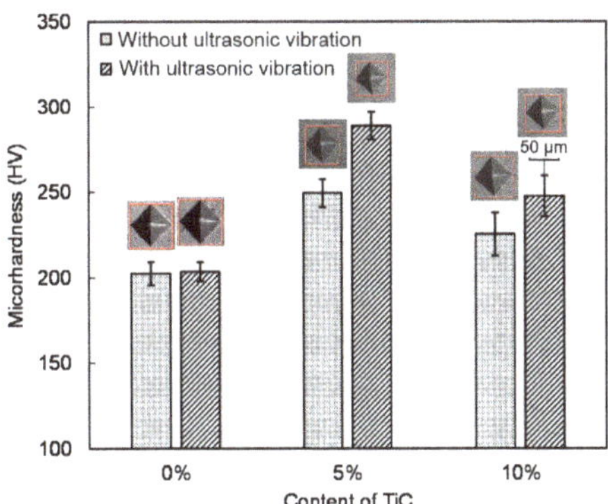

Figure 8. The effects of ultrasonic vibration and TiC content on microhardness.

3.5.2. Wear Resistance

The sliding width of the material removal trail and wear rate are shown in Figure 9. Both the sliding width and wear rate had a negative relationship with the wear resistance. With the assistance of ultrasonic vibration, the wear resistance increased. The significant reduction of the fabrication defects provided a smoother interface and reduced the friction coefficient of the coating layers [49]. Under the same test condition, the friction force was reduced, resulting in a smaller material removal volume and higher wear resistance. In addition, the reinforcement aggregation was reduced by ultrasonic vibration. It contributed to the formation of the uniform anti-wear protective layer during the dry sliding tests to further increase the wear resistance.

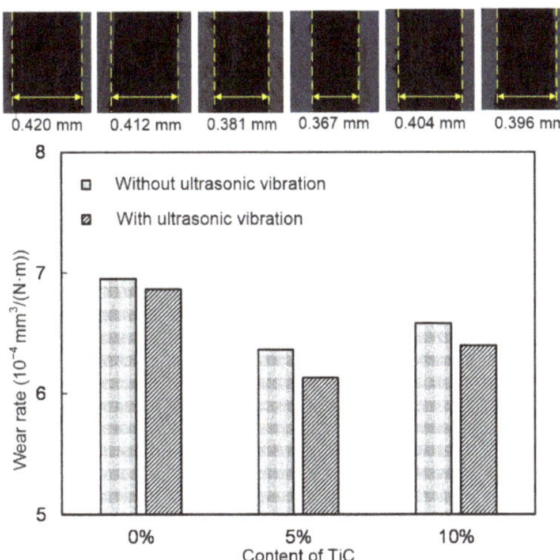

Figure 9. The effects of ultrasonic vibration and TiC content on microhardness.

Compared with CP-Ti, TiC-TMC coatings showed a higher wear resistance. In addition, the friction force was measured during the dry sliding tests. The average friction coefficients of each combination of input were calculated and are listed in Table 4. The major reason was that the hardness of TiC was extremely high, which increased the hardness of the TiC-TMC coatings. The adhesive wear mode was changed to abrasive wear, which could significantly increase the wear resistance [50,51]. In addition, during the dry sliding tests, the SiC ball would leave a rough worn surface on the samples. The spherical TiC particles might have the function of self-lubrication [52,53]. With the increase of the TiC content to a high level (10 wt.%), the wear resistance was reduced. The major reason was that the greater number of fabrication defects increased the friction coefficient.

Table 4. The average friction coefficient of the CP-Ti and TiC-TMC coatings.

Conditions	Friction Coefficient
CP-Ti without UV	0.322
CP-Ti with UV	0.324
5 wt.% TiC-TMC coatings without UV	0.201
5 wt.% TiC-TMC coatings with UV	0.129
10 wt.% TiC-TMC coatings without UV	0.296
10 wt.% TiC-TMC coatings with UV	0.268

4. Conclusions

In this study, TiC-TMC coatings with different TiC contents were fabricated by an ultrasonic vibration-assisted LDED process. The effects of ultrasonic vibration and TiC content on the phase composition, TiC aggregation, bonding quality, fabrication defects and mechanical properties were investigated.

With the assistance of ultrasonic vibration, the liquid materials were mixed and stirred by the actions of acoustic streaming and transient cavitation in liquid material solidification, which promoted the dissolution of TiC(O) and the precipitation of refined $Ti_xC_y(P)$. In the TiC-TMC coatings fabricated by LDED without ultrasonic vibration, TiC(O) particles preferred to aggregate in a titanium matrix. In contrast, the process with ultrasonic vibration could significantly reduce the TiC(O) aggregation by decreasing the amount of TiC(O) and improving the distribution. In addition, extra heat energy was generated by the ultrasonic vibration assistant, which could increase the bonding quality and reduce the fabrication defects. The improved TiC(O) aggregation and fabrication defects were effective in enhancing the microhardness and wear resistance of the ultrasonic vibration-assisted LDED fabricated TiC-TMC coatings.

With the increase of TiC content, the phase transformation from the TiC(O) phase to the $Ti_xC_y(P)$ phase was suppressed. The major reason was that the higher TiC content suppressed the further dissolution of TiC(O) in liquid titanium. At a high content of TiC, there were more undissolved TiC(O) particles existing in the TiC-TMC coatings after the fabrication, which aggravated the reinforcement aggregation. In addition, the higher TiC content exacerbated the generation of a lack of fusion defects as the TiC powders needed more energy to be melted than Ti. The generation of both the TiC(O) aggregation and fabrication defects decreased the mechanical properties of microhardness and wear resistance of the TiC-TMC coatings with a higher TiC content.

Author Contributions: Data curation, Y.L. and D.Z.; Formal analysis, Y.L.; Investigation, Y.L.; Methodology, Y.L.; Resources, Y.L.; writing—original draft, Y.L.; Writing—review & editing, Y.L., D.Z., H.W. and W.C. All authors have read and agreed to the published version of the manuscript.

Funding: This research received no external funding.

Institutional Review Board Statement: Not applicable.

Informed Consent Statement: Not applicable.

Data Availability Statement: Not applicable.

Conflicts of Interest: The authors declare no conflict of interest.

References

1. Peters, M.; Kumpfert, J.; Ward, C.H.; Leyens, C. Titanium Alloys for Aerospace Applications. *Adv. Eng. Mater.* **2003**, *5*, 419–427. [CrossRef]
2. Hartman, A.D.; Gerdemann, S.J.; Hansen, J.S. Producing lower-cost titanium for automotive applications. *JOM* **1998**, *50*, 16–19. [CrossRef]
3. Attar, H.; Bönisch, M.; Calin, M.; Zhang, L.C.; Scudino, S.; Eckert, J. Selective laser melting of in situ titanium-titanium boride composites: Processing, microstructure and mechanical properties. *Acta Mater.* **2014**, *76*, 13–22. [CrossRef]
4. Guo, C.; Zhou, J.; Chen, J.; Zhao, J.; Yu, Y.; Zhou, H. Improvement of the oxidation and wear resistance of pure Ti by laser cladding at elevated temperature. *Surf. Coat. Technol.* **2010**, *205*, 2142–2151. [CrossRef]
5. Dhanda, M.; Haldar, B.; Saha, P. Development and characterization of hard and wear resistant MMC coating on Ti-6Al-4V substrate by laser cladding. *Procedia Mater. Sci.* **2014**, *6*, 1226–1232. [CrossRef]
6. Cui, Z.; Yang, H.; Wang, W.; Wu, H.; Xu, B. Laser cladding Al-Si/Al_2O_3-TiO_2 composite coatings on AZ31B magnesium alloy. *J. Wuhan Univ. Technol. Mater. Sci. Ed.* **2012**, *27*, 1042–1047. [CrossRef]
7. Song, S.X.; Ai, X.; Zhao, J.; Huang, C.Z. Al_2O_3/Ti($C_{0.3}N_{0.7}$) cutting tool material. *Mater. Sci. Eng. A* **2003**, *356*, 43–47. [CrossRef]
8. Selamat, M.; Watson, L.; Baker, T. XRD and XPS studies on surface MMC layer of SiC reinforced Ti-6Al-4V alloy. *J. Mater. Process. Technol.* **2003**, *142*, 725–737. [CrossRef]
9. Yoon, C.S.; Kang, S.; Kim, D.Y. Dissolution and reprecipitation behavior of TiC-TiN-Ni cermets during liquid-phase sintering. *Korean J. Ceram.* **1997**, *3*, 124–128.
10. Fan, Z.; Miodownik, A.; Chandrasekaran, L.; Ward-Close, M. The Young's moduli of in situ Ti/TiB composites obtained by rapid solidification processing. *J. Mater. Sci.* **1994**, *29*, 1127–1134. [CrossRef]
11. Yuan, P.; Gu, D. Molten pool behaviour and its physical mechanism during selective laser melting of TiC/AlSi10Mg nanocomposites: Simulation and experiments. *J. Phys. D Appl. Phys.* **2015**, *48*, 035303. [CrossRef]
12. Liu, W.; Dupont, J.N. Fabrication of functionally graded TiC/Ti composites by laser engineered net shaping. *Scr. Mater.* **2003**, *48*, 1337–1342. [CrossRef]
13. Rambo, C.R.; Travitzky, N.; Zimmermann, K.; Greil, P. Synthesis of TiC/Ti–Cu composites by pressureless reactive infiltration of TiCu alloy into carbon preforms fabricated by 3D-printing. *Mater. Lett.* **2005**, *59*, 1028–1031. [CrossRef]
14. Man, H.C.; Zhang, S.; Cheng, F.; Yue, T. Microstructure and formation mechanism of in situ synthesized TiC/Ti surface MMC on Ti-6Al-4V by laser cladding. *Scr. Mater.* **2001**, *44*, 2801. [CrossRef]
15. Gu, D.; Meng, G.; Li, C.; Meiners, W.; Poprawe, R. Selective laser melting of TiC/Ti bulk nanocomposites: Influence of nanoscale reinforcement. *Scr. Mater.* **2012**, *67*, 185–188. [CrossRef]
16. Joseph, J. Direct laser fabrication of compositionally complex materials: Challenges and prospects. *Addit. Manuf. Appl. Met. Compos.* **2020**, 147–163. [CrossRef]
17. Hu, Y.; Cong, W. A review on laser deposition-additive manufacturing of ceramics and ceramic reinforced metal matrix composites. *Ceram. Int.* **2018**, *44*, 20599–20612. [CrossRef]
18. Farshidianfar, M.H.; Khajepour, A.; Gerlich, A.P. Effect of real-time cooling rate on microstructure in laser additive manufacturing. *J. Mater. Process. Technol.* **2016**, *231*, 468–478. [CrossRef]
19. Pei, Y.; Zuo, T. Gradient microstructure in laser clad TiC-reinforced Ni-alloy composite coating. *Mater. Sci. Eng. A* **1998**, *241*, 259–263. [CrossRef]
20. Yang, S.; Liu, W.; Zhong, M.; Wang, Z. TiC reinforced composite coating produced by powder feeding laser cladding. *Mater. Lett.* **2004**, *58*, 2958–2962. [CrossRef]
21. Kathuria, Y. Nd-YAG laser cladding of Cr_3C_2 and TiC cermets. *Surf. Coat. Technol.* **2001**, *140*, 195–199. [CrossRef]
22. Wu, W. Dissolution precipitation mechanism of TiC/Ti composite layer produced by laser cladding. *Mater. Sci. Technol.* **2010**, *26*, 367–370. [CrossRef]
23. Ding, H.; Liu, X.; Yu, L.; Zhao, G. The influence of forming processes on the distribution and morphologies of TiC in Al-Ti-C master alloys. *Scr. Mater.* **2007**, *57*, 575–578. [CrossRef]
24. Wu, X. In situ formation by laser cladding of a TiC composite coating with a gradient distribution. *Surf. Coat. Technol.* **1999**, *115*, 111–115. [CrossRef]
25. Cong, W.; Ning, F. A fundamental investigation on ultrasonic vibration-assisted laser engineered net shaping of stainless steel. *Int. J. Mach. Tools Manuf.* **2017**, *121*, 61–69. [CrossRef]
26. Wang, H.; Hu, Y.; Ning, F.; Cong, W. Ultrasonic vibration-assisted laser engineered net shaping of Inconel 718 parts: Effects of ultrasonic frequency on microstructural and mechanical properties. *J. Mater. Process. Technol.* **2020**, *276*, 468–478. [CrossRef]
27. Li, M.; Zhang, Q.; Han, B.; Song, L.; Cui, G.; Yang, J.; Li, J. Microstructure and property of Ni/WC/La2O3 coatings by ultrasonic vibration-assisted laser cladding treatment. *Opt. Lasers Eng.* **2020**, *125*, 105848. [CrossRef]
28. Ning, F.; Hu, Y.; Cong, W. Microstructure and mechanical property of TiB reinforced Ti matrix composites fabricated by ultrasonic vibration-assisted laser engineered net shaping. *Rapid Prototyp. J.* **2019**, *25*, 581–591. [CrossRef]
29. Noltingk, B.E.; Neppiras, E.A. Cavitation produced by ultrasonics. *Proc. Phys. Soc. Sect. B* **1950**, *63*, 674. [CrossRef]
30. Schneider, C.A.; Rasband, W.S.; Eliceiri, K.W. NIH Image to ImageJ: 25 years of image analysis. *Nat. Methods* **2012**, *9*, 671–675. [CrossRef]

31. Kacher, J.; Landon, C.; Adams, B.L.; Fullwood, D. Bragg's Law diffraction simulations for electron backscatter diffraction analysis. *Ultramicroscopy* **2009**, *109*, 1148–1156. [CrossRef]
32. Ya, B.; Zhou, B.; Yang, H.; Huang, B.; Jia, F.; Zhang, X. Microstructure and mechanical properties of in situ casting TiC/Ti6Al4V composites through adding multi-walled carbon nanotubes. *J. Alloys Compd.* **2015**, *637*, 456–460. [CrossRef]
33. Zhang, X.; Song, F.; Wei, Z.; Yang, W.; Dai, Z. Microstructural and mechanical characterization of in-situ TiC/Ti titanium matrix composites fabricated by graphene/Ti sintering reaction. *Mater. Sci. Eng. A* **2017**, *705*, 153–159. [CrossRef]
34. Komarov, S.V.; Kuwabara, M.; Abramov, O.V. High power ultrasonics in pyrometallurgy: Current status and recent development. *ISIJ Int.* **2005**, *45*, 1765–1782. [CrossRef]
35. Chrysanthou, A.; Chen, Y.; Vijayan, A.; O'sullivan, J. Combustion synthesis and subsequent sintering of titanium-matrix composites. *J. Mater. Sci.* **2003**, *38*, 2073–2077. [CrossRef]
36. Ranganath, S.; Subrahmanyam, J. On the in situ formation of tic and ti2c reinforcements in combustion-assisted synthesis of titanium matrix composites. *Met. Mater. Trans. A* **1996**, *27*, 237–240. [CrossRef]
37. Wanjara, P.; Drew, R.; Root, J.; Yue, S. Evidence for stable stoichiometric Ti2C at the interface in TiC particulate reinforced Ti alloy composites. *Acta Mater.* **2000**, *48*, 1443–1450. [CrossRef]
38. Yuan, P.; Gu, D.; Dai, D. Particulate migration behavior and its mechanism during selective laser melting of TiC reinforced Al matrix nanocomposites. *Mater. Des.* **2015**, *82*, 46–55. [CrossRef]
39. Li, Y.; Hu, Y.; Cong, W.; Zhi, L.; Guo, Z. Additive manufacturing of alumina using laser engineered net shaping: Effects of deposition variables. *Ceram. Int.* **2017**, *43*, 7768–7775. [CrossRef]
40. Nazari, K.A.; Rashid, R.R.; Palanisamy, S.; Xia, K.; Dargusch, M.S. A novel Ti-Fe composite coating deposited using laser cladding of low cost recycled nano-crystalline titanium powder. *Mater. Lett.* **2018**, *229*, 301–304. [CrossRef]
41. Wang, L.; Felicelli, S.; Gooroochurn, Y.; Wang, P.; Horstemeyer, M. Optimization of the LENS® process for steady molten pool size. *Mater. Sci. Eng. A* **2008**, *474*, 148–156. [CrossRef]
42. Khosrofian, J.M.; Garetz, B. Measurement of a Gaussian laser beam diameter through the direct inversion of knife-edge data. *Appl. Opt.* **1983**, *22*, 3406–3410. [CrossRef]
43. Lang, R.J. Ultrasonic atomization of liquids. *J. Acoust. Soc. Am.* **1962**, *34*, 6–8. [CrossRef]
44. Fan, Z.; Zhao, Y.; Tan, Q.; Mo, N.; Zhang, M.-X.; Lu, M.; Huang, H. Nanostructured Al_2O_3-YAG-ZrO_2 ternary eutectic components prepared by laser engineered net shaping. *Acta Mater.* **2019**, *170*, 24–37. [CrossRef]
45. Zhang, B.; Li, Y.; Bai, Q. Defect formation mechanisms in selective laser melting: A review. *Chin. J. Mech. Eng.* **2017**, *30*, 515–527. [CrossRef]
46. Rashid, R.R.; Palanisamy, S.; Attar, H.; Bermingham, M.; Dargusch, M.S. Metallurgical features of direct laser-deposited Ti6Al4V with trace boron. *J. Manuf. Process.* **2018**, *35*, 651–656. [CrossRef]
47. Wang, R.; Gu, D.; Chen, C.; Dai, D.; Ma, C.; Zhang, H. Formation mechanisms of TiB_2 tracks on Ti6Al4V alloy during selective laser melting of ceramic-metal multi-material. *Powder Technol.* **2020**, *367*, 597–607. [CrossRef]
48. Shen, Y.-L.; Williams, J.; Piotrowski, G.; Chawla, N.; Guo, Y. Correlation between tensile and indentation behavior of particle-reinforced metal matrix composites: An experimental and numerical study. *Acta Mater.* **2001**, *49*, 3219–3229. [CrossRef]
49. Liu, P.; Zhang, Y. A theoretical analysis of frictional and defect characteristics of graphene probed by a capped single-walled carbon nanotube. *Carbon* **2011**, *49*, 3687–3697. [CrossRef]
50. Yang, J.; Gu, W.; Pan, L.; Song, K.; Chen, X.; Qiu, T. Friction and wear properties of in situ $(TiB_2 + TiC)/Ti_3SiC_2$ composites. *Wear* **2011**, *271*, 2940–2946. [CrossRef]
51. Li, J.; Yu, Z.; Wang, H. Wear behaviors of an $(TiB + TiC)/Ti$ composite coating fabricated on Ti6Al4V by laser cladding. *Thin Solid Films* **2011**, *519*, 4804–4808. [CrossRef]
52. Deng, J.; Liu, L.; Yang, X.; Liu, J.; Sun, J.; Zhao, J. Self-lubrication of Al_2O_3/TiC/CaF_2 ceramic composites in sliding wear tests and in machining processes. *Mater. Des.* **2007**, *28*, 757–764. [CrossRef]
53. Feng, Y.; Zhang, J.; Wang, L.; Zhang, W.; Tian, Y.; Kong, X. Fabrication techniques and cutting performance of micro-textured self-lubricating ceramic cutting tools by in-situ forming of Al_2O_3-TiC. *Int. J. Refract. Met. Hard Mater.* **2017**, *68*, 121–129. [CrossRef]

metals

MDPI

Article

3D-Printed Connector for Revision Limb Salvage Surgery in Long Bones Previously Using Customized Implants

Jong-Woong Park [1,2], Hyun-Guy Kang [1,2,*], June-Hyuk Kim [1] and Han-Soo Kim [3]

[1]   Orthopaedic Oncology Clinic, National Cancer Center, Goyang 10408, Korea; jwpark82@ncc.re.kr (J.-W.P.); docjune@ncc.re.kr (J.-H.K.)
[2]   Division of Convergence Technology, National Cancer Center, Goyang 10408, Korea
[3]   Department of Orthopaedic Surgery, Seoul National University Hospital, Seoul 03080, Korea; hankim@snu.ac.kr
*   Correspondence: ostumor@ncc.re.kr; Tel.: +82-31-920-1665

Abstract: In orthopedic oncology, revisional surgery due to mechanical failure or local recurrence is not uncommon following limb salvage surgery using an endoprosthesis. However, due to the lack of clinical experience in limb salvage surgery using 3D-printed custom-made implants, there have been no reports of revision limb salvage surgery using a 3D-printed implant. Herein, we present two cases of representative revision limb salvage surgeries that utilized another 3D-printed custom-made implant while retaining the previous 3D-printed custom-made implant. A 3D-printed connector implant was used to connect the previous 3D-printed implant to the proximal ulna of a 40-year-old man and to the femur of a 69-year-old woman. The connector bodies for the two junctions of the previous implant and the remaining host bone were designed for the most functional position or angle by twisting or tilting. Using the previous 3D-printed implant as a taper, the 3D-printed connector was used to encase the outside of the previous implant. The gap between the previous implant and the new one was subsequently filled with bone cement. For both the upper and lower extremities, the 3D-printed connector showed stable reconstruction and excellent functional outcomes (Musculoskeletal Tumor Society scores of 87% and 100%, respectively) in the short-term follow-up. To retain the previous 3D-printed implant during revision limb salvage surgery, an additional 3D-printed implant may be a feasible surgical option.

Keywords: extremity; revision; limb salvage surgery; 3D printing; customized; implant

Citation: Park, J.-W.; Kang, H.-G.; Kim, J.-H.; Kim, H.-S. 3D-Printed Connector for Revision Limb Salvage Surgery in Long Bones Previously Using Customized Implants. *Metals* **2021**, *11*, 707. https://doi.org/10.3390/met11050707

Academic Editor: Atila Ertas

Received: 2 April 2021
Accepted: 22 April 2021
Published: 26 April 2021

Publisher's Note: MDPI stays neutral with regard to jurisdictional claims in published maps and institutional affiliations.

Copyright: © 2021 by the authors. Licensee MDPI, Basel, Switzerland. This article is an open access article distributed under the terms and conditions of the Creative Commons Attribution (CC BY) license (https://creativecommons.org/licenses/by/4.0/).

1. Introduction

Custom-made implants for limb salvage surgery pre-date modular implants; however, their extensive manufacturing duration remains a major drawback [1]. Modular endoprostheses are the most commonly-used implants that have been used to overcome the disadvantages of early custom-made implants [1–3]. They can be used by orthopedic oncology surgeons to provide surgical flexibility, and standardization of the implant pieces facilitates manufacturing. However, there are anatomic sites where a modular endoprosthesis cannot be used. Furthermore, the adjacent normal joint is often sacrificed during reconstruction using the modular implant due to limited bone stock for implant fixation. Recently, three-dimensional (3D)-printing has been introduced for medical applications, wherein custom-made implants can be fabricated in a few days or weeks. These 3D-printed custom-made implants have shown promising short- and mid-term surgical outcomes in orthopedic oncology [4–7].

For limb salvage surgery of the long bones, the modular endoprosthesis system is available for most anatomical locations. However, the 3D-printed custom-made implant has advantages in long bone surgery for preservation of normal adjacent joints, and in limb salvage surgeries for specific anatomical locations without an existing conventional tumor endoprosthesis. When limb salvage surgery with a conventional endoprosthesis including an artificial joint is required, intercalary replacement with joint-saving of the long bone is often possible with a 3D-printed custom-made implant.

Limb salvage surgery with an endoprosthesis is more vulnerable to mechanical failure than implants for arthroplasty without replacement of a major bone defect [8–11]. One of the concerns associated with custom-made implants is the higher probability of mechanical failure as the implant design is not standardized [12]. While the mechanical strength of the full solid-body implant fabricated by 3D-printing is comparable to those fabricated by forged, wrought, or casting [13], the lattice structure, that is an intended porotic structure to facilitate tissue integration around the implant, requires mechanical and clinical verification [6,7,14]. In this situation, the biggest concern in practice is overcoming failure of bone union or fracture of implant after performing limb salvage surgery with the 3D-printed implant. As most 3D-printed custom-made implants are not modular but rather a single unit, it is difficult to replace only the broken parts or those where the tumor has recurred nearby.

Herein, we present a surgical method that utilizes an additional 3D-printed implant in revision limb salvage surgery to preserve the previous 3D-printed custom-made titanium alloy implants in long bones.

2. Materials and Methods

Two representative patients underwent revision surgeries using a 3D-printed connector to connect the previous 3D-implant to the host bone. The reasons for revision surgeries after the 3D-printed limb salvage surgeries were tumor recurrence and implant fracture. The previous implants achieved bony union radiographically at the junctions between the implant and the host bone, and neither patient underwent revision surgery due to non-union at the junction. Rather than removing all implants, the revision surgeries were performed in a manner that preserved the previous implant as much as possible. The 3D-printed customized implants were designed using the Materialise Interactive Medical Image Control System (Materialise; Leuven, Belgium) and Magics 22 (Materialise; Leuven, Belgium) and fabricated using an electron beam melting-type (EBM) 3D-printer (ARCAM A1, Arcam AB, Mölndal, Sweden) with Ti6Al4V (Ti6Al4V-ELI Per ASTM 136). The custom-made implant was created by the MEDYSSEY Company (Jecheon, Korea) and certified by the Ministry of Food and Drug Safety. All study participants provided informed consent, and the study design was approved by the appropriate ethics review board (NCC2017-0129).

For the 3D-printed connector, two junctional parts for the previous implant and remaining host bone were designed first. The connector body was subsequently designed for the most functional position or angle by twisting or tilting. The junction to the previous metal implants and the bodies were mainly solid structure, while the junction to the remaining host bone partly had lattice structure for contact surface with the bone. The lattice structure had 750 um pores [7,15]. The main concept of the 3D-printed connector is that it mimics that of a modular prosthesis, with an assembly mechanism to insert one piece into another (Modular Universal Tumor and Revision System, Implantcast, Buxtehude, Germany; Global Modular Replacement System, Stryker, Kalamazoo, MI, USA). The previous 3D-printed implant was tapered on the inside, and the 3D-printed connector encased the outside of the previous implant. The 1.5-mm gap between the previous implant and the new one was filled with bone cement. The metal artifact around the previous 3D-printed implant that was seen in the computed tomography (CT) images was removed and smoothed as much as possible, though there was a risk of error during the modelling process. For both patients, the design of the implant used in the first surgery was saved as an STL file; in the CT images taken for the new implant, the previous implant part was not modelled by dealing with metal artifacts but was replaced by fusion of the implant design data.

The first patient reported here was a 40-year-old man who underwent treatment for a desmoplastic fibroma, a locally aggressive intermediate tumor, in his forearm [14]. He had undergone orthopedic surgery on the same side of both forearm bones and had used an external fixator for fracture fixation five years ago. It is not clear whether the tumor had originated from the fracture site or had entered the screw holes of the previous external fixator, but the tumor involved both forearm bones. There was no distant metastasis. Bilateral limb salvage surgery for the forearm bones was performed using two 3D-printed custom-made implants for the radius and the ulna (Figure 1A–E). One year postoperatively, local tumor recurrence was detected at the proximal junction of the ulnar implant. The radial implant showed good fixation in both proximal and distal junctions. The first revision surgery was performed to remove the recurred tumor and to make a single forearm bone by fusion of the proximal and distal remnant bones. However, fusion at the proximal part failed and a second revision surgery was planned (Figure 1F–H).

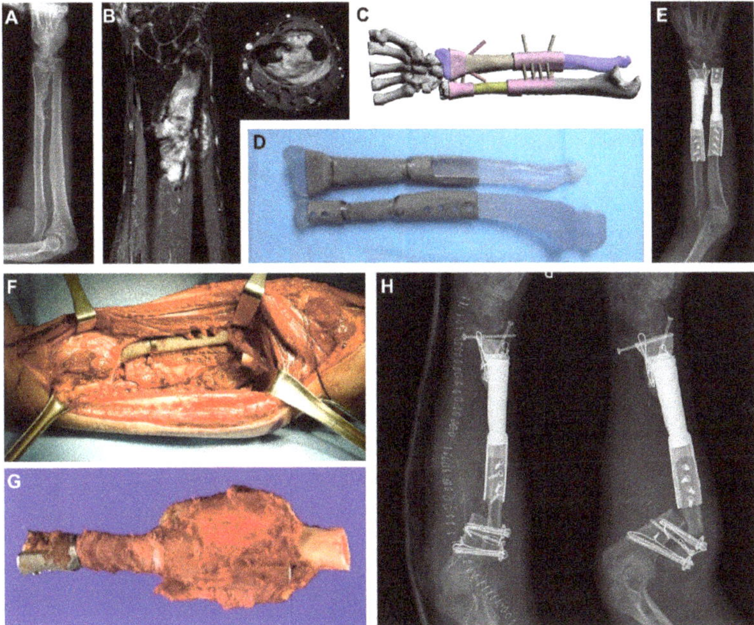

Figure 1. Failure of the 3D-printed implants in both forearm bones [14]. (**A**) A preoperative plain radiography and (**B**) gadolinium-enhanced T1-weighted magnetic resonance image showing desmoplastic fibroma arising from both forearm bones. (**C**) A graphic design and (**D**) photograph of 3D-printed implants. (**E**) A postoperative plain radiograph after limb salvage surgery. Photographs showing (**F**) removal of the ulnar implant with a recurrent tumor and (**G**) a specimen. (**H**) Plain radiographs showing immediate postoperative status after the first revision surgery (**left**) and mechanical failure 4 months later (**right**).

The second representative patient was a 69-year-old woman who had chondrosarcoma at the left distal half of the femur without distant metastasis. She had a limb length discrepancy of 10 cm due to a childhood history of osteomyelitis in the contralateral femur. The patient underwent limb salvage surgery using a 3D-printed customized implant for the distal part of her right femur with retention of the natural knee joint, and acute shortening of the right femur for partial correction of limb length discrepancy (Figure 2A–C). The patient could ambulate independently three months postoperatively. However, six months postoperatively, she slipped, and the proximal part of the implant was broken between the implant body and the fixation plate. The unexpected implant fracture was analyzed and the main causes of the fracture were stress concentration at the broken junction and internal defects of the 3D-implant. The distal junction to the knee joint was well-fixed. The first revision surgery was performed with additional dual plates with expectation of bony union or extracortical bone bridge to the porous structure of the implant body. However, 21 months later, mechanical failure occurred and a second revision surgery was planned (Figure 2D–F).

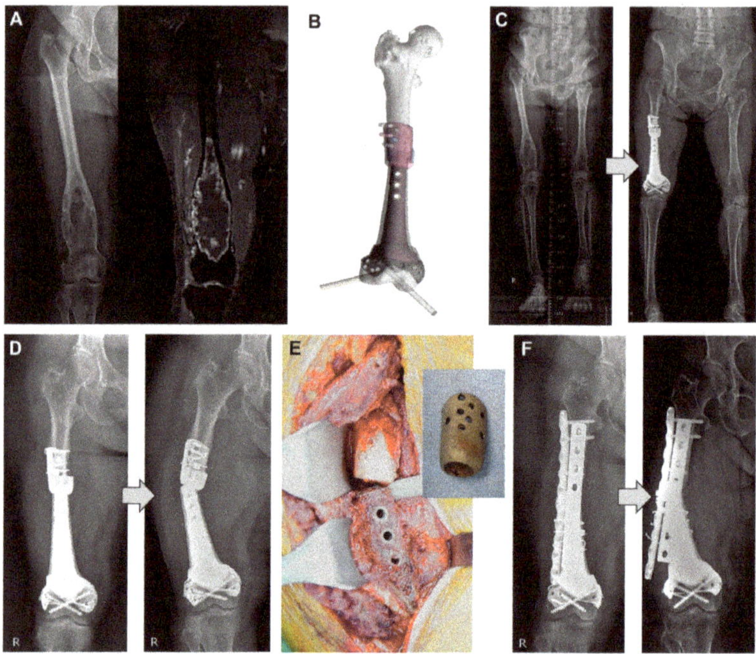

Figure 2. Failure of the 3D-printed implants in the right distal femur. (**A**) A preoperative plain radiography and gadolinium-enhanced T1-weighted magnetic resonance images showing chondrosarcoma in the right distal femur. (**B**) A graphical design of the 3D-printed implant. (**C**) A preoperative (**left**) and postoperative teleradiogram (**right**) after limb salvage surgery. (**D**) Photographs showing implant fracture at 6 months postoperatively. (**E**) Intraoperative photographs and the removed broken proximal part. (**F**) Plain radiographs showing immediate postoperative status after the first revision surgery (**left**) and another mechanical failure 21 months later (**right**).

3. Results

3.1. Forearm

For the patient who underwent single forearm bone surgery, the proximal part of the previous radial implant and the proximal ulna needed to be connected while maintaining the distal part, including the wrist joint. The wrist and hand function in the salvaged forearm was normal; however, thumb extension was limited due to soft tissue adhesion of the thumb flexor muscles. The elbow joint function was originally limited from 0 to 90 degrees due to radial head dislocation secondary to a childhood trauma. The proximal part of the previous radial implant was half the cylindrical plate that originally wrapped the proximal radial shaft. The junctional part of the 3D-printed connector up to the previous radial implant was designed to have a metal insert replace the space of the original radial shaft and the remaining half of the cylindrical plate from the opposite direction of the previous implant. The second junctional part was designed for the proximal ulna near the elbow joint, and this comprised two-thirds of a cylindrical plate with multiple holes for the screws and suture. The functional position of the forearm rotation was set at 15 degrees toward pronation from the neutral position. The body of the 3D-printed connector was twisted to the functional rotational position (Figure 3A–D).

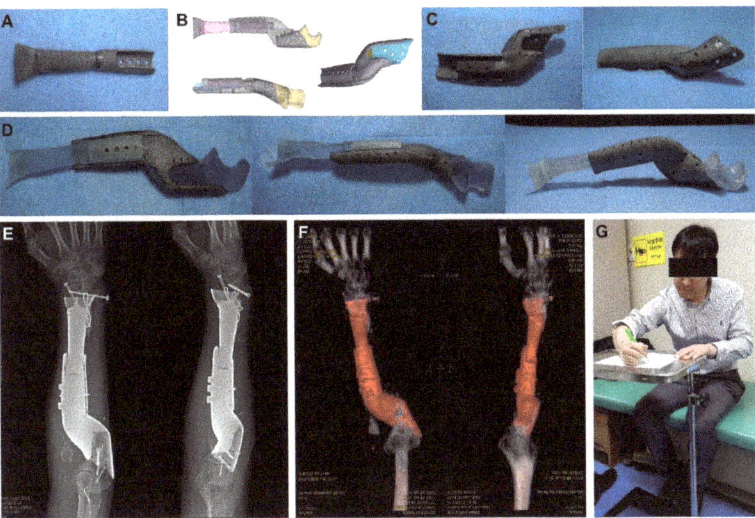

Figure 3. Revision limb salvage surgery using 3D-printed connector in the forearm. (**A**) A photograph of the previous 3D-printed implant for the radius. (**B**) Graphic designs of the 3D-printed connector and (**C,D**) photographs of the 3D-printed connector for single forearm bone surgery preserving the previous 3D-printed radial implant. Postoperative (**E**) plain radiographs and (**F**) computed tomography reconstruction images after second revision surgery using the 3D-printed connector. (**G**) A photograph showing functional position of the forearm during writing.

In the second revision surgery, the broken screws and the remaining part of the proximal radius were removed. The wrist joint was stable after the first revision surgery, and the host bones near the wrist were firmly bonded to the previous 3D-printed implant. The proximal part of the previous 3D-printed implant and the ulnar remnant near the elbow joint were fixed using the 3D-printed connector. The fixation between the old and new 3D-printed implants was achieved by winding three wires through a matched hole with bone cement. The 3D-printed connector and the proximal ulna showed conforming stability after inserting the bone into the implant, which was designed to sufficiently enclose the bone and fixed using screws (Figure 3E–F). The patient recovered the original range of motion of his elbow (0–90 degrees) at 10 weeks postoperatively and was able to write with his forearm in the 15-degree pronation position (Figure 3G). The wrist and elbow function were well-maintained without mechanical failure until the last follow-up at 26 months after the second revision surgery. The Musculoskeletal Tumor Society (MSTS) score at the last visit was 26 (87%).

3.2. Femur

For the patient who underwent revision surgery using the 3D-printed connector for the femur, the femoral shaft of the previous implant body and proximal femur needed to be connected while maintaining the distal part, including the knee joint. The knee joint had an extension lag of 10 degrees due to acute shortening of the femur, though independent ambulation without pain was possible. Furthermore, partial correction of the limb length discrepancy led to an improvement in gait, and the left heel that did not touch the ground during ambulation before the limb length correction, could touch the ground while walking after the correction (Figure 4D). The junctional part of the 3D-printed connector to the previous broken 3D-printed custom-made implant was a full cylinder that surrounded the remaining previous implant. Another junctional part was designed as a one-third plate with screw holes to fix the proximal femur near the hip joint. The body of the 3D-printed connector was designed to restore the tilting angle of the femur from the mechanical axis of the lower limb (Figure 4A,B).

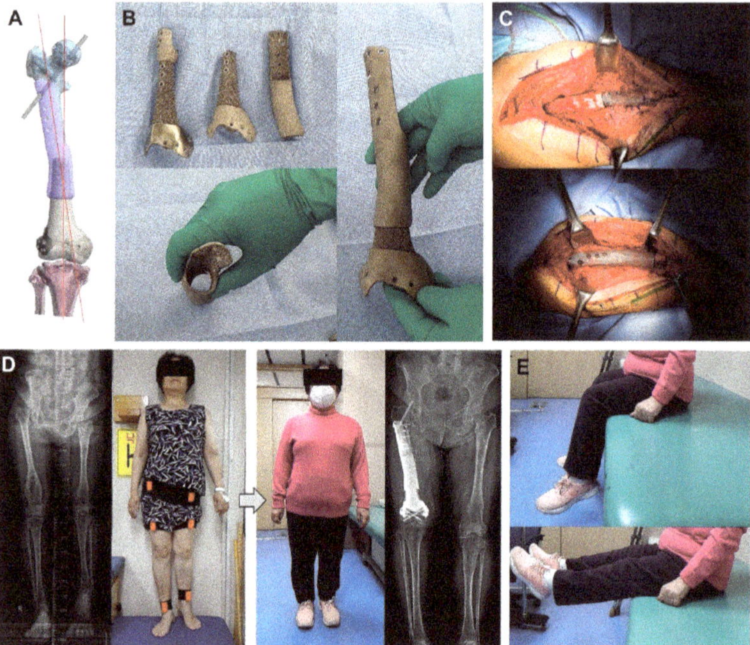

Figure 4. Revision limb salvage surgery using 3D-printed connector in the femur. (**A**) A graphic design of the 3D-printed connector. (**B**) Photographs of the previous 3D-printed implant (duplicated), broken implant (duplicated), and the 3D-printed connector. (**C**) Intraoperative photographs before and after application of the 3D-printed connector. (**D**) Preoperative teleradiograph and photograph showing the chondrosarcoma of the right distal femur and limb length discrepancy of 10 cm (**left**), and final postoperative pictures (**right**). (**E**) Photographs showing normal knee joint function.

In the second surgery, all the metal plates and screws were removed, leaving the knee joint below the shaft of the broken 3D-printed customized implant used in the first surgery. The 3D-printed connector was connected by inserting the shaft of the previous 3D implant into the 3D-printed connector. The gap between the old and new 3D-printed implant was filled with bone cement. The remaining proximal femur on the hip side was fixed by multiple screws. The patient resumed walking on crutches at 2 weeks after the second operation and was able to walk without crutches without pain at 10 weeks postoperatively. The range of motion of the knee joint was normal without extension lag (Figure 4C–E). At 10 months after the second revision surgery, the MSTS score was 30 (100%).

4. Discussion

When a megaprosthesis is fabricated by 3D printing, it is often difficult to implement a modular-type; therefore, it is fabricated as a single mass. After limb salvage surgery using a megaprosthesis, revision surgeries are often required for a variety of reasons, including local recurrence and mechanical failure. One of the concerns regarding the use of a 3D-printed customized implant in limb salvage surgery is that it makes revision surgery difficult. In this study, an additional 3D-printed implant was introduced during revision without removing all of the previously inserted 3D-printed implant. The surgical results using the additional 3D-printed connector implant were acceptable in the short-term follow-up.

The metal-to-metal fixation mechanism is still a problem in the 3D-printed implant. The locking screw mechanism has not been introduced in 3D-printed implants as the male and female screw threads are hard to implement due to insufficient precision and roughness of the electron beam melting-type 3D-printed product. Therefore, cemented fixation to the intended gap between the implants was mainly utilized in these patients.

This study has some limitations. It was a preliminary study based on a small number of patients with a short follow-up. Although the short-term surgical results were promising, they are difficult to generalize. These results imply that the 3D-printed connector implant is a surgical option to consider when revision surgery after previous limb salvage surgery using a 3D-printed customized implant is needed, and retaining the previous 3D-printed implant totally/partially would be beneficial. Regarding the mechanical property of the 3D-printed implants, long-term follow-up clinical data and biomechanical data is required. There was no serious mismatch between the 3D connector and previous implant or remaining bone. However, deformation of the 3D-printed customized implant by residual stress and contraction was not accurately measured in this paper.

In conclusion, to retain the previous 3D-printed implant during revision limb salvage surgery, an additional 3D-printed implant may be a feasible surgical option.

Author Contributions: J.-W.P. conceptualized, interpreted the patient data and implant designs and was a major contributor in writing the manuscript. H.-G.K. was a primary implant designer and visualized the illustrations. J.-H.K. and H.-S.K. contributed to validation of the implant designs. All authors have read and agreed to the published version of the manuscript.

Funding: This research was supported by the Industrial Strategic Technology Development Program ('P0008805'), funded by the Ministry of Trade, Industry & Energy (MOTIE, Korea).

Institutional Review Board Statement: The study was conducted according to the guidelines of the Declaration of Helsinki, and approved by the Institutional Review Board of National Cancer Center, Korea (NCC2017-0129, 1 June 2017).

Informed Consent Statement: Informed consent was obtained from all subjects involved in the study.

Data Availability Statement: The datasets used and/or analyzed during the current study are available from the corresponding author on reasonable request.

Conflicts of Interest: The authors declare no conflict of interest.

References

1. Malawer, M.M.; Wittig, J.C.; Bickels, J.; Wiesel, S.W. *Operative Techniques in Orthopaedic Surgical Oncology*, 2nd ed.; Wolters Kluwer: Philadelphia, PA, USA, 2016.
2. Biermann, J.S. *Orthopaedic Knowledge Update: Musculoskeletal Tumors 3*; American Academy of Orthopaedic Surgeons, Lippincott: New Tork, NY, USA, 2013.
3. Sim, F.H.; Choong, P.F.; Weber, K.L. *Master Techniques in Orthopaedic Surgery: Orthopaedic Oncology and Complex Reconstruction*; Lippincott Williams & Wilkins: Philadelphia, PA, USA, 2011.
4. Angelini, A.; Trovarelli, G.; Berizzi, A.; Pala, E.; Breda, A.; Ruggieri, P. Three-dimension-printed custom-made prosthetic reconstructions: From revision surgery to oncologic reconstructions. *Int. Orthop.* **2019**, *43*, 123–132. [CrossRef] [PubMed]
5. Ji, T.; Yang, Y.; Tang, X.; Liang, H.; Yan, T.; Yang, R.; Guo, W. 3D-printed modular hemipelvic endoprosthetic reconstruction following periacetabular tumor resection: Early results of 80 consecutive cases. *J. Bone Jt. Surg Am.* **2010**, *102*, 1530–1541. [CrossRef] [PubMed]
6. Park, J.W.; Kang, H.G.; Kim, J.H.; Kim, H.S. The application of 3D-printing technology in pelvic bone tumor surgery. *J. Orthop. Sci.* **2020**. [CrossRef] [PubMed]
7. Park, J.W.; Kang, H.G.; Kim, J.H.; Kim, H.S. New 3-dimensional implant application as an alternative to allograft in limb salvage surgery: A technical note on 10 cases. *Acta Orthop.* **2020**, *91*, 489–496. [CrossRef] [PubMed]
8. Theil, C.; Röder, J.; Gosheger, G.; Deventer, N.; Dieckmann, R.; Schorn, D.; Hardes, J.; Andreou, D. What is the likelihood that tumor endoprostheses will experience a second complication after first revision in patients with primary malignant bone tumors and what are potential risk factors? *Clin. Orthop. Relat. Res.* **2019**, *477*, 2705–2714. [CrossRef] [PubMed]
9. Shehadeh, A.; Noveau, J.; Malawer, M.; Henshaw, R. Late complications and survival of endoprosthetic reconstruction after resection of bone tumors. *Clin. Orthop. Relat. Res.* **2010**, *468*, 2885–2895. [CrossRef] [PubMed]
10. Jeys, L.M.J.; Kulkarni, A.; Grimer, R.J.; Carter, S.R.; Tillman, R.M.; Abudu, A. Endoprosthetic reconstruction for the treatment of musculoskeletal tumors of the appendicular skeleton and pelvis. *J. Bone Jt. Surg Am.* **2008**, *90*, 1265–1271. [CrossRef] [PubMed]
11. Capanna, R.; Scoccianti, G.; Frenos, F.; Vilardi, A.; Beltrami, G.; Campanacci, D.A. What was the survival of megaprostheses in lower limb reconstructions after tumor resections? *Clin. Orthop. Relat. Res.* **2015**, *473*, 820–830. [CrossRef] [PubMed]

12. Schwartz, A.J.; Kabo, J.M.; Eilber, F.C.; Eilber, F.R.; Eckardt, J.J. Cemented distal femoral endoprostheses for musculoskeletal tumor: Improved survival of modular versus custom implants. *Clin. Orthop. Relat. Res.* **2010**, *468*, 2198–2210. [CrossRef] [PubMed]
13. Liu, S.; Shin, Y.C. Additive manufacturing of Ti6Al4V alloy: A review. *Mater. Des.* **2019**, *164*, 107552. [CrossRef]
14. Park, J.W.; Song, C.A.; Kang, H.G.; Kim, J.H.; Lim, K.M.; Kim, H.S. Integration of a three-dimensional-printed titanium implant in human tissues: Case study. *Appl. Sci.* **2020**, *10*, 553. [CrossRef]
15. Park, J.W.; Kang, H.G. New 3-dimensional implant application as an alternative to allograft in limb salvage surgery: A technical note on 10 cases. *Acta Orthop.* **2020**, *91*, 617–619. [CrossRef] [PubMed]

Article

In Vitro Physical-Chemical Behaviour Assessment of 3D-Printed CoCrMo Alloy for Orthopaedic Implants

Radu Mirea [1,*], Iuliana Manuela Biris [2], Laurentiu Constantin Ceatra [1], Razvan Ene [3,4], Alexandru Paraschiv [1], Andrei Tiberiu Cucuruz [5], Gabriela Sbarcea [6,*], Elisa Popescu [3] and Teodor Badea [1]

[1] Romanian Research and Development Institute for Gas Turbines—COMOTI, 220D Iuliu Maniu Blvd., 061126 Bucharest, Romania; laurentiu.ceatra@comoti.ro (L.C.C.); alexandru.paraschiv@comoti.ro (A.P.); teodor.badea@comoti.ro (T.B.)

[2] University Emergency Hospital, 169 Splaiul Independentei, 050098 Bucharest, Romania; ema.biris@gmail.com

[3] Bucharest Emergency Clinical Hospital, 8 Floreasca Street, 014461 Bucharest, Romania; razvan77ene@yahoo.com (R.E.); elisa_zukie@yahoo.com (E.P.)

[4] Orthopedics and Traumatology Department, "Carol Davila" University of Medicine and Pharmacy, 8 Eroii Sanitari Blvd, 050474 Bucharest, Romania

[5] SC Cromatec Plus SRL, 1 Petre Ispirescu Street, 077167 Snagov, Romania; andrei.cucuruz@gmail.com

[6] National Research and Development Institute for Electrical Engineering—ICPE-CA, 313 Splaiul Unirii, 030138 Bucharest, Romania

* Correspondence: radu.mirea@comoti.ro (R.M.); gabriela.sbarcea@icpe-ca.ro (G.S.); Tel.: +40-724-977-646 (R.M.); +40-724-794-258 (G.S.)

Citation: Mirea, R.; Biris, I.M.; Ceatra, L.C.; Ene, R.; Paraschiv, A.; Cucuruz, A.T.; Sbarcea, G.; Popescu, E.; Badea, T. In Vitro Physical-Chemical Behaviour Assessment of 3D-Printed CoCrMo Alloy for Orthopaedic Implants. *Metals* **2021**, *11*, 857. https://doi.org/10.3390/met11060857

Academic Editors: Atila Ertas and Aleksander Lisiecki

Received: 29 April 2021
Accepted: 22 May 2021
Published: 24 May 2021

Publisher's Note: MDPI stays neutral with regard to jurisdictional claims in published maps and institutional affiliations.

Copyright: © 2021 by the authors. Licensee MDPI, Basel, Switzerland. This article is an open access article distributed under the terms and conditions of the Creative Commons Attribution (CC BY) license (https://creativecommons.org/licenses/by/4.0/).

Abstract: In this study, a CoCrMo-based metallic alloy was manufactured using a 3D-printing method with metallic powder and a laser-based 3D printer. The obtained material was immersed in a simulated body fluid (SBF) similar to blood plasma and kept 2 months at 37 °C and in relative motion against the SBF in order to mimic the real motion of body fluids against an implant. At determined time intervals (24, 72, 168, 336, and 1344 h), both the metallic sample and SBF were characterized from a physical-chemical point of view in order to assess the alloy's behaviour in the SBF. Firstly, the CoCrMo based metallic sample was characterized by scanning electron microscopy (SEM) for assessing surface corrosion and X-ray diffraction (XRD) for determining if and/or what kind of spontaneous protective layer was formed on the surface; secondly, the SBF was characterized by pH, electrical conductivity (EC), and inductively coupled plasma mass spectroscopy (ICP-MS) for assessing the metal ion release. We determined that a 3D-printed CoCrMo alloy does not represent a potential biological hazard in terms of the concentration of metal ion releases, since it forms, in a relatively short period of time, a protective CoCr layer on its exposed surface.

Keywords: 3D printing; powder metallurgy; simulated body fluid; biomaterial

1. Introduction

"Traditional" orthopaedic implants are titanium (Ti)-based alloys since they provide a series of characteristics suitable for such applications. They can be used in hip joint replacement, tibia rod, clavicle plate, etc. Nevertheless, there are other metallic alloys used in orthopaedic implants that are not Ti based. Cobalt-chromium (CoCr)-based alloys represent a viable solution and molybdenum (Mo) is often added to enhance the inner structure in terms of increasing the ductility and strength of the alloy [1]. Although casting and forging are, and continue to be, the main methods for obtaining both CoCr- and Ti-based alloys for orthopaedic implants, 3D printing is gradually gaining its position within this field. In recent years, 3D printing (both polymer and metal printing) has opened several exciting possibilities to create customized orthopaedic implants [2]. There are three rapid prototyping techniques used to produce metallic implants, i.e., selective laser melting (SLM), electron beam melting (EBM), and laser engineered net shaping (LENS), and each of them having their advantages and disadvantages. The SLM technique uses high power lasers to melt metallic powders in order to achieve the required shape of the implant, while the EBM technique uses a focused high-energy electron beam for direct solidification of metal powders. The LENS technique creates a molten metal pool on a substrate, and then the metal powder is injected [3]. A 3D printing process usually uses digital design software and/or a 3D digital scanner in order to achieve the virtual version of the metallic implant to be produced. Then, the obtained model is transformed into an .STL file format enabling the exact spatial coordinates (xyz) of a model's surfaces. Next, the model is "sliced" by using another software, each "slice" being 25–100 μm thick. By stacking these "slices", the final shape of the implant is obtained. A 3D printer can use different types of materials starting from various plastics to different metallic powder mixtures. About 50 metallic alloys can be used in 3D printing and they are mostly Ti, Ni, Al, stainless steel, CoCr, etc., alloys and more than 80% of them are used as implants [4].

According to the ASTM standard F2792-12a [5], the SLM technique is classified as powder bed fusion technology [6]. Therefore, in this paper, we describe the use of this technique and characterize the obtained CoCr-based material used for implants. Thus, an in vitro behaviour assessment of the obtained metallic material used for orthopaedic implants has been performed during 60 days of immersion in a simulated body fluid (SBF) similar to blood plasma.

For a metallic material to be suitable as biomaterial, it should meet a series of desired properties ranging from mechanical (matching the material's elasticity modulus with that of the bone, high strength, high wear, corrosion resistance, etc.) to biocompatibility (not toxic to environmental cells and tissues, quickly integrated in the body, etc.) [7,8].

CoCr-based alloys have higher wear resistance than Ti-based alloys; therefore, they are especially used for implants that, over time are subjected to extensive wear, such as hip joints, etc. Clinically, CoCrMo alloys are used the most, since they are the perfect combination between high strength and high ductility, but the main drawback is their high elasticity modulus, which may lead to greater stress in the bone. This drawback may be overcome by 3D printing of CoCr-based alloys. Although EBM is the most used technique for 3D printing of CoCr alloys [8], the SLM technique can also be used. When prepared by using the SLM technique, Mo must be added to the powder since it enables high temperature gradients during melting and subsequent rapid cooling and, especially, it enables the formation of the fine granular structure of the implant [9].

Thus, it is possible to produce implants that have tailored mechanical properties [10] but the issue of the material's biocompatibility still remains. One of the main problems is metal ion dissociation from the metallic material into the body fluids, which are in close contact with the metal. It should be mentioned that corrosion is the most frequent phenomena that occurs in the case of an implant and one of the most important ones to be avoided. It is well known that, from a biocompatibility point of view, a suitable material performs with an appropriate host response (e.g., minimum disruption of normal body function) and does not cause any allergic and/or inflammatory response when placed in vivo [11].

The first requirement of any material that is placed in the human body is that it should be biocompatible and not cause any adverse reaction. Corrosion and surface oxide film dissolution are the two mechanisms which introduce additional ions into the body. Extensive release of ions from implants can result in adverse biological reactions. Corrosion is the first consideration for a material of any type that is to be used in the body because metal ion release takes place mainly due to corrosion of surgical implants [12]. The two main corrosion types that often appear are spot and pitting corrosion, as described in [13,14].

In the particular case of orthopaedic implants, micro-motions are known to occur at points of fixation, while corrosion is caused by the body fluids, which contain various inorganic and organic molecules [15].

In this paper, we present the results obtained by immersing a CoCr-based 3D-printed material in vitro (being the "next best thing") into an SBF for 60 days at 37 °C with the fluid being in relative motion against the material. We assessed the appearance of surface corrosion, metal ion dissociation, and the appearance of a spontaneous protective layer on the exposed surface.

Considering that even a small concentration of metal ion release into the body could be very aggressive due to their migration and accumulation in different organs, sometimes far away from the point of release, it is important to quantify the concentration of ion release in various environments; thus, ICP-MS investigations were conducted [16].

Material corrosion may lead to implant failure as [17] has assessed and may cause an allergic and/or toxic response of the body.

A complete understanding of 3D-printed metallic implants is needed, nevertheless, what has been assessed so far is genuinely promising. Some of the advantages of 3D-printed implants as compared with the "traditional" manufactured ones, are highlighted in [18], among them, we highlight the following: the final shape of the implant can be obtained in a single phase of production; as many as needed can be produced, even if only one piece; the mechanical properties are better than cast or forged implants; the design and geometry can be as complicated as needed; porosity can be designed as needed for the elasticity module to match that of a bone; and the pores can be designed to be opened or inter-correlated in order to ensure maximum osseointegration.

2. Materials and Methods

The experimental part of this study was to manufacture a 3D-printed CoCrMo metallic alloy that could be used for an orthopaedic implant. The CoCr alloy was chosen as a good alternative to Ti-based alloys given its wear resistance [7]. Molybdenum was added to the mixture in order to increase the ductility and to increase the strength of the alloy.

2.1. Method for Producing the CoCrMo Alloy

2.1.1. 3D Manufacturing

The metallic implant was manufactured using a selective laser melting (SLM) TruPrint 1000 (TRUMPF Inc., Ontario, ON, Canada) printer and CoCrMo metal powder (Starbond CoS powder 30 from S&S Schefter GmbH, Mainz, Germany) within the 10–30 μm grain size and with the chemical composition presented in Table 1.

Table 1. Chemical composition of the metallic powder.

Element	Co	Cr	W	Mo	Other: C, Fe, Mn, N, Si
wt.%	59	25	9.5	3.5	Less than 1.5

For 3D manufacturing of the metallic implant, the following process parameters were used: 90 W laser power, 600 mm/s laser speed, 20 μm layer thickness, and 55 μm laser spot. An Ar (99.995%) flow system was used to reduce the oxygen level to less than 300 ppm. The material was mechanically removed from the building platform and was tested in a simulated body fluid solution without any post-processing operations. Moreover, its hardness was measured to be 463.5 (HV 0.5). Then, the sample was polished by using a rotating polishing machine, until its roughness dropped below 0.05 μm (N2). The used polishing paper had a roughness varying from 240 to 1200. Then, the sample was surfaced using diamond powder suspensions of 3 and 1 μm, respectively.

2.1.2. SEM Assessment

The micro-structural analysis was performed by using an FEI F50 Inspect (FEI Company, Brno, Czech Republic) with a magnification of 4000×, HV = 20.00 kV, WD 11.6 mm, spot 3.0, and dwell of 30 μs. The method has been used to emphasize deep corrosion and surface transformations of the metallic samples [17,19].

2.1.3. Mass Variation Assessment

The aim of the mass loss/gain assessment was to determine, the corrosion rate and correlated with XRD the formation of protective layer on sample's surface.

2.1.4. XRD Assessment

The XRD analysis was conducted to establish if and/or what kind of oxides had formed on the sample's surface. A Bruker D8 Discover diffractometer (Bruker, Billerica, MA, USA) was used. The diffractometer settings were as follows: primary optics used a Cu tube (λ = 1.540598 Å) and a Göebel mirror, while secondary optics used a 1D LynxEye detector (Bruker, Billerica, MA, USA). The plots were recorded at 0.04° angle and 1s/step scanning speed. They were indexed using the ICDD Release 2015 database.

2.2. Method for Preparing the SBF

The SBF solution was prepared, as shown in [20]; it was an updated version of the simulated body fluid submitted, in 2003, with detailed instructions for its preparation to the Technical Committee ISO/TC 150 of the International Organization for Standardization as a solution for in vitro measurements of implant materials. Table 2 shows the concentrations of reagents to be used for 1 L of SBF. The SBF solution was prepared in the lab by mixing extra pure substances procured from the market from WWR Chemicals. The sample was immersed in the prepared SBF in an Erlenmeyer polypropylene (PP) recipient. The recipient was immersed in a thermostatic bath equipped with a movable device that allowed a frequency of 60 movements/minute with a distance of ±1 cm right/left in order to mimic the relative motion of body fluid against an implant. A PP wire was used to ensure that the sample was kept immersed in the SBF, however, did not come in contact with the recipient.

Table 2. Reagents for preparing the updated simulated body fluid [20]. (reprinted from ref. [20] with permission from Elsevier).

Reagent	Amount for 1 L of SBF
Sodium chloride	8.035 g
Sodium bicarbonate	0.355 g
Potassium chloride	0.225 g
Potassium phosphate dibasic tri-hydrate	0.231 g
Magnesium chloride hexahydrate	0.311 g
1M hydrochloric acid	39 mL
Calcium chloride	0.292 g
Sodium sulphate	0.072 g
Tris(hydroxyl-methyl) amino methane	6.118 g

2.2.1. PH and Electrical Conductivity Assessment

The pH and electrical conductivity characterizations were performed in order to determine the redox phenomena that occurred and the variation of ions in the SBF. A Mettler-Toledo pH/conduct meter (Mettler-Toledo, Greifensee, Switzerland) was used.

2.2.2. ICP-MS Assessment

Inductively coupled plasma mass spectrometry (ICP-MS) [21] was used for determining the concentration of metal ions dissociated in the SBF. A Dirac Elan II ICP-MS spectrometer (Perkin-Elmer, Toronto, ON, Canada) was used.

3. Results and Discussions

The conducted experiments focused on both the material and the SBF. The surface corrosion of the material was assessed by scanning electron microscopy (SEM), XRD to determine if and/or what kinds of oxides had formed on its surface while immersed, and mass variation. The SBF was assessed by its pH, electrical conductivity variation, and by ICP-MS analysis to determine the concentration of metal ion dissociated from the metallic material.

In order to better assess the ion dissociation and surface corrosion, the protective oxide layer of the sample was removed on one side and the exposed material was polished until its roughness dropped below 0.05 μm (N2). All the above-mentioned assessment that refers to the material was performed on the exposed surface.

3.1. Material Characterization

3.1.1. SEM Assessment

The SEM technique uses electrons for images; thus, it is considered to be the most suitable technique for high-resolution imaging of surfaces. The basic principle is similar to OM, but electrons are used instead of light, allowing a much higher magnification (higher than 100,000×).

The aim of this type of assessment was to investigate even deeper the corrosion that can occur on the surface. A magnification of 4000× was considered to be adequate in this case, and therefore the SEM assessment was performed by means of this magnification and the covered area was 80 μm × 80 μm. The analysis was performed on the exposed surface, aimed at highlighting the formation of the protective layer as a replacement for the layer removed.

The SEM images (Figure 1) show surface corrosion of the sample as pitting in the first two weeks of exposure and, during the next two weeks, the formation of porous structures. According to [22], this porous structure may not be pitting (a type of corrosion that often appears) but porous single crystals of Cr_2O_3 and Co_3O_4. This hypothesis is strongly supported by ICP-MS that does not show any increase in metal ions into the SBF and pH and EC plots that show an increase in their values.

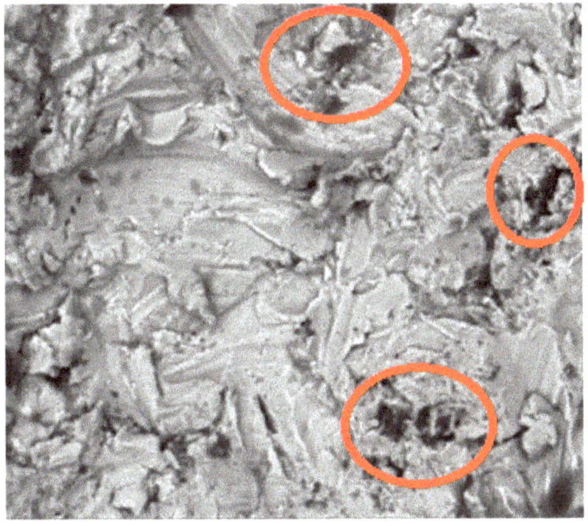

(a) Initial

(b) 15 days

Figure 1. *Cont.*

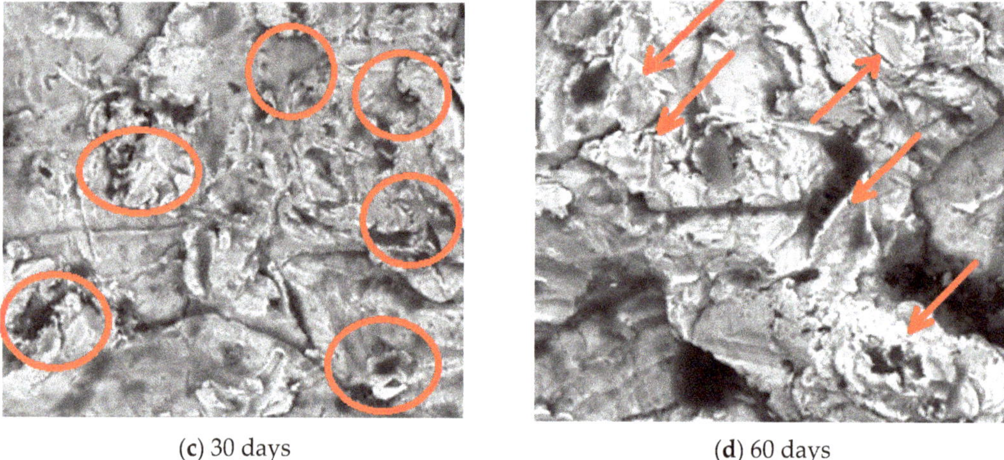

(**c**) 30 days   (**d**) 60 days

Figure 1. SEM images showing surface corrosion of 3D-printed CoCrMo metallic implant during SBF exposure. (**a**) Initial (0 h); (**b**) after 15 days; (**c**) after 30 days; (**d**) after 60 days.

Thus, the proposed formation mechanisms consist of Cr and Co ions, under the influence of the acidic environment of the SBF (mainly due to HCl) forming porous single crystals of Cr_2O_3 and Co_3O_4, as has been studied by [22].

The surface analysis was performed, as in [23], and showed that, during the first two weeks of exposure, the material became porous due to the metal ion release into the SBF and, as a direct consequence, the porous surface increased almost 150% (Figure 1b). In the next two weeks, the formation of the oxide layer can be observed since most of the pores shrink almost 60% (Figure 1c). In the second month of exposure, the formation of porous single crystals of Cr_2O_3 and Co_3O_4 can be observed on the sample's surface (Figure 1d).

3.1.2. XRD Assessment

An XRD assessment is usually used in materials science for determining the crystallographic structure of a material, and it is one of the most used techniques in the field. It uses X-rays to irradiate the material. Firstly, incident X-rays are sent towards the material and the reflected X-rays are measured in terms of intensities and angles.

In this study, X-ray diffraction (XRD) was used to establish if and/or what kind of oxides formed on the sample's surface. As stated before, a D8 Discover diffractometer (Bruker, Billerica, MA, USA) was used for sample analysis.

Figure 2 shows the existence of a protective layer on the polished surface during sample exposure to the SBF. On the basis of each sample's composition, different oxide layers form on the surface.

As shown in Figure 2, the cobalt-chromium (CoCr) protective layer, which is actually an alloy, is formed on the sample's surface. The layer does not have a crystalline structure (peak dimensions are rather modest) but it is well represented leading to the conclusion that it takes a while, but after its formation, the layer offers a protective interface between the SBF and the metallic implant.

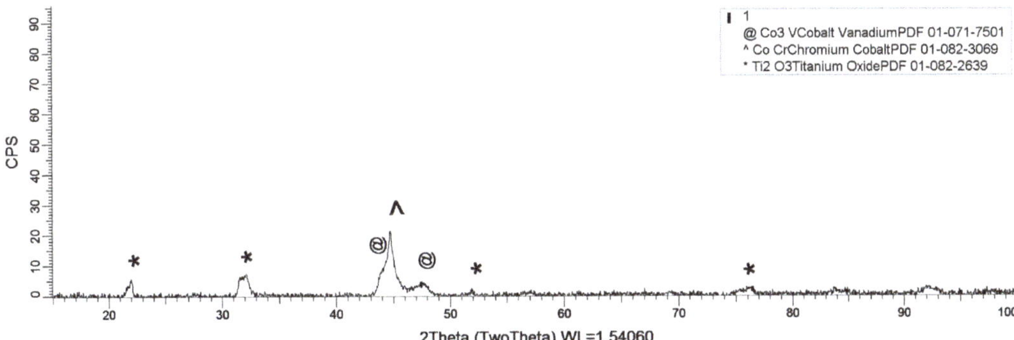

Figure 2. XRD plot for the 3D-printed CoCrMo metallic implant.

3.1.3. Mass Variation Assessment

The mass variation assessment as shown in Figure 3 was conducted by a gravimetric comparison of the sample's mass. Thus, the sample was weighed at its initial state, and then after 15, 30, and 60 days of exposure in the SBF. The aim of the assessment is to highlight the mass loss and/or gain while the sample is in the SBF. On the one hand, mass loss can be correlated with metal ion release, but this assessment is made by using ICP-MS, as described in Section 3.2.2. On the other hand, mass gain may occur due to the formation of metal oxides on the sample's surface, as emphasized in Section 3.1.2.

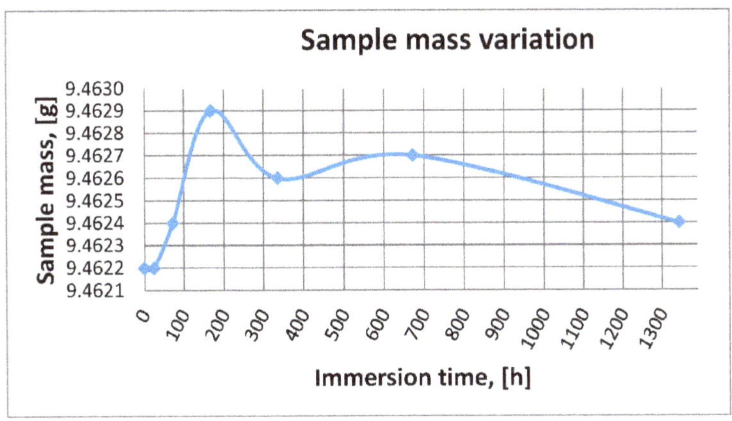

Figure 3. 3D-printed CoCrMo metallic implants mass vs. immersion time variation.

3.2. SBF Characterization

3.2.1. SBF pH and Electrical Conductivity Assessment

The pH value refers to the hydrogen ion concentration in a solution. The hydrogen concentration is closely related to corrosion phenomena, since a low pH value indicates an acidic environment suitable for surface corrosion. In addition, electrical conductivity shows the ion concentration in a solution; the higher the value, the higher the number of ions that are dissociated.

Therefore, pH and electrical conductivity monitoring represent an important assessment in order to evaluate the redox phenomena that occur while the sample is immersed in the SBF.

It can be observed in Figures 4 and 5 that the variation of pH values and electrical conductivity are closely interdependent, as well as with mass variation shown in Figure 3.

Thus, observing the curve's allure which indicates pH and EC variation during the immersion period, in the first week of exposure, the SBF's pH varies significantly while the electrical conductivity increases. During this period, the sample gains mass due to the formation of an oxide layer on its surface. Redox phenomena during the first week of exposure are intense, as seen in Figure 4, and the concentration of metal ions is increased, as can be observed in Table 3.

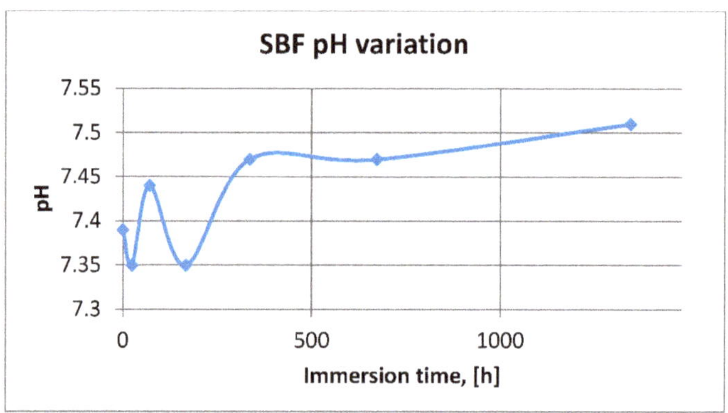

Figure 4. SBF pH vs. immersion time variation.

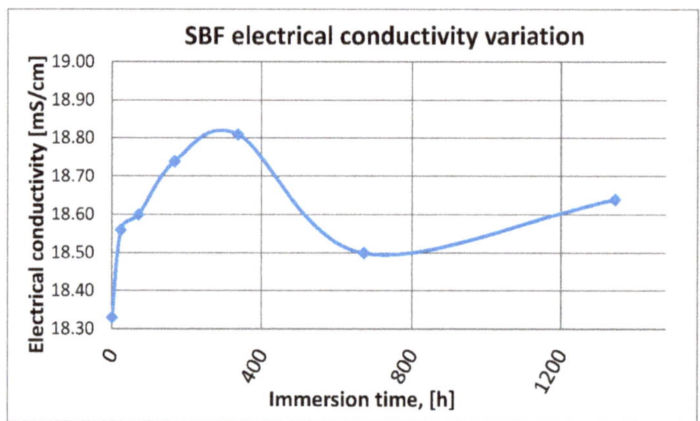

Figure 5. SBF electrical conductivity vs. immersion time variation.

Table 3. Measured metal ion concentrations in the SBF.

Element Time	Co (µg/L)	Cr (µg/L)	Mo (µg/L)	Mn (µg/L)
0 h—Initial	0	0	0	0
24 h	5.346	7.39	0.453	1.374
72 h	9.771	6.852	0.729	1.824
168 h	14.646	6.801	1.108	2.147
336 h	20.32	6.58	1.611	2.295
672 h	30.058	6.48	2.413	2.551
1334 h	30.325	5.149	2.598	2.145

Furthermore, this trend continues during the second week of exposure when the sample's mass decreases as metal ions are dissociated in the SBF, correlated with a continuous increase in electrical conductivity and pH.

After the second week of exposure, the amplitude of redox phenomena diminishes, as the protective layer forms on the sample's surface.

The most probable hypothesis regarding the allure of the plots shown in Figures 3–5 can be described as follows: The newly formed oxide layers of Cr_2O_3 and Co_3O_4 are still reacting to each other in order to form the protective layer of Co-Cr highlighted by the XRD plot. Thus, O_2 is released back into the SBD in its ion form and quickly reacts with ionic H_2 leading to a decrease in pH, observed in the last section of the plot shown in Figure 4. Nevertheless, metal ions do not dissociate since ICP-MS measurements show this aspect. In fact, Cr continues to be consumed since its concentration decreases (as shown in Table 3). The sample's mass decreases in the final interval of the plot shown in Figure 3, since O_2 is released from the initial oxide layers of Cr_2O_3 and Co_3O_4 and also due to the continuous (even at slower rate) release of Co and Mo ions into the SBF.

3.2.2. Metal Ion Release Assessment

ICP-MS is a powerful method for determining trace metal in given environments, due to its capability to measure very low concentrations of metal ions dissolved in fluids. This approach is well known, widely used by various research teams, and reported in frequently quoted scientific journals.

Basically, the ICP-MS method consist of determining the concentration of metal ions in a given solution. Thus, as required by the experimental design, 50 mL of the SBF used for sample immersion was collected and the solution was used for determining the nature and concentration of metal ions that might have dissociated from the sample into the SBF.

Given the data provided by Table 1, the following metal ions were monitored: Co, Cr, Mo, and Mn, since it is well known that large concentrations can lead to allergic and/or toxic reactions.

Table 3 shows the variation of metal ions dissociated from the sample in the SBF during the exposure period. Given the fact that the sample's exposed surface is relatively large, being almost impossible for a metallic implant to lose such a large area of protective layer, the immediate effect is the high concentration of metal ions released into the SBF. The aim of this assessment is to highlight the formation of a new protective oxide layer soon enough after the implant's surface is damaged (for various reasons, e.g., poor manipulation during implant surgery, mechanical shock, etc.) and to assess the concentration of metal ions dissociated into the SBF.

It can be observed that the formation of the oxide protective layer occurs after just 1 month of exposure. The correlation with pH, electrical conductivity, and mass variation plots highlights the fact that 3D-printed CoCr-based alloy may be used as orthopaedic implant. More mechanical and biocompatibility testing must be performed in the near future, but steps forward have been made.

Even though the metallic sample used in this study had a protective film on its surface, sample preparation removed it before immersing the sample in the SBF. It should be mentioned that the exposed surface was 909.65 mm^2, and based on this, the corrosion rate and ion release rate could be calculated.

Table 4 shows the metal ion release rate from each sample and their corrosion rate. The values can be calculated using the following formula:

$$c_i = \frac{m_{i,\,sol}}{m_{i,sample} \times S \times 16} \tag{1}$$

where c_i is the ion concentration, $m_{i,sol}$ is the ion mass in solution, $m_{i,sample}$ is the ion mass in sample, S is the exposed surface and, 16 is the number of weeks of exposure.

Table 4. Calculated corrosion and ion release rates.

Sample	Corrosion Rate $\{g/(mm^2\ Week)\}$	Ion Release $(\{mg/L\}/(mm^2\ Week))$				
		Co	Cr	Mo	Mn	W
3D-printed CoCrMo	$6.87\,e^{-5}$	$1.4\,e^{-5}$	$5.13\,e^{-5}$	$1.5\,e^{-5}$	$14.9\,e^{-5}$	$10.3\,e^{-5}$

The corrosion rate can be calculated by using the following formula:

$$C_r = \frac{m_m}{m_i \times S \times 16} \tag{2}$$

where m_m is the measured mass of the sample, m_i is the initial mass of the sample, S is the exposed surface and, 16 is the number of weeks of exposure

The corrosion rate is almost insignificant (e^{-5}) since its measurement is grams over mm^2 and week. Most implant materials are kept in the body for a maximum of 2 years (almost 100 weeks in bad cases), and the overall corrosion rate can be declared as 4.3‰. The more important aspect is the release rate of Mn, but as Table 3 shows, this release is only in the first 72 h of exposure, then, the trend is constant, and therefore it can be considered that Mn turns passive.

As a comparison, Table 5 lists the maximal allowed concentrations in the human body and the biological effects.

Table 5. Biological effects and maximal concentrations of metal ion accumulation in the human body.

Metal	Effect
Nickel (Ni)	It is the main cause of contact dermatitis. The main biological parameter is the amount of metal released on the skin during direct contact and exposure to human sweat. The limit is 0.5 mg/cm^2 x week, of which an insignificant part of Ni sensitive subjects will react. It has a toxic effect by creating cellular lesions and large cellular cultures. It is dangerous for bones and tissues, although less dangerous than Co or V, and it has cancer potency. The normal level of Ni in the blood is 5 mg/L [1].
Cobalt (Co)	Its function limits the role of vitamin B12 [1], by diminishing the adsorption of Fe in the blood stream [24]. The normal concentration of Co in human fluids is 1.5 mg/L.
Chromium (Cr)	It causes ulceration and central nerve system disorders [24]. The maximal concentration in the blood stream should be 28 mg/L. Its compounds are adsorbed only after oral ingestion. Cr (III) is usually deposited in reticular systems in the cell, while Cr (IV) can penetrate cellular membrane in both directions [1].
Aluminium (Al)	It provokes epileptic episodes and Alzheimer's disease [24]. The maximal concentration in the blood stream should be 30 mg/L.
Vanadium (V)	It is very toxic in its elementary state [24], therefore, the maximum concentration should not exceed 0.5 µg/L.
Molybdenum (Mo)	It is an essential element use by specific enzymes. It is easily adsorbed through the intestines, and its normal concentration in the blood stream should be 1–3 ppm. It is very toxic and sometimes lethal in large doses, regular symptoms are diarrhoea, coma, heart failure, and inhibitor for some essential enzymes. In addition, large concentration of Mo can interfere with Ca and P metabolism [1].

4. Conclusions

After being immersed for 60 days in a blood plasma SBF and having a large area depleted from its protective oxide layer, the sample shows pitting corrosion in the first two weeks of exposure which is visible by SEM.

Although the material has a large area in direct contact with SBF, a new protective layer of CoCr alloy is formed relatively quickly, as highlighted by the XRD assessment.

Nevertheless, during this period, metal ions are released in the SBF and the more concerning ones are Co and Cr.

The pH, electrical conductivity, and sample mass variation give a general idea about redox phenomena that occur.

A 3D printing technique was used for producing a CoCrMo-based alloy to be used as an orthopaedic implant by using the SLM technique and the incipient conclusion is that the produced CoCrMo alloy is suitable for such applications.

The concentration of metal ions dissociated from the biomaterial into the SBF solution was assessed by using the ICP-MS technique, which highlighted that the concentration of some metal ions varied during SBF solution exposure. The main conclusion is that those ions form a protective layer on the sample's surface mainly due to their reactivity and lead to the passivation of the exposed surface, thus, minimizing the concentration of ions that can dissociate into real body fluid.

Author Contributions: Conceptualization, R.M. and I.M.B.; methodology, R.M., L.C.C., R.E., and G.S.; software, A.P. and G.S.; validation, R.M., I.M.B., and R.E.; analysis, L.C.C., A.T.C., G.S., T.B., and E.P.; investigation, R.M., L.C.C., A.T.C., A.P., G.S., E.P., and T.B.; resources, R.M.; writing—original draft preparation R.M., R.E., and G.S.; writing—review and editing, R.M., R.E., and G.S., funding acquisition, R.M. All authors have read and agreed to the published version of the manuscript.

Funding: This research was carried out within POC-A1-A1.2.3-G-2015, ID/SMIS code P_40_422/105884, and the "TRANSCUMAT" Project, grant no. 114/09.09.2016 (subsidiary contract no. 1/D.1.5/114/24.10.2017), a project supported by the Romanian Minister of Research and Innovation.

Institutional Review Board Statement: Not applicable.

Informed Consent Statement: Not applicable.

Data Availability Statement: Data sharing not applicable.

Conflicts of Interest: The authors declare no conflict of interest. The funders had no role in the design of the study; in the collection, analyses, or interpretation of data; in the writing of the manuscript, or in the decision to publish the results.

References

1. Baht, S.V. *Biomaterials*; Narosa Publishing House: New Delhi, India, 2002.
2. Benmassauod, M.M.; Kohama, C.; Kim, T.W.B.; Kadlowec, J.A.; Foltiny, B.; Mercurio, T.; Ranganathan, S.I. Efficiency of eluted antibiotics trough 3D printed femoral implants. *Biomed. Microdevices* **2019**, *21*, 1–10.
3. El-Hajje, A.; Kolos, E.C.; Wang, J.K.; Malekaseedi, S.; He, Z.; Wiria, F.E.; Choong, C.; Ruys, A.J. Physical and mechanical characterization of 3D-printed porous titanium for biomedical applications. *J. Mater. Sci.* **2014**, *25*, 2471–2481.
4. Attarilar, S.; Ebrahimi, M.; Djavanroodi, F.; Fu, Y.; Wang, L.; Yang, J. 3D printing technologies in metallic implants: A thematic review of the techniques and procedures. *Int. J. Bioprinting* **2021**, *7*, 21–46. [CrossRef]
5. *ASTM Standard F2792-12a Standard Terminology for Additive Manufacturing Technologies*; ASTM International: West Conshohocken, PA, USA, 2012.
6. Sing, S.L.; An j Yeong, W.Y.; Wiria, F.E. Laser and electron-beam powder-bed additive manufacturing of metallic imlants: A review of processes, materials and design. *J. Orthop. Res.* **2016**, *34*, 369–385. [CrossRef]
7. Hussein, M.A.; Mohammed, A.S.; Al-Aqeeli, N. Wear characteristics of metallic biomaterials: A review. *Materials* **2015**, *8*, 2749–2768. [CrossRef]
8. Popescu, D.; Ene, R.; Popescu, A.; Cîrstoiu, M.; Sinescu, R.; Cîrstoiu, C. Total Hip Joint Replacement in Young Male Patient With Osteoporosis, Secondary To Hypogonadotropic Hypogonadism. *Acta Endocrinol.* **2015**, *11*, 109–113. [CrossRef]
9. Prasad, K.; Bazaka, O.; Chua, M.; Rochford, M.; Fredrick, L.; Spoor, J.; Synes, R.; Tieppo, M.; Collins, C.; Cao, A.; et al. Metallic biomaterials: Current chalanges and opportunities. *Materials* **2017**, *10*, 884. [CrossRef]
10. Gawlik, M.M.; Wiese, B.; Desharnis, V.; Ebel, T.; Willumeit-Romer, R. The effect of surface treatments on the degradation of biomedical Mg alloys—A review. *Materials* **2018**, *11*, 2561. [CrossRef] [PubMed]
11. Eliaz, N. Corrosion of metallic biomaterials: A review. *Materials* **2019**, *12*, 407. [CrossRef]
12. Bidhendi, H.R.A.; Pouranvari, M. Corrosion study o metallic biomaterials in simulated body fluid. *Metalurjia-MJoM* **2011**, *17*, 13–22. [CrossRef]
13. Baino, F.; Yamaguchi, S. The use of simulated body fluid (SBF) for assessing material bioactivity in the context of tissue engineering: Review and chalanges. *Biomimetics* **2020**, *5*, 57. [CrossRef]

14. Petkovic, D.S.; Mandrino, D.; Sarler, B.; Horky, J.; Ojdanic, A.; Zehetbauer, M.J.; Orlov, D. Surface analysis of biodegradable Mg-alloys after immersion in simulated body fluid. *Materials* **2020**, *13*, 1740.
15. Diomidis, N.; Mischler, S.; More, N.S.; Manish, R. Tribo-electrochemical characterization of metallic biomaterials for total joint replacement. *Acta Biomater.* **2012**, *8*, 852–859. [CrossRef] [PubMed]
16. Popa, M.; Demetrescu, I.; Vasilescu, E.; Drob, P.; Ionita, D.; Vasilescu, C. Corrosion Resistance of Some Thermo-mechanically Treated Titanium Bioalloys Depending on pH of Ringer Solution. *Rev. Roum. Chim.* **2009**, *60*, 241–247.
17. Nica, M.; Cretu, B.; Ene, D.; Antoniac, I.; Gheorghita, D.; Ene, R. Failure Analysis of Retrieved Osteosynthesis Implants. *Materials* **2020**, *13*, 1201. [CrossRef]
18. Wong, K.-C.; Scheinemann, P. Additive manufactured metallic implants for orthopaedic applications, Adv. In Metallic Biomaterials. *Sci. China Mater.* **2018**, *61*, 440–454. [CrossRef]
19. Available online: https://www.mee-inc.com/hamm/scanning-electron-microscopy-sem/ (accessed on 22 January 2021).
20. Marques, M.R.C.; Loedberg, R.; Almukainzi, M. Simulated biological fluids with possible applicaation in dissolution testing. *Technologies* **2011**, *18*, 15–26.
21. Mirea, R.; Ceatra, L.; Cucuruz, A.T.; Ene, R.; Popescu, E.; Biris, I.; Cretu, M. Advanced experimental investigation of used metallic biomaterials. *Rom. J. Mater.* **2019**, *59*, 138–145.
22. Dickinson, C.; Zhou, W.; Hodgkins, R.P.; Shi, Y.; Zhao, D.; He, H. Formation mechanism of porous single-crystal Cr_2O_3 and Co_3O_4 templated by mesoporous silica. *Chem, Matter.* **2006**, *18*, 3088–3095. [CrossRef]
23. Gorji, N.E.; O'Connor, R.; Brabazon, D. XPS, XRD and SEM characterization of the virgin and recycled powders for 3D printing applications. *IOP Conf. Ser. Mater. Sci. Eng.* **2019**, *591*, 012016. [CrossRef]
24. Aksakal, B.; Yildirim, Ö.S.; Gul, H. Metallurgical failure analysis of various implant materials used in orthopedic applications. *J. Fail. Anal. Prev.* **2004**, *4*, 17–23. [CrossRef]

 *metals*

MDPI

Article

Fatigue Assessment of Selective Laser Melted Ti-6Al-4V: Influence of Speed Manufacturing and Porosity

Unai Segurajauregi [1], Adrián Álvarez-Vázquez [2], Miguel Muñiz-Calvente [2,*], Íker Urresti [1] and Haydee Naveiras [1]

[1] Ikerlan Technology Research Centre, Basque Research and Technology Alliance (BRTA), Paseo J.M. Arizmendiarrieta 2, 20500 Arrasate-Mondragon, Spain; usegurajauregi@ikerlan.es (U.S.); iurresti@ikerlan.es (Í.U.); nphaydee@gmail.com (H.N.)

[2] Department of Construction and Manufacturing Engineering, University of Oviedo, 33203 Gijón, Spain; alvarezvadrian@uniovi.es

[*] Correspondence: munizcmiguel@uniovi.es

Abstract: Additive Manufacturing represents a promising technology as an alternative to the conventional manufacturing process, with rapid and economic product development, as well as a significant weight reduction and a freeform design. Although the mechanical properties of additively manufactured metals, such as the Ti-6Al-4V alloy, are well-established, a complete understanding of the fatigue performance is still a pending aspiration due to its inherent stochastic complexity and the influence of several manufacturing factors. This paper presents a study of the influence of speed manufacturing and porosity in the fatigue behaviour of a Ti-6Al-4V alloy. To this aim, a numerical simulation of the expected porosity at different laser velocities is performed, together with a simulation of the residual stresses. These numerical results are compared with experimental measurements of residual stresses and a qualitative analysis of the porosities. Then, fatigue strength is experimentally obtained for two different laser speeds and fitted by a probabilistic model. As a result, the probabilistic S–N fields for different laser velocities are found to be similar, with scatter bands nearly coincident, drawing the conclusion that this effect is negligible in comparison with other concurrent ones, such as roughness or surface defects from manufacturing conditions, promoting crack initiation and premature fatigue failure.

Keywords: additive manufacturing; fatigue; titanium; selective laser melting

Citation: Segurajauregi, U.; Álvarez-Vázquez, A.; Muñiz-Calvente, M.; Urresti, Í.; Naveiras, H. Fatigue Assessment of Selective Laser Melted Ti-6Al-4V: Influence of Speed Manufacturing and Porosity. *Metals* **2021**, *11*, 1022. https://doi.org/10.3390/met11071022

Academic Editors: Atila Ertas, Adam Stroud and Matteo Benedetti

Received: 3 May 2021
Accepted: 22 June 2021
Published: 25 June 2021

Publisher's Note: MDPI stays neutral with regard to jurisdictional claims in published maps and institutional affiliations.

Copyright: © 2021 by the authors. Licensee MDPI, Basel, Switzerland. This article is an open access article distributed under the terms and conditions of the Creative Commons Attribution (CC BY) license (https://creativecommons.org/licenses/by/4.0/).

1. Introduction and Motivation

Additive Manufacturing (AM), formerly known as Rapid Prototyping (RP) (see ASTM F2792-12a [1]), has undoubtedly been increasing over the last two decades as a technology that is disrupting current manufacturing processes, and attracts interest from both industrial and academic perspectives [2–5]. Known also as 3-D printing, AM consists of a progressive consolidation of raw materials, such as powder or wire, in a layer-by-layer fashion, in an opposite approach to traditional manufacturing processes, which are typically based on machining block parts, that is, the subtraction or removal of material [6]. Moreover, this novel technology has several important advantages compared with traditional methods: an agile development product from Computer Aided Design (CAD) to fabrication; a significant reduction of weight in the final design (with potential reductions of up to nearly 50% [6,7]); and a geometric freedom that allows the production of parts otherwise not possible with conventional methods. Additionally, AM may lead to the reduction of carbon emissions compared with traditional manufacturing processes, due to the use of lighter weight parts [6,8].

Polymeric materials were originally preferred for producing additively manufactured parts [5], but nowadays both non-metallic (composites, ceramics) and metallic materials are usually employed. The titanium (Ti) alloys, in particular, are of great interest because of their increasing use in the aerospace industry due to their weight saving, operating temperature, corrosion resistance and compatibility with biological and composite materials. Unfortunately, their higher cost compared with other alternatives hitherto represents the most important limitation [6,9–12].

Three different technologies were developed to produce metal additively manufactured parts: Laser Beam Melting (LBM), Laser Metal Deposition (LMD) and Selective Laser Melting (SLM), the latter being the most commonly used [13,14]. In this process, the energy of the laser source is applied to melt powder, as a raw material, within a powder-bed layer. Then, the 3D geometrical design is built up by recoating a new powder-layer and subsequent melting [5]. Nowadays, these manufacturing technologies for developing structural alloys are particularly useful, since the intrinsic heat can be directly used to trigger the chemical reactions, such as those implied in precipitation hardening alloys [15,16].

There are different works in the literature that have focused on researching the influence of different additive manufacturing parameters on the fatigue performance of Ti-6Al-4V, such as the microstructure, the build direction, the residual stresses and the porosity. In the first case, Nalla et al. [17] have investigated the influence of the microstructure on both bimodal and coarser lamellar types, concluding that the latter improved the fatigue behaviour in the HCF zone, whereas Thijs et al. [18] studied the influence of the scanning parameters and scanning strategy on the microstructure during the SLM process. In the second case, Edwards et al. [6,19] presented a study on the effect of the build direction, revealing that the cracks oriented perpendicular to the build layers provide enhanced fatigue crack growth behaviour. Regarding the residual stresses, several researchers [20,21] have found that a high temperature pre-heating during the additive manufacturing process may reduce thermal gradients. Lastly, [22] evidenced that the failure initiation in SLM or EBM manufactured titanium alloys is governed by porosity and lack of fusion. Nevertheless, previous works have not investigated the influence of speed manufacturing, which would be conducted with different porosities and could imply different fatigue behaviours.

The aim of this paper is to study the influence of different laser velocities on the porosity of additively manufactured specimens of the Ti-6Al-4V alloy and to evaluate the fatigue performance associated with those porosities. Firstly, numerical simulations were developed to study the expected porosity considering different laser velocities, together with residual stresses inherent to the manufacturing process. Secondly, these numerical results were compared with experimental measurements of residual stresses and a qualitative analysis of the porosities. Thirdly, a tensile test was conducted on specimens produced at two different laser velocities in order to evaluate its influence on the mechanical properties. After that, a fatigue experimental campaign was carried out on specimens at two different laser speeds and the results were evaluated according to a probabilistic S–N model developed by [23], contrary to the fatigue assessment methodologies performed in the literature, based on deterministic S–N models [24–26], through the inherent and non-negligible scatter titanium fatigue tests [6,27–31].

The paper is structured as follows: in Section 2, the material and methods employed in this study are detailed, including the material and geometry selected (Section 2.1), the manufacturing conditions (Section 2.2) and the experimental procedures followed in the testing (Section 2.3). Section 3 details the numerical study of the porosity and the residual stresses together with the experimental results obtained. Section 4 is focused on both the tensile (Section 4.1) and fatigue characterization of two different laser velocities (Section 4.2). Section 5 presents an interpretation of the experimental results and, finally, Section 6 summarises the main conclusions drawn from this work.

2. Materials and Methods

This section describes the material and geometry of the specimens to be used in the experimental campaign in this work. Then, the manufacturing conditions are also detailed, distinguishing different batches of samples fabricated for porosity, tensile and fatigue characterization. Finally, the experimental testing procedures are exposed.

2.1. Material and Geometry

The specimens were produced using the SLM technique employing a titanium-based alloy, Ti-6Al-4V. The dimensions of the specimen are indicated in Figure 1. Note that the build direction Z-axis is indicated, starting at the support.

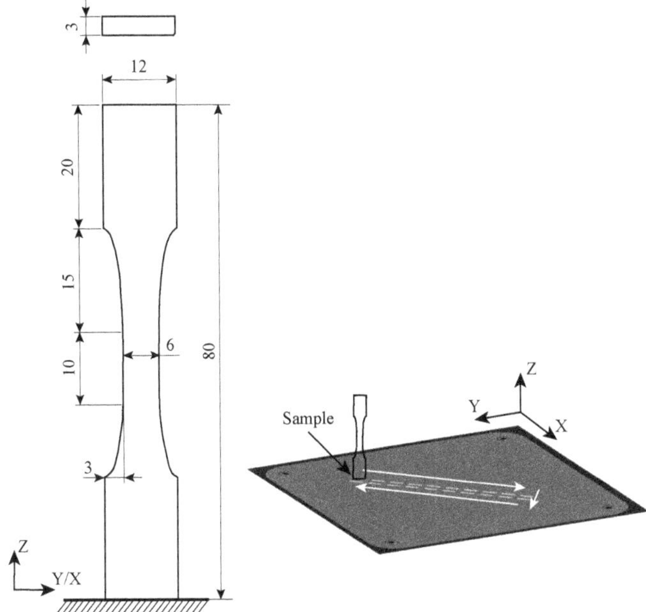

Figure 1. Dimensions (in mm.) of the Ti-6Al-4V specimens with build direction in Z and the support.

2.2. Manufacturing Conditions

All the specimens were provided by the manufacturer Optimus3D (Vitoria–Gasteiz, Spain) in the same orientation. The SLM parameters selected are detailed in Table 1. Three different batches of additively manufactured Ti-6Al-4V samples were produced with the following purposes:

- Batch 1: Porosity characterization. Fourteen specimens were fabricated with 7 different velocities to evaluate their influence on the porosity (2 samples per velocity): {800, 1100, 1200, 1300, 1500, 1700, 1900} mm/s.
- Batch 2: Tensile characterization. Two specimens were manufactured with laser velocities of 1200 and 1900 mm/s (1 sample per velocity) for tensile characterization of the material.
- Batch 3: Fatigue characterization. Fifteen specimens were manufactured with laser velocities of 1200 (7 samples) and 1900 mm/s (8 samples) for fatigue characterization.

Table 1. SLM parameters selected.

Layer Thickness [mm]	Hatch Spacing [mm]	Power [W]
0.03	0.1	200

2.3. Testing and Characterization Procedures

2.3.1. Porosity Measurement Procedure

The samples from *Batch 1* were sliced using a diamond disc cutter by removing a layer thickness of no less than 1 mm, in order to avoid the effect of the cutting process on the microstructure. Silicon carbide papers with different grades of 80, 240, 600, 1200 and 2500 (in this order) were used for the grinding with a continuous water stream for flushing the loose and abrasive particles. Then, a manual polishing process was applied to the surfaces using a diamond suspension of 9, 3 and 1 microns of particle size on a Remet LS1.

It is worth mentioning that neither the heat treatment nor the machining process were conducted on specimens before the experimental campaign, since the aim of this work was to study only the laser velocity, that is, without any additional varying concomitant effect that could mislead the interpretation of the experimental results. After that, the porosity was qualitatively analysed using both optical and scanning electron microscope (SEM). The entire surface of the specimens was analyzed with a resolution of 500 μm in order to identify the zones with larger pores. After that, different scanning zooms (10 and 50 μm) were applied to focus on the zones where pores had been observed. In cases where the resolution of 500 μm was not enough to identify pores in any part of the specimen, 50 μm was used to check the entire surface. It is important to remark that smaller pores could not be identified by the applied resolution. Still, the authors assumed that the influence of those pores on the fatigue life could be disregarded, compared to the pores identified in this study.

2.3.2. Residual Stresses Measurement Procedure

The measurement of the residual stresses was performed by way of the hole drilling strain gage method according to ASTM E837–13a [32] using an MTS-300 RS measurement machine supplied by SINT Technology, as can be seen in Figure 2. The parameters selected included a drilling speed of 0.2 mm/min, a drill delay of 2–3 s and an acquisition delay set to 5–10 s. The RESTAN software (SINT Technology) was used to obtain the residual stresses.

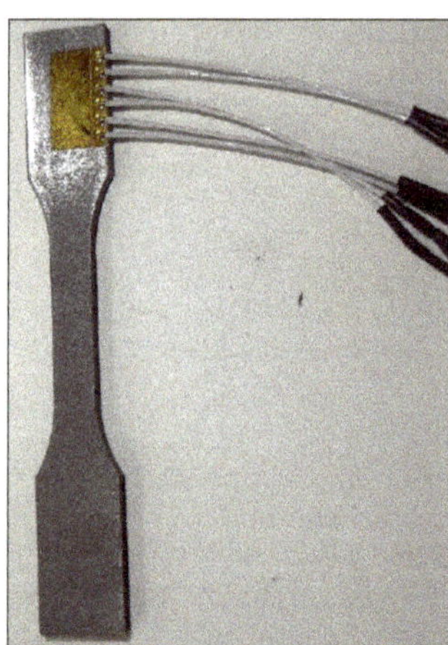

Figure 2. Strain gage method in hole drilling for measurement of residual stresses.

2.3.3. Tensile and Fatigue Characterization Procedure

The tensile tests were developed according to EN 2002 [33]. A strain rate of 0.05 mm/min was used for yield stress σ_{ys} at 0.2 % and 2 %/min for tensile strength σ_r. The displacement was measured using DIC equipment.

The fatigue tests were conducted for sinusoidal load at $R = 0.1$ and at ambient temperature, according to ASTM E466-07 [34], at a frequency of 6 Hz with stress ranging from 100 to 600 MPa, in a servohydraulic MTS Bionix. As previously mentioned, a total of 7 samples were tested for laser velocity $v = 1200$ mm/s and 8 samples for laser velocity $v = 1900$ mm/s.

3. Study of Porosity and Residual Stresses

This section presents the results obtained from numerical and experimental studies related to the estimation and measurement of specimens porosity and residual stresses.

3.1. Numerical Study: Expected Porosity and Residual Stresses

Two different numerical studies were conducted: a porosity simulation for each of the different laser velocities considered in *Batch 1*, and a finite element simulation for the estimation of the residual stresses.

In the first case, the tool known as "Additive Science", from the Additive Suite developed by ANSYS [35], was used to simulate the expected porosity with different laser velocities. The following constraints in the dimensions of the melt pool were introduced as an input (see Figure 3):

- Reference depth ($D \geq 0.045$). It ensures that half of the third layer is passed in the material fusion.
- Ratio depth/width ($D/W < 0.95$). It ensures the depth is nearly the same as the width.
- Ratio length/width ($L/W < 4$). It ensures the length is not larger than the width.

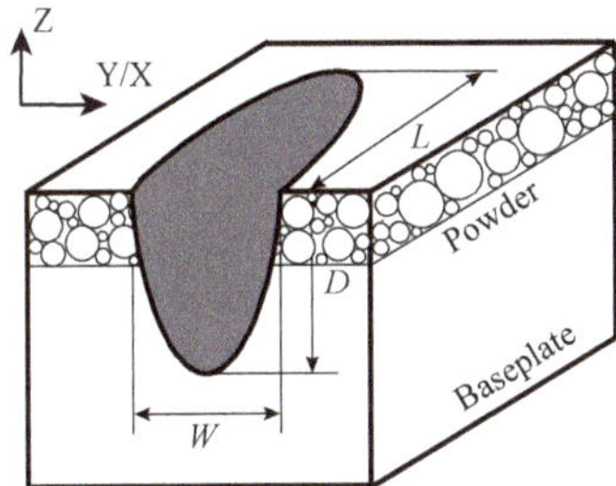

Figure 3. Dimensions of the melt pool.

Based on the constraints on the melt pool, the geometrical mean results provided by Additive Science for each velocity are summarised in Table 2.

Table 2. Numerical results of the geometrical dimensions of the melt pool for different velocities.

Scan Speed [mm/s]	$\overline{D}$ [mm]	$\overline{L}$ [mm]	$\overline{W}$ [mm]
800	0.115	0.447	0.157
1100	0.082	0.446	0.136
1200	0.074	0.442	0.130
1300	0.067	0.439	0.125
1500	0.056	0.428	0.115
1700	0.047	0.417	0.108
1900	0.039	0.400	0.101

Once the dimensions of the melt pool had been estimated, the software provided the expected porosity for each of the different laser velocities considered, as shown in Table 3, together with the corresponding energy density in the SLM process. The porosity was directly estimated by ANSYS as a function of the energy density applied, which depends on the power, velocity and the dimensions of the melt pool. It is also worth mentioning that this software only considered the porosity that was due to a lack of fusion, which was predominant at high velocities, and the porosity that was due to spherical vaporization was discarded. Then, in order to identify the kind of porosity to be experimentally found, the causal relationship between the energy density and the kind of porosity proposed by Dilip et al. [36] was used: spherical pores correspond with energy density >60 J/mm^3, a lack of pores (or completely net) corresponds with energy density in the interval 55–60 J/mm^3 and sharper pores correspond with energy density <55 J/mm^3. Then, according to these limits, only the velocity 800 mm/s was expected to produce spherical pores, but with a lower percentage of porosity. On the contrary, those velocities higher than 1300 mm/s were expected to give sharper pores, with the porosity percentage increasing as much as the velocity increases. According to the simulation performed and the results obtained, middle velocities ranging from 1100 to 1200 were not expected to produce pores, but completely net structures. Finally, it is worth mentioning that the values reported in Table 2 regarding the percentage of porosity are not related to experimental measurements, but to results obtained by the Additive Suite tool developed by ANSYS, in combination with the work reported by Dilip et al. [36], which studied the influence of processing parameters on the evolution of melt pool and porosity in Ti-6Al-4V alloy parts fabricated by selective laser melting.

Table 3. Numerical results of the porosity simulations for different velocities.

Scan Speed [mm/s]	Energy Density [J/mm^3]	Porosity [%]
800	83.33	0.00
1100	60.61	0.00
1200	55.56	0.00
1300	51.28	0.00
1500	44.44	0.01
1700	39.22	0.08
1900	35.09	0.36

In the second case, the Additive Science tool allowed the residual stress to be numerically simulated, as can be seen in Figure 4, where the equivalent von Mises stress is depicted along the sample.

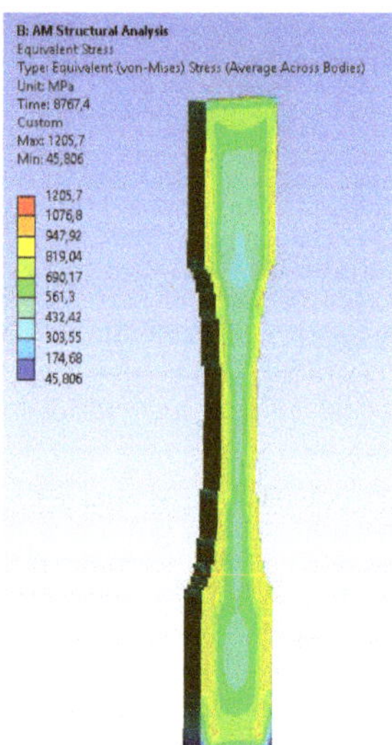

Figure 4. Residual stresses simulated upon the removal of the support.

3.2. Experimental Results

3.2.1. Porosity

The porosity was qualitatively analysed using both optical and scanning electron microscopes (SEM) for each of the laser speeds considered in Batch 1, as can be seen in Figure 5. As is well known, the geometrical forms of the pores are expected to be heterogeneous depending on the laser velocity, which occurred in this case; ranging from spherical (Figure 5a) at low velocities and high laser power, to irregular and sharper (Figure 5d–f) for large velocities and low laser power, which is in accordance with the simulated results in Section 3.1. The former are known to be caused due to improper settings or processing parameters [12,37,38], while the latter are usually related in the literature to the argon gas entrapped during the manufacturing process [39–41]. Furthermore, the middle velocities were expected to produce a negligible porosity in comparison with the other velocities (see Table 2), which is corroborated in the micrographs in Figure 5b,c).

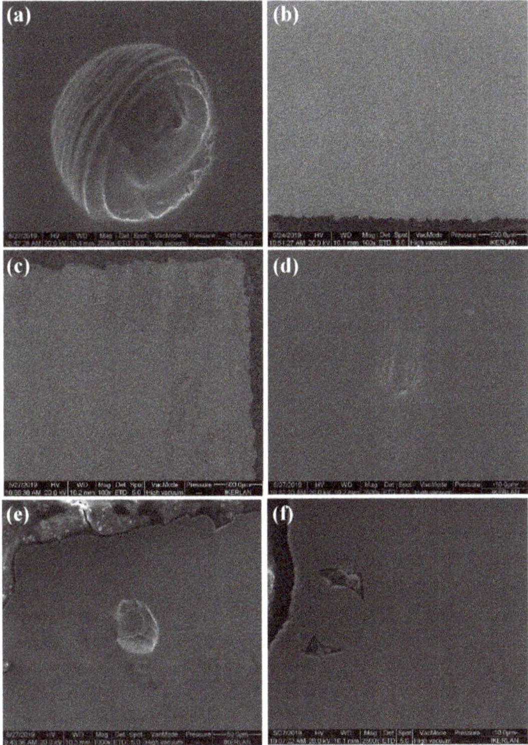

Figure 5. SEM micrographs for different laser velocities: (**a**) $v = 800$ mm/s, (**b**) $v = 1100$ mm/s, (**c**) $v = 1200$ mm/s, (**d**) $v = 1500$ mm/s, (**e**) $v = 1700$ mm/s, (**f**) $v = 1900$ mm/s.

3.2.2. Residual Stress Measurement

Figure 6 illustrates the experimental results of the maximum and minimum residual stresses along the distance for both velocities considered in Batch 1. The maximum values of the residual stresses for the higher velocity evolve steadily along the distance, while for the lower velocity, a peak is present at the 0.1 mm distance. The same behaviour is observed in the case of minimum residual stress but at a lower order of magnitude. In general terms, there is an inverse trend between the development of residual stresses and the laser speed, that is, the values for both maximum and minimum residual stresses are higher for the larger velocity until a certain distance of almost 0.6 mm, where both trends tend to be equal.

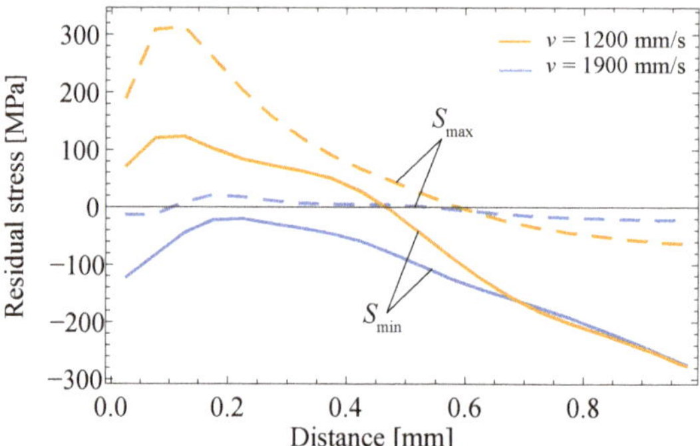

Figure 6. Experimental values of maximum and minimum residual stresses for two different laser velocities: $v = 1200$ mm/s and $v = 1900$ mm/s.

4. Study of Tensile and Fatigue Behaviour

Once the study of the expected porosity and residual stresses and its comparison with the experimental results was performed, the tensile and fatigue characterization was conducted, and is now described in this section.

4.1. Tensile Behaviour

Engineering stress–strain curves for both velocities considered in Batch 2 are illustrated in Figure 7. As can be seen, the linear-elastic regime is approximately the same in both cases, but with a larger yield strength the lower the laser speed. In the plastic zone, however, the samples manufactured at a lower laser velocity produce a better tensile performance for the same strain value. Table 4 summarises the mechanical constants of the additively manufactured specimens for both velocities.

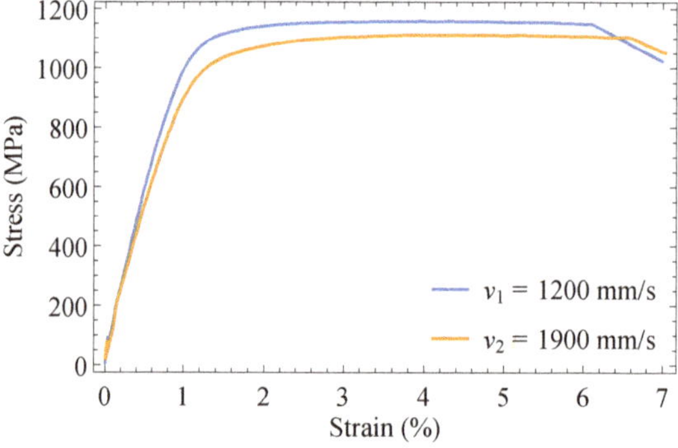

Figure 7. Engineering stress vs. strain curves for Ti-6Al-4V for two velocities: 1200 and 1900 mm/s.

Table 4. Tensile properties for additively manufactured Ti-6Al-4V obtained from the stress–strain curves in Figure 7.

Velocity (mm/s)	σ_{ys} (MPa)	σ_r (MPa)	ε_f (%)
1200	1068.62	1158.96	6.12
1900	990.69	1114.76	6.58

4.2. Fatigue Behaviour

Finally, the fatigue assessment of manufactured Ti-6Al-4V samples from Batch 3 was conducted according to the probabilistic model developed by [23]. In this model, the *p*-percentile curves in the S–N field are given as the following Weibull distribution:

$$p = 1 - \exp\left[-\left(\frac{(\log \Delta\sigma - C)(\log N - B) - \lambda}{\delta}\right)^{\beta}\right], \tag{1}$$

with B as the horizontal asymptote for the stress, that is, the fatigue strength, C as the vertical asymptote for the lifetime, that is, no fatigue failure will occur below this limit, and λ, δ and β as the location, scale and shape Weibull parameters, such that $(\log \Delta\sigma - C)(\log N - B) > \lambda$.

Figure 8 depicts the estimated S–N fields for both velocities considered in this batch, which were estimated easily with ProFatigue software [42]. As can be seen, the inherent scatter of fatigue results is non-negligible, thus a probabilistic model is more suitable than a deterministic one. The fatigue performance on both velocities exhibits the same behaviour with no differences in the lifetime cycle for different given stress ranges, and the scatter bands are approximately similar. For this reason, speed manufacturing seems to not have a significant effect on the fatigue performance of additively manufactured Ti-6Al-4V.

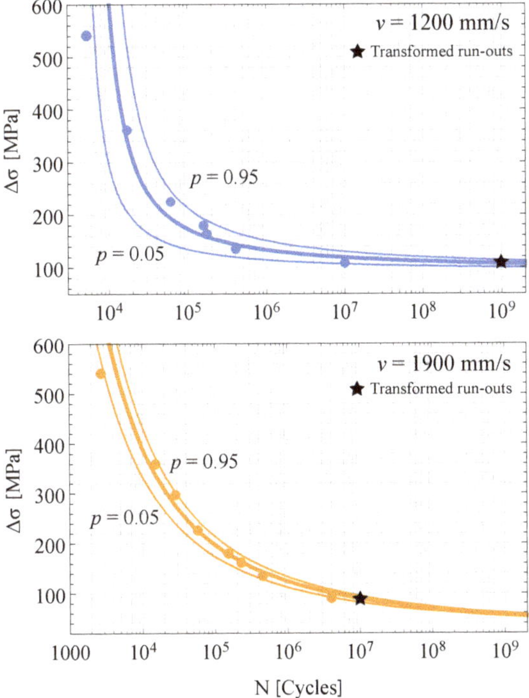

Figure 8. Resulting S–N curves from the Castillo–Canteli model for Ti-6Al-4V with velocities 1200 mm/s and 1900 mm/s.

Finally, the experimental campaign retrieved from [6], corresponding with a lower laser velocity of $v = 200$ mm/s, was also estimated according to the Castillo–Canteli model and is superposed in Figure 9 for comparison purposes. As a result, a wide range of laser speeds were considered from 200 to 1900 mm/s, providing robustness to the final conclusions drawn from this work. Indeed, though having a lower laser velocity, the resulting

S–N field is approximately the same with both previous velocities at 1200 and 1900 mm/s, enhancing the conclusion that there was a negligible effect on the fatigue performance.

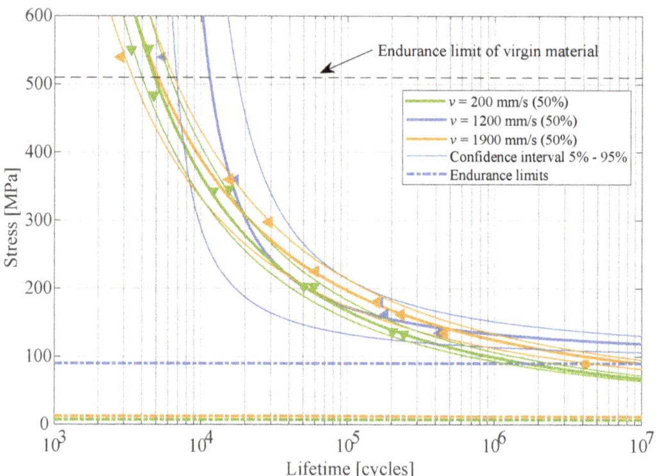

Figure 9. Comparison of S–N fields for different laser velocities: $v = 200$ mm/s (data from Edwards and Ramulu [6]), $v = 1200$ mm/s and $v = 1900$ mm/s.

5. Discussion

In this work, different laser speeds were considered in additively manufactured Ti-6Al-4V alloy to evaluate their influence on the fatigue performance, where any other additional concurrent effect was avoided (heat treatment, machining surfaces, etc.). The experimental fatigue results were estimated according to a probabilistic model developed by [23], and the resulting S–N curves were approximately the same with the scatter bands being nearly coincident. In other words, the laser speed effect is negligible and no influence is found on the fatigue lifetime of Ti-6Al-4V specimens. Then, in order to enhance this conclusion, an external experimental campaign retrieved from [6]—at a lower velocity from those previously considered, that is, $v = 200$ mm/s—was also estimated and compared with previous results. Though having a wide range of laser velocities, from 200 to 1900 mm/s, the resulting S–N fields are approximately the same and the scatter bands are nearly coincident. However, this evidence does not allow for the conclusion that the effect of laser speed is negligible in terms of the fatigue lifetime of the specimens, since other concurrent effects could be covering the effects associated with the laser velocity and misleading the conclusions. The authors postulate that the high surface roughness obtained in the additive manufacturing process ($R_a = 3.27$, $R_z = 15.83$), together with the few differences in porosities depending on both laser speeds, lead to all cracks originating from the surfaces of the specimens, misleading conclusions about the potential effects of the porosities on the fatigue lifetime. Future work could consider the prior polishing of the specimens in order to improve the surface roughness, thus inducing the failure to occur from the pores.

The authors want to remark that the tensile and fatigue properties presented in this paper are related to only one SLM building direction. Taking into account that the material properties of metals manufactured by SLM cannot be considered isotropic— that is, dependent on the testing direction—the conclusions of this paper must be taken into consideration only for the manufacturing direction in which the specimens were made, since different results could be achieved for other directions. Furthermore, the relationship between speed manufacturing and the resulting mechanical properties is not straightforward in additively manufactured samples, as other variables are implied, such as the melt pool depth and temperature required for complete melting or evaporation; the effects of these variables were not considered in this study but are proposed as the basis for further work.

6. Conclusions

- The experimental values of the residual stresses increase for lower laser speeds for both maximum and minimum values.
- The expected porosity was simulated for different laser velocities, establishing limiting energy densities to identify the kind of pores: spherical, sharper or absent.
- The porosity was qualitatively analysed for seven different velocities, corroborating that, for lower speeds, the pores are spherical while for larger speeds they are sharper and more irregular. On the contrary, for middle velocities no pores were detected.
- Tensile experimental results at two different laser speeds showed an influence on the mechanical properties of Ti-6Al-4V alloys, especially in the plastic regime.
- A probabilistic model was used to estimate the fatigue lifetime for two different velocities, concluding that this effect is negligible in comparison with other concurrent variables, such as surface defects or roughness.
- The influence of speed manufacturing will only be non-negligible when other concurrent effects, such as those caused by the machining process or heat treatments, are softened or relaxed.

Author Contributions: Conceptualization, U.S., M.M.-C., Í.U. and H.N.; methodology, U.S., M.M.-C., Í.U. and H.N.; software, U.S., M.M.-C. and H.N.; validation, M.M.-C. and A.Á.-V.; formal analysis, U.S., M.M.-C. and A.Á.-V.; investigation, U.S., M.M.-C. and Í.U.; resources, U.S., Í.U. and H.N.; data curation, U.S. and H.N.; writing—original draft preparation, U.S. and A.Á.-V.; writing—review and editing, A.Á.-V., U.S. and M.M.-C.; visualization, A.Á.-V. and M.M.-C.; supervision, All authors; project administration, I.U. and M.M.-C.; funding acquisition, U.S. and Í.U. All authors have read and agreed to the published version of the manuscript.

Funding: This research received no external funding.

Institutional Review Board Statement: Not applicable.

Informed Consent Statement: Not applicable.

Data Availability Statement: Not applicable.

Conflicts of Interest: The authors declare no conflict of interest.

References

1. ASTM F2792-12a. *Standard Terminology for Additive Manufacturing Technologies*; Technical Report; American Society for Testing and Materials: Philadelphia, PA, USA, 2012.
2. Choi, D.S.; Lee, S.H.; Shin, B.S.; Whang, K.H.; Song, Y.A.; Park, S.H.; Jee, H.S. Development of a direct metal freeform fabrication technique using CO_2 laser welding and milling technology. *J. Mat. Proc. Technol.* **2001**, *113*, 273–279. [CrossRef]
3. Thompson, S.M.; Bian, L.; Shamsaei, N.; Yadollahi, A. An overview of direct laser deposition for additive manufacturing; part i: Transport phenomena, modeling and diagnostics. *Addit. Manuf.* **2015**, *8*, 36–62. [CrossRef]
4. Huang, Y.; Leu, M.C. *Frontiers of Additive Manufacturing Research and Education*; Technical Report; University of Florida: Gainesville, FL, USA, 2016.
5. Gibson, I.; Rosen, D.; Stucker, B. *Additive Manufacturing Technologies. Rapid Prototyping to Direct Digital Manufacturing*, 3rd ed.; Springer: Berlin/Heidelberg, Germany, 2021.
6. Edwards, P.; Ramulu, M. Fatigue performance evaluation of selective laser melted Ti-6Al-4V. *Mater. Sci. Eng. A* **2014**, *598*, 327–337. [CrossRef]
7. Edwards, P.; O'Conner, A.; Ramulu, M. Electron Beam Additive Manufacturing of Titanium Components: Properties and Performance. *J. Manuf. Sci. Eng.* **2013**, *135*, 061016. [CrossRef]
8. Sreenivasan, R.; Goel, A.; Bourell, D.L. Sustainability issues in laser-based additive manufacturing. *Phy. Proc.* **2010**, *5*, 81–90. [CrossRef]
9. Boyer, R.R. An overview on the use of titanium in the aerospace industry. *Mater. Sci. Eng.* **1996**, *213*, 103–114. [CrossRef]
10. Peters, M.; Kumpfert, J.; Ward, C.H.; Leyens, C. Titanium alloys for aerospace applications. *Adv. Eng. Mater.* **2003**, *5*, 419–427. [CrossRef]
11. Banerjee, D.; Williams, J.C. Perspectives on titanium science and technology. *Acta Mater.* **2013**, *61*, 844–879. [CrossRef]
12. Boyer, R.R. A review of the fatigue properties of additively manufactured Ti-6Al-4V. *JOM* **2018**, *70*, 349–357.

13. Abe, F.; Osakada, K.; Shiomi, M.; Uematsu, K.; Matsumoto, M. The manufacturing of hard tools from metallic powders by selective laser melting. *J. Mater. Process. Technol.* **2001**, *111*, 210–213. [CrossRef]
14. Santos, E.C.; Shiomi, M.; Laoui, T.; Osakada, K. Rapid manufacturing of metal components by laser forming. *Int. J. Mach. Tools. Manuf.* **2006**, *46*, 1459–1468. [CrossRef]
15. Kürnsteiner, P.; Wilms, M.; Weisheit, A.; Barriobero-Vila, P.; Gault, B.; Jägle, E.; Raabe, D. In-process precipitation during laser additive manufacturing investigated by atom probe tomography. *Microsc. Anal.* **2017**, *23*, 694–695. [CrossRef]
16. Simonelli, M.; McCartney, D.G.; Barriobero-Vila, P.; Aboulkhai, N.T.; Tse, Y.Y.; Clare, A.; Hague, R. The influence of iron in minimizing the microstructural anisotropy of Ti-6Al-4V produced by laser powder-bed fusion. *Metall. Mater. Trans. A* **2020**, *51*, 2444–2459. [CrossRef]
17. Nalla, R.K.; Ritchie, R.O.; Boyce, B.L.; Campbell, J.P.; Peters, J.O. Influence of microstructure oon high-cycle fatigue of Ti-6Al-4V: bimodal vs. lamella structures. *Metall. Mater. Trans. A* **2002**, *33*, 899–918. [CrossRef]
18. Thijs, L.; Verhaeghe, F.; Craeghs, T.; VanHumbeeck, J.; Kruth, J.P. A study of the microstructural evolution during selective laser melting of Ti-6Al-4V. *Acta Mater.* **2010**, *58*, 3303–3312. [CrossRef]
19. Edwards, P.; Ramulu, M. Effect of build direction on the fracture toughness and fatigue crack growth in selective laser melted Ti-6Al-4V. *Fatigue Fract. Eng. Mater. Struct.* **2015**, *38*, 1228–1236. [CrossRef]
20. Van Zyl, I.; Yadroitsava, I.; Yadroitsev, I. Residual stresses in Ti6Al4V objects produced by direct metal laser sintering. *Addit. Manuf.* **2016**, *27*, 134–141.
21. Ali, H.; Ma, L.; Ghadbeigi, H.; Mumtaza, K. In-situ residual stress reduction, martensitic decomposition and mechanical properties enhancemente through high temperature powder bed pre-heating of selective laser melted Ti6Al4V. *Mater. Sci. Eng. A* **2017**, *695*, 211–220. [CrossRef]
22. Günther, J.; Krewerth, D.; Lippmann, T.; Leuders, S.; Tröster, T.; Weidner, A.; Biermann, H.; Niendorf, T. Fatigue life of additively manufactured Ti-6Al-4V in the very high cycle fatigue regime. *Int. J. Fatigue* **2017**, *94*, 236–245. [CrossRef]
23. Castillo, E.; Fernández-Canteli, A. *A Unified Statistical Methodology for Modeling Fatigue Damage*; Springer: Dordrecht, The Netherlands, 2009.
24. Spierings, A.B.; Starr, T.L.; Wegener, K. Fatigue perfomance of additive manufactured metallic parts. *Rapid Prototyp. J.* **2013**, *19*, 88–94. [CrossRef]
25. Kasperovich, G.; Hausmann, J. Improvement of fatigue resistance and ductility of TIAl6V4 processed by selective laser melting. *J. Mater. Process. Technol.* **2015**, *220*, 202–214. [CrossRef]
26. Walker, K.F.; Liu, Q.; Brandt, M. Evaluation of fatigue crack propagation behaviour in Ti-6Al-4V manufactured by selective laser melting. *Int. J. Fatigue* **2017**, *104*, 302–308. [CrossRef]
27. Le, V.; Pessard, E.; Morel, F.; Edy, F. Influence of porosity on the fatigue behaviour of additively fabricated Ta6V alloys. In Proceedings of the 12th International Fatigue Congress (FATIGUE 2018), Potiers, France, 27 May–1 June 2018; Volume 165, p. 02008.
28. Greitemeier, D.; Palm, F.; Syassen, F.; Melz, T. Fatigue performance of additive manufactured TiAl6V4 using electron and laser beam melting. *Int. J. Fatigue* **2017**, *94*, 211–217. [CrossRef]
29. Wycisk, E.; Emmelmann, C.; Siddique, S.; Walther, F. High cycle fatigue (HCF) performance of Ti-6Al-4V alloy processed by selective laser melting. *Adv. Mater. Res.* **2013**, *816*, 134–139. [CrossRef]
30. Leuders, S.; Thöne, M.; Riemer, A.; Niendorf, T.; Tröster, T.; Maier, H.J. On the mechanical behaviour of titanium alloy TiAl6V4 manufactured by selective laser melting: Fatigue resistance and crack growth performance. *Int. J. Fatigue* **2013**, *48*, 300–307. [CrossRef]
31. Li, P.; Warner, D.H.; Fatemi, A.; Phan, N. Critical assessment of the fatigue performance of additivelymanufactured Ti-6Al-4V and perspective for future research. *Int. J. Fatigue* **2016**, *85*, 130–143. [CrossRef]
32. ASTM E837-13a. *Standard Test Method for Determining Residual Stresses by the Hole-Drilling Strain-Gage Method*; Technical Report; American Society for Testing and Materials: Philadelphia, PA, USA, 2020.
33. EN 2002. *Aerospace Series–Metallic Materials–Test Methods–Part 1: Tensile Testing at Ambient Temperature*; Technical Report; European Standards: Brussels, Belgium, 2005.
34. ASTM E466-07. *Standard Practice for Conducting force Controlled Constant Amplitude axial Fatigue Tests of Metallic Materials*; Technical Report; American Society for Testing and Materials: Philadelphia, PA, USA, 2007.
35. ANSYS. *Additive Print and Additive Science User Guide*; ANSYS, Inc.: Canonsburg, PA, USA, 2019.
36. Dilip, J.J.S.; Zhang, S.; Ten, C.; Zeng, J.; Robinson, C.; Pal, D.; Stucker, B. Influence of processing parameters on the evolution of melt pool, porosity, and microstructures in Ti-6Al-4V alloy parts fabricated by selective laser melting. *Prog. Addit. Manuf.* **2017**, *2*, 157–167. [CrossRef]
37. Wang, P.; Nai, M.L.S.; Aw, B.; Wei, L.J. Effect of processing parameters on microstructure and mechanical properties of Ti-6Al-4V made by selective electron beam melting additive manufacturing. In Proceedings of the Annual International Solid Freeform Fabrication Symposium (SFF Symp 2016), Austin, TX, USA, 8–10 August 2016.
38. Wang, P.; Nai, M.L.S.; Tan, X.; Vastola, G.; Raghavan, S.; Sin, W.J.; Tor, S.B.; Pei, Q.X.; Wei, J. Recent progress of additive manufactured Ti-6Al-4V by electron beam melting. In Proceedings of the Annual International Solid Freeform Fabrication Symposium (SFF Symp 2016), Austin, TX, USA, 8–10 August 2016.

39. Mok, S.H.; Bi, G.; Folkes, J.; Pashby, I. Deposition of Ti-6Al-4V using a high power diode laser and wire, Part I: Investigation on the process characteristics. *Surf. Coat. Technol.* **2008**, *202*, 3933–3939. [CrossRef]
40. Tammas-Williams, S.; Zhao, H.; Léonard, F.; Derguti, F.; Todd, I.; Prangnell, P. XCT analysis of the influence of melt strategies on defect population in Ti-6Al-4V components manufactured by selective electron beam melting. *Mater. Charact.* **2015**, *102*, 47–61. [CrossRef]
41. Carroll, B.E.; Palmer, T.A.; Beese, A.M. Anisotropic tensile behavior of Ti-6Al-4V components fabricated with directed energy deposition additive manufacturing. *Acta Mater.* **2015**, *87*, 309–320. [CrossRef]
42. Fernández-Canteli, A.; Przybilla, C.; Nogal, M.; López-Aenlle, M.; Castillo, E. Profatigue: A software program for probabilistic assessment of experimental fatigue data sets. *Procedia Eng.* **2014**, *74*, 236–241. [CrossRef]

metals

MDPI

Article

Deposition of Nickel-Based Superalloy Claddings on Low Alloy Structural Steel by Direct Laser Deposition

André Alves Ferreira [1,2,*], Rui Loureiro Amaral [2], Pedro Correia Romio [1], João Manuel Cruz [3], Ana Rosanete Reis [1,2] and Manuel Fernando Vieira [1,2,*]

[1] Faculty of Engineering, University of Porto, R. Dr. Roberto Frias, 4200-465 Porto, Portugal; pedro.romio@gmail.com (P.C.R.); areis@inegi.up.pt (A.R.R.)
[2] LAETA/INEGI—Institute of Science and Innovation in Mechanical and Industrial Engineering, R. Dr. Roberto Frias, 4200-465 Porto, Portugal; ramaral@inegi.up.pt
[3] SERMEC-Group, R. de Montezelo 540, 4425-348 Porto, Portugal; joaocruz@sermecgroup.pt
* Correspondence: andreferreira@fe.up.pt (A.A.F.); mvieira@fe.up.pt (M.F.V.); Tel.: +351-910-461-480 (A.A.F.)

Abstract: In this study, direct laser deposition (DLD) of nickel-based superalloy powders (Inconel 625) on structural steel (42CrMo4) was analysed. Cladding layers were produced by varying the main processing conditions: laser power, scanning speed, feed rate, and preheating. The processing window was established based on conditions that assured deposited layers without significant structural defects and a dilution between 15 and 30%. Scanning electron microscopy, energy dispersive spectroscopy, and electron backscatter diffraction were performed for microstructural characterisation. The Vickers hardness test was used to analyse the mechanical response of the optimised cladding layers. The results highlight the influence of preheating on the microstructure and mechanical responses, particularly in the heat-affected zone. Substrate preheating to 300 °C has a strong effect on the cladding/substrate interface region, affecting the microstructure and the hardness distribution. Preheating also reduced the formation of the deleterious Laves phase in the cladding and altered the martensite microstructure in the heat-affected zone, with a substantial decrease in hardness.

Keywords: direct laser deposition; Inconel 625; parametrisation; microstructure; microhardness; preheating

Citation: Ferreira, A.A.; Amaral, R.L.; Romio, P.C.; Cruz, J.M.; Reis, A.R.; Vieira, M.F. Deposition of Nickel-Based Superalloy Claddings on Low Alloy Structural Steel by Direct Laser Deposition. *Metals* **2021**, *11*, 1326. https://doi.org/10.3390/met11081326

Academic Editors: Atila Ertas and Adam Stroud

Received: 16 July 2021
Accepted: 20 August 2021
Published: 22 August 2021

Publisher's Note: MDPI stays neutral with regard to jurisdictional claims in published maps and institutional affiliations.

Copyright: © 2021 by the authors. Licensee MDPI, Basel, Switzerland. This article is an open access article distributed under the terms and conditions of the Creative Commons Attribution (CC BY) license (https://creativecommons.org/licenses/by/4.0/).

1. Introduction

The laser-based additive manufacturing (LBAM) technologies applied in the production and repair of industrial components emerged in the late 1990s. Their use continues to extend to many industrial sector applications for components that operate in extreme conditions. LBAM technologies are unique and versatile in the manufacturing of parts with complex geometry, functionally graded or customised, producing an improvement in properties that can be used for a variety of industrial applications, such as within the aerospace, metallurgy, energy, and automotive industries [1,2]. Direct laser deposition (DLD) is an LBAM technology used for the additive manufacturing of metal parts, reconstructions, and repairs. DLD consists of the supply, through a nozzle, of metallic powder (or wire) processed by a focused laser, creating a melt pool on the surface of a metallic substrate. Several processing variables directly or indirectly affect the quality and structural integrity of components, dictated by solidification and metallurgical bonding [3].

DLD involves interactions between the laser beam, powder, and substrate in an environment with local protection from inert gas. Laser power, scanning speed, beam size, and powder feed rate are parameters that play a dominant role in cladding geometry (height, width, and length), dilution, and metallurgical properties. Clad overlapping, gas flow rate, powder flow profile, powder quality (size, shape, and density), and preheating are important secondary parameters [4,5].

The success of DLD depends on the selection of processing conditions that guarantee an effective bonding of the deposited material. This proper bonding produces adequate thermal delivery control, dense layers, a small heat-affected zone (HAZ), low dilution, minimal distortion, and good surface quality, with an attractive set of mechanical properties as well as resistance to wear and corrosion [6–8]. A controlled DLD process can replace conventional processes (i.e., electric arc welding and thermal spraying) in order to repair industrial components. Traditional approaches present drawbacks in component repairs, such as the time required, the limited thickness of the deposition layers, the low metallurgical bond, the formation of porosities and cracks, and the distortion of substrates (caused by overheating of the components). Therefore, it is of industrial interest to develop high-efficiency and -precision repair technologies to increase component life.

This additive manufacturing technology is considered the best strategy for reconstructing and repairing damaged components in terms of environmental benefit and economic feasibility. However, it is regarded as a complex process due to uncertainties in the quality and reliability of recovered industrial components [9], requiring further investigation to consolidate the results reported in this area. Although the equipment cost is high, DLD has successfully repaired dies, moulds, turbines, and gears. Adaptability for automation, ease of assembly of the laser on a CNC machine or robotic arm, and lower post-processing requirements are additional advantages of the DLD process [10,11].

DLD still has a way to go for broader industrial applications. Theoretical and experimental studies have developed relevant information about DLD; however, there are still many challenges, such as process optimisation, 3D reconstruction of highly complex structures, and substrate preheating effects, which need to be clarified. The production of wear-resistant claddings on low and medium carbon steel substrates is an application that can have many industrial applications, both in component repair and protective coating with a thick resistant layer.

Nickel-based superalloys are an excellent option for producing this wear-resistant layer. These alloys have been adopted in multiple applications due to their properties, such as mechanical behaviour at high temperatures, hardness, mechanical resistance, and good fatigue resistance, creep, and corrosion [12,13]. These properties are conferred by the structure and chemical composition of the alloy, mainly by elements such as molybdenum (Mo) and niobium (Nb), which form a solid solution in a nickel–chromium matrix [14]. While conventional manufacturing with these high-performance alloys has been difficult due to excessive tool wear and low material deposition rates, LBAM technologies can overcome these constraints, improving delivery times, and reducing manufacturing costs [15].

The use of nickel-based superalloys in DLD must consider the high cooling rates of the process, promoted by the localised thermal delivery induced by the laser beam, which can lead to the formation of metastable phases and the segregation of elements. These microstructural effects reduce the toughness and hardness of the coated components [16–19]. Preheating (PHT) is essential to control the cooling rate, minimising this effect. Increasing PHT temperature promotes the growth of the melt pool (depth and width), melting more substrate, thus increasing dilution [20–24]. PHT also prevents cladding delamination or cracking and reduces distortion and residual stresses due to the lower thermal gradient between the cladding and substrate [7,25,26].

In this study, Inconel 625, a nickel-based superalloy, powder was deposited on 42CrMo4 steel, while the process parameters were varied. 42CrMo4 steel is often used to produce components, such as gears and main shafts, and Inconel 625 is employed in the repair/remanufacturing of these components by SERMEC Group. Single layers were produced to evaluate the metallurgical bonding with a substrate; the influence of several processing parameters, such as laser power, scanning speed, and powder feed rate on the cladding quality, was evaluated considering the absence of cracks and structural imperfections. Preheating was performed on an optimised cladding condition in order to moderate the microstructure and mechanical responses. Microhardness profiles of claddings were obtained and correlated with the microstructures.

2. Materials and Methods

2.1. DLD System Setup

A laser system, LDF 3000–100, was used to produce DLD claddings. The system has a high-power fibre-coupled laser diode (wavelength 900–1030 nm, depending on the power), with a nominal beam power of 6000 W. The machine concept is a KUKA KR90 R3100 industrial robot, based on a 6-axis industrial robot. All axes are connected to the robot and laser control units, which command the temperature of the melt pool as well as the laser power. The powder was fed during the deposition process by a coaxial feeding system, as illustrated in Figure 1. The substrate was preheated (PHT) to 300 °C with a manual gas system. The temperature control of the preheated substrate was done by a digital pyrometer, for verification of the uniformity of the substrate surface temperature distribution. Preheating is intended to decrease the cooling rate in the melt pool and HAZ regions as well as eliminate moisture. Tests were performed on substrates after the production of clads, with and without PHT, to evaluate susceptibility to cracking and eventual formation of metastable phases.

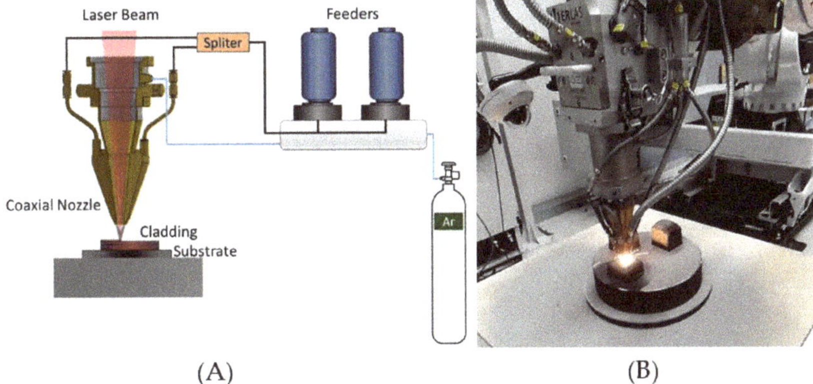

(A)

(B)

Figure 1. (**A**) Schematic representation of the direct laser deposition system with two feeders; (**B**) system in operation.

2.2. Feedstock Powder and Substrate

A nickel-based superalloy (MetcoClad 625 from Oerlikon), similar to Inconel 625, produced by the gas-atomised process, was used in this study. This powder was developed specifically for laser processing and presents a spherical morphology as well as particle sizes ranging between 45 and 90 µm. Figure 2 shows scanning electron microscopy (SEM) images of the morphology of the MetcoClad 625 powder.

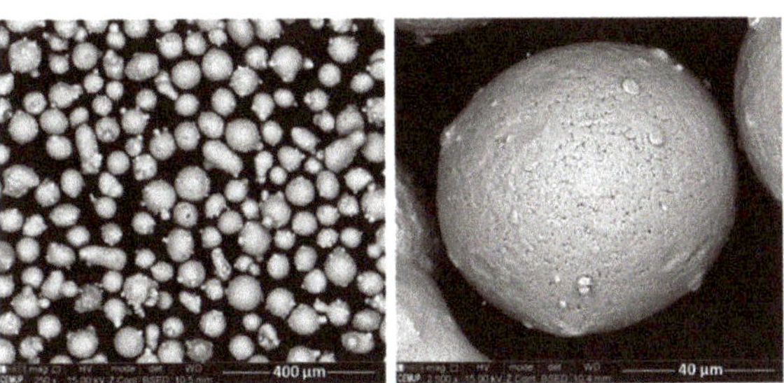

Figure 2. SEM images of MetcoClad 625 powder.

42CrMo4 steel, in the quenched and tempered condition, was used as a substrate for the DLD deposition. Specimens with dimensions of 100 mm × 120 mm × 15 mm were prepared for depositions. This steel is classified as a low alloy structural steel with high mechanical strength and toughness as well as good fatigue resistance and machinability, being widely used in the manufacturing of critical industrial components, such as gears, automotive components, wing generators, and drilling joints for deep wells [27]. Its mechanical and chemical properties are described in standard EN 10,269 [28]. The chemical composition of the MetcoClad 625 (M625) powder and 42CrMo4 steel are shown in Table 1.

Table 1. Chemical composition of M625 powder and 42CrMo4 steel (wt.%).

Material	Ni	Cr	Mo	Nb	Si	Mn	C	Fe
M625	60.8	21.3	9.2	4.6	-	-	-	4.1
42CrMo4	-	1.1	0.2	–	0.3	0.7	0.4	97.3

2.3. Process Parameters

The main process parameters (laser power, scanning speed, and feed rate) were considered to evaluate the effect of processing conditions on clads. The evaluation of the clad quality depends on clad characteristics, namely, absence of cracks or pores, good metallurgical bond (interfaces without microcracks, pores, and fragile phases, exhibiting good wettability), and dilution between 15 and 30%. The values of the tested parameters are shown in Table 2. All results are representative of single-layer samples. The terminology ILP_SS_FR was used to identify the samples, being I—M625 powder (Inconel powder); LP—laser power (kW); SS—scanning speed (mm/s); FR—feed rate (g/min). Eighteen combinations of different processing conditions were tested, with and without preheating (see Table 3). Before deposition, the substrates were cleaned with pure acetone.

Table 2. Process parameters tested for M625 deposition on 42CrMo4 substrate.

Process Parameters	Values
Laser power (LP)	1.0, 1.5, 2.0, 2.5, and 3.0 kW
Scanning speed (SS)	2.0, 4.0, 6.0, and 10.0 mm/s
Feed rate (FR)	10, 15, and 20 g/min

In all tests, a spot size of 2.5 mm and an offset in the Z-axis of 0.2 mm were used. High-purity argon (99.99%), with a 5.5 L/min flow rate, was used as the shielding gas for minimising contamination of the melt pool during the DLD process. Samples with and without PHT were cooled in air.

2.4. Mechanical and Microstructural Characterisation

Samples from each deposition were cut for microstructural and mechanical characterisation using a metallographic cut-off machine with refrigeration in order to avoid substrate and cladding overheating. Samples were mounted in resin and polished down to a 1 μm diamond suspension. Kalling's N°. 2 chemical etching ($CuCl_2$—5 g, hydrochloric acid—100 mL, and ethanol—100 mL) was used to reveal the microstructures.

The measurements of the height, depth, and width of the claddings produced by the DLD technique were performed using a Leica DVM6 A 2019 digital microscope (DM) (Wetzlar, Germany). The Leica DM 4000 M optical microscope (OM) (Wetzlar, Germany) allowed for a microstructural analysis at low magnifications to evaluate, for example, the size of the heat-affected zone. A scanning electron microscope, FEI Quanta 400 FEG (ESEM, Hillsboro, OR, USA), equipped with energy-dispersive X-ray spectroscopy (EDX) (EDAX Genesis X4M, Oxford Instrument, Oxfordshire, UK) and electron backscatter diffraction (EBSD) (EDAX-TSL OIM EBSD, Mahwah, NJ, USA) was used for higher magnification observation and phase identification. For EBSD evaluation, the samples went through an additional polishing step, using a 0.06 μm silica colloidal suspension, for a superior surface finish and to remove polishing-induced plastic deformations, allowing Kikuchi patterns to be obtained [29]. EBSD allows for the obtaining of information on microstructural characteristics with a small interaction volume and a high resolution, for which TSL OIM Analysis 5.2 software was used. For all raw data obtained by EBSD, a dilatation clean-up routine was performed, with a grain tolerance angle of 15° and a minimum grain size of 10 points.

Quantitative image analysis was employed on optical images using the ImageJ software, version 1.51p (National Institutes of Health, Bethesda, MD, USA).

Vickers microhardness tests gave the mechanical characterisation. The tests were performed using a test force of 300 g for 15 s in a Struers Duramin 5 (Struers Inc., Cleveland, OH, USA) Vickers hardness tester. Each hardness value corresponds to the average of three indentations.

3. Results and Discussion

3.1. Processing Effects

The quality of cladding was first evaluated by inspection with a digital microscope (DM). The microstructural analysis of all the claddings produced did not detect cracks, pores, or inclusions of significant dimensions, an essential requirement for obtaining high-performance deposits.

SEM characterisation confirmed the observations made by the DM. Figure 3 shows an SEM image of the cross-section of an M625 clad deposited on 42CrMo4 steel, representing the geometric aspects of cladding: height (h), width (w), depth (d), clad area (AC), melting area (AM), and wetting angle (θ). These geometric aspects were measured on all claddings using the ImageJ software. The results are shown in Table 3.

Table 3. Dimensional analysis of claddings produced by DLD.

Sample	Cladding Dimensional Analysis						
	W (mm)	H (mm)	D (mm)	θ (°)	AC (mm^2)	AM (mm^2)	D (%)
I1_2_15	2.79	3.07	0.25	121	9.9	0.3	2.6
I1_6_15	3.07	1.22	0.06	64	3.0	0.0	1.3
I1.5_10_10	3.09	0.71	0.43	47	5.8	2.1	26.5
I1.5_10_15	3.11	1.03	0.23	64	2.3	0.2	7.3
I2_2_15	3.62	3.66	1.28	107	15.2	3.0	16.6
I2_4_15	3.50	2.47	1.12	88	7.2	2.4	25.0
I2_6_10	3.38	1.39	1.16	53	3.3	2.3	40.6
I2_6_15	3.61	1.47	0.51	65	4.5	1.3	22.0
I2_6_20	3.63	1.71	0.56	74	4.9	1.2	19.5
I2_10_10	3.20	1.20	0.72	58	2.7	1.1	29.1
I2_10_15	3.36	0.81	0.88	40	1.9	1.5	45.1
I2.5_10_10	3.33	0.86	1.04	49	2.1	2.3	52.1
I2.5_10_15	3.36	1.23	1.00	51	2.9	1.9	39.8
I3_2_15	5.05	3.40	2.06	99	15.4	6.6	29.9
I3_4_15	4.64	2.06	0.89	70	7.5	2.5	25.3
I3_6_10	3.89	1.85	1.08	74	6.2	2.8	30.8
I3_6_15	4.71	0.78	0.68	37	2.8	2.6	48.0
I3_6_20	3.94	1.55	1.34	69	4.6	3.3	41.6

This analysis of the geometry of the single-track deposits is critical as it provides information on process yield and cladding performance. For example, the contact angle is an essential parameter in assessing the quality of the cladding [30,31]. Higher beads can promote low wettability during the deposition of multi-tracks, hindering overlapping and generating discontinuities in the hatch spacing of the overlapping deposits, thus facilitating crack propagation. Typically, a contact angle greater than 90° is associated with lower-quality claddings [32]. Table 3 shows that depositions with higher heights have a high contact angle, as samples I1_2_15, I2_2_15, and I3_2_15 demonstrate.

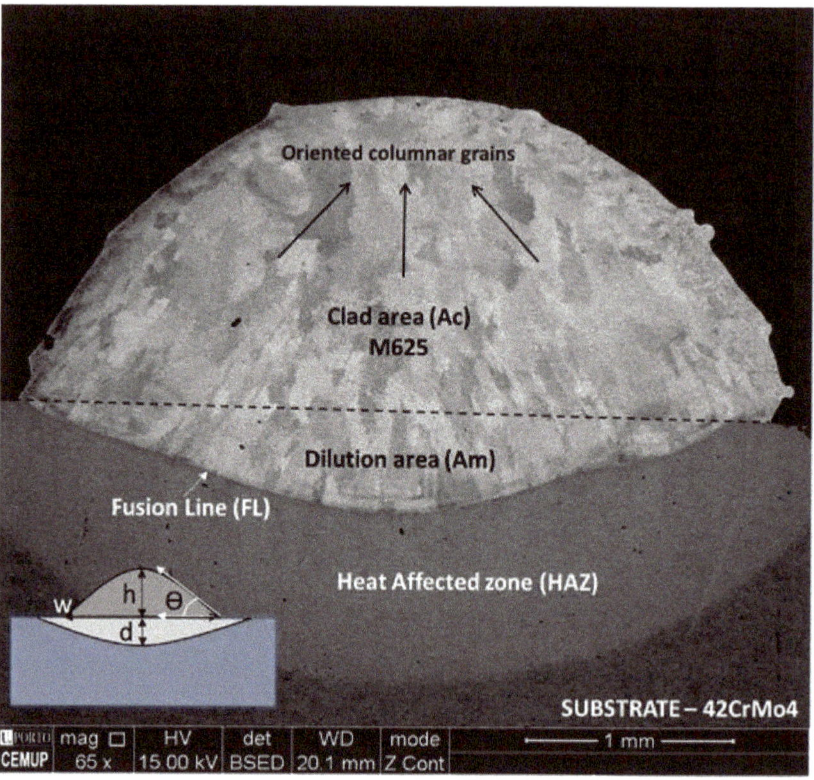

Figure 3. Cross-section of a single layer of M625 deposited on 42CrMo4: d—depth; h—height; w—width; θ—contact angle; AC—clad area; and AM—melting area.

The results presented in Table 3 show that the analysed processing parameters (laser power, scanning speed, powder feed rate, and preheating) strongly influence the production of the cladding and its bonding to the substrate. The selection of a processing window that guarantees a metallurgically bonded clad with good material yield is a fundamental task. This selection is difficult since the mutual interaction of the various parameters is complex. Thus, it is common to apply combined parameters in the DLD process to obtain a more accurate relationship between processing parameters and the clad characteristics [33].

One of the most used complex parameters is powder deposition density (PDD), which expresses the combined effect of feed rate, scanning speed, and laser spot size (φ) (Equation (1)) [34–36]. Figure 4 exposes the linear growth of the cladding area with the increase in the PDD parameter.

$$PDD \ (g/mm^2) = FR/(SS \times \varphi) \tag{1}$$

This parameter shows that by increasing the feed rate or decreasing the scanning speed, we can obtain claddings with a larger area, which was expected since both situations result in more powder supply in the same period of time. PDD is a valuable parameter, but this relationship is only valid for the cladding area and not for the total area of the deposit, including the area of the substrate that has been melted. This last area is vital because it affects the quality of the cladding.

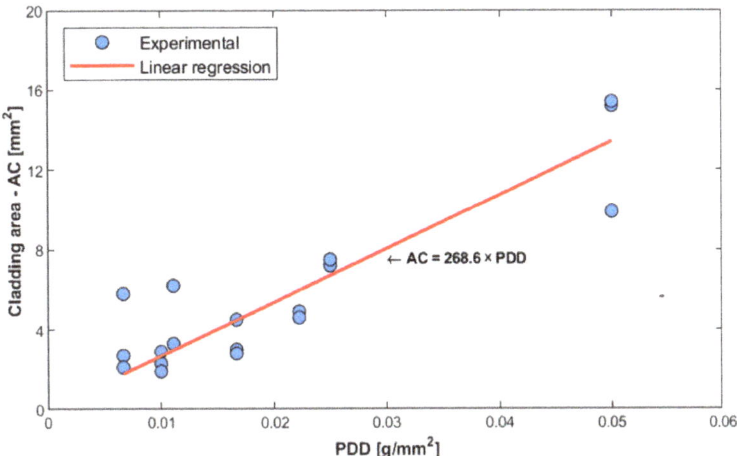

Figure 4. Dependence of cladding area on the value of the PDD complex parameter.

Additionally, with the measured areas (Table 3), it was possible to calculate the dilution, which quantifies the relative amount of melted substrate during laser processing, according to Equation (2).

$$\text{Dilution (\%)} = \text{AM}/(\text{AC} + \text{AM}) \times 100 \tag{2}$$

A dilution ratio between 15 and 30% is sufficient to allow a good metallurgical bond between the substrate and the cladding. Higher dilution, which means a greater melting of the substrate, is undesirable since it reduces the deposition yield and induces considerable changes in the chemical composition of the deposited material, modifying the expected properties of the cladding.

The results obtained, and presented in Table 3, indicate that dilution increases with increased laser power, keeping the other processing variables constant. A laser power of 1.5 kW is enough to guarantee a dilution higher than 15% for almost all conditions (the only exception is the I1.5_10_15 sample).

Despite this apparent direct relation between laser power and dilution, the effect of other critical processing variables, namely the scanning speed and the feed rate, makes the establishment of relationships between processing condition and dilution complex. To overcome this difficulty, it is common to apply complex parameters, empirically adjusted, to the clad/substrate set under analysis to define the processing window [33,37,38].

Figure 5 shows a process window map, which associates laser power with the scanning speed and feeding rate ratio, as well as the dilution that is correlated with the laser power through complex parameter LP $(SS/FR)^{0.5}$; it is represented by two curves, one for 15% and the other for 30%. Additionally, as was also considered in the map, a vertical line that corresponds to the acceptable limit for the wetting angle is present.

As seen in Figure 5, the shaded area, delimited by the previous conditions (dilution range, acceptable wetting angle, and process parameters), reveals the desirable practical manufacturing processing window of Inconel 625. In addition, the dilution increases proportionally with the increase in laser power and decreases with the ratio between scanning speed and feeding rate [39]. The flow of liquid metal in the melt pool is dominated by Marangoni's convection effect, caused by the surface tension gradient. As the temperature of the substrate increases, this effect becomes more evident. However, in practice, the surface tension gradient (γ) dγ/dT depends on both temperature (T) and composition. In this case, the most significant influence factor is the thermal gradient promoted between laser beam and substrate, as Le et al. [40] demonstrated, where the increase in the substrate temperature becomes more evident.

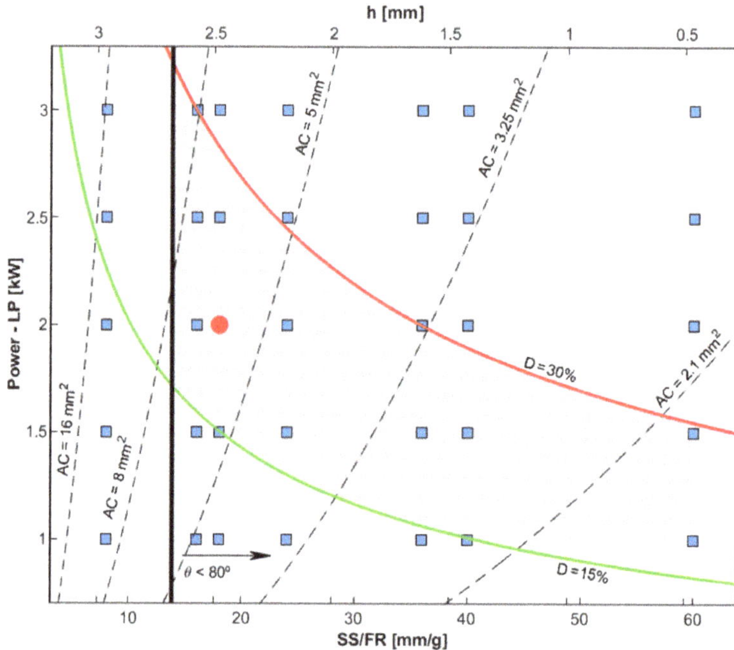

Figure 5. DLD process window map of M625 for additive manufacturing of single tracks. Red point represents the experimental I2_6_20 condition.

Laser power control allows the required good metallurgical integrity and dilution to be achieved [41], producing a cladding with a good metallurgical bond and uniform properties. Increasing the power of the laser promotes increased energy density as well as greater dilution and mixing between the substrate and the deposited powder, which is characteristic of the laser alloy process [42].

Considering the analyses performed, the I2_2_15, I2_4_15, I2_6_15, I2_6_20, I2_10_10, I3_2_15, and I3_4_15 conditions showed dilutions within the established range, between 16.6% and 29.9%, but a lower dilution value is preferable. Nevertheless, the I2_2_15, I2_4_15, and I3_2_15 samples presented poor wettability angles (107°, 88°, and 99°, respectively) that may lead, in future, to claddings with overlapping defects between strands. On the other hand, the four remaining conditions presented good wetting angles, but taking into account not only the quality of the process but also its efficiency, condition I2_6_20 (Figure 5 red point) is the only one that allows for the possibility of manufacturing a larger cladding area and consequently a higher cladding high, which has an inverse linear relationship with the SS/FR ratio. Considering this evaluation, the I2_6_20 condition will be used to perform the analyses throughout the following sections.

3.2. Microstructures and Mechanical Characterisation of the DLD Samples

Figure 6 shows the cladding microstructure formed adjacent to the substrate by the deposition of M625 on 42CrMo4 steel. This microstructure consists of columnar grains, mainly dendrites, and cellular morphologies in a few regions.

As shown in Figure 6B, on solidification a continuous thin layer, < 10 μm, consisting of planar grains formed in the vicinity of the substrate. This morphology evolves into columnar grains with continued solidification of the cladding. This microstructure is consistent with the results of a similar study in which martensitic stainless steel is deposited [43]. The very high thermal gradient in the contact zone of the melt pool with the cold substrate contributed to the formation of planar grains. The microstructure evolves into a columnar/dendritic structure due to a rapid decrease in the thermal gradient when more material solidifies. Moreover, columnar grains grow perpendicular to the substrate/solidified material, i.e., in the opposite direction of the primary heat dissipation source, as is usual in DLD solidification [44]. As seen in Figure 7, this columnar/dendritic structure is the characteristic microstructure of the cladding.

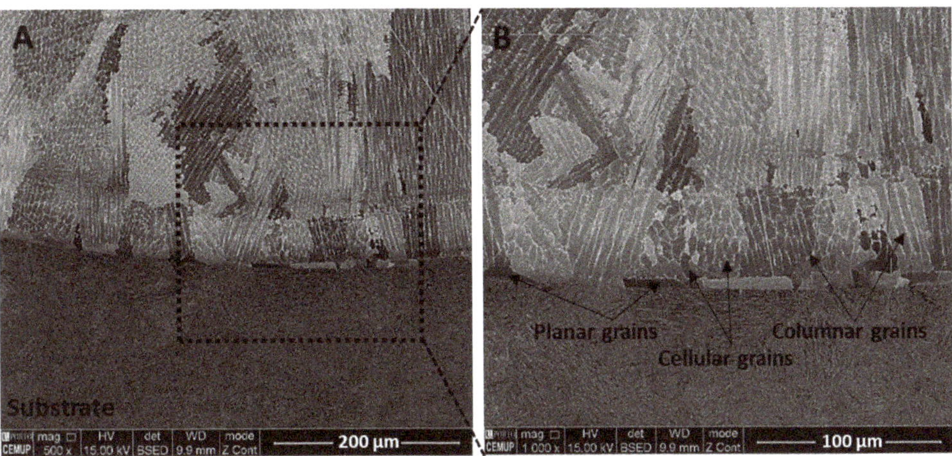

Figure 6. Microstructure of cladding produced by the I2_6_20 cladding condition on the preheated substrate, showing (**A**) substrate interface cladding zone. (**B**) A higher magnification illustration of the interface.

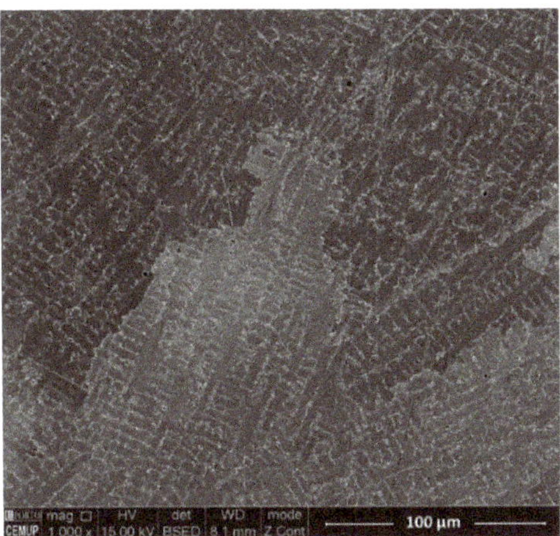

Figure 7. SEM image of the I2_6_20 cladding microstructure.

Figure 8 shows a high-magnification SEM image and elemental mapping of the microstructure of the I2_6_20 cladding. This microstructure consists of the γ matrix, bright zones (surrounded by a segregation zone), and rounded dark particles. The elemental mapping indicates that the bright zones are rich in Mo and Nb, indicating the formation of the Laves phase located at the interdendritic region. Rounded dark particles are complex oxides dispersed in the γ matrix. Table 4 shows the chemical composition obtained by EDX analysis of the zones indicated in Figure 8.

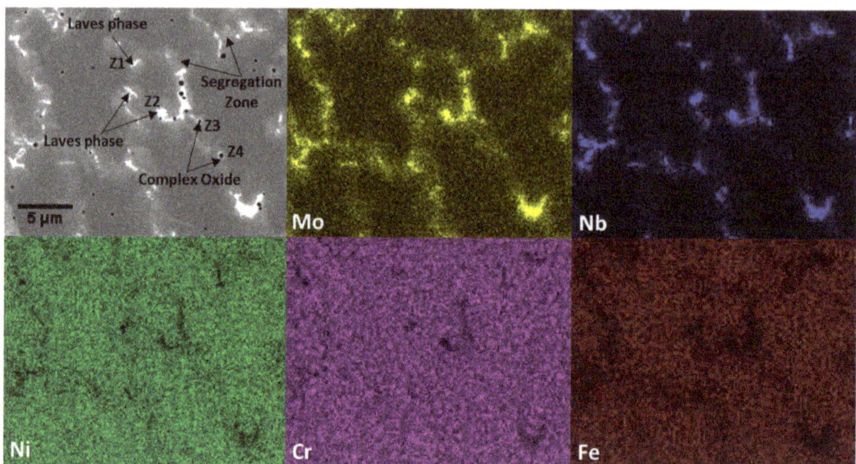

Figure 8. SEM image and the elemental maps obtained by EDX microanalysis showing segregation zones associated with the Laves phase and complex oxides.

Table 4. Chemical composition (%) of the zones indicated in Figure 8.

Phases	Zones	Ni	Cr	Mo	Nb	Si	Fe	Mn	Al	Ca	C	O
Laves	Z1	42.16	17.39	10.97	10.88	3.17	15.43	-	-	-	-	-
phase	Z2	42.29	17.91	10.62	10.34	2.63	16.21	-	-	-	-	-
Oxide	Z3	0.33	4.24	-	1.11	13.00	-	2.44	15.83	4.70	2.58	55.77
	Z4	0.71	4.38	-	1.18	14.71	-	1.78	14.58	6.91	-	55.75

The γ dendrites are formed during the solidification of nickel-based superalloys processed by DLD, segregating Nb and Mo into the liquid, thus creating the local conditions for forming the Laves phase. The final stage of non-equilibrium solidification thus gives rise to this microstructure consisting mainly of the γ matrix and Laves phase. A similar microstructure has been found in other nickel-based superalloy solidification studies [45–51].

Laves phase formation promotes the initiation and propagation of cracks, with a detrimental effect in mechanical response, reducing ductility, ultimate tensile strength, fracture resistance, and fatigue life [52,53]. Thus, this phase reduces the performance of Inconel 625, requiring control of morphology and distribution in the cladding.

A similar analysis was performed at the cladding/substrate interface, Figure 9. In the continuous thin layer adjacent to the substrate, characterized by planar grains (see Figure 6), the thermal gradient and growth rate are significantly different from those of the dendritic region, and the Laves phase was not detected.

Figure 9 shows that DLD deposition of M625 on the 42CrMo4 substrate led to cladding with a flat interface, with a thin continuous layer of planar γ grains, followed by γ dendrites and a dispersed Laves phase. As already mentioned, the appearance of the Laves phase in this region has a detrimental effect and should be minimized. Therefore, another cladding was performed with the substrate preheated to 300 °C. This procedure leads to a decrease in the thermal gradient of the deposited layer, influencing the microstructure. Preheating (PHT) increases the interdiffusion of constituents of M625 and 42CrMo4 at the bonding interface and, by decreasing the substrate cooling rate, can also affect the microstructure and properties in the heat-affected zone.

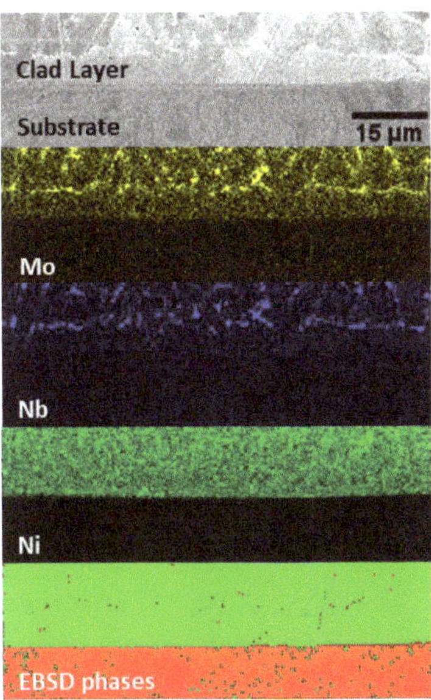

Figure 9. SEM image and elemental maps obtained by EDX microanalysis of M625 deposited on 42CrMo4 substrate. An EBSD image shows the phase distribution in this region: FCC phases in green and BCC phases in red.

The influence of PHT application on the cladding/substrate interface is shown in Figure 10. The thin layer of planar grains was not formed. Substrate heating significantly reduced the thermal gradient in the initial solidification phase of the melt pool, leading to the formation of dendrites throughout the cladding. Furthermore, it appears that PHT slightly increases the interdendritic spacing from 5–7 µm to 6–9 µm, measured by ImageJ software in Figure 6 and Figure 10, respectively. This observation confirms the decrease in the cooling rate allowing the growth of the interdendritic spacing. This feature is consistent with a study that indicates that lower cooling rates promote dendrite growth and decrease cellular grain formation in the initial stages of solidification [54].

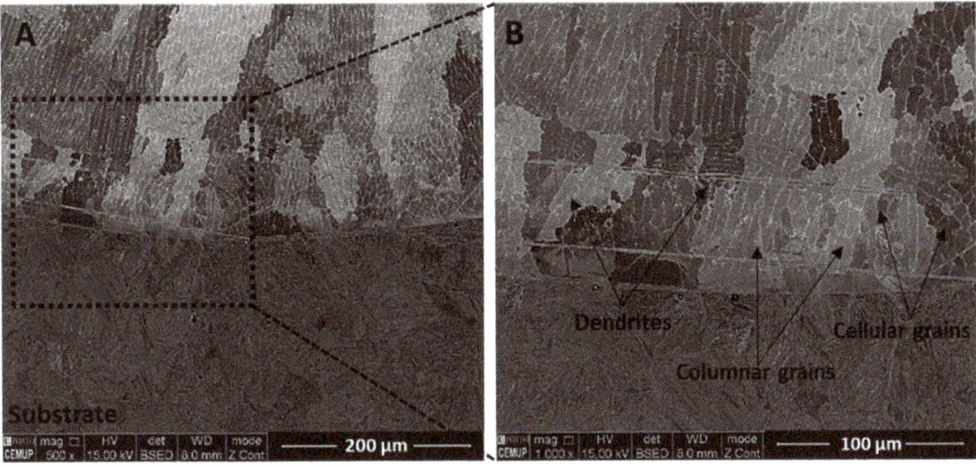

Figure 10. Microstructure of cladding produced by the I2_6_20 cladding condition on the preheated substrate, showing (**A**) substrate interface cladding zone. (**B**) A higher magnification illustration of the interface.

Besides, the application of PHT has led to the reduction in Laves phases, this effect highlighted at the interface zone, as seen in Figure 11. A lower cooling rate effectively maintained the Nb and Mo in the γ matrix, avoiding segregation. Moreover, the elemental mapping and EBSD image in Figure 11 confirm a more intense interdiffusion at the interface, which is not as plane as it was without PHT, Figure 9. Diffusion of nickel, which is a gamma-phase stabilizer, to steel is associated with a greater amount of residual austenite in this region. Thus, the application of preheating seems helpful for reducing the deleterious Laves phase and for enhancing metallurgical bonding.

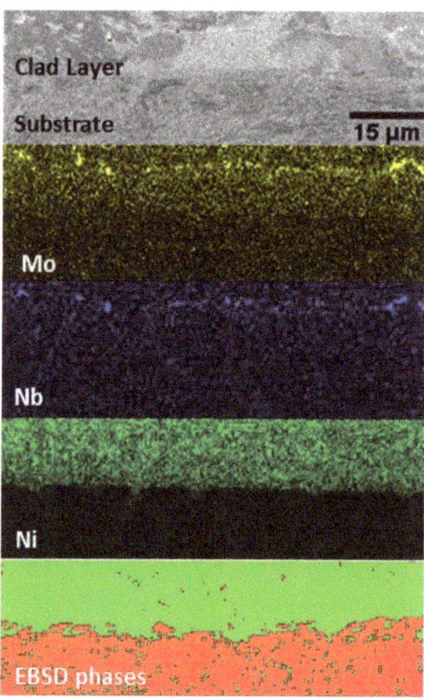

Figure 11. SEM image and elemental maps obtained by EDX microanalysis of M625 deposited on preheated (PHT) 42CrMo4 substrate. An EBSD image shows the phase distribution in this region: FCC phases in green and BCC phases in red.

The HAZ of 42CrMo4 steel is also a critical region, as substrate heating by the laser followed by rapid cooling leads to martensite formation, which can create cracks and allow for rapid crack propagation. Microstructural differences in HAZ caused by PHT were analysed by EBSD, as illustrated in Figure 12. This figure shows that PHT affects the HAZ microstructure, with larger (longer and wider) martensite laths caused by a slower cooling rate. Martensite with wider laths is associated with lower mechanical strength which, together with the more significant amount of residual austenite determined when using PHT, can reduce the brittleness of this region.

Figure 12 also shows that no preferential crystalline orientation was formed in the HAZ of claddings without or with PHT.

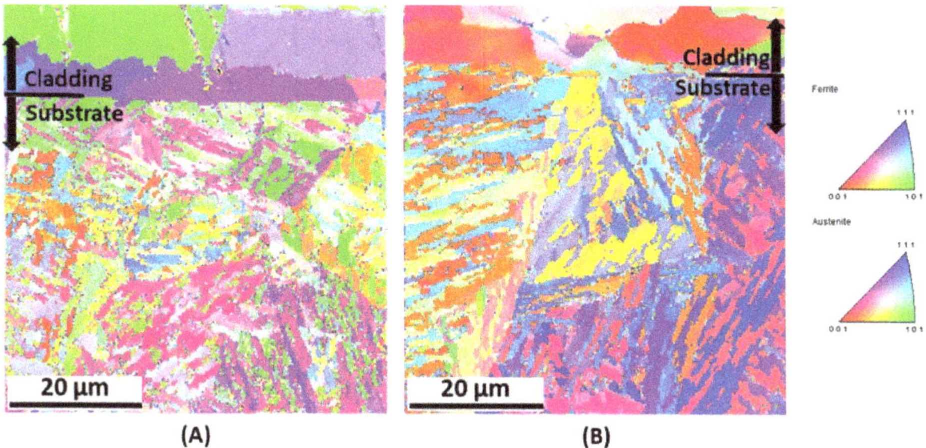

Figure 12. EBSD images of the I2_6_20/substrate interface region. (**A**) Without PHT; (**B**) with PHT.

3.3. Microhardness Measurements

The microhardness profile shown in Figure 13 shows that the hardness distribution in the cladding zone is uniform and independent of PHT, with an average hardness of 258 ± 2 HV and 253 ± 2 HV for the samples with and without preheating, respectively. A sharp transient zone with a pronounced hardness increase is measured in the HAZ of the sample without PHT (maximum hardness of 491 ± 23 HV). The application of PHT to the substrate before deposition produced a more uniform distribution of hardness in the HAZ (368 ± 25 HV), with a less sharp transient near the interface. Furthermore, the hardness peak has been eliminated, and the hardness values show less dispersion. These differences in hardness are attributed to changes in the microstructure induced by PHT, as seen in Figure 12, and its influence on chemical composition distribution, seen in Figures 9 and 11, and indicate that the HAZ region is less prone to cracking.

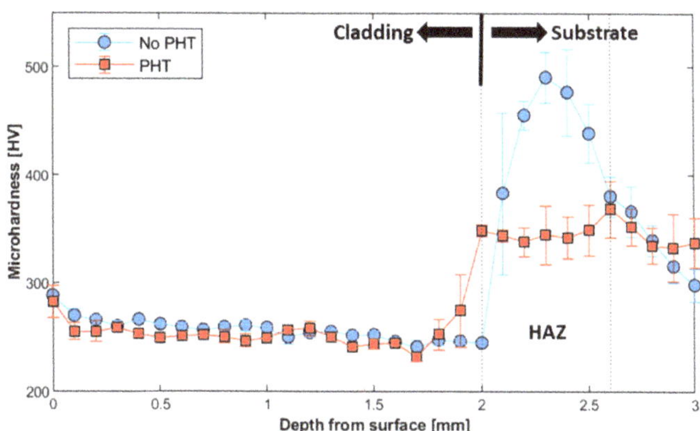

Figure 13. Microhardness profile of cladded samples applying the I2_6_20 condition with and without PHT.

4. Conclusions

The deposition of Inconel 625 claddings onto a 42CrMo4 steel substrate was performed using direct laser deposition (DLD) and varying processing conditions: laser power, scanning speed, feed rate, and preheating. Macro- and microstructural analysis, in addition to the hardness measurements, led to the following main conclusions:

- A DLD process window map considering processing variables shows that several combinations can be used. However, the cladding produced with 2 kW of laser power, a scanning speed of 6 mm/s, and a 20 g/min feed rate presented adequate dilution and wettability.
- The deposited layers were produced without significant structural defects such as cracks, pores, or other types of discontinuities.
- Substrate preheating to 300 °C influences the microstructure of the cladding/substrate interface, reducing the formation of the deleterious Laves phase.
- PHT also alters the hardness profile, mainly in the heat-affected zone, due to modification of the martensite microstructure and increased residual austenite.

Author Contributions: A.A.F., M.F.V., and A.R.R. proposed the methodology and results data analyses followed in this research; A.A.F., J.M.C., P.C.R., and R.L.A. carried out the experimental tests. All authors participated in the discussion of results and writing of the manuscript. All authors have read and agreed to the published version of the manuscript.

Funding: This research was funded by FEDER through the Operational Programme for Competitiveness and Internationalization (COMPETE 2020), Projetos em Copromoção (project POCI-01-0247-FEDER-039848).

Institutional Review Board Statement: Not applicable.

Informed Consent Statement: Not applicable.

Data Availability Statement: Not applicable.

Acknowledgments: The authors give their thanks to the CEMUP (Centro de Materiais da Universidade do Porto) for expert assistance with SEM.

Conflicts of Interest: The authors declare no conflict of interest.

References

1. Bian, L.; Shamsaei, N.; Usher, J.M. *Laser-Based Additive Manufacturing of Metal Parts*, 1st ed.; CRC Press: Boca Raton, FL, USA, 2018.
2. Ur Rahman, N.; Matthews, D.T.; de Rooji, M.; Khorasani, A.M.; Gibson, I.; Cordova, L.; Römer, G. An Overview: Laser-Based Additive Manufacturing for High Temperature Tribology. *Front. Mech. Eng.* **2019**, *5*, 1–15. [CrossRef]
3. Thompson, S.M.; Bian, L.; Shamsaei, N.; Yadollahi, A. An overview of Direct Laser Deposition for additive manufacturing; Part I: Transport phenomena, modeling and diagnostics. *Addit. Manuf.* **2015**, *8*, 36–62. [CrossRef]
4. Song, L.; Bagavath-Singh, V.; Dutta, B.; Mazumder, J. Control of melt pool temperature and deposition height during direct metal deposition process. *Int. J. Adv. Manuf. Technol.* **2012**, *58*, 247–256. [CrossRef]
5. Petrat, T.; Graf, B.; Gumenyuk, A.; Rethmeier, M. Laser metal deposition as repair technology for a gas turbine burner made of inconel 718. *Phys. Procedia* **2016**, *83*, 761–768. [CrossRef]
6. Alimardani, M.; Fallah, V.; Khajepour, A.; Toyserkani, E. The effect of localised dynamic surface preheating in laser cladding of Stellite 1. *Surf. Coatings Technol.* **2010**, *204*, 3911–3919. [CrossRef]
7. Leunda, J.; García Navas, V.; Soriano, C.; Sanz, C. Effect of laser tempering of high alloy powder metallurgical tool steels after laser cladding. *Surf. Coatings Technol.* **2014**, *259*, 570–576. [CrossRef]
8. Dass, A.; Moridi, A. State of the art in directed energy deposition: From additive manufacturing to materials design. *Coatings* **2019**, *9*, 418. [CrossRef]
9. Lahrour, Y.; Brissaud, D. A Technical Assessment of Product/Component Re-manufacturability for Additive Remanufacturing. *Procedia CIRP* **2018**, *69*, 142–147. [CrossRef]
10. Kush, M. *Advanced Manufacturing Technologies Modern Machining Advanced Joining Sustainable Manufacturing*; Springer International Publishing: New York, NY, USA, 2018; p. 294. [CrossRef]
11. Pinkerton, A.J.; Wang, W.; Li, L. Component repair using laser direct metal deposition. *Proc. Inst. Mech. Eng. B J. Eng. Manuf.* **2008**, *222*, 827–836. [CrossRef]
12. Vilar, R.; Almeida, A. Repair and manufacturing of single crystal Ni-based superalloys components by laser powder deposition—A review. *J. Laser Appl.* **2015**, *27*, S17004. [CrossRef]
13. Abioye, T.E.; McCartney, D.G.; Clare, A.T. Laser cladding of Inconel 625 wire for corrosion protection. *J. Mater. Process. Technol.* **2015**, *217*, 232–240. [CrossRef]
14. Marchese, G.; Colera, X.G.; Calignano, F.; Lorusso, M.; Biamino, S.; Minetola, P.; Manfredi, D. Characterization and Comparison of Inconel 625 Processed by Selective Laser Melting and Laser Metal Deposition. *Adv. Eng. Mater.* **2017**, *19*, 1–9. [CrossRef]

15. Gonzalez, J.A.; Mireles, J.; Stafford, S.W.; Perez, M.A.; Terrazas, C.A.; Wicker, R.B. Characterization of Inconel 625 fabricated using powder-bed-based additive manufacturing technologies. *J. Mater. Process. Technol.* **2019**, *264*, 200–210. [CrossRef]

16. Long, Y.T.; Nie, P.L.; Li, Z.G.; Huang, J.; Li, X.; Xu, X.M. Segregation of niobium in laser cladding Inconel 718 superalloy. *Trans. Nonferrous Met. Soc. China* **2016**, *26*, 431–436. [CrossRef]

17. Xiao, H.; Li, S.; Han, X.; Mazumder, J.; Song, L. Laves phase control of Inconel 718 alloy using quasi-continuous-wave laser additive manufacturing. *Mater. Des.* **2017**, *122*, 330–339. [CrossRef]

18. Singh, G.; Vasudev, H.; Bansal, A.; Vardhan, S.; Sharma, S. Microwave cladding of Inconel-625 on mild steel substrate for corrosion protection. *Mater. Res. Express* **2020**, *7*. [CrossRef]

19. Feng, K.; Feng, K.; Chena, Y.; Denga, P.; Li, Y.; Zhao, H.; Lu, F.; Li, R.; Jian, H.; Li, Z. Improved high-temperature hardness and wear resistance of Inconel 625 coatings fabricated by laser cladding. *J. Mater. Process. Technol.* **2017**, *243*, 82–91. [CrossRef]

20. Farahmand, P.; Kovacevic, R. Laser cladding assisted with an induction heater (LCAIH) of Ni-60%WC coating. *J. Mater. Process. Technol.* **2015**, *222*, 244–258. [CrossRef]

21. Wang, Z.; Zhao, J.; Zhao, Y.; Zhang, H.; Shi, F. Microstructure and microhardness of laser metal deposition shaping K465/stellite-6 laminated material. *Metals* **2017**, *7*, 512. [CrossRef]

22. Bennett, J.; Dudas, R.; Cao, J.; Ehmann, K.; Hyatt, G. Control of Heating and Cooling for Direct Laser Deposition Repair of Cast Iron Components. In Proceedings of the International Symposium on Flexible Automation, ISFA 2016, Cleveland, OH, USA, 1–3 August 2016.

23. The Laser Repair Process of High-Speed Gear Transmission Body with Better Wear Resistance and Higher Hardness. Patent CN106222651A, 14 December 2016.

24. Sadhu, A.; Choudhary, A.; Sarkar, S.; Nair, A.M.; Nayak, P.; Pawar, S.D.; Muvvala, G.; Pal, S.K.; Nath, A.K. A study on the influence of substrate preheating on mitigation of cracks in direct metal laser deposition of NiCrSiBC-60%WC ceramic coating on Inconel 718. *Surf. Coatings Technol.* **2020**, *389*, 125646. [CrossRef]

25. Shim, D.S.; Baek, D.S.; Lee, S.B.; Yu, J.H.; Choi, Y.S.; Park, S.H. Influence of heat treatment on wear behavior and impact toughness of AISI M4 coated by laser melting deposition. *Surf. Coatings Technol.* **2017**, *328*, 219–230. [CrossRef]

26. He, W.; Shi, W.; Li, J.; Xie, H. In-situ monitoring and deformation characterisation by optical techniques; part I: Laser-aided direct metal deposition for additive manufacturing. *Opt. Lasers Eng.* **2019**, *122*, 74–88. [CrossRef]

27. Sun, C.; Fu, P.X.; Liu, H.W.; Liu, H.H.; Du, N.Y. Effect of tempering temperature on the low temperature impact toughness of 42CrMo4-V steel. *Metals* **2018**, *8*, 232. [CrossRef]

28. *Steels and Nickel Alloys for Fasteners with Specified Elevated and/or Low Temperature*; BSIStandard: London, UK, 2013; ISBN 978 0 580 76784 5.

29. Engler, O.; Randle, V. *Introduction to texture analysis: Macrotexture, Microtexture and Orientation Mapping*, 2rd ed.; CRC Press: Boca Raton, FL, USA, 2010.

30. Abioye, T.E.; Folkes, J.; Clare, A.T. A parametric study of Inconel 625 wire laser deposition. *J. Mater. Process. Technol.* **2013**, *213*, 2145–2151. [CrossRef]

31. Toyserkani, E.; Khajepour, A.; Corbin, S. *Laser Cladding*, 1st ed.; CRC Press: Boca Raton, FL, USA, 2004.

32. Li, C.; Guo, Y.B.; Zhao, J.B. Interfacial phenomena and characteristics between the deposited material and substrate in selective laser melting Inconel 625. *J. Mater. Process. Technol.* **2017**, *243*, 269–281. [CrossRef]

33. Ferreira, A.A.; Darabi, R.; Sousa, J.P.; Cruz, J.M.; Reis, A.R.; Vieira, M.F. Optimization of direct laser deposition of a martensitic steel powder (Metco 42c) on 42CrMo4 steel. *Metals* **2021**, *11*, 672. [CrossRef]

34. Emamian, A.; Corbin, S.F.; Khajepour, A. In-Situ Deposition of Metal Matrix Composite in Fe-Ti-C System Using Laser Cladding Process. *Met. Ceram. Polym. Compos. Var. Uses* **2011**. [CrossRef]

35. Farshidianfar, M.H.; Khajepour, A.; Gerlich, A.P. Effect of real-time cooling rate on microstructure in Laser Additive Manufacturing. *J. Mater. Process. Technol.* **2016**, *231*, 468–478. [CrossRef]

36. Emamian, A.; Corbin, S.F.; Khajepour, A. Effect of laser cladding process parameters on clad quality and in-situ formed microstructure of Fe-TiC composite coatings. *Surf. Coatings Technol.* **2010**, *205*, 2007–2015. [CrossRef]

37. Ocelík, V.; de Oliveira, U.; de Boer, M.; de Hosson, J.T.M. Thick Co-based coating on cast iron by side laser cladding: Analysis of processing conditions and coating properties. *Surf. Coatings Technol.* **2007**, *201*, 5875–5883. [CrossRef]

38. de Oliveira, U.; Ocelík, V.; De Hosson, J.T.M. Analysis of coaxial laser cladding processing conditions. *Surf. Coatings Technol.* **2005**, *197*, 127–136. [CrossRef]

39. Liu, J.; Li, J.; Cheng, X.; Wang, H. Effect of dilution and macrosegregation on corrosion resistance of laser clad AerMet100 steel coating on 300M steel substrate. *Surf. Coatings Technol.* **2017**, *325*, 352–359. [CrossRef]

40. Le, T.N.; Lo, Y.L. Effects of sulfur concentration and Marangoni convection on melt-pool formation in transition mode of selective laser melting process. *Mater. Des.* **2019**, *179*, 107866. [CrossRef]

41. Marya, M.; Singh, V.; Hascoet, J.-Y.; Marya, S. Metallurgical Investigation of the Direct Energy Deposition Surface Repair of Ferrous Alloys. *J. Mater. Eng. Perform.* **2018**, *27*, 813–824. [CrossRef]

42. Ocelík, V.; Hemmati, I.; De Hosson, J.T.M. The influence of processing speed on the properties of laser surface deposits. *Surf. Contact Mech. Incl. Tribol. XII* **2015**, *1*, 93–103. [CrossRef]

43. Hemmati, I.; Ocelík, V.; De Hosson, J.T.M. Microstructural characterisation of AISI 431 martensitic stainless steel laser-deposited coatings. *J. Mater. Sci.* **2011**, *46*, 3405–3414. [CrossRef]

44. Dinda, G.P.; Dasgupta, A.K.; Mazumder, J. Laser aided direct metal deposition of Inconel 625 superalloy: Microstructural evolution and thermal stability. *Mater. Sci. Eng. A* **2009**, *509*, 98–104. [CrossRef]
45. Naghiyan Fesharaki, M.; Shoja-Razavi, R.; Mansouri, H.A.; Jamali, H. Microstructure investigation of Inconel 625 coating obtained by laser cladding and TIG cladding methods. *Surf. Coatings Technol.* **2018**, *353*, 25–31. [CrossRef]
46. Xiao, H.; Li, S.M.; Xiao, W.J.; Li, Y.Q.; Cha, L.M.; Mazumder, J.; Song, L.J. Effects of laser modes on Nb segregation and Laves phase formation during laser additive manufacturing of nickel-based superalloy. *Mater. Lett.* **2017**, *188*, 260–262. [CrossRef]
47. Knorovsky, G.A.; Cieslak, M.J.; Headley, T.J. Inconel 718 A Solidification Diagram. *Metall. Transactions A* **1989**, *20*, 2149–2158. [CrossRef]
48. Wang, L.; Dong, J.; Tian, Y.; Zhang, L. Microsegregation and Rayleigh number variation during the solidification of superalloy Inconel 718. *Mineral. Metall. Mater.* **2008**, *15*, 594–599. [CrossRef]
49. Dupont, J.N. Solidification of an Alloy 625 Weld Overlay. *Metall. Mater. Trans. A* **1996**, *27*, 3612–3620. [CrossRef]
50. Cieslak, M.J.; Headley, T.J.; Kollie, T.; Romig, A.D. A Melting and Solidification Study of Alloy 625. *Met. Mater. Trans. A* **1988**, *19*, 2319–2331. [CrossRef]
51. Nie, P.; Ojo, O.A.; Li, Z. Numerical modeling of microstructure evolution during laser additive manufacturing of a nickel-based superalloy. *Acta Mater.* **2014**, *77*, 85–95. [CrossRef]
52. Xie, H.; Yang, K.; Li, F.; Sun, C.; Yu, Z. Investigation on the Laves phase formation during laser cladding of IN718 alloy by CA-FE. *J. Manuf. Process.* **2020**, *52*, 132–144. [CrossRef]
53. Chen, Y.; Guo, Y.; Xu, M.; Ma, C.; Zhang, Q.; Wang, L.; Yao, J.; Li, Z. Study on the element segregation and Laves phase formation in the laser metal deposited IN718 superalloy by flat top laser and gaussian distribution laser. *Mater. Sci. Eng. A* **2019**, *754*, 339–347. [CrossRef]
54. Porter, D.A.; Easterling, K.E.; Sherif, M.Y. *Phase Transformations in Metals and Alloys*, 3rd ed.; CRC Press: Boca Raton, FL, USA, 2009.

metals

MDPI

Article

Comparative Study on Microstructure and Corrosion Resistance of Al-Si Alloy Cast from Sand Mold and Binder Jetting Mold

María Ángeles Castro-Sastre [1,*], Cristina García-Cabezón [2], Ana Isabel Fernández-Abia [1], Fernando Martín-Pedrosa [2] and Joaquín Barreiro [1]

[1] Department of Mechanical, Informatics and Aerospace Engineering, Campus de Vegazana, University of León, 24071 León, Spain; aifera@unileon.es (A.I.F.-A.); jbarg@unileon.es (J.B.)

[2] Materials Engineering, E.I.I., Universidad de Valladolid, 47011 Valladolid, Spain; crigar@eii.uva.es (C.G.-C.); fmp@eii.uva.es (F.M.-P.)

* Correspondence: macass@unileon.es; Tel.: +34-98-72-93-588

Abstract: This investigation is focused on the corrosion evaluation of an as-cast Al-Si alloy, obtained by two different casting methods: traditional sand casting and three-printing casting, using a binder jetted mold. The experimental results are discussed in terms of chemical composition, microstructure, hardness, and corrosion behavior of two different casting parts. The microstructure and composition of the sample before and after the corrosion tests was analyzed using light microscopy (OM), scanning electron microscopy (SEM), and energy-dispersive X-ray spectroscopy (EDX) and X-ray diffraction (DRX). The corrosion of the two processed castings was analyzed using anodic polarization (PA) test and electrochemical impedance spectroscopy (EIS) in an aerated solution of 3.5% by weight NaCl, similar to the seawater environment. After the corrosion process, the samples were analyzed by inductively coupled plasma/optical emission spectrometry (ICP/OES); the composition was used to determine the chloride solution after immersion times. The sample processed by binder jetting mold showed higher corrosion resistance with nobler potentials, lower corrosion densities, higher polarization resistance, and more stable passive layers than the sample processed by sand casting. This improvement of corrosion resistance could be related to the presence of coarse silicon particles, which decrease of cathodic/anodic ratio and the number of micro-galvanic couples, and the lower amount of intermetallic β Al-Fe-Si phase observed in cast alloy solidified in binder jetting mold.

Keywords: binder jetting; sand casting; aluminum alloy; microstructure; corrosion

Citation: Castro-Sastre, M.Á.; García-Cabezón, C.; Fernández-Abia, A.I.; Martín-Pedrosa, F.; Barreiro, J. Comparative Study on Microstructure and Corrosion Resistance of Al-Si Alloy Cast from Sand Mold and Binder Jetting Mold. *Metals* **2021**, *11*, 1421. https://doi.org/10.3390/met11091421

Academic Editor: Atila Ertas

Received: 26 July 2021
Accepted: 31 August 2021
Published: 8 September 2021

Publisher's Note: MDPI stays neutral with regard to jurisdictional claims in published maps and institutional affiliations.

Copyright: © 2021 by the authors. Licensee MDPI, Basel, Switzerland. This article is an open access article distributed under the terms and conditions of the Creative Commons Attribution (CC BY) license (https://creativecommons.org/licenses/by/4.0/).

1. Introduction

Techniques based on layer-by-layer fabrication have expanded enormously over the past two decades due to the availability of a variety of materials (metals, ceramics, polymers, and sands) and post-treatment procedures [1]. These technologies began for the manufacture of prototypes due to the geometric freedom and validation of physical prototypes during the product development cycle, reducing delivery time. [2,3]. However, several sectors (aerospace, automotive, foundry) have included this new technology in their production to make functional parts. In particular, casting industries immediately identified the benefits of additive manufacturing for the production of molds [4]. The ability to make parts layer by layer allows the designer to achieve one-piece molds, integrating sprue, runner, gate, and feeder into the mold design, and these unique geometries and complex designs are difficult to achieve through traditional casting processes. In addition, manufacturing time is reduced since the production of patterns, use of tools, and machining processes is avoided [5]. These aspects generate enormous advantages considering that more than 70% of all metal castings are produced by the sand casting process [6]. Moreover, sand consumption can be significantly reduced with the additive manufacturing process, which can be helpful for energy saving and reducing environmental pollution as it allows the selective deposition of an acid catalyst (e.g., furan).

Binder jetting is one of the most suitable additive manufacturing technologies for the manufacture of casting molds. Several studies have focused on improving this process, applied to different materials. Tang et al. [7] optimized the binder jetting process for inkjet printing Ti6Al4V parts with excellent properties. Mariani et al. [8] demonstrated the feasibility of binder jetting technology to print WC–Co parts. This study revealed the importance of the characteristics of the initial powder and its influence on the mechanical properties of the printed parts. Other studies have focused on improving inkjet technology. Cheng et al. [9] investigated the process of driving waveform design for the multi-nozzle piezoelectric printhead. The developed procedure allows reducing the number of experimental tests to find suitable waveform parameters for different 3D printers and new materials. Other studies [10] have focused on improving the roller-type leveling mechanism in order to achieve desired layer thickness, smooth surface, and good color quality. Li et al. [11] conducted a study to solve the jamming problem of the printhead and to improve the dimensional accuracy of binder jetting technology. For this purpose, several water-based binders and gypsum composite powders were prepared, and the optimum binder-powder assembly was then determined through elementary adhesive testing and roller paving testing.

In the bibliography, there are several works focused on validating the advantages of applying additive manufacturing technologies to the manufacture of casting molds. Sneling et al. [12] took advantage of using 3D printed molds to cast parts with complex geometries, such as cellular structures and sandwich panels. In this way, they would avoid joints in the final part that would act as stress concentrators. Walker et al. [13] used binder injection technology to manufacture molds with built-in sensors that allowed monitoring the casting process by recording different parameters (temperature, pressure, humidity, gas chemistry) in different parts of the mold during the casting process.

Despite the advantages of applying additive manufacturing in the foundry industry, the process is not yet ready to completely replace the traditional casting process. A deeper understanding of the properties of printed molds and their effect on the quality of castings is still needed. The quality of castings depends on various factors related to the casting process, as well as the properties of the mold used. With regard to the casting process, it is necessary to consider the rate of heat removal from the metal to the mold, the pouring temperature, and the chemical composition of the starting alloy. Regarding the properties of the mold, it is necessary to know the mechanical properties, the surface finish, and the permeability fundamentally. Most of the studies have focused on studying the properties of printed molds using gypsum powder and sand in order to obtain castings of light alloys such as aluminum. McKenna et al. [14] analyzed the effects of temperature and curing time on the mechanical properties of the printed mold. The authors concluded that due to the loss of volatile compounds, permeability increased with increasing curing time. Vaezi et al. [15] studied the effect of layer thickness on mold properties. They observed that by decreasing the thickness of the layer and increasing the amount of binder added between particles bond, an improvement in tensile strength was achieved, although worse results were produced in the surface quality of the mold. Edwin et al. [16] studied the feasibility of printing plaster molds to produce aluminum metal castings. To improve the mechanical and thermal properties of the mold, they performed post-processing operations such as infiltration. The infiltrated molds had good resistance to heat, although the mold did not have enough mechanical strength to withstand the metallostatic pressure due to the molten metal.

There have been some studies that analyzed the quality of the castings obtained from 3d printed molds but the number of studies is limited. Snelling et al. [17,18] conducted a comparative study using two different materials to produce the 3DP mold—ZCorp's plaster vs. ExOne's silica—in order to obtain A356 aluminum alloy casting. In this study, they focused on analyzing various properties of the casting, including roughness, density, hardness, porosity, and microstructure. They concluded that the material of the mold significantly influences the quality of the final part. Different factors, such as the amount of binder used to bind the particles of the mold, the grain size, the permeability, and the composition of the mold material, modify the material properties of resultant castings. These properties are related to the microstructural aspects of the final casting matrix (secondary dendrite arm separation, grain size), secondary intermetallic phases (size, morphology, distribution, and quantity), and porosity [19]. Specifically, the microstructure of the Al-Si alloys, the object of study in this article, present in their microstructure a matrix of α-aluminum dendrites as the main component and several phases. Among the phases present are eutectic Si particles, several secondary intermetallic phases such as Al_2Cu, Mg_2Si, and several Fe-bearing phases (AlFeSi, AlFeSiMn). The secondary phases (eutectic Si, Al_2Cu, Mg_2Si, AlFeSi, and AlFeSiMn) provide specific properties to the alloy and can affect the mechanical and corrosion behavior of the material.

Al-Si-Cu alloys are passive materials, however, they show high susceptibility to pitting corrosion in chloride solution; this has been related to the presence of secondary phases because most of these particles are more noble than Al-matrix [20]. These micro- galvanic cells cause the dissolution of the aluminum matrix.

Part of this secondary phase are Si particles. The low solubility of silicon in the aluminum matrix originates the precipitation of Si particles. The size, morphology, and distribution of these particles play an important role in the mechanical and corrosion behavior of Al-Si-Cu alloys [21]. The cathodic behavior of Si within the aluminum-rich matrix acting as an anode results in the formation of micro-galvanic couples that cause localized corrosion [22]. The influence of silicon morphology and its effect on the area ratio in galvanic corrosion has been analyzed by Tahamtan et al. [23]. Thick flake Si particles with sharp edges that increase the cathode/anode ratio greatly increase the density of the corrosion stream [24].

Also, the electrochemical properties of Al-Si-Cu alloys are affected by morphologies, type, and distribution of intermetallic phases that are, in turn, a function of alloy composition and cooling rate [25,26]. Intermetallic particles such as Al_2CuMg, and Mg_2Si, are anodic with respect to the Al matrix and preferably corrode with respect for the surrounding Al matrix [27]. However, the particles of intermetallic phases such as Al_2Cu and especially the iron-rich phases (β-Al-Fe-Si and α-Al-Fe-Mn-Si) are mostly cathodic to the Al and cause the dissolution of surrounding eutectic Al matrix adjacent to these intermetallic phases [28]. The iron content determines the type and morphology of these iron-rich intermetallic phases. The plate shape phase, such as βAl_5FeSi, is the most abundant, and the greater the amount of iron, the longer and wider the needles are. This phase is very hard and fragile and induces a high susceptibility to localized corrosion. The presence of manganese favors the "Chinese script" form, such as $\alpha Al_{15}(Fe,Mn)_3Si_2$, which improves corrosion resistance compared with those that are manganese-free [29–31].

The studies at the microstructural level, mechanical properties and corrosion behavior of Al-Si alloys produced by the traditional sand casting process are known. However, little information on these aspects has been studied for castings using molds fabricated by additive manufacturing technology. This work aims to focus on this gap to investigate the microstructure and corrosion behavior of the AlSiCu alloy. For this purpose, the alloys were solidified in traditional sand molds and binder jetting molds.

2. Materials and Methods

2.1. Materials and Molds Manufacturing

For this study, the aluminum alloy Al-Si-Cu (EN-AC 4600) was used. The chemical composition was analyzed by optical emission spectroscopy (OES), and the results are shown in Table 1.

Table 1. Alloy composition in wt. (%).

Si	Fe	Cu	Mn	Mg	Zn	Ni	Cr	Pb	Sn	Ti	Bi	Cd	Co	V	Zr	Al
9.53	1.1	2.05	0.21	0.09	0.9	0.07	0.067	0.04	0.0115	0.08	0.045	0.0006	0.001	0.008	0.008	Bal

The metal pouring conditions were the same for both molds. The aluminum smelting was carried out in a Hobersal PR/400 (Hobersal, Barcelona, Spain) furnace at 770 °C, and the pouring temperature was 750 °C.

The material used to print the binder jetting mold was a commercial powder plaster, $CaSO_4.1/2H_2O$ (VisiJet PXL Core, from 3DSystems, Rock Hill, SC, USA) with a purity of 80–90%, according to the manufacturer and a commercial binder solution (VisiJet PXL Clear) based on water with 2% Pyrrolidone. The characterization of the powder and the binder were carried out by the authors in a previous work [32]. The mold was manufactured using binder jetting technology with a Project CJP 660Pro machine (3D Systems, Rock Hill, SC, USA). Binder Jetting technology is a powder-based process that uses a liquid binder to selectively bind the powder particles that are spread on a work platform. This process is repeated layer by layer until the part is completed according to the CAD design.

To manufacture the sand casting mold, commercial silica sand (Petrobond) was used. A pattern, including the part and filling system, was manufactured by fused deposition modeling (FDM) technology, using Ultimaker 2+ machine (Ultimaker B.V., Utrecht, Dutch). Figure 1 shows the geometry of the part and the methodology used for the manufacture of both molds, obtaining the casting parts, and metallographic and corrosion study. The part has a rounded geometry with a sufficient area-volume ratio to analyze changes in the internal structure due to heat transfer and cooling rate.

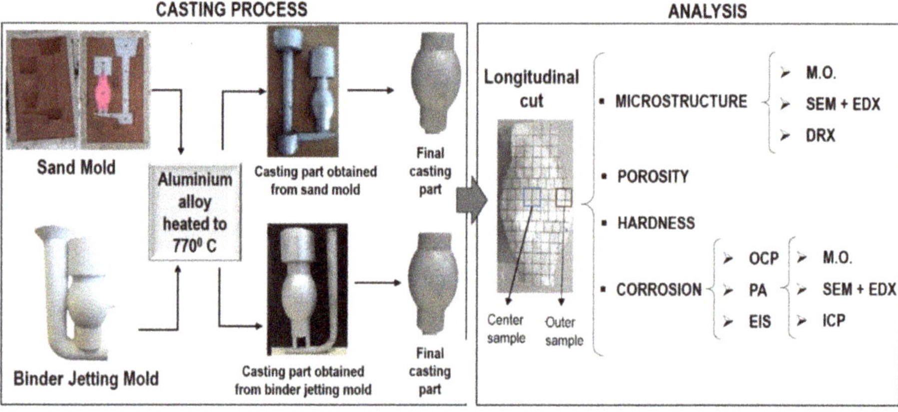

Figure 1. The methodology used for the casting process and analysis of the casting samples (metallographic and corrosion study).

2.2. Microstructural Characterization and Porosity

The metallographic study of both castings was carried out. First, the castings were longitudinally cut and then smaller samples were cut from two different zones of the casting parts, the center and the outer zone. The samples obtained were cold encapsulated and polished (as standard). To reveal the microstructure of the samples, the polished specimens were then etched using Keller's reagent (2.5 mL HNO_3; 1.5 mL HCl, 1 mL HF and 95 mL H_2O) for 12 s. Once polished and attacked, the study of different parameters was carried out, resorting to the use of different techniques.

For the microstructure and porosity study, optical and electronic microscopy was used. The images were taken from each sample (center sample and outer sample) corresponding to the castings obtained from both molds (sand mold and binder jetting printed mold). For the porosity study, a total of 30 images were taken with light microscopy (LM) randomly distributed on the surface of each sample, and then ImageJ 2016-0205 software (National Institute of Health, Bethesda, MD, USA) was used to determine the area occupied by the pores.

The morphological and chemical analysis of the phases was carried out by means of secondary electron imaging (SEI) obtained using a JEOL 6480 scanning electron microscope (SEM) (JEOL, Tokyo, Japan), equipped with an energy dispersive X-ray spectroscopy (EDS) technique was used by means of the AZtec (Matsuzawa Seike Co., Ltd., Tokyo, Japan) analysis software. We also used X-ray diffraction analysis (XRD) using a Bruker D8 Discover diffractometer equipped with a Cu anode (=1.5418 Å).

2.3. Microhardness

A microhardness Vickers test was carried out using Matsuzawa MXT70 microdurometer(EG&G Princeton Applied Research, potentiostat, NJ, USA) with a load of 100 g for 30 s. Hardness samples were extracted from the center and outer locations (Figure 1). Indentations were made until at least 10 indentations were acceptable and each hardness value was calculated by taking the average of the ten identations. The microhardness was evaluated as the arithmetical mean value, and standard deviation was determined.

2.4. Corrosion Testing and Samples Characterization

The corrosion resistance of the casting Al-Si-Cu alloy obtained using a binder jetting mold in comparison with traditional sand casting mold, used as reference, was evaluated from three electrochemical methods: open circuit potential (OCP), anodic polarization measurement (PA), and electrochemical impedance spectroscopy (EIS). The electrochemical testings were carried out in a three-electrode cell using a saturated calomel electrode (SCE) as the reference electrode, graphite as the counter-electrode, and the Al-Si-Cu alloy as the working electrode. The potentiostat 273A EG&G PAR (Princeton Applied Research, Princeton, NJ, USA) and the impedance analyzer Solartron SI 1260 (Princeton Applied Research, Princeton, NJ, USA)were used for monitoring the tests. The corrosion tests were carried out in solution with chlorides (3.5% NaCl at 25 °C). The samples were wet ground with SiC abrasive papers, followed by polishing with 1 mm diamond aqueous suspension. All the electrochemical tests were repeated three times.

The open-circuit potential was measured for 3600 s of immersion of samples in the chloride solution. Potentiodynamic anodic potential curves were made following the ASTM G-5 [33]. Once a nearly quasi-stable potential has been reached, the anodic potentiodynamic scan started at 250 mV below VOCP, reaching $1V_{SCE}$, using 50 mV/min as the potential scan rate. The corrosion potential and current densities were obtained through Tafel's analysis. Electrochemical impedance spectroscopy (EIS) was carried out to evaluate the corrosion polarization resistance (RP). Impedance spectra were performed at open circuit potential after a stabilization step in the open circuit during 1800 s with a signal amplitude of 10 mV at OCP and with a frequency range from 10 MHz to 1Hz. Observations by optical microscopy were done on the corroded samples to understand how the mold casting influences the corrosion performance in Al-Si-Cu alloy.

2.5. Inductive Coupling Plasma (ICP) with Optical Emission Spectrophotometer (OES)

In order to know which phases were dissolved after the corrosion process and to keep the piece in the same medium for 14 days, inductive coupling plasma (ICP) with optical emission spectrophotometer (OES) (ICP-OES) was used. The model was Perkin Elmer Optima 2000DV. The samples were measured directly, undiluted. The ICP equipment (Agilent Technologies Australia, Made in Malaysia) was calibrated with the following standards: 0, 0.01, 0.05, 0.1, 0.5, 1, and 10 ppm.

3. Results and Discussion

3.1. Microstructural Characterization of Aluminum Alloy Castings

Figure 2 shows the microstructures solidified in both molds. Figure 2a–f represents the solidified microstructures in a binder jetting mold, while Figure 2g–l represents the solidified microstructures in a sand mold. As can be seen, when comparing these figures, different mold materials produce some changes in the morphology and size of some phases.

Regarding the size of the phases, it is evident that they are much smaller in the parts obtained by sand mold, especially in the outer area of the parts, which could be associated with the higher cooling speed in this area. As for the shape, it can be seen that elongated shapes predominate in the sand mold. However, the phases present in the outer zone of the binder jetting mold appear as thick phases with an irregular geometric and elongated shape in most of the phases. The phases of Si, in these cases, are more elongated.

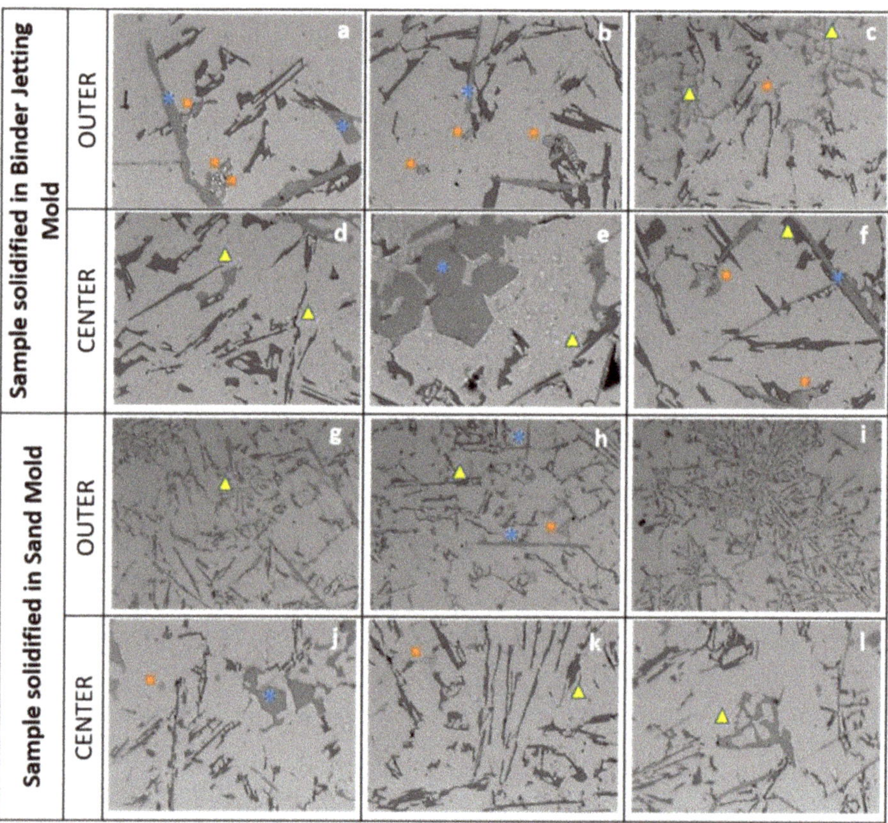

Figure 2. Optical micrographs of the samples: obtained in the outer area (**a–c**) from binder jetting mold and (**g–i**) from sand mold, and obtained in the center area (**d–f**) from binder jetting mold and (**j–l**) from sand mold. * Si phase; ■ Al$_2$Cu phase; △ Al-Fe-Si phase.

It is known that the microstructure in aluminum alloy contains a matrix of α-aluminum dendrites as the main component, which is surrounded by the eutectic Al-Si aggregate and various secondary phases. Among the phases present are eutectic Si particles, several secondary intermetallic phases such as Al$_2$Cu, Mg$_2$Si, and several Fe-bearing phases (AlFeSi, AlFeSiMn). In our case, the microstructure that results from the melting in both molds consisted of an aluminum matrix, in which some phase is distributed. Coarse pro-eutectic Si particles (Figure 2e–j) and also Si needles particles (Figure 2b–f,h) can be observed. Iron phases can generally be divided into three different morphologies: polyhedral crystals, Chinese script, or thin platelets. The plate-like phases, such as β-Al$_5$FeSi, are considered more harmful than script-type phases and polyhedral crystals, such as α-Al$_{15}$- (Fe,Mn)$_3$Si$_2$ [29,30,34]. All of them appear in the two castings obtained in both types of molds (Figure 2c–h,k,l). Al$_2$Cu appears with irregular rounded shapes, most of them near to the Si and Fe phases, indicating that these phases may act as a nucleation site (Figure 2a–c,f,h,j,k). This phase seems to be mostly found in the cast part obtained in a mold made with Binder Jetting.

Using SEM, it was possible to obtain Secondary Electron Imaging (SE), in order to know the composition of the different phases identified by OM (Figure 2). The results are shown in Figure 3. In this Figure, some images are collected, and one of them corresponds to the image of secondary electrons (SE) of the selected area, five of them to high-resolution X-ray spatial distribution maps corresponding to the elements present in the SE area. Iron is represented by green point density, which patterns distribution mappings is closely correlated with yellow point density corresponding to Mn and in some points with aluminum. This correlation between this element is indicating the presence of Al-Fe-Mn phase. The silicon distribution is not close to any element, so this indicates the presence of eutectic silicon particles.

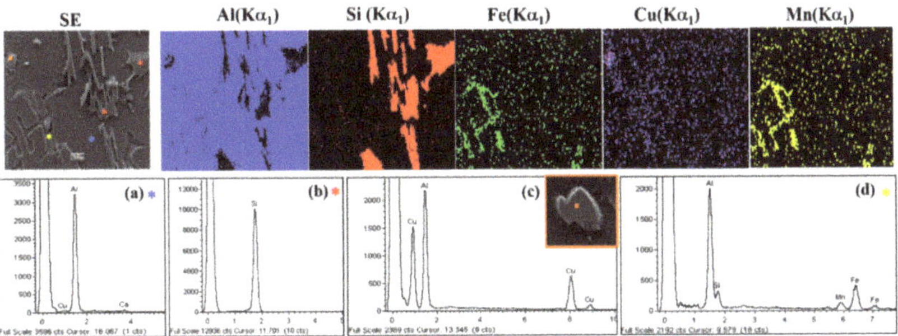

Figure 3. Mapping analysis of an area on the surface of the casting specimen, and 3 (a–d) X-ray spectra obtained at different points in the area as a function of the distribution of the elements.

The ED X-ray spectrum is derived from different areas on the SE image, depending on the distribution of the elements shown in the X-ray images. This has allowed corroborating the presence of different phases: silicon (Figure 3b), Al_2Cu (Figure 3c), and Al-Fe-Si-Mn (Figure 3d) in that area.

The secondary iron phase (Al-Fe-Mn-Si) appeared in both molds with two different defined morphologies (Figure 4). On the one hand, particles with the shape of Chinese script appeared (Figure 4a), and on the other hand, particles with irregular shapes (Figure 4b).

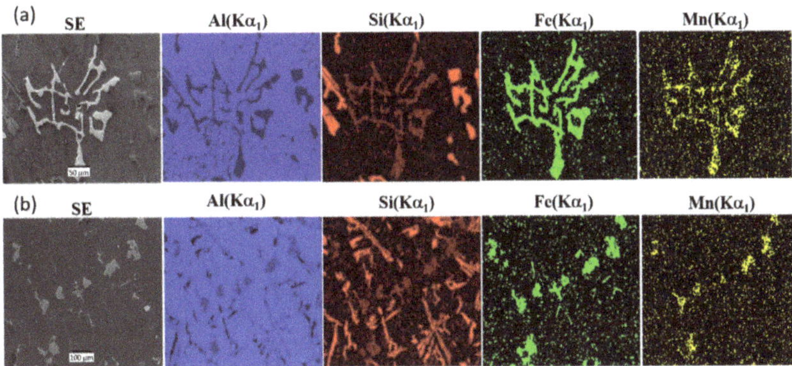

Figure 4. Mapping analysis of two different phases of iron. Chinese script (a), and irregular shape (b).

Figure 5 and Table 2 show the experimental results of the XRD patterns of the four samples studied (sand mold and binder jetting mold in the outer and central zones). As it can be observed, the phase proportions vary slightly both in the analyzed position (center-exterior) and when changing the mold. The XRD analysis revealed the presence of Al, and Si eutectic, as the major peaks. In addition, small diffraction peaks of $CuAl_2$ intermetallic phase were detected, which proved that the Cu particles had reacted with the Al matrix-forming $CuAl_2$ intermetallic, as suggested by the EDS analysis. In Figure 5, peaks at ~8°, 44°, 64°, 78°, and 83° correspond to the Bragg planes (111), (200), (220), (311), and (222), respectively, indicating the presence of aluminum. On the other hand, the peaks that appear at ~28°, 48°, 56°, 77°, and 88° indicate the presence of the eutectic Si phase in the Bragg planes (111), (220), (311), (331), and (422). These results are consistent with other investigations of AlSi alloys as well as sintered compounds of these [35–37].

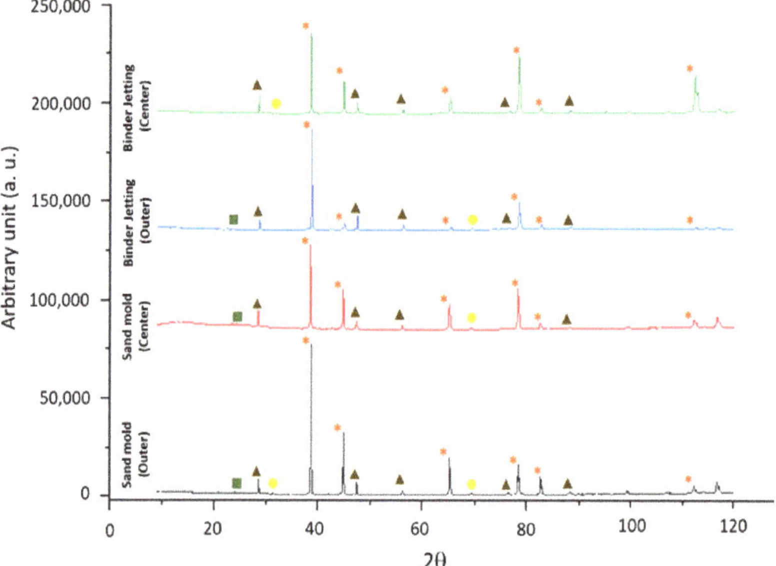

Figure 5. Experimental XRD patterns of the aluminum-silicon alloy processed by sand and binder jetting mold. (⁎ Al; ▲ Si; ● Al_2Cu; ■ AlFeMn).

Table 2. Semi-quantitative results for the samples solidifies in both molds (sand mold and binder jetting mold).

Phases	Binder Jetting		Sand Casting	
	Outer	Center	Outer	Center
Al	42	76.1	30.6	80.3
Si	7.6	0.7	5.1	1.4
Al-Fe-Si	28.1	18.4	52.2	12.4
Al_2Cu	10.9	2.6	7.5	2.0
Al-Fe-Mn-Si	11.3	1.4	1.0	2.5
Mg_2Si	2.7	0.8	1.1	1.3

Weaker peaks at approximately 41°, 42°, and 69° corresponding to the (220) and (222) Bragg planes indicating the minority presence of Al_2Cu [38].

The XRD results reveal that the composition of the alloy is similar in both cases, although the content of each phase is different for the binder jetting and sand mold. The results of a semi-quantitative analysis of these samples are shown in Table 2. According to the results, it is possible to conclude that the silicon phase is the majority in the exterior zone for both molds, however, it appears in a greater proportion for samples obtained by binder jetting mold. In addition, for both molds, the secondary phases predominate in the outer area of the piece, and they are also the majority in binder jetting mold. However, the aluminum phase is predominant for both molds in the central zone, although this phase is the majority in the case of the sand mold.

In addition to the identification of phases and their distribution, in this work, the dendritic size was also determined in both areas of the castings (exterior and center) because it is an important parameter to take into account when evaluating the corrosion behavior.

From the OM micrographs, the secondary dendrite arm separation (SDAS) was measured (Figure 6) and the results are collected in Table 3. In both areas, exterior and center, higher SDAS values for the sample were obtained with the binder jetting mold. From it, it was determined to establish the cooling speeds of the solidified alloys in the two molds and in the two studied areas. For this purpose, Equation (1) was used [39].

$$\log R = -2.5 \log \Lambda s + 4.5 \tag{1}$$

where Λs is the secondary dendrite arm separation (µm) and R is the cooling speed (°C/s).

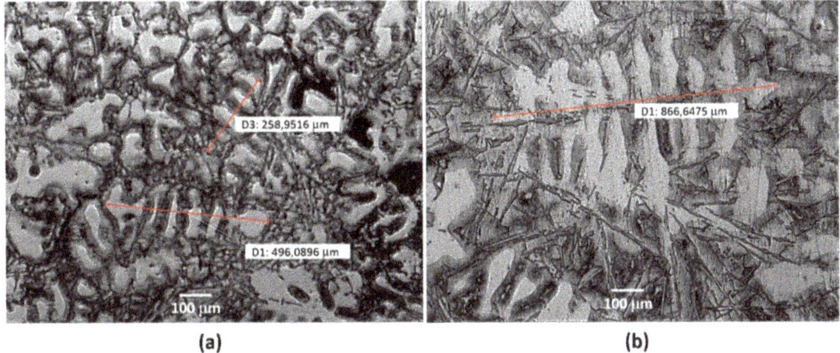

Figure 6. OM micrographs used to evaluate the SDAS (**a**) dendrites solidified in sand mold, and (**b**) dendrites solidified in binder jetting mold.

Table 3. Cooling rates obtained from Equation (1) and the SDAS values.

Parameters	Binder Jetting		Sand Casting	
	Outer	Center	Outer	Center
SDAS (µm)	96.21	120.37	70.62	81.48
Cooling Rate (°C/s)	0.35	0.19	0.75	0.52

Table 3 shows the cooling rates, calculated using Equation (1), taking into account the determined secondary experimental dendrite arm spacing (SDAS). In the investigated alloy, it can be seen that the highest solidification speed occurred in the outer areas of the cast part, according to the lowest value of SDAS.

These compositional and morphological differences observed in the samples analyzed, notably influence the corrosion behavior, as indicated in the following sections.

3.2. Porosity and Hardness Study of Solidified Samples

To analyze porosity, a longitudinal cut was made in the castings, as indicated in Section 2.2. For both cases, sand casting and binder jetting casting, two types of porosity were detected. As can be seen in Figure 7, in both casting processes, the same type of porosity appeared in the samples: (1) spherical pores as a consequence of gas trapping (Figure 7a,c), and (2) dendritic porosity due to the shrinkage that occurs during the solidification process (Figure 7b,d).

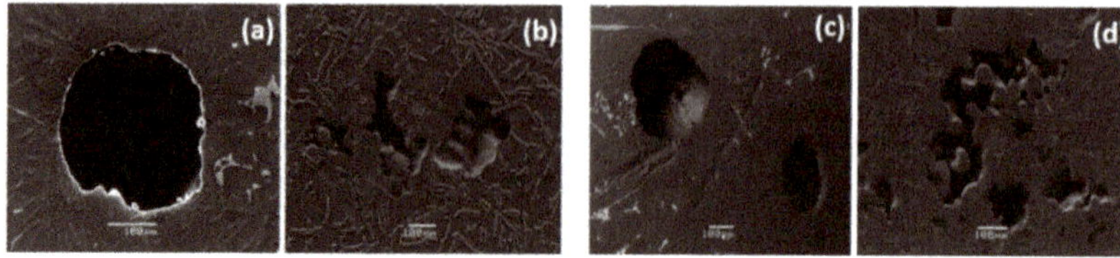

Figure 7. Micrographs obtained by SEM revealing the porosity present in the samples solidified in binder jetting mold (**a**,**b**) and in sand mold (**c**,**d**).

Although the same type of porosity was observed with both processes, it should be noted that the presence of pores due to gas trapping was higher when the sample was obtained from the mold processed by additive manufacturing. This may be because the 3D mold is processed by additive manufacturing; it contains higher moisture, volatile substances, and impurities compared to the traditional mold. This high content of volatile substances means that the castings from the mold processed by 3D Binder Jetting present a slight increase in both the number and the size of the pores.

The porosity percentage on the surface of the casting part was also determined (Table 4). The area occupied by pores was significantly higher in the specimen obtained with binder jetting mold, with this difference being more accused of the center specimen.

Table 4. % Porosity of casting specimens solidified in sand mold and binder jetting mold.

Type of Mold	% Porosity	
	Outer	Center
Sand mold	3.39	3.02
Binder jetting mold	3.68	4.35

The mean Vickers microhardness for all samples are indicated in Table 5. The microhardness values do not present significant differences neither with the type of mold, nor with the region analyzed in the samples.

Table 5. Hardness values of casting specimens: binder jetting and sand casting.

Type of Mold	HV 0.1/30	
	Outer	Center
Sand Mold	65.54	62.86
Binder Jetting Mold	63.26	62.70

3.3. Corrosión Behavior

Al-Si-Cu alloy processed by sand casting and by binder jetting casting corresponding to the central and outer zones were immersed in the 3.5 wt.% NaCl solution at 20 °C during 3600 s and the open circuit potential was recorded, Figure 8a. For both sections and for both casting processes, the potential slightly increased over time, thus, it is indicative of a passive layer formation process on the surface. In all cases, there were notable potential fluctuations that reveal the difficulties of forming a stable passive layer. The potentials are slightly higher for binder jetting casting, indicating that thermodynamically the corrosion resistance is somewhat higher than for sand casting. The differences between the center and outer sections are minimal for both samples.

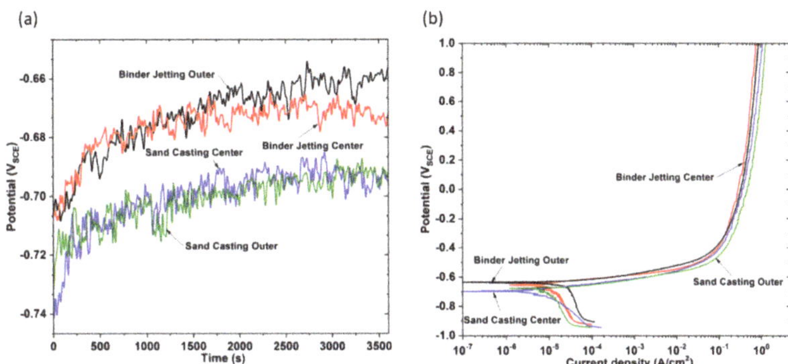

Figure 8. OCP evolution of binder jetting and of sand mold (**a**) and anodic polarization scans in the binder jetting and in the sand mold (**b**) in 3.5% NaCl, center and outer sections.

The effect of the molding on the corrosion rate of the AlSiCu alloy sample was also studied using the anodic polarization technique. Figure 8b shows the solidified potentiodynamic polarization curves for binder jetting mold and sand mold corresponding to the center and outer zones. The morphology of the curves coincides in all cases after the cathodic branch is observed; the zone of corrosion potential is slightly higher for binder jetting casting, which is consistent with the previous open circuit potential values. Then the current density increases until reaching the zone of passivity without observing in any case pitting potential, although as is known, pitting corrosion is one of the most common corrosion mechanisms in these Al alloys [40]. The curves are slightly displaced to the left for alloy solidified on the binder jetting mold, which indicates lower current density and therefore higher corrosion resistance.

Corrosion parameters such as corrosion potential (E_{corr}) and corrosion current density (i_{corr}) obtained from Tafel's analysis, are tabulated in Table 6, also included passive current density (i_{pass}) at $0.5V_{SCE}$ and polarization resistance (R_P) values. The results indicate the decrease in corrosion and passive current density with the use of binder jetting mold for the center section. However, the differences are negligible for the outer sections, which show in both cases the lowest polarization resistances. The resistance of the central section of the binder jetting casting alloy was three times higher than that obtained in the sand casting, so the kinetics of the corrosion process expressed is also substantially improved in addition to increasing thermodynamic stability with the use of the binder jetting mold.

Table 6. Electrochemical parameters estimated by Tabel method.

Sample		E_{corr} (V)	i_{corr} (μA/cm^2)	R_p (Ω/cm^2)	I_{pas} (A/cm^2)
Sand mold	Center	−0.695	90.1	289.29	0.71
	Outer	−0.683	43.0	605.54	0.90
Binder jetting mold	Center	−0.652	14.9	1752.6	0.52
	Outer	−0.643	40.5	643.31	0.61

In order to study the effect of the casting mold on the nature of the passive layer and to obtain information on the corrosion mechanisms, electrochemical impedance spectroscopy tests were carried out, which reveal greater differences between the alloys depending on the mold used in the solidification. In the Nyquist diagram (Figure 9a), a larger semi-circle is clearly observed in the binder jetting alloys, and the radius of the semi-circle is indicative of the resistance to electronic transfer and is commonly related to corrosion resistance, this coincides as shown in the Bode diagram (Figure 9b), with a higher impedance modulus in the lower frequency zone for the solidified alloy in binder jetting mold. The sample corresponding to the outer section solidified in sand casting clearly showed the lowest corrosion resistance.

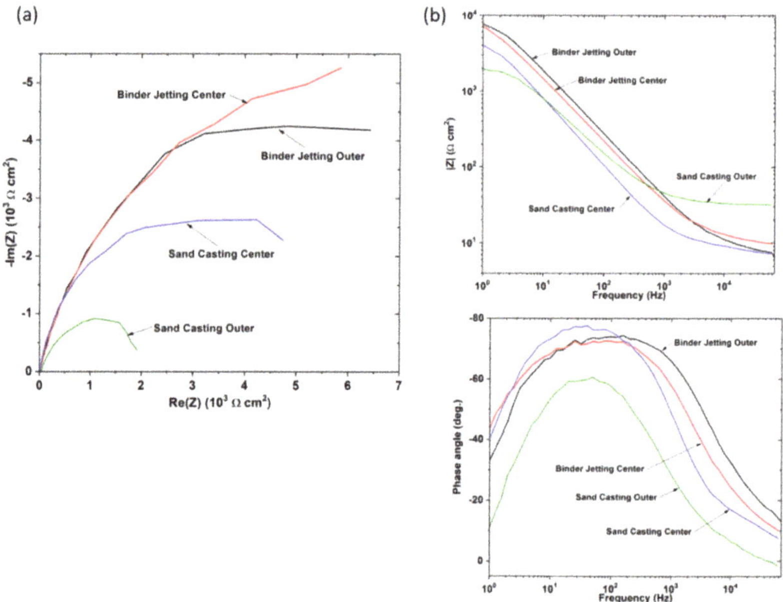

Figure 9. Nyquist (**a**) and Bode (**b**) plots of binder jetting and sand casting samples in 3.5% NaCl, center and outer sections.

The three electrochemical testing results indicate the best behavior of the samples processed by additive manufacturing mold, which must be correlated with the structural differences observed previously. From the electrochemical point of view, the most of intermetallic phases are more noble the aluminum matrix, making the alloy system highly susceptible to localized corrosion [41].

Silicon in coarse flake shapes precipitates as a consequence of the low solubility of this alloy element in the aluminum-rich matrix. Silicon is cathodic with respect to the eutectic aluminum matrix, which may lead to the formation of micro-galvanic couples. The amount of silicon [42], the morphology of silicon particles as well as the cathodic/anodic ratio [43] have important effects on galvanic corrosion between silicon and aluminum solid solution. Other microstructural features as dendrite cell size, secondary dendrite arm spacing, and grain size of aluminum phase also can affect galvanic corrosion, and it is concluded that a thicker dendritic structure tends to give the material a higher corrosion resistance compared to thinner structures where the distance between the arms is smaller [44].

The enhanced corrosion resistance of alloy solidified in the binder jetting mold compared with the alloy solidified in sand mold could be associated, in the first instance, with the reduced area ratio of eutectic silicon particles to eutectic aluminum phase [24,43] but on the other hand, with the presence of coarse silicon particles visualized in the eutectic of binder jetting alloy that can originate a lower number of micro-galvanic couples. Nevertheless, it is known that the effect of silicon is not so important due to the low corrosion current densities [34] resulting from the high polarization of silicon particles [45].

On the other hand, it is known that the presence of Fe in Al-Si-Cu alloys originates the β-Al-Fe-Si phase in the form of longer and thicker needles, with the higher the iron content and the higher the cooling speed [46]. This phase has a nobler potential than the aluminum matrix and therefore enhances the susceptibility to localized corrosion [40]. The presence of Mn resulting in α-Al-Fe-Mn-Si in the form of the so-called "Chinese script", with lower potential reducing the galvanic corrosion compared with the β-Al-Fe-Si intermetallic phase [47]. The Fe-intermetallic compounds are more visible in outer sections of both samples, but the sample processed by sand casting had a higher amount of needles and larger size while in the sample processed by binder jetting, a greater amount of script phase was visible, it can explain the enhance corrosion behavior observed for outer ceramic casting sample. These results are in agreement with results observed by other authors [40,42,48]. These studies concluded that the iron-intermetallics compounds acted as more effective cathodes than Si on Al-Si-Cu in 3.5% NaCl solution.

Finally, the content of the Mg_2Si phase in the sample processed by binder jetting mold was higher than processed by sand one, which may enhance corrosion resistance [49] because it is anodic with respect to the aluminum matrix.

3.4. Microstructural Characterization of the Molten Samples after the Corrosion Test

After the corrosion test, a clear dissolution of phases was observed (Figure 10). The OM micrographs show that there has been a greater dissolution of the phases in the sand than in the binder jetting mold, regardless of the area studied. This may be due to the fact that the size of the phases corresponding to the sample processed by sand mold were smaller than the phase size corresponding to the samples processed by binder jetting mold.

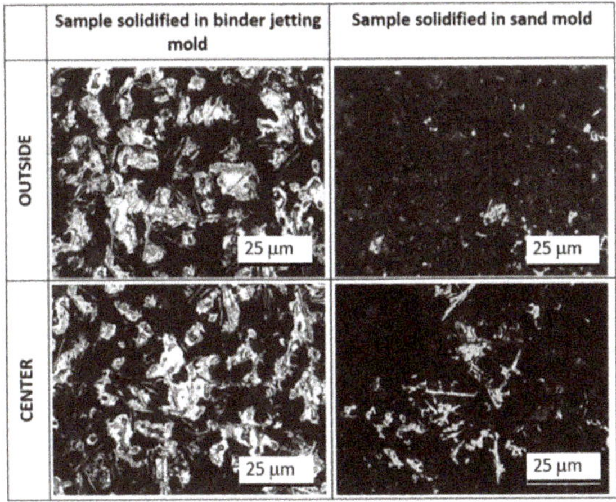

Figure 10. OM micrographs of corroded surface products immersed in 3.5% NaCl solution for the specimen in the binder jetting mold and sand casting mold.

The samples were left in the corrosive medium (3.5% NaCl) for 75 days in order to study the behavior of the oxide layer formed, as well as the effect on the existing phases. Figure 11 shows six images of an area for each sample processed with both molds (A) and (B). One of the images (SE) was derived from secondary electrons, while the other five images are high-resolution X-ray spatial distribution maps corresponding to the elements present in the SE area. The relative abundance of each element is represented by the spatial distribution of the density of dots of different colors depending on the element that is represented in each image. For both cases, there is a correlation distribution between aluminum (green image) and oxygen (red image), although in the case of the sand mold, a lower intensity and number of spots can be observed in the case of oxygen with respect to aluminum. However, in the binder jetting specimen, there is a homogeneous distribution between both elements. This could be indicative that the oxide layer formed in the sample manufactured by binder jetting is more resistant to the corrosive medium. The other elements (Si, Fe, Cu) continue to appear in the molten samples.

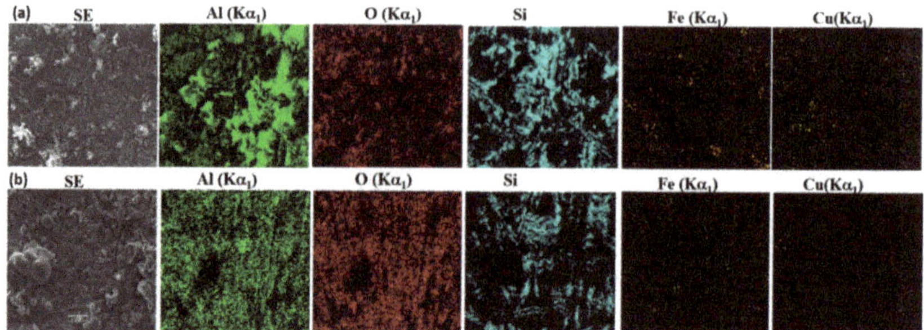

Figure 11. Element distribution mapping for Al-Si Alloy processed by sand mold (**a**) and binder jetting mold (**b**).

In order to know the number of dissolved elements after the 75 days of immersion, the concentration of the elements present in the liquid was determined by the technique of optical emission spectrometry in plasma inductively coupled.

The concentration obtained for each element after 75 days of immersion in 3.5% NaCl is presented in Table 7. In view of the results presented in the table, we can conclude that there is a greater dissolution of Al, Cu, and Si in the sample obtained with sand mold. This is in agreement with the results obtained with OM and SEM-EDX.

Table 7. Concentration (wt.%) of the elements in the 3.5% NaCl after 75 days of immersion.

Element (ppm)	Reference Material	Binder Jetting		Sand Casting	
		Center	Outer	Center	Outer
Al	<0.10	0.20	2.28	2.57	2.36
Cu	<0.01	0.03	0.06	0.20	0.04
Fe	<0.02	0.02	0.02	0.03	0.02
Mg	<0.01	0.19	0.20	0.25	0.15
Mn	<0.01	0.01	0.01	0.01	0.01
Ni	<0.02	0.02	0.02	0.02	0.02
Pb	<0.05	0.05	0.05	0.05	0.05
Si	<0.10	0.28	0.32	0.51	0.34
Sn	<0.10	0.10	0.10	0.10	0.10
Ti	<0.01	0.01	0.01	0.01	0.01
Zn	<0.02	0.02	0.08	0.33	0.02

4. Conclusions

The present study is directed to investigate the influence of two different molds on the microstructure and corrosion behavior of an aluminum alloy. The major conclusions are:

The corrosion behavior depends on the mold, but it is irrespective of the aluminum cast zone.

The OCP results revealed the formation of the passive layer for the alloy processed with the two molds. Although the slightly higher potentials in OCP, the lower current densities in the PA test, as well as the decrease in corrosion and passive current density in the Tafel test obtained for samples processed by additive manufacturing, reveal that in this case, the aluminum alloy is more corrosion-resistant.

The variation of corrosion behavior with both molds is related to the differences found in the microstructure. The binder jetting-processed alloy microstructure shows galvanic pairs of silicon and α-Al-Fe-Mn-Si intermetallic precipitates that limit corrosion resistance. While with the sand mold, the microstructure is rich in needle-like intermetallic compounds (β-Al-Fe-Si) that make corrosion accelerate.

Author Contributions: Conceptualization, M.Á.C.-S. and C.G.-C.; methodology, M.Á.C.-S.; validation, F.M.-P.; formal analysis, M.Á.C.-S. and C.G.-C.; investigation, C.G.-C.; data curation, C.G.-C.; writing—original draft preparation, M.Á.C.-S. and C.G.-C.; writing—review and editing, A.I.F.-A. and F.M.-P.; supervision, J.B.; funding acquisition, A.I.F.-A. and J.B. All authors have read and agreed to the published version of the manuscript.

Funding: This research was funded by the Ministry of Science, Innovation, and Universities of Spain, grant number DPI2017- 89840-R.

Institutional Review Board Statement: Not applicable.

Informed Consent Statement: Not applicable.

Data Availability Statement: Not applicable.

Conflicts of Interest: The authors declare no conflict of interest.

References

1. Yılmaz, M.S.; Özer, G.; Öter, Z.C.; Ertugrul, O. Effects of hot isostatic pressing and heat treatments on structural and corrosion properties of direct metal laser sintered parts. *Rapid Prototyp. J.* **2021**, *27*, 1059–1067. [CrossRef]
2. Cohen, A.L. From rapid prototyping to the second industrial revolution. In Proceedings of the Third International Conference on Rapid Prototyping, Dayton, OH, USA, 23–25 June 1992; pp. 189–192.
3. Low, Z.X.; Chua, Y.T.; Ray, B.M.; Mattia, D.; Metcalfe, I.S.; Patterson, D.A. Perspective on 3D printing of separation membranes and comparison to related unconventional fabrication techniques. *J. Membr. Sci.* **2017**, *523*, 596–613. [CrossRef]
4. Attar, M.; Aydin, S.; Arabaci, A.; Mutlu, I. Mechanical meta-material-based polymer skin graft production by rapid prototyping and replica method. *Rapid Prototyp. J.* **2021**, *27*, 278–287. [CrossRef]
5. Wang, J.; Sama, S.R.; Manogharan, G. Re-Thinking Design Methodology for Castings: 3D Sand-Printing and Topology Optimization. *Int. J. Met.* **2019**, *13*, 2–17. [CrossRef]
6. EPA (Environmental Protection Agency). *Profile of the Metal Casting Industry*; Office of Compilance Sector Notebook Project; EPA (Environmental Protection Agency): Washington, DC, USA, 1997.
7. Tang, Y.; Huang, Z.; Yang, J.; Xie, Y. Enhancing the Capillary Force of Binder-Jetting Printing Ti6Al4V and Mechanical Properties under High Temperature Sintering by Mixing Fine Powder. *Metals* **2020**, *10*, 1354. [CrossRef]
8. Mariani, M.; Goncharov, I.; Mariani, D.; De Gaudenzi, P.G.; Popovich, A.; Lecis, N.; Vedani, M. Mechanical and microstructural characterization of WC-Co consolidated by binder jetting additive manufacturing. *Int. J. Refract. Met. Hard Mater.* **2021**, *100*, 105639. [CrossRef]
9. Cheng, Y.-L.; Tseng, T.-W. Study on driving waveform design process for multi-nozzle piezoelectric printhead in material-jetting 3D printing. *Rapid Prototyp. J.* **2021**, *27*, 1172–1180. [CrossRef]
10. Cheng, Y.-L.; Chang, C.-H.; Kuo, C. Experimental study on leveling mechanism for material-jetting-type color 3D printing. *Rapid Prototyp. J.* **2020**, *26*, 11–20. [CrossRef]
11. Li, J.; Yan, R.; Yang, Y.; Xie, F. Water-based binder preparation and full-color printing implementation of a self-developed 3D printer. *Rapid Prototyp. J.* **2021**, *27*, 530–536. [CrossRef]
12. Snelling, D.; Li, Q.; Meisel, N.; Williams, C.B.; Batra, R.C.; Druschitz, A.P. Lightweight Metal Cellular Structures Fabricated via 3D Printing of Sand Cast Molds. *Adv. Eng. Mater.* **2015**, *17*, 923–932. [CrossRef]
13. Walker, J.; Harris, E.; Lynagh, C.; Beck, A.; Lonardo, R.; Vuksanovich, B.; Thiel, J.; Rogers, K.; Conner, B.; MacDonald, E.; et al. 3D Printed Smart Molds for Sand Casting. *Int. J. Met.* **2018**, *12*, 785–796.
14. George, S.; Roy, A.; McKenna, N.; Singamneni, S.; Diegel, O.; Singh, D.; Neitzert, T.; St George, J.; Roy Choudhury, A.; Yarlagadda, P. Direct Metal casting through 3D printing: A critical analysis of the mould characteristics. In Proceedings of the 9th Global Congress on Manufacturing and Management (GCMM2008), Gold Coast, QLD, Australia, 12–14 November 2008; pp. 1–5.
15. Vaezi, M.; Chua, C.K. Effects of layer thickness and binder saturation level parameters on 3D printing process. *Int. J. Adv. Manuf. Technol.* **2011**, *53*, 275–284. [CrossRef]
16. Garzón, E.O.; Alves, J.L.; Neto, R.J. Study of the viability of manufacturing ceramic moulds by additive manufacturing for rapid casting. *Ciência Tecnologia Materiais* **2017**, *29*, e275–e280. [CrossRef]
17. Snelling, D.A.; Williams, C.B.; Druschitz, A.P. Mechanical and material properties of castings produced via 3D printed molds. *Addit. Manuf.* **2019**, *27*, 199–207. [CrossRef]
18. Snelling, D.; Blount, H.; Forman, C.; Ramsburg, K.; Wetzel, A.; Williams, C.; Druschitz, A. The Effects of 3D Printed Molds on Metal Castings. In Proceedings of the Solid Freeform Fabrication Symposium, Austin, TX, USA, 12–14 August 2013; pp. 827–845.
19. Barresi, J.; Kerr, M.J.; Wang, H.; Couper, M.J. Effect of magnesium, iron and cooling rate on mechanical properties of Al-7Si-Mg foundry alloys. *AFS Trans.* **2000**, *108*, 563–570.
20. Szklarska-Smialowska, Z. Pitting corrosion of aluminum. *Corros. Sci.* **1999**, *41*, 1743–1767. [CrossRef]
21. Guo, W.; Li, J.; Qi, M.; Xu, Y.; Ezatpour, H.R. Effects of heat treatment on the microstructure, mechanical properties and corrosion resistance of AlCoCrFeNiTi0.5 high-entropy alloy. *J. Alloys Compd.* **2021**, *884*, 161026. [CrossRef]

22. Davis, J.R. *Alloying: Understanding the Basics*; ASM International: Metals Park, OH, USA, 2001; p. 647.
23. Fadavi Boostani, A.; Tahamtan, S. Fracture behavior of thixoformed A356 alloy produced by SIMA process. *J. Alloys Compd.* **2009**, *481*, 220–227. [CrossRef]
24. Tahamtan, S.; Fadavi Boostani, A. Evaluation of pitting corrosion of thixoformed A356 alloy using a simulation model. *Trans. Nonferrous Met. Soc. China* **2010**, *20*, 1702–1706. [CrossRef]
25. Ye, H. An overview of the development of Al-Si-alloy based material for engine applications. *J. Mater. Eng. Perform.* **2003**, *12*, 288–297. [CrossRef]
26. Chen, C.L.; West, G.D.; Thomson, R.C. Characterisation of Intermetallic Phases in Multicomponent Al-Si Casting Alloys for Engineering Applications. *Mater. Sci. Forum.* **2006**, *519*, 359–364. [CrossRef]
27. Donatus, U.; Thompson, G.E.; Omotoyinbo, J.A.; Alaneme, K.K.; Aribo, S.; Agbabiaka, O.G. Corrosion pathways in aluminium alloys. *Trans. Nonferrous Met. Soc. China* **2017**, *27*, 55–62. [CrossRef]
28. Li, J.; Dang, J. A summary of corrosion properties of Al-Rich solid solution and secondary phase particles in al alloys. *Metals* **2017**, *7*, 84. [CrossRef]
29. Couture, A. Iron in aluminium casting alloys—a literature survey. *AFS Int. Cast Met. J.* **1981**, *6*, 9–17.
30. Crepeau, P.N. Effect of Iron in Al-Si Casting Alloys A Critical Review. *AFS Trans.* **1995**, *103*, 361–366.
31. Cao, X.; Campbell, J. Morphology of β-Al$_5$FeSi Phase in Al-Si Cast Alloys. *Mater. Trans.* **2006**, *47*, 1303–1312. [CrossRef]
32. Castro-Sastre, M.A.; Fernández-Abia, A.I.; Rodríguez-González, P.; Martínez-Pellitero, S.; Barreiro, J. Characterization of materials used in 3D-printing technology with different analysis techniques. *Ann. DAAAM Proc.* **2018**, *29*, 947–954.
33. ASTM International. *ASTM G-5-87, Standard Reference Test Method for Making Potentiostatic and Potentiodynamic Anodic Polarization Measurements*; ASTM International: West Conshohocken, PA, USA, 1993.
34. Bonsack, W. Effects of minor alloying elements on aluminium casting alloys: Part II. Aluminum–silicon alloys. *ASTM Bull.* **1943**, *124*, 41–51.
35. Liu, J.; Xiu, Z.; Liang, X.; Li, Q.; Hussain, M.; Qiao, J.; Jiang, L. Microstructure and properties of Sip/Al-20 wt.% Si composite prepared by hot-pressed sintering technology. *J. Mater. Sci.* **2014**, *49*, 1368–1375. [CrossRef]
36. Bonatti, R.S.; Siqueira, R.R.; Padilha, G.S.; Bortolozo, A.D.; Osório, W.R. Distinct Alp/Sip composites affecting its densification and mechanical behavior. *J. Alloys Compd.* **2018**, *757*, 434–447. [CrossRef]
37. Bonatti, Y.A.; Bortolozo, A.D.; Costa, D.; Osório, W.R. Morphology and size effects on densification and mechanical behavior of sintered powders from Al-Si and Al-Cu casting alloys. *J. Alloy. Compd.* **2019**, *786*, 717–732. [CrossRef]
38. Tash, M.M.; Mahmoud, E.R.I. Development of in-Situ Al-Si/CuAl$_2$ metal matrix composites: Microstructure, hardness, and wear behavior. *Materials* **2016**, *9*, 442. [CrossRef] [PubMed]
39. Drouzy, M.; Richard, M. Effet des conditions de solidification sur la qualité des alliage de fonderie de A-U5 GT et A-S7 G, estimation des caractéristiques mécaniques. *Fonderie* **1969**, *285*, 49–56.
40. Arrabal, R.; Mingo, B.; Pardo, A.; Mohedano, M.; Matykina, E.; Rodríguez, I. Pitting corrosion of rheocast A356 aluminium alloy in 3.5 wt.% NaCl solution. *Corros. Sci.* **2013**, *73*, 342–355. [CrossRef]
41. Jain, S. *Corrosion and Protection of Heterogeneous Cast Al-Si (356) and Al-Si-Cu-Fe (380) Alloys by Chromate and Cerium Inhibitors*; The Ohio State University: Columbus, OH, USA, 2006.
42. Osório, W.R.; Goulart, P.R.; Garcia, A. Effect of silicon content on microstructure and electrochemical behavior of hypoeutectic Al-Si alloys. *Mater. Lett.* **2008**, *62*, 365–369. [CrossRef]
43. Tahamtan, S.; Boostani, A.F. Quantitative analysis of pitting corrosion behavior of thixoformed A356 alloy in chloride medium using electrochemical techniques. *Mater. Des.* **2009**, *30*, 2483–2489. [CrossRef]
44. Oztürk, I.; Agaoglu, G.H.; Erzi, E.; Dispinar, D.; Orhan, G. Effects of strontium addition on the microstructure and corrosion behavior of A356 aluminum alloy. *J. Alloys Compd.* **2018**, *763*, 384–391. [CrossRef]
45. Davis, J.R. *ASM Specialty Handbook: Aluminum and Aluminum Alloys*; ASM International: Metals Park, OH, USA, 1993.
46. Kucharikova, L.; Liptáková, T.; Tillová, E.; Kajánek, D.; Schmidová, E. Role of chemical composition in corrosion of aluminum alloys. *Metals* **2018**, *8*, 581. [CrossRef]
47. Samuel, A.M.; Doty, H.W.; Valtierra, S.; Samuel, F.H. βAl$_5$FeSi phase platelets-porosity formation relationship in A319.2 type alloys. *Int. J. Met.* **2018**, *12*, 55–70.
48. Mingo, B.; Arrabal, R.; Pardo, A.; Matykina, E.; Skeldon, P. 3D study of intermetallics and their effect on the corrosion morphology of rheocast aluminium alloy. *Mater. Charact.* **2016**, *112*, 122–128. [CrossRef]
49. Yasakau, K.A.; Zheludkevich, M.L.; Lamaka, S.V.; Ferreira, M.G.S. Role of intermetallic phases in localized corrosion of AA5083. *Electrochim. Acta.* **2007**, *52*, 7651–7659. [CrossRef]

metals

Article

Experimental Investigation of Pressure Drop Performance of Smooth and Dimpled Single Plate-Fin Heat Exchangers

Kanishk Rauthan *, Ferdinando Guzzomi, Ana Vafadar, Kevin Hayward and Aakash Hurry

School of Engineering, Edith Cowan University, Joondalup, WA 6027, Australia; f.guzzomi@ecu.edu.au (F.G.); a.vafadarshamasbi@ecu.edu.au (A.V.); Kevin.Hayward@ecu.edu.au (K.H.); a.hurry@ecu.edu.au (A.H.)
* Correspondence: k.rauthan@ecu.edu.au; Tel.: +61-8-6304-4690

Abstract: Passive heat exchangers (HXs) form an inseparable part of the manufacturing industry as they provide high-efficiency cooling at minimal overhead costs. Along with the aspects of high thermal cooling, it is essential to monitor pressure loss while using plate-fin HXs because pressure loss can introduce additional power costs to a system. In this paper, an experimental study was conducted to look at the effects of dimples on the pressure drop characteristics of single plate-fin heat exchangers. To enable this, different configurations of National Advisory Committee for Aeronautics (NACA) fins with smooth surfaces and 2 mm-diameter dimples, 4 mm-diameter dimples and 6 mm-diameter dimples were designed and 3D printed using fused deposition modelling (FDM) of ABS plastic. The depth to diameter ratio for these dimples was kept constant at 0.3 with varied diameters and depths. These were then tested using a subsonic wind tunnel comprised of inlet and outlet pressure taps as well as a hot wire velocimeter. Measurements were taken for pressure differences as well as average velocity. These were then used to calculate friction factor values and to compare the smooth fin to the dimpled fins in relation to their relative pressure drop performance. It was observed that for lower velocities the 4 mm dimples provided minimum pressure drop, with a difference of 58% when compared to smooth fins. At higher velocities, 6 mm dimples increased the pressure drop by approximately 34% when compared to smooth fins. It can also be concluded from the observed data in this study that shallower dimples produce lower pressure drops compared to deeper dimples when the depth to diameter ratio is kept constant. Accordingly, deeper dimples are more effective in providing drag reduction at lower velocities, whereas shallower dimples are more effective for drag reduction at higher velocities.

Keywords: pressure drop; heat exchanger; additive manufacturing; surface textures; dimples; drag reduction

Citation: Rauthan, K.; Guzzomi, F.; Vafadar, A.; Hayward, K.; Hurry, A. Experimental Investigation of Pressure Drop Performance of Smooth and Dimpled Single Plate-Fin Heat Exchangers. *Metals* **2021**, *11*, 1757. https://doi.org/10.3390/met11111757

Academic Editor: Atila Ertas

Received: 1 October 2021
Accepted: 28 October 2021
Published: 1 November 2021

Publisher's Note: MDPI stays neutral with regard to jurisdictional claims in published maps and institutional affiliations.

Copyright: © 2021 by the authors. Licensee MDPI, Basel, Switzerland. This article is an open access article distributed under the terms and conditions of the Creative Commons Attribution (CC BY) license (https://creativecommons.org/licenses/by/4.0/).

1. Introduction

When discussing heat exchangers (HXs), a primary or direct contact surface is defined as a surface that separates two fluids at different temperatures. As noted by Thulukkanam [1] and Shah and Sekulic [2], additional surfaces can be attached to a primary surface to increase the heat transfer area and further improve the heat transfer characteristics of an HX. These secondary appendages are referred to as fins and aid with convective heat transfer.

With recent advancements in brazing and welding technology, most currently manufactured compact heat exchangers (CHX) and plate-fin heat exchangers (PFHX) involve the brazing, semi-welding and all-welding of plates to a core [3]. Diffusion-bonded HXs have also been applied in high-pressure industrial contexts [4].

Another important development in HX manufacturing is the introduction of additive manufacturing (AM), which allows for complex geometries and matrices to be fabricated [5]. As a result, HXs have become more compact and more efficient [4]. For instance, in a study conducted by Wong et al. [6], it was concluded that an AM-fabricated lattice-structured HX provided minimal resistance to airflow as compared to traditional plate-fin and pin-fin HXs.

However, the presence of a large number of voids within the lattice structure was shown to not allow the adequate transfer of heat, hindering the heat transfer capabilities of the HX. Another example of AM-fabricated HXs is the topology-optimised pin fin heat exchanger that has been developed by Dede, Joshi and Zhou, as shown in Figure 1 [7].

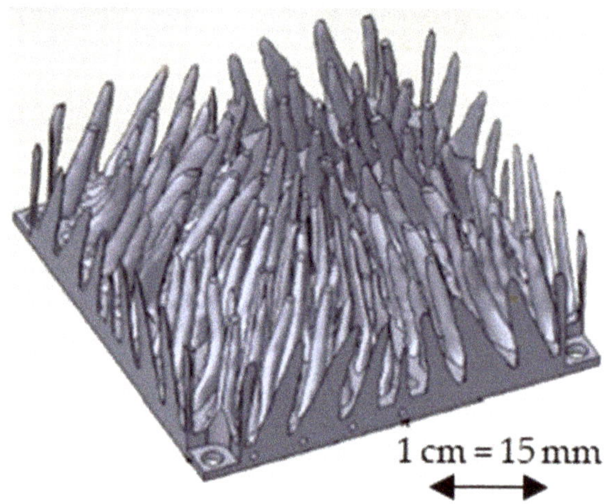

Figure 1. Topology-optimised AM-fabricated Pin-Fin HX CAD model. Reprinted with permission from [7]. Copyright 2015 by American Society of Mechanical Engineers (ASME).

The study conducted by McDonough [8] presents multiple examples of additively manufactured heat exchangers, highlighting the complexity in design along with the ease of fabrication of such complex shapes. Interested readers are highly encouraged to refer to the above-mentioned study to gain in-depth details of applications of additive manufacturing.

Further details of this are presented in a recently published paper by Vafadar et al. [9]. This above study discusses the application of metal AM in the development of complex geometries, for different industries, including the civil, electrical, oil and gas sectors. AM is presented in the above paper as a viable alternative to conventional machining (CM) in terms of reducing both costs and energy consumption [10].

The high thermal performance of CHXs (up to 98%) in combination with an ever-increasing demand for higher rates of heat transfer has prompted the research and development of heat transfer augmentation techniques that have minimal pressure drop penalties via the aid of forced convection. Forced convection, as defined by Sheikholeslami and Ganji [11], refers to the mechanism of fluid transport where fluid is brought into motion using external sources, such as pumps, fans, blowers, etc. To achieve this, there are two methods of heat transfer augmentation that are widely applied in forced convection: using complex shapes and using surface textures on extended surfaces.

One of the most commonly used types of HXs are PFHXs, which have external fins that increase the overall heat transfer area of an exchanger. These fins can be further modified to introduce turbulence in the flow, which enhances heat transfer [12]. These modifications may occur in the form of surface textures that are classified as passive methods of heat transfer enhancement, since they do not require any external source of power. Often, this enhancement in heat transfer is accompanied by an increase in pressure drop that requires additional pumping power for fans/blowers [12]. Hence, investigating surface textures to enhance heat transfer with a reduced pressure drop has become a major area of research for engineers.

Surface textures are an extension of passive methods and include features that are embossed/engraved on the surfaces of HXs to improve thermal and pressure drop performances. One surface texture that has been extensively studied by a range of different researchers is dimples. A number of numerical and experimental studies have been conducted that deal with dimples as a means of improving heat transfer while reducing drag/pressure drop. For instance, Burgess and Ligrani [13] have investigated the effects of dimple depth on the Nusselt number and friction factor, considering the dimple print diameter to dimple depth ratios (ε) of 0.1, 0.2 and 0.3 whilst keeping the print diameter constant at 50.8 mm with constant streamwise and spanwise pitches of 82.2 and 41.1 mm, respectively. Streamwise pitch (s) reflects the spacing between the dimples in the direction of the flow, whereas spanwise pitch (p) defines the spacing between dimples perpendicular to the flow of the fluid.

Their study compared the experimental values obtained for a depth to diameter ratio of 0.1 with other studies possessing depth to diameter ratios of 0.2, 0.3, 0.28 and 0.19 for friction factor values. Their study concluded that the values of friction factors increased with either an increase in depth to diameter ratio or an increase in the number of dimples [13]. Investigations conducted by Rao, Wan and Xu [14] on pin-fin dimpled channels with various dimple depths have revealed the dependency of pressure loss behaviour on dimple depth to diameter ratios. Further, their study included dimple depth to diameter ratios of 0.1, 0.2 and 0.3, where the obtained results showed that pin-fin dimple channels with shallower dimples exhibited a reduction of up to 17.6% for the values of friction factors [14]. Experiments on various density patterns of dimples along with different dimple depths have been conducted by Nesselrooij et al. [15] to show the sensitivity of drag reduction to the direction of fluid flow and flow conditions such as the bulk velocity. Their study looked at dimple depths of 0.025 and 0.05 with 20 mm and 60 mm dimple print diameters, respectively. Multiple dimple orientations, such as flow aligned and staggered, were also investigated.

Nesselrooij et al. [15] concluded that for both low-density and high-density patterns, increasing the depth to diameter ratio from 0.025 to 0.05 increased the overall drag at all velocities. The reason for this was attributed to the interactions between the boundary layer and the spanwise velocity component [15]. In turn, shallow dimples with low-density patterns produced a drag coefficient that was 4% less than the drag coefficient for flat plates [15]. Although drag reduction for the high-density deeper dimples improved with the Reynolds number, it still produced an 8% increase in drag performance [15]. Rao and Feng [16] conducted experimental and numerical studies for spherical and teardrop dimples with a depth to diameter ratio of 0.2. Their study concluded that compared to smooth flat plate, both dimple geometries increased the friction factor with the teardrop dimpled channel having relatively higher friction factor values than those of spherical dimples [16].

Moon, O'Connell and Glezer [17] experimentally investigated the effect of the fluid flow channel height on the heat transfer coefficient and friction factor. They tested four different channel heights (6, 13, 19 and 25 mm) while keeping the channel width and length constant. The bottom surface of the test setup was machined to incorporate the dimples.

Their study revealed that heat transfer improved by approximately two-fold, whilst friction factors ranged from 1.6 to 2 times relative to a smooth channel [17]. Moon, O'Connell and Glezer [17] also noted that studies conducted by other authors on dimples showed a lower pressure drop in comparison to other turbulators (such as continuous ribs) whilst significantly improving the thermal performance by a margin of 38%. Several other authors, including Abbas et al. [18], Zhong et al. [19] and Yan, Yang and Wang [20], have also conducted studies on the drag reduction properties of dimples on golf balls and cylinders, concluding that the addition of dimples has a significant effect on reducing the overall drag observed.

It is evident from the above review that much research and many experiments have been conducted using dimpled surfaces on a flat plate. Although flat plates provide a good indication of the thermal and pressure drop performances of dimpled HXs, they do not provide information regarding their implementation for plate-fin compact heat exchangers.

This is due to the difference in the orientation of PFHXs when compared to flat plates. In practical applications of PFHXs, multiple sides of the fins are exposed to the fluid flow and the fins are indeed perpendicular to the flow. Additionally, there is much disparity within the current literature on dimpled channel heat exchangers in terms of results relating to the thermal and pressure drop performances, a gap which the current study seeks to bridge. The effects of variable dimple diameters and dimple depths on the pressure drop performance of single elliptical plate-fin HXs will be tested. These will be compared to experimental results and conclusions provided by various authors. Based on the results of this current study, future work can be conducted on the thermal performances of various dimple configurations.

It should be noted here that the scope of the current study is limited to the investigation of passive surface textures, as they do not require any additional machines or materials to be formed/created. One of these surface modifications makes use of the dimples as turbulators to improve heat transfer while introducing a minimal pressure drop.

2. Experimental Study

2.1. Experimental Setup

For the experimental setup, the 'AF1125 Subsonic Wind Tunnel' (TECQUIPMENT, Nottingham, UK) was utilised, since it is able to replicate forced convection conditions as seen in the real-world applications of PFHXs. The following Figure 2 shows an overview of the experimental setup.

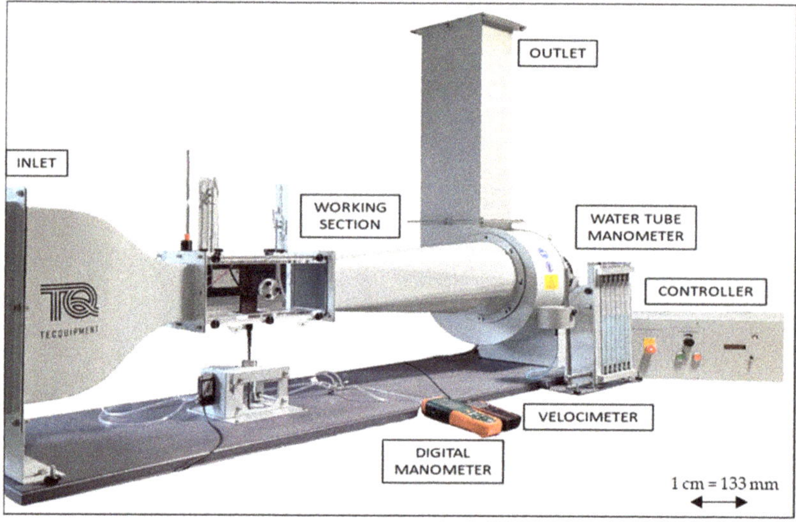

Figure 2. AF1125 subsonic wind tunnel.

The AF1125 wind tunnel consists of an inlet air duct followed by a rectangular working section, a diverging outlet duct and a blower that sits at the end of the diverging duct. The fluid (air) is drawn into the wind tunnel by the blower. Before entering the working section, the fluid encounters the flow straightener which sits right at the entrance of the inlet. The flow straightener minimises the lateral movement of fluid caused by the suction of the fluid and the swirling motion of the air by allowing the air to pass through an array of honeycomb-shaped cells. This ensures that the fluid reaching the working section has no lateral components of velocity and that the flow is as straight as possible.

The rectangular working section measures at a cross section of 125 mm × 125 mm, with a length of 350 mm. Further, it is equipped with four manometer probes that measure the static and dynamic pressure in the inlet and outlet sections of the wind tunnel. Static pressure or mechanical pressure, as defined by Shaughnessy, Katz and Schaffer [21], is the force measured perpendicular to a surface of interest that is exposed to fluid. This form of pressure represents the average of the normal stress at a certain point within the fluid. Accordingly, if the kinetic energy of a fluid particle was converted into pressure potential energy, the increase in the pressure due to this conversion is represented by dynamic pressure [21]. These probes were used in this study to measure the dynamic pressure at the inlet and outlet of the working section with the aid of HD755 digital differential manometer (EXTECH, Nashua, NH, USA). The manometer measures the pressure difference between two points. Notably, it was connected to measure the difference between the pitot tube pressure and the static pressure values at the inlet and outlet, respectively. Aquarium valves were used to switch between the inlet and outlet pressure taps in this study, as is shown in Figure 3. Pressure values were also used as performance indicators for assessing the pressure drop performances of the varied HXs that were tested. Finally, VT110 velocimeter (KIMO Instruments, Mumbai, India) was installed near the inlet section to measure the velocity of the inlet flow. The measured velocity was then used to calculate the Reynolds number of the free stream flow.

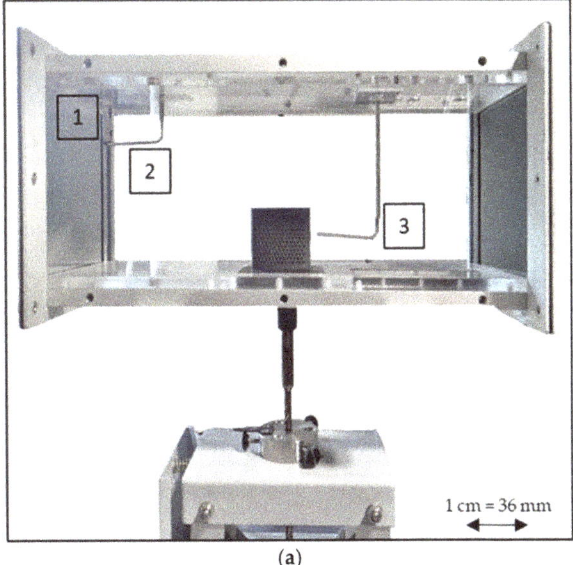

(a)

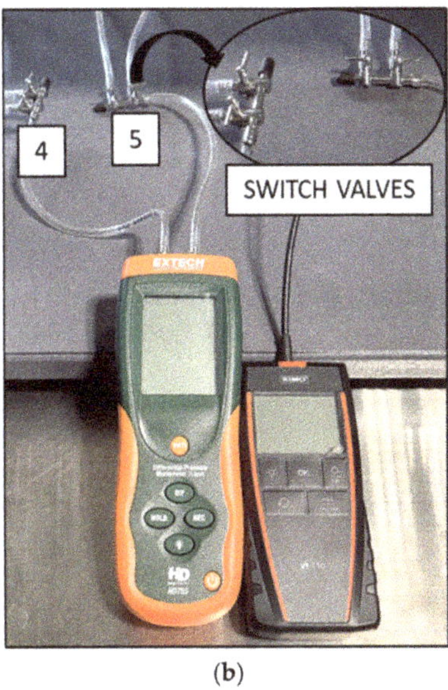

(b)

Figure 3. (a) Subsonic wind tunnel working section with the fin; (b) digital manometer and hot wire anemometer together with the switch valve assembly.

Table 1 summarises the apparatus and the various measuring components installed within the wind tunnel, and these are further highlighted in Figure 3.

Table 1. Component list for pressure testing in subsonic wind tunnel.

Sr. No.	Component
1	Velocimeter
2	Inlet Pressure Probes
3	Outlet Pressure Probes
4	Inlet Differential Pressure Valve
5	Outlet Differential Pressure Valve

2.2. Pressure Performance Criteria

There are several parameters that can be used to evaluate the pressure drop performances of heat exchangers, such as the coefficient of drag and friction factor. For the purpose of the current study, friction factor was chosen as the evaluation criteria for pressure drop performance, as measuring the coefficient of drag was not feasible. The experimental setup illustrated in Section 2.1 has been designed specifically to measure the inlet and outlet dynamic pressure, as well as the free stream velocity. Friction factor emerges as the preferred choice in representing the head loss within the test section due to friction, since it is directly proportional to the pressure drop, as shown in Equation (1). Friction factor values allow for results to be presented in a non-dimensional form, which can be readily evaluated against results from similar studies in relation to the coefficient of drag.

2.2.1. Friction Factor

The friction factor or Darcy friction factor can be used to estimate pressure loss due to friction in pipe flow or open-channel flow [22]; it is a non-dimensional quantity and can be expressed using the general equation shown below, adapted from Barker [22].

$$f = \frac{\Delta P 2 D_h}{\rho L V^2} \tag{1}$$

where f is the dimensionless Darcy friction factor, ρ (kg/m^3) is the fluid density, L (m) is the characteristic length of the fin, V (m/s) is the fluid velocity, D_h (m) is the hydraulic diameter and ΔP (Pa) is the pressure difference between the inlet and the outlet of the working section.

The hydraulic diameter can be calculated using the equation given below, which has been adapted from Genium Publishing Corporation [23].

$$D_h = \frac{2WH}{(H + W)} \tag{2}$$

where W (m) and H (m) are the width and height of the wind tunnel working section, respectively.

2.2.2. Reynolds Number

The Reynolds number can be used to characterise the inertial and viscous properties of a moving fluid [24], where it can be calculated using the equation given below that has been adapted from Vafadar, Guzzomi and Hayward [25].

$$Re = \frac{\rho V D_h}{\mu} \tag{3}$$

where μ (kg/m·s) is the dynamic viscosity of the fluid.

2.3. Uncertainty

Since the results are presented relatively using the same exact test equipment, uncertainty was calculated using single sample standard deviation in order to outline inaccuracies in measurements and to describe the reliability of the results. Standard deviation was calculated for the 5 individual runs to make sure the experimental data and the equipment did not show any error. A maximum standard deviation of 8% was noted for the smooth NACA fin at a velocity of 12 m/s. Standard deviation for the remainder of the tests was below 5%. These uncertainties are within a reasonable range of accuracy, where similar uncertainties have been noted by Burgess and Ligrani [13] as well as Rao, Li and Feng [16]. Individual runs for all geometry profiles were conducted consecutively within a period of 120 min. This was to ensure the ambient conditions within the lab were as consistent as possible for all individual runs, without introducing bias from external factors such as change in temperature, pressure and humidity.

2.4. Test Methodology

Thermal probes were removed from within the test section to avoid hindering the path of the fluid and to ensure the pressure drop was based solely on the geometry being tested. The bottom section of the wind tunnel was redesigned in a way to accommodate the base of the fin and to make it flush with the bottom panel of the wind tunnel. In turn, this excludes the pressure drop of the plate itself and provides measurements for only the pressure drop performance of the fin and the dimples.

Pilot tests were conducted to establish the range of velocities to be tested, where velocities from 2 m/s to 30 m/s were tested. Since the range of the velocimeter reached its maximum at 30 m/s, pressure readings were measured at velocities starting at 12 m/s and going up in increments of 2 m/s thereafter until 28 m/s. This made it possible to obtain nine measurements to evaluate the pressure drop. Pressure measurements at 30 m/s were not measured since the velocimeter failed to provide consistent readings, where subsequent experimental data showed a relatively large standard deviation. Similarly, velocities below 12 m/s showed inconsistent pressure readings for the NACA fins and therefore, were not included in these tests. The main reason for this inconsistency in the measurements was the limited range of the measuring equipment which failed to provide adequate measurements for velocities below 12 m/s and beyond 28 m/s.

Experiments were conducted five times for each of the geometries being tested. Initially, ten pilot tests were conducted, where standard deviation was calculated for five runs and ten runs, respectively. Average standard deviations of 0.29 Pa and 0.27 Pa were observed for the five and ten runs, respectively. Accordingly, it was decided that five runs were sufficient to ensure accurate results without increasing the uncertainty of the measurements. An average of the five collective runs was used for the final calculation of friction factor as well as Reynolds number.

During the experiments, static inlet/outlet pressure, dynamic inlet/outlet pressure, ambient room temperature and fluid velocity were recorded. These were further used to calculate friction factor for different fin geometries.

A boundary layer thickness of 3 mm was estimated based on flat plate approximation of the elliptical fin along the length of the entire NACA profile. As a result, the pitot tube within the wake/outlet section of the fin was positioned 6 mm off the trailing edge of the fin in order to not capture the boundary layer effects and measure the pressure drop based on the wake of the fin. An equation suggested by Çengel and Cimbala [26] was used to calculate the approximate thickness of the turbulent boundary layer based on flat plate approximation.

$$\delta \cong x \times \frac{0.38}{(Re)^{1/5}} \tag{4}$$

where δ is the boundary layer thickness (mm), x is the characteristic length from the leading edge (mm) and Re is the Reynolds number (dimensionless).

A digital manometer was used to provide a rolling average of the pressure drop for the inlet and outlet sections of the tested fins. Similarly, a velocimeter was used at the inlet section to calculate the average of the velocity. Both the manometer and velocimeter do not have the capability to provide values over a data acquisition system, whereby the rolling average was taken for the pressure and velocity. The overall pressure drop was then calculated by obtaining the difference between the inlet and outlet's averaged experimental pressure values, whilst the Reynolds number was calculated using the average of the velocity distribution.

2.5. Test Parameters

Different configurations of dimples, as shown in Figure 4, were tested. The aim was to investigate the effect of varying dimple diameters and depths on pressure drop performance while keeping the depth to diameter ratio, also known as surface roughness (ε), constant at 0.3.

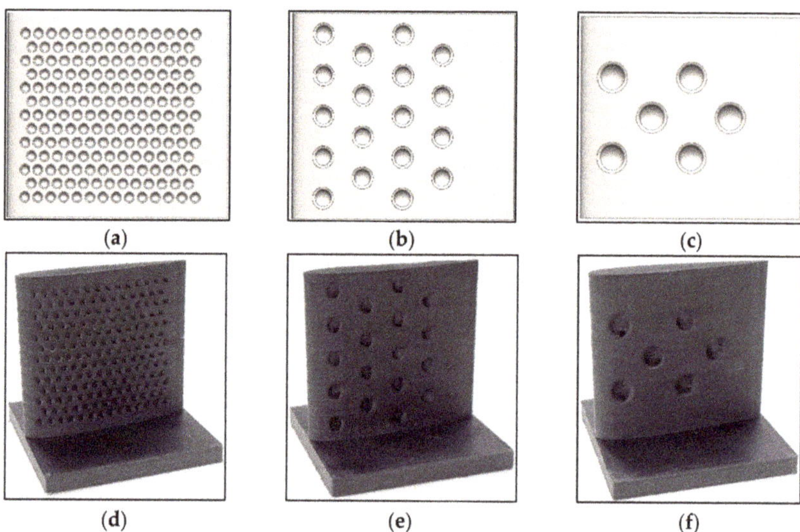

Figure 4. (**a**) HX 1–2 mm dimples; (**b**) HX 2–4 mm dimples; (**c**) HX 3–6 mm dimples; (**d**) 3D-printed HX 1; (**e**) 3D-printed HX 2; (**f**) 3D-printed HX 3.

Test fins were fabricated using 3D printing of ABS plastic with Zortrax M200 3D printer (Zortrax, Olsztyn, Poland). Three-dimensional models were developed for the plate and the fin as a single solid using Solidworks (Dassault Systèmes, Waltham, MA, USA) and these were then converted into appropriate file formats to process using the Zortrax M200 (Zortrax, Olsztyn, Poland). Three-dimensional-printed fins were then inspected for surface defects and irregularities. No post-processing of the fins was required and the finished samples are shown below in Figure 4.

Initial experiments investigated the effects of varying the dimple diameter and depth on pressure drop performance; the tested dimple geometries are shown in Figure 4.

These dimpled fins' pressure drop performances were compared to the smooth NACA 63-015 profile fin. Measurements for the smooth fin were used as the base measurements to indicate the effective change in pressure drop after implementing dimples. The smooth fin is shown below in Figure 5.

Figure 5. NACA 63-015-inspired smooth 3D-printed PFHX.

As stated in the literature review, a larger diameter to depth ratio has been shown/proven to provide a better thermal performance. A large diameter to depth ratio can introduce stagnant air pockets and turbulence, which may increase the pressure drop. Therefore, dimple diameters of 2 mm, 4 mm and 6 mm were chosen with a D/d ratio of 0.3. The selection of the ratio was based upon the previous studies while the selection of the dimple diameters was limited by the geometry of the fin and the manufacturing equipment. All fins were printed using the Zortrax M200 3D printer which provides a resolution of 0.4 mm for a single printable point. Taking this into consideration and the curvature of the dimples, it was decided to use 2 mm as the minimum dimple size which corresponds to a depth of 0.6 mm. Dimple diameters lower than 2 mm would have depths lower than 0.4 mm, which when manufactured would lose the details of the curvature of the dimple. The number of dimples on the fins for the above-mentioned diameters were 352, 36 and 12, respectively. The number and configuration of dimples were normalised using the volume of material removed, where all efforts were made to keep the removed volume as consistent as possible. This is shown in Table 2.

Table 2. Smooth vs. dimpled fins' mechanical properties.

Fin	Fin Volume (mm^3)	Fin Surface Area (mm^2)	Δ Volume (mm^3)	Δ S.A (mm^2)
HX Smooth	11,571	5149	-	-
HX 1–2 mm Dimples	11,191	5475	380	326
HX 2–4 mm Dimples	11,260	5277	311	128
HX 3–6 mm Dimples	11,216	5241	356	91

Figure 6a,b present the geometrical dimensions used to create the varied dimple profile and the array of dimples on the fin, respectively.

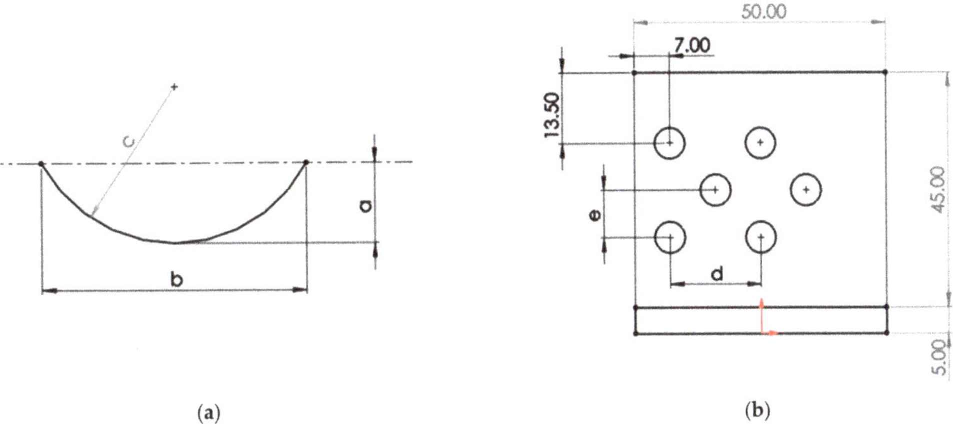

Figure 6. Geometrical dimensions for (**a**) dimple profile; (**b**) fin with dimples. Dimensions are in millimeters (mm).

Table 3 below presents the summary of the samples that were tested in the current study, including the geometrical configurations of dimples as well as their spanwise/streamwise spacing. All dimensions are in millimeters.

Table 3. Geometric configuration of HX samples.

Fin	a	b	c	d	e	a/b
HX Smooth	-	-	-	-	-	-
HX 1–2 mm Dimples	0.6	2	1.13	3	3	0.3
HX 2–4 mm Dimples	1.2	4	2.27	18	4.5	0.3
HX 3–6 mm Dimples	1.8	6	3.4	18	9	0.3

3. Results

Figure 7a presents the average of the pressure drop measurements for the smooth NACA 63-015 fin whereas Figure 7b shows the results of five individual runs that were undertaken for the NACA fin. The trendline corresponds to a second-order polynomial equation, the coefficient for which is shown in Figure 7a.

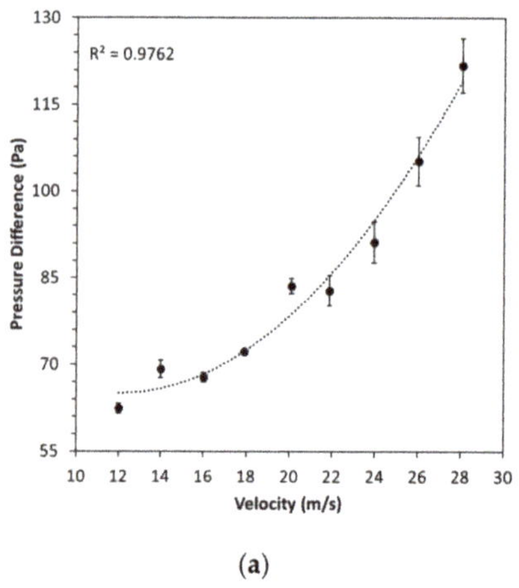

(a)

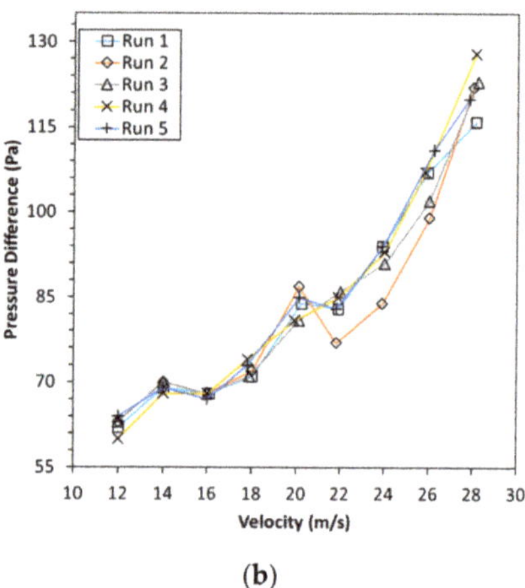

(b)

Figure 7. Experimental results for smooth NACA/elliptical profile; (a) average of 5 runs and (b) 5 individual runs.

Figure 8a presents the average of the pressure drop measurements for the 2 mm-dimpled fin whereas Figure 8b provides the pressure measurements for the five individual runs. The trendline corresponds to a second-order polynomial equation, the coefficient for which is shown in Figure 8a.

Figure 9a shows the pressure drop measurements for the 4 mm-dimpled fin and Figure 9b provides the results for the five individual measurements undertaken. The trendline corresponds to a second-order polynomial equation, the coefficient for which is shown in Figure 9a.

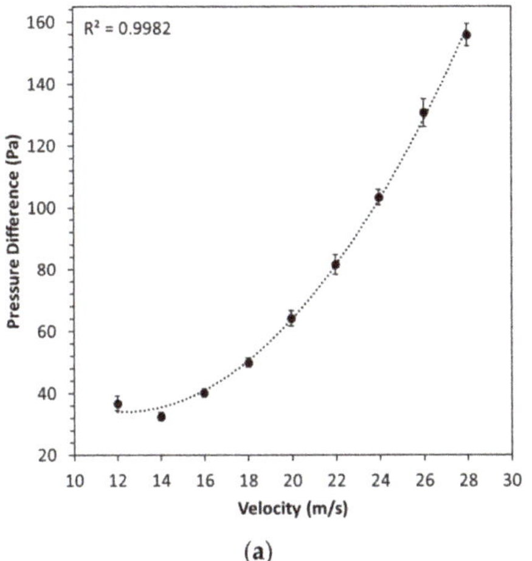
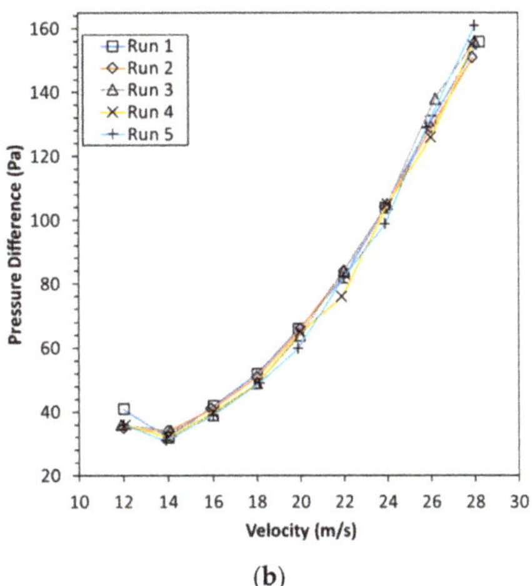

Figure 8. Experimental results for 2 mm dimples engraved on the NACA/elliptical Profile; (**a**) average of 5 runs; (**b**) 5 individual runs.

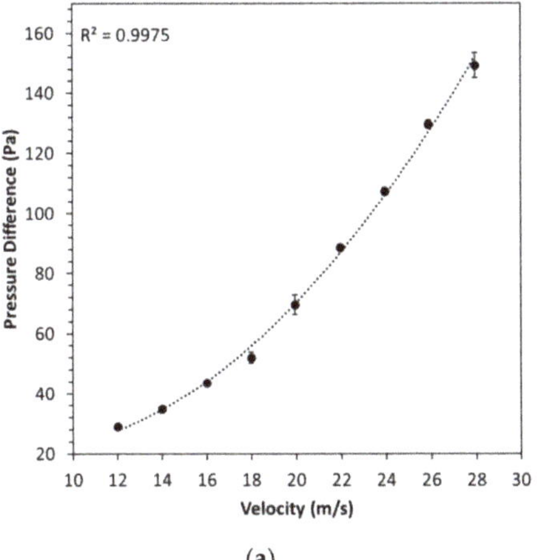
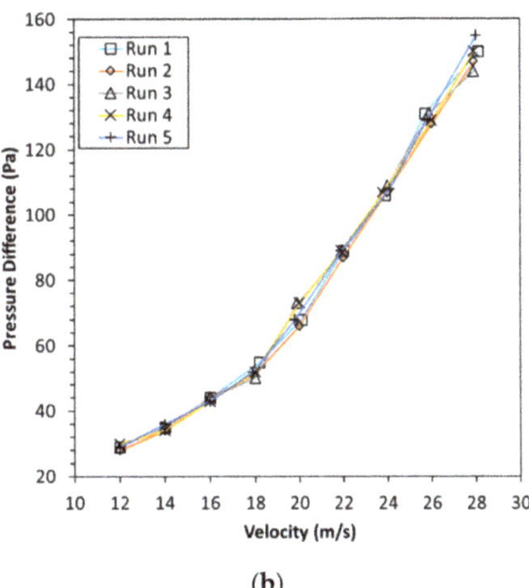

Figure 9. Experimental results for 4 mm dimples engraved on NACA/elliptical profile; (**a**) average of 5 runs; (**b**) 5 individual runs.

Figure 10a highlights the average pressure drop for the 6 mm-dimpled fin whereas Figure 10b presents the results for the five individual runs undertaken. The trendline corresponds to a second-order polynomial equation, the coefficient for which is shown in Figure 10a.

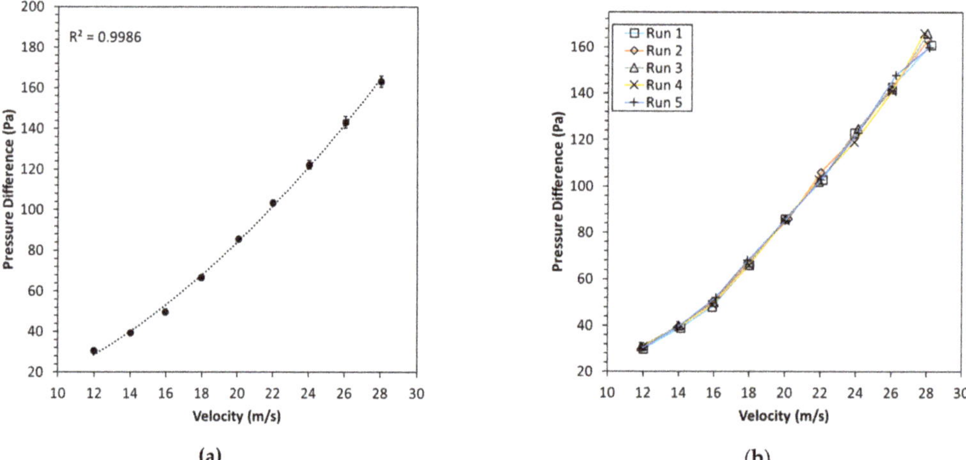

(**a**) (**b**)

Figure 10. Experimental results for 6 mm dimples engraved on NACA/elliptical profile; (**a**) average of 5 runs; (**b**) 5 individual runs.

Figures 11 and 12 compares the pressure drop performances of the different dimpled fins (2 mm, 4 mm, 6 mm) against the smooth NACA fin as well as their friction factor characteristics.

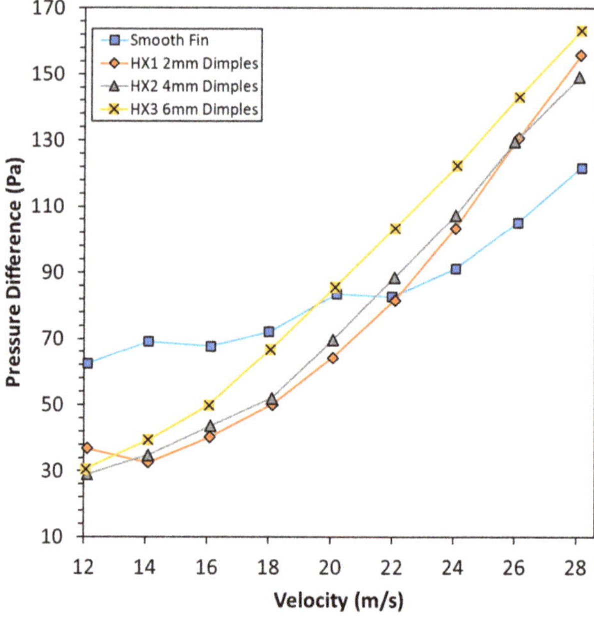

Figure 11. Comparison of pressure drop between smooth, 2 mm, 4 mm and 6 mm dimples engraved on NACA/elliptical profile.

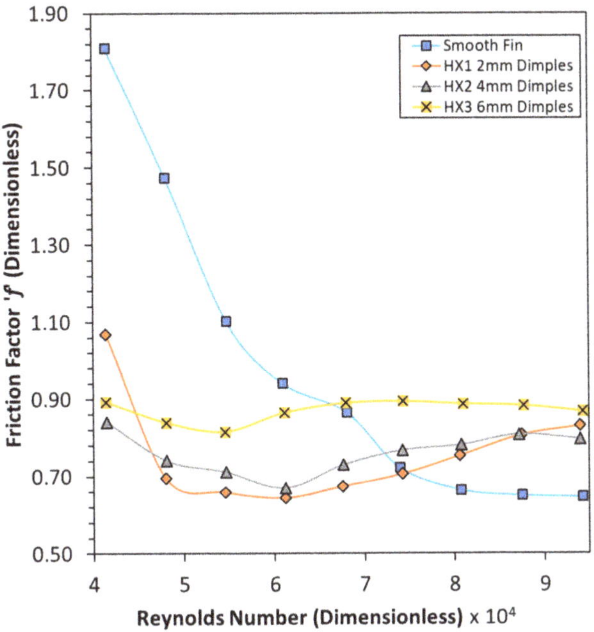

Figure 12. Comparison of friction factor value between smooth, 2 mm, 4 mm and 6 mm dimples engraved on NACA/elliptical profile.

4. Discussion of Results

The results from the wind tunnel testing of smooth NACA/elliptical fin, and fins with 2 mm, 4 mm and 6 mm dimples, are presented in Figure 7, Figure 8, Figure 9, Figure 10, respectively. These include the five individual runs to determine the accuracy of the measurement as well as the average of these five runs. These average runs were then compared to each other, as seen in in Figure 11, for their relative pressure drop performances. It can be observed in Figure 11 that with an increase in diameter as well as depth of the dimples, there was a noticeable increase in the pressure drop of the fins. A similar trend has been observed by Ting [27] in relation to golf balls, where initially increasing the surface roughness of golf balls via dimples reduced the coefficient of drag. The above study also observed that after a certain limit is reached for dimple depth, a further increase in the depth of dimples had the opposite effect, where the coefficient of drag increased significantly. In the current study, fins with dimples outperformed the smooth fin in terms of pressure drop performance up to velocities of 20 m/s, after which the pressure drop for the dimpled fins increased significantly. These observations can be attributed to the fact that the addition of dimples delays the boundary layer separation, which subsequently induces a narrower wake behind the fin leading to reduced form drag. These dimples introduce localised turbulence within and around the dimple cavity, which re-energizes the boundary layer and delays boundary layer separation. As a result of this localised turbulence, the boundary layer sticks to the surface for a longer duration and reduces the overall wake behind the fin. This contributes to a reduction in form drag, which is seen as the initial decline in the friction factor or the lower pressure drop observed with a low Reynolds number. Form drag is highly dependent on flow separation, whereas friction drag is a function of shear stresses observed on the surface of a geometry [28]. With an increase in the Reynolds number, there is an increase in shear stresses, which in turn increases the overall friction drag. For the case of dimpled fins, shear stresses are further magnified due to the introduction of localised turbulence by dimples, where as a result there is a significant increase in the pressure drop for dimpled fins starting at a velocity of 20–22 m/s, which is equivalent to a Reynolds number of approximately 7×10^4–7.5×10^4. Experiments conducted by Chowdhury et al. [29] have shown that increasing the depth of dimples lowers the critical Reynolds number, where the transition to turbulent flow happens earlier within the flow in comparison to smooth fins. Even though this shift in transition has the possibility of increasing the coefficient of drag in the transcritical regime, as observed by Chowdhury et al. [29], it also creates local turbulence within dimples. This local turbulence can be controlled via the depth of the dimples and used to energise the boundary layer, which delays the aforementioned separation. Accordingly, as can be observed in Figure 11, the fin with 2 mm dimples showed a higher pressure drop at 12 m/s relative to 4 mm and 6 mm dimples, both of which experienced negligible differences in the pressure drop performance at 12 m/s. Starting at 14 m/s, the fin with 2 mm dimples showed a consistently lower pressure drop, followed by fins with 4 mm and 6 mm dimples, respectively. This trend could be observed up until 26 m/s, after which the fin with 2 mm dimples surpassed the 4 mm-dimpled fin and showed an increase in pressure drop. It is worth noting here that the difference in pressure drop between 2 mm- and 4 mm-dimpled fins was quite insignificant when compared with 6 mm dimples. As suggested by various authors, including Ge, Fang and Liu [30], this could be attributed to the threshold of dimple depth and diameter, whereby if the depth and diameter of dimples is increased beyond a limit, the favourable effect of pressure reduction is negated and an increase in pressure drop can consequently be observed.

Figure 12 shows the friction factor calculated using Equation (1) provided in the theory section. The friction factor allows for the comparison of head loss within an open-channel flow, where flow can be observed in a pipe measured over a specific distance [31]. In this study, friction factor values were plotted against the Reynolds number for a smooth fin and dimpled fins (2 mm, 4 mm and 6 mm). This allowed the authors to determine the relationship between the head loss observed within the wind tunnel for various dimpled and non-dimpled fins relative to the Reynolds number. It is evident from the trends observed in Figure 12 that the friction factor for the smooth NACA fin was the highest up to a Reynolds number of 6.5×10^4, whereas the 2 mm dimples provided the lowest friction factor for Reynolds numbers ranging from 4.6×10^4 to 7.2×10^4. Experiments conducted by Choi, Jeon and Choi [32] found that with an increase in the Reynolds number, a sharp decrease in the coefficient of drag can be observed for dimpled spheres as compared to smooth spheres. A similar trend is also visible in Figure 12 for the 2 mm-dimpled fin and the smooth fin, where a sharp decrease in the friction factor can be observed for the 2 mm-dimpled fin between Reynolds numbers of 4×10^4 and 5×10^4, and for the smooth fin between the range of 4×10^4 and 6×10^4. The same study by Choi, Jeon and Choi [32] further concluded that the coefficient of drag hit a constant value after the sharp decline region. Their study examined dimpled spheres with varied surface roughness values (ε), with a higher surface roughness indicating deeper dimples. The results published in the above study further summarised that the coefficient of drag for higher surface roughness (ε) spheres showed a sharp decline at a lower Reynolds number while maintaining a higher constant value than spheres with lower surface roughness values (ε). Similar observations are evident in Figure 12 where, on the one hand, the equipment could not capture the sharp decline of the friction factor for deeper dimples (4 mm and 6 mm), as was observed for the 2 mm-dimpled fin and the smooth fin. On the other hand, it can be clearly observed that the friction factor starts to plateau at a Reynolds number of approximately 6.8×10^4, with 6 mm dimples showing the largest head loss, and 2 mm dimples exhibiting the lowest head loss. The overall trend for the coefficient of drag as observed by Choi, Jeon and Choi [32] in their experimental studies can also be observed in Figure 12 for the plot of the friction factor. This similarity in the overall trend can be explained by the formation of a separation bubble within the dimples resulting in delayed boundary layer separation, as mentioned earlier, leading to a reduction in the overall drag and consequently reducing the overall pressure drop.

These results also coincide with the trends observed by Rao, Wan and Xu [14], Nesselrooij et al. [15] and Patel and Borse [33] where the introduction of dimples was shown to reduce pressure drop relative to a flat plate. Further, a study conducted by Rao, Wan and Xu [14] also concluded that shallower dimples perform better at reducing the pressure drop as compared to deeper dimples, since shallower dimples reduce the velocity near the upstream half of the dimples which consequently reduces turbulent mixing in the main flow. With deeper dimples, even though turbulent mixing is reduced in the upstream half of the dimples, there is a strong flow impingement near the downstream rim of the dimples [14]. Interested readers may refer to the above-mentioned study by Rao, Wan and Xu [14] for additional insight regarding the pressure loss due to flow impingement.

This flow impingement introduces additional pressure loss within the flow. This is visible in Figure 11, where the fin with 2 mm dimples provided the lowest pressure drop, followed by fins with 4 mm and 6 mm dimples, respectively. At low velocities of up to 22 m/s, a maximum pressure drop reduction of 58% was observed using 4 mm dimples relative to the smooth NACA fin, whereas at high velocities above 22 m/s, a 34% increase in pressure drop was observed with 6 mm dimples relative to the smooth NACA fin. Figure 13 presents the percentage difference in pressure for all three configurations of dimples tested against the smooth fin.

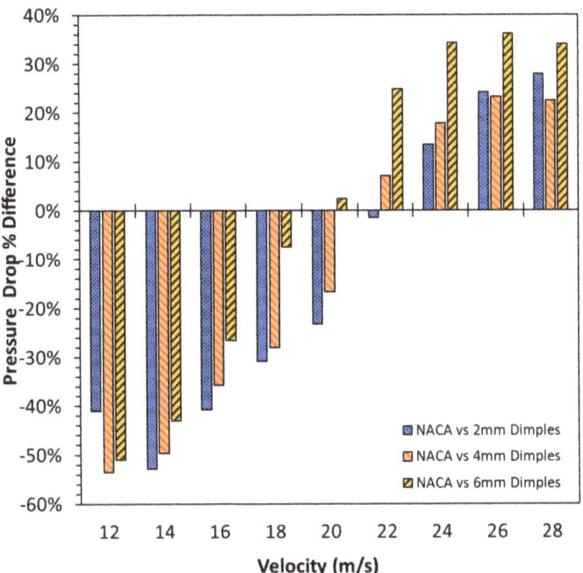

Figure 13. Pressure drop % difference for NACA fin vs. 2 mm-, 4 mm- and 6 mm-dimpled fins.

It should be observed here that the results obtained by several authors including Choi, Jeon and Choi [32] as well as Abbas et al. [18] for drag coefficients coincide with the trends observed in Figure 12 for friction factor values. As a result, it can be summarised from Figure 12 that deeper dimples induce drag reduction at a lower Reynolds number, which can be seen in the sharp decline in the friction factor value for 2 mm dimples. This sudden decrease in friction factor could not be captured in the 4 mm and 6 mm dimples due to the limit of the measuring equipment and the reasons outlined in Section 2.4 of the current study. However, from the similarity of the trends observed in the literature and Figure 12, the critical regime of the sudden decrease in drag for 4 mm and 6 mm dimples should lie beyond a Reynolds number $<4 \times 10^4$. This is evident from the values for a friction factor of 4 mm- and 6 mm-dimpled fins, as they are observed to increase when going from a Reynolds number of 5×10^4 to 4×10^4, suggesting a linear increase in overall drag at velocities lower than 12 m/s. Consequently, it can also be concluded from Figure 12 that deeper dimples induce a higher pressure drop at a higher Reynolds number of >7 × 10^4 when compared to shallower dimples. This is evident from the trend where friction factor values start to plateau within the transcritical regime (*Re*, 6.5 × 10^4—9 × 10^4), whereby at that stage the 6 mm dimpled fins have the highest friction factor values, followed by the 4 mm- and 2 mm-dimpled fins. The smooth NACA fin provided the lowest frictional resistance. In summary, it is apparent that deeper dimples are preferred for reducing drag at lower velocities with a larger pressure drop an in the transcritical regime. Shallower dimples are preferred for medium to high velocities with a smaller pressure drop in the transcritical regime.

5. Conclusions

In this study, we have experimentally investigated the pressure drop characteristics of NACA/elliptical fins with smooth surfaces and with 2 mm, 4 mm and 6 mm engraved dimples. The experimental results for the smooth and dimpled fins were obtained using wind tunnel testing and compared relative to each other for velocities ranging from 12 m/s to 28 m/s, translating to a Reynolds number range of 4 × 10^4–9.2 × 10^4. The results show a clear distinction in pressure drop and friction factor for smooth and dimpled fins. Dimpled fins outperformed the smooth fin for velocities of 12 m/s to approximately 22 m/s, with 2 mm-dimpled fins providing the lowest pressure drop and the least friction amongst the compared fins. Another important observation was the increase in the pressure drop and friction factor with relative increases in the dimple diameter and depth. With the depth to diameter ratio being constant at 0.3 for all dimpled fins, it was evident that shallower dimples performed better and induced a lower pressure drop compared to dimples with larger diameters and depths. Dimples with 4 mm diameters led to a reduction of 58% in pressure drop at a velocity of 12 m/s when compared to the smooth fins. Dimples with 2 mm diameters showed a 53% reduction in pressure drop at a velocity of 16 m/s relative to the smooth fin. Dimples with the largest diameter of 6 mm showed a reduction of 55% at a velocity of 12 m/s when compared to the smooth fin and the 2 mm dimples. This occurred due to the fact that a higher surface roughness or deeper dimples shift the critical Reynolds number upstream, which reduces the drag at low velocities while the constant value of the drag coefficient within the transcritical regime increased significantly. A smaller surface roughness or shallower dimples provided a larger drag reduction at higher velocities with the constant value of drag coefficient within the transcritical regime being lower when compared to deeper dimples. As a result, applications that require drag reduction at lower velocities should opt for deeper dimples, whilst applications that require significant drag reduction at high velocities should utilise shallower dimples.

Future Work

The current study focuses on the investigation of the pressure drop characteristics of dimpled fins relative to a smooth NACA fin. Even though favourable results have been obtained with the use of dimples to reduce the pressure drop, a framework is still missing in order to optimise dimple design for specific applications. As a result of the current investigation, the following future works are proposed, which seek to further establish dimples and the use of turbulators as a means of increasing the efficiency of PFHXs.

- Heat transfer studies should be conducted on the geometries investigated in the current study. Even though a significant pressure drop reduction was observed with the introduction of dimples, how well these dimples allowed the transfer of heat will eventually be a decisive factor in the relevance of the use of additive manufacturing and the use of turbulators in improving the efficiency of HXs.
- Mathematical models should be investigated that allow the design of dimples based on common parameters observed in the application of HXs. These include velocity, drag, Reynolds number, surface roughness and other factors that contribute to increasing the pressure drop efficiency of HXs. The current literature regarding dimples has mainly focused on experimental results, where the design of the dimples is either cho-

sen arbitrarily based on existing literature or is influenced by their existing applications in golf balls. These designs have been verified for their effectiveness based on these experimental parameters. A framework or a mathematical model that provides insight into the design of dimple parameters, such as the optimal depth and diameter for specific velocities or specialised applications, will be highly advantageous in removing the need for extensive testing of various arbitrarily chosen dimple designs.

Author Contributions: Conceptualization, A.V., F.G. and K.H.; Methodology, K.R. and A.H.; Investigation, K.R.; Resources, A.V. and F.G.; Data curation, K.R.; Writing—original draft preparation, K.R.; Writing—review and editing, K.R., A.V., F.G., K.H. and A.H.; Supervision, F.G., A.V. and K.H.; Project administration, K.R., F.G., A.V., K.H. and A.H.; Funding acquisition, F.G. and A.V. All authors have read and agreed to the published version of the manuscript.

Funding: This study was supported by Edith Cowan University (ECU) under Grant G1004423. Additionally this manuscript received an open access publication grant from the School of Engineering, Edith Cowan University, Australia.

Institutional Review Board Statement: Not applicable.

Informed Consent Statement: Not applicable.

Data Availability Statement: Not applicable.

Acknowledgments: The authors would like to acknowledge and thank the School of Engineering, Edith Cowan University, Australia, for providing and administering the resources and the requirements for the successful completion of this research. The authors would also like to thank the technical and literacy staff including Adrian Davis, Guanliang Zhou, Yatt Yap and Michael Stein for their support during the course of the research in technical needs and writing requirements.

Conflicts of Interest: The authors declare that there is no conflict of interest.

References

1. Thulukkanam, K. *Heat Exchanger Design Handbook*; Taylor & Francis Group: Boca Raton, FL, USA, 2013.
2. Shah, R.K.; Sekulic, D.P. *Fundamentals of Heat Exchanger Design*; John Wiley & Sons: Hoboken, NJ, USA, 2003.
3. Li, Q.; Flamant, G.; Yuan, X.; Neveu, P.; Luo, L. Compact heat exchangers: A review and future applications for a new generation of high temperature solar receivers. *Renew. Sustain. Energy Rev.* **2011**, *15*, 4855–4875. [CrossRef]
4. Hesselgreaves, J.E.; Law, R.; Reay, D.A. *Compact Heat Exchangers*; Elsevier: Oxford, UK, 2017.
5. Oliveira, J.P.; LaLonde, A.; Ma, J. Processing parameters in laser powder bed fusion metal additive manufacturing. *Mater. Des.* **2020**, *193*, 108762. [CrossRef]
6. Wong, M.; Owen, I.; Sutcliffe, C.J.; Puri, A. Convective heat transfer and pressure losses across novel heat sinks fabricated by selective laser melting. *Int. J. Heat Mass Transf.* **2009**, *52*, 281–288. [CrossRef]
7. Dede, E.M.; Joshi, S.N.; Zhou, F. Topology Optimization, Additive Layer Manufacturing, and Experimental Testing of an Air-Cooled Heat Sink. *J. Mech. Des.* **2015**, *137*, 111403. [CrossRef]
8. McDonough, J.R. A perspective on the current and future roles of additive manufacturing in process engineering, with an emphasis on heat transfer. *Therm. Sci. Eng. Prog.* **2020**, *19*, 100594. [CrossRef]
9. Vafadar, A.; Guzzomi, F.; Rassau, A.; Hayward, K. Advances in Metal Additive Manufacturing: A Review of Common Processes, Industrial Applications, and Current Challenges. *Appl. Sci.* **2021**, *11*, 1213. [CrossRef]
10. Frazier, W.E. Metal Additive Manufacturing: A Review. *J. Mater. Eng. Perform.* **2014**, *23*, 1917–1928. [CrossRef]
11. Sheikholeslami, M.; Ganji, D.D. Nanofluid Forced Convection Heat Transfer. In *Applications of Nanofluid for Heat Transfer Enhancement*; Elsevier: Amsterdam, The Netherlands, 2017; pp. 127–193.
12. Alam, T.; Kim, M.H. A comprehensive review on single phase heat transfer enhancement techniques in heat exchanger applications. *Renew. Sustain. Energy Rev.* **2018**, *81*, 813–839. [CrossRef]
13. Burgess, N.; Ligrani, P. Effects of dimple depth on channel nusselt numbers and friction factors. *J. Heat Transf.* **2005**, *127*, 839–847. [CrossRef]
14. Rao, Y.; Wan, C.; Xu, Y. An experimental study of pressure loss and heat trasnfer in the pin fin-dimple channels with various dimple depths. *Int. J. Heat Mass Transf.* **2012**, *55*, 6723–6733. [CrossRef]
15. van Nesselrooij, M.; Veldhuis, L.L.M.; van Oudheusden, B.W.; Schrijer, F.F.J. Drag reduction by means of dimpled surfaces in turbulent boundary layers. *Exp. Fluids* **2016**, *57*, 142. [CrossRef]
16. Rao, Y.; Li, B.; Feng, Y. Heat transfer of turbulent flow over surfaces with spherical dimples and teardrop dimples. *Exp. Therm. Fluid Sci.* **2015**, *61*, 201–209. [CrossRef]
17. Moon, H.; O'Connell, T.; Glezer, B. Channel height effect on heat transfer and friction in a dimple passage. *J. Eng. Gas Turbines Power* **2000**, *122*, 307–313. [CrossRef]

18. Abbas, Z.; Shah, S.I.A.; Javed, A.; Jami, M.L. Influence of dimple design on aerodynamic drag of Golf balls. In Proceedings of the 2019 Sixth International Conference on Aerospace Science and Engineering (ICASE), Islamabad, Pakistan, 12–14 November 2019.
19. Zhong, X.; Shuzhao, L.; Xiaoyan, W.; Xijiang, Z. Research on Drag Reduction Effect of Concave Non-smooth Surface in Air. *Adv. Mater. Res.* **2011**, *299–300*, 7–11.
20. Yan, F.; Yang, H.; Wang, L. Study of the Drag Reduction Characteristics of Circular Cylinder with Dimpled Surface. *Water* **2021**, *13*, 197. [CrossRef]
21. Shaughnessy, E.J.; Katz, I.M.; Schaffer, J.P. 8.4 Static, Dynamic, Stagnation and Total Pressure. In *Introduction to Fluid Mechanics*; Oxfor University Press: New York, NY, USA, 2005; pp. 490–491.
22. Barker, G. *The Engineer's Guide to Plant Layout and Piping Design for the Oil and Gas Industries*; Gulf Professional Publishing: Burlington, MA, USA, 2018.
23. Genium Publishing Corporation. 402.1—Guidelines for the Use of Data in Section 402. In *Fluid Flow—Data Book*; Genium Publishing Corporation: New York, NY, USA, 1993; p. 5.
24. Layton, W. 5.4 The Reynolds Number. In *Introduction to the Numerical Analysis of Incompressible Viscous Flows*; Society for Industrial and Applied Mathematics: Philadelphia, PA, USA, 2008; pp. 83–86.
25. Vafadar, A.; Guzzomi, F.G.; Hayward, K. Experimental Investigation and Comparison of the Thermal Performance of Additively and Conventionally Manufactured Heat Exchangers. *Metals* **2021**, *11*, 574. [CrossRef]
26. Çengel, Y.A.; Cimbala, J.M. Chapter 10—Approximate Solutions of the Navier-Stokes Equations: Turbulent Flat Plate Boundary Layer. In *Fluid Mechanics: Fundamentals and Applications*; McGraw-Hill Higher Education: New York, NY, USA, 2017; p. 1117.
27. Ting, L.L. Effects of Dimple Size and Depth on Golf Ball Aerodynamic Performance. In Proceedings of the ASME/JSME 2003 4th Joint Fluids Summer Engineering Conference, Honolulu, HI, USA, 6–10 July 2009.
28. Schuetz, T. 4.3.1 Pressure and Friction Drag. In *Aerodynamics of Road Vehicles*; SAE International: Pittsburgh, PA, USA, 2016; pp. 220–221.
29. Chowdhury, H.; Loganathan, B.; Wang, Y.; Mustary, I.; Alam, F. A study of dimple characteristics on golf ball drag. *Procedia Eng.* **2016**, *147*, 87–91. [CrossRef]
30. Ge, M.-w.; Fang, L.; Liu, Y.-q. Drag reduction of wall bounded incompressible turbulent flow based on active dimples/pimples. *J. Hydrodyn.* **2017**, *29*, 261–271. [CrossRef]
31. Chanson, H. *Hydraulics of Open Channel Flow*; Butterworth Heinemann: Oxford, UK, 2004.
32. Choi, J.; Jeon, W.-P.; Choi, H. Mechanism of drag reduction by dimples on a sphere. *Phys. Fluids* **2006**, *18*, 041702. [CrossRef]
33. Patel, I.H.; Borse, S.L. Experimental Investigation of Heat Transfer Enhancement over the Dimpled Surface. *Int. J. Eng. Sci. Technol.* **2012**, *4*, 3666–3672.

MDPI

St. Alban-Anlage 66

4052 Basel

Switzerland

Tel. +41 61 683 77 34

Fax +41 61 302 89 18

www.mdpi.com

Metals Editorial Office

E-mail: metals@mdpi.com

www.mdpi.com/journal/metals

www.ingramcontent.com/pod-product-compliance
Lightning Source LLC
LaVergne TN
LVHW071542170726
843515LV00008B/2200

9 783036 539034